CYBER TERRORISM AND INFORMATION WARFARE

I. Assessment of Challenges

Yonah Alexander
and
Michael S. Swetnam

Fifth Volume, Second Series
Terrorism: Documents of International & Local Control

Yonah Alexander
Donald J. Musch
General Editors

Oceana Publications, Inc. / Dobbs Ferry, New York

You may order this or any other Oceana publications by visiting Oceana's website at http://www.oceanalaw.com

Terrorism: Documents of International and Local Control, Second Series

9780379214154
9780379214161

Library of Congress Catalog Card Number: 99-75945

Manufactured in the United States of America on acid-free paper.

Table of Contents

Preface

About the Editors

White House

1. Critical Foundations Protecting America's Infrastructures, the Report of the President's Commission on Critical Infrastructure Protection, October 13, 1997 1

2. The President's Commission on Critical Infrastructure Protection, Statement of Robert T. Marsh before the Subcommittee on Technology, Terrorism and Government Information, Committee on the Judiciary, United States Senate, November 5, 1997 151

3. White Paper—the Clinton Administration's Policy on Critical Infrastructure Protection. Presidential Decision Directive 63, May 1998 163

4. Protecting America's Critical Infrastructures: PDD 63, The White House, Office of the Press Secretary 179

5. Summary of Presidential Decision Directives 62 and 63, The White House, Office of the Press Secretary 181

6. The President's National Security Telecommunications Advisory Committee, Information Infrastructure GrouP Report, September 1998 183

7. Statement by the Press Secretary, Administration Updates Encryption Policy, The White House, Office of the Press Secretary . . 201

8. Press Briefing by the Vice President, Deputy Chief of Staff John Podesta, Principal Associate Deputy Attorney General Robert Litt, Assistant Director of the FBI Carolyn Morris, Under Secretary of Commerce William Reinsch, Deputy Secretary of Defense John Hamre, and Deputy National Security Advisor Jim Steinberg, The White House, Office of the Press Secretary, September 16, 1998 205

9. President's Export Council, Subcommittee on Encryption (Pecsenc), September 18, 1998 217

10. Remarks by Richard A. Clarke, National Coordinator for Security, Infrastructure Protection, and Counter-terrorism, American Bar Association Committee on Law and National Security, November 12, 1998 221

11. President Clinton and Vice President Gore: Keeping America Secure for the 21st Century, January 22, 1999 227

12. Funding for Domestic Preparedness and Critical Infrastructure Protection. The White House, Office of the Press Secretary 229

13. Keeping America Secure for the 21st Century: Computer Security and Critical Infrastructure, The White House Office of the Press Secretary, January 22, 1999 231

14. Remarks by the President on Keeping America Secure for the 21st Century, National Academy of Sciences, Washington, D.C., The White House, Office of the Press Secretary 233

15. Press Briefing by Attorney General Janet Reno, Secretary of HHS Donna Shalala, and Richard Clarke, President's National Coordinator for Security, Infrastructure and Counterterrorism, The White House, Office of the Press Secretary 239

16. Information Technology for the Twenty-first Century: A Bold Investment in America's Future, The White House, Office of the Vice President 247

17. President's Information Technology Advisory Committee, Report to the President, Information Technology Research: Investing in Our Future, February 1999 253

18. The Electronic Intrusion Threat to National Security and Emergency Preparedness (NS/EP) Telecommunications, An Awareness Document, National Communications System, March 1999 343

FBI Documents

19. Cybernation: The American Infrastructure in the Information Age, A Technical Primer on Risks and Reliability 461

20. Statement of Louis J. Freeh, Director Federal Bureau of Investigation, July 25, 1996, before the Committee on Commerce, Science, and Transportation, United States Senate, Impact of Encryption on Law Enforcement and Public Safety 497

21. Electronic Surveillance—Report on the FBI's Publication of the Second Notice of Capacity, Statement of FBI Assistant Director James K. Kallstrom, Washington DC, FBI National Press Office, January 14, 1997 503

22. Speech by Louis J. Freeh, Director of the FBI, 1997 International Computer Crime Conference, New York, New York March 4, 1997 515

23. Encryption, Statement of Louis J. Freeh, Director Federal Bureau of Investigation, before the Senate Judiciary Committee Hearing on Encryption, United States Senate, Washington, D.C., July 9, 1997 . . . 523

24. The Impact of Encryption on Public Safety, Statement of Louis J. Freeh, Director Federal Bureau of Investigation, before the Permanent Select Committee on Intelligence, United States House of Representatives, Washington, D.C., September 9, 1997 529

25. Threats to U. S. National Security, Hearing on Current and Projected Threats to U. S. National Security, Statements for the Record before the Senate Select Committee on Intelligence, Washington, D.C., January 28, 1998 537

26. Cybercrime, Transnational Crime, and Intellectual Property Theft, Statement for the Record, Michael A. Vatis, Deputy Assistant Director and Chief, National Infrastructure Protection Center, Federal Bureau of Investigation before the Congressional Joint Economic Committee, Washington, DC March 24, 1998 545

27. National Infrastructure Protection Center Cybernotes, January 5, 1999 557

28. Encryption: Impact on Law Enforcement, Laboratory Division, Engineering Research Facility, Quantico, Virginia, January 8, 1999 and June 3, 1999 571

29. National Infrastructure Protection Center Cybernotes, January 20, 1999 587

30. National Infrastructure Protection Center Cybernotes, February 3, 1999 599

31. National Infrastructure Protection Center Cybernotes, February 17, 1999 615

32. National Infrastructure Protection Center Cybernotes, March 3, 1999 639

33. Critical Infrastructure Protection and Information Warfare Issues, Statement for the Record of Michael A. Vatis, Director, National Infrastructure Protection Center, Federal Bureau of Investigation, before the Senate Armed Service Committee, Subcommittee on Emerging Threats and Capabilities, Washington, D. C., March 16, 1999 659

34. National Infrastructure Protection Center Cybernotes, March 17, 1999 669

35. National Infrastructure Protection Center Cybernotes, March 31, 1999 687

36. National Infrastructure Protection Center Cybernotes, April 14, 1999 705

37. National Infrastructure Protection Center Cybernotes,
April 28, 1999 721

38. National Infrastructure Protection Center Cybernotes,
May 12, 1999 741

39. National Infrastructure Protection Center Cyber Notes,
May 26, 1999 755

Cyber Terrorism and Information Warfare: Threats and Responses

Four Volume Set

PREFACE

Terrorism—the calculated employment or the threat of violence by individuals, subnational groups, and state actors to attain political, social, and economic objectives in the violation of law, intended to create an overwhelming fear in a target area larger than the victims attacked or threatened—is as old as civilization itself. Yet unlike their historical counterparts, present-day terrorists have introduced into contemporary life a new scale of violence in terms of conventional and unconventional threats and impact. The internationalization and brutalization of modern terrorism make it clear that we have entered into an Age of Super and Cyber Terrorism with its serious implications to national, regional, and global security concerns.

Perhaps the most significant dangers are those relating to the safety, welfare, and rights of ordinary people; the stability of the state system; the health of economic development; the expansion of democracy; and perhaps even the survival of civilization itself.

Thus, despite the end of the Cold War and the evolving era of the New World Order, terrorism remains as threatening as ever. Undoubtedly, the conflicts emerging from ideological, religious, and national animosities will continue to make terrorism a global problem well into the twenty-first century. Ensuring the safety of its citizens at home and abroad will, therefore, continue to be every government's paramount responsibility in the coming months and years.

Future Prospects: From Conventional to Super Terrorism

There are several specific reasons why terrorism will grow in the future. First, terrorism has proved very successful in attracting publicity, disrupting the activities of government and business, and causing significant death and destruction. Second, arms, explosives, supplies, financing, and secret communications are readily available. Third, an international support network of groups and states exists that greatly facilitates the undertaking of terrorist activities.

Future attacks are sure to be characterized by both continuity and change. Groups that are small and unsophisticated can be expected to rely mostly on bombings and small arms. Those with more skills and state support will probably attempt to carry out more complex operations, such as kidnappings, assassinations, and facility attacks. Moreover, technological developments offer new targets and new capabilities for terrorist groups.

Consequently, future incidents could be much more costly in terms of human lives and property than they were in the past. Thus, the prospect of the use of mass-destruction weapons (biological, chemical, and nuclear terrorism) could inflict large-scale damage to persons and property.

Two factors suggest this likely development of more destructive forms of terrorism well into the twenty-first century. First, ironically, bringing terrorism under substantial control in the foreseeable future through national and international legislation, increased security and enforcement measures, as well as preemptive and punitive military strikes might, in fact, hasten the advent of more daring types of terrorism. As a result of this trend, other vulnerable targets created by technological advances of contemporary society are likely to become more attractive to terrorists.

A second distinct consideration that might encourage escalated terrorism is the fact that ideological and political violence is usually a means to an end: it progresses in proportion to the aims envisioned. If the goals are higher, then the level of terrorism must necessarily be higher.

It is possible, therefore, that certain conditions could provide terrorists with an incentive to escalate their attacks dramatically. A relevant example is the fanaticism that might, in the view of the terrorists, be "sanctioned" by God or "justified" by ideological or political "rationalizations."

In short, the advances of science and technology are slowly turning all modern society into a potential victim of mass destruction terrorism. As such, there is no immunity for the noncombatant segment nor for nations without a direct connection to political violence motivated by particular conflicts. "Super terrorism," then, is likely to have an unprecedented serious impact on the future quality of life and on orderly, civilized existence.

Cyber Terrorism and Information Warfare

An emerging trend of international life is the growing threat of information warfare and cyber terrorism. The U.S. Chairman of the Joint Chiefs of Staff Instruction (CJCSI, Number 3210.01, dated January 2, 1996) provides several relevant unclassified definitions related to the former danger. The first definition describes "Information Systems" as the entire infrastructure, organization, personnel, and components that collect, process, store, transmit, display, disseminate, and act on information." The second definition states "Information Warfare" consists of "actions taken to achieve information superiority by affecting adversary information, information-based processes, information systems, and computer-based networks while defending one's own information, information-based processes, information systems, and computer-based networks."

Other approaches to describe "Information Warfare" are articulated invariably in the expanding academic and professional literature on the subject. For example, Winn Schwartau in his book *Information Warfare: Cyberterrorism—Protecting Your Personal Security in the Electronic Age*, 2nd ed. (New York: Thunder's Mouth Press, 1996) classifies "Information Warfare" on the basis of three categories. The first class is "Personal Information Warfare," dealing with "attacks against an individual's electronic privacy," including "the disclosure of digital records and database entries wherever information is stored." The second class, "Corporate Information Warfare," focuses on the competition or the "war" between business corporations worldwide. And the third class, "Global Information Warfare" is a mode of "cyber warfare" waged against "industries, global economic forces, or against entire countries or states."

In short, "Information Warfare," as reflected in the manifold definitions offered as well as the coverage of the media, consists of a broad spectrum of threats ranging from electronic jamming to psychological operations underscoring the perpetrators' deliberate exploitation of military and civilian information systems' inherent vulnerabilities and thereby adversely affecting national and global security.

Consider for example, the following cases, selected at random from thousands of reported events, as illustrations of society's vulnerability to attacks:

- During 1986-1989, a ring of West German "hackers" had stolen passwords, programs, and other data for the KGB after the ring pierced military, scientific, and industry computers in the United States, Western Europe, and Japan.
- In 1994, "hackers" gained compete access to all U.S. Air Force classified networks in Rome, New York, laboratories and breached other classified sites, such as the South Korean Atomic Research Institute, through access to the Rome Laboratory's system.
- In 1997-1998, an unemployed former part-time computer science student in Canada, gained illegal access to computer systems at the U.S. National Aeronautics and Space Administration (NASA), U.S. National Oceanographic and Atmospheric Administration, and several U.S. and Canadian universities.
- In February 1998, Ehud Tenenbaum, an Israeli "hacker" (better known as "The Analyzer"), jointly with two young collaborators from California, mounted cyber attacks against the Pentagon's systems, a nuclear weapons research lab, and other significant targets.
- Several months later in 1998, a "hacker" group, calling themselves Masters of Downloading/2016216 (MOD) and claiming to consist of 15 members (8 in the United States, 5 in the United Kingdom, and 2 in Russia), penetrated a U.S. Department of Defense highly protected computer.

- Intruders, suspected "cyber spooks" from the Russian Academy of Sciences (a government-supported organization that interacts with Russia's top military laboratories), had stolen "sensitive information" about defense technical research matters during 1998-1999.
- In Spring 1999, during the Kosovo War, Serb entities attempted to overwhelm NATO computers by "ping attacks" which establishes communication with a target computer and occupies its functions by continuing to stay linked and feeding it information.
- In June 1999, tens of thousands of computer systems in several major U.S. companies (including AT&T, Boeing, and General Electric) had their files infected, damaged, or destroyed by a software virus, Explore.Zip worm. This file-destroying bug, first detected in Israel, arrived as an e-mail message and utilized MAPI commands and Microsoft Outlook on Windows systems to propagate itself. It has been estimated that damages from the Explore.Zip worm and other viruses (e.g., Melissa macro virus) topped $7 billion during the first two quarters of 1999. By the end of the year, computer intrusions will cost private industry more than $20 billion.

What is of particular concern is the prevailing assessment of intelligence agencies, strategic thinkers, and scientists that not only "hackers" and "crackers" (criminal "hackers") but also terrorists (individuals, groups, and state sponsors) are likely to intensify the exploitation of the new "equalizer" weapon as a form of "electronic warfare." It is estimated, for instance, that hostile low-risk perpetrators launching a well-coordinated attack with about thirty computer experts strategically placed around the globe and with a budget of approximately ten million dollars, could bring the United States, the only superpower, to its knees.

It is evident that the threat of electronic "non-explosive" terrorist assaults is growing with every passing day. Three contributing factors account for this reality. First, the "globalization" of the Internet users (currently numbering over 120 million with an estimated 1 billion people by the year 2005) makes governmental efforts to control Internet attacks much more challenging than ever before.

Second, there are now approximately 30,000 "hacker-oriented" sites on the Internet, thus resulting in "democratization" of the tools to be used for disruption and destruction. With the step-by-step Internet "cookbooks," the utilization of works, Trojan horses, logic bombs, and other modus operandi alternatives are becoming a permanent fixture of international life.

And third, with the Cold War now behind us, terrorist organizations have broken away from their place within the formerly bipolar world, to become multidirectional, causing further complications to our technologically vulnerable societies. These new political realities, coupled with easily accessible cyber weapons, have enhanced the threats and capabilities of terrorist groups to the degree in which they could forever alter our planet's existence.

Thus, the emergence of the term cyber terrorism has been accepted as a new mode of information warfare that explicitly adopts methods utilized by terrorists by means of the Internet. Interestingly, Kodak put its first camera on the market about 100 years ago, with the slogan "You push a button, we do the rest." At the close of the 20th century, terrorism at the push of the button has already happened.

Currently the Internet already serves as an arena for propaganda and psychological warfare. Ideological extremists such as Neo-Nazi groups have called for ethnic, racial, and religious violence by their sympathizers. Traditional terrorist organizations, like the Hezbollah, supported by Iran and Syria, maintains on its web site a daily record of "heroic" battles of its fighters in Southern Lebanon. And Afghanistan, the newest state sponsor of terrorism, publishes "on-line" its radical brand of Islam.

Terrorists have also used their laptops to store operation plans. Ramzi Ahmed Yusuf (who is serving a life sentence for his role in perpetrating the World Trade Center bombing in New York in 1993 and for other terrorist crimes) developed on his computer the plot to blow-up some dozen American airlines over the Pacific Ocean. And terror "networks," such as the underground infrastructure of Osama bin Laden, the "international jackal" (accused of destroying the American embassies in Kenya and Tanzania in 1998), are sustained via personal computers with satellite uplinks and encrypted messages.

Is the worst yet to come? It is only a matter of time. Consider the possibility of waking to the news one morning to hear that a group of terrorists employing electronic "sniffers" sabotaged the global financial system by disrupting international fund-transfer networks, causing an unprecedented stock plunge on the New York, London, and Tokyo exchanges.

Clearly, there are numerous other devastating scenarios, including altering formulas for medication at pharmaceutical plants, "crashing" telephone systems, misrouting passenger trains, changing pressure in gas pipelines to cause valve failure, disrupting operations of air traffic control, triggering oil refinery explosions and fires, scrambling the software used by emergency services, "turning off" power grids, and detonating simultaneously hundreds of computerized bombs around the world.

In sum, this new medium of communication, command and control, supplemented by unlimited paralyzing and destructive keyboard attacks on civilian and military critical infrastructure nerve centers, forces us to think about the "unthinkable" with grave concern.

Governmental and Nongovernmental Responses

To be sure, the growing concern with the threats of information warfare and cyber terrorism has already resulted in an expansion of security measures undertaken by governmental, intergovernmental and business communities. Suffice to mention that the U.S. Government has for the past several years been engaged in developing a comprehensive strategy to protect its critical infrastructures, including telecommunications, energy, banking, finance, water systems, government operations, and emergency services. By Executive Order in July 1996, President Bill Clinton established the Commission on Critical Infrastructure Protection headed by General Robert T. Marsh, U.S. Air Force (Ret.). As a joint government and private sector endeavor, the Commission developed a national policy and implementation strategy for protecting the eight critical infrastructures.

Additionally, in February 1998, a National Infrastructure Protection Center (NIPC) located at the Federal Bureau of Investigation's (FBI) headquarters in Washington, D.C., was established for the purpose of serving as the U.S. Government focal point for threat assessment, warning, investigation, and response for threats or attacks against U.S. critical infrastructures. NIPC is directed by Michael Vatis, Assistant Director of the FBI. And in October 1999, the U.S. Department of Defense established a Cyber Warfare Center to be based at the Space Command in Colorado Springs under General Richard Myers of the Air Force. The mission of this new structure is to defend the U.S. systems from electronic attacks and disable the enemy's computer systems.

Clearly, these policies and actions, coupled with other U.S. governmental moves as well as a multitude of private security measures, signify a positive trend to address the new threats to national and global security in the information age. The academic community also has an obligation to identify the nature of the cyber threats and to recommend possible strategies to reduce the risks. One step in this direction is the formation of the Inter-University Center for Cyber Terrorism and Information Warfare. This structure, established in 1997 as a consortium of universities and think tanks coordinated by the Center for Counter Terrorism Studies, Potomac Institute for Policy Studies, located in Virginia. Several academic efforts have been taken, including the sponsorship of two international seminars on "Cyber Terrorism and Information Warfare: Threats and Responses" (April 16, 1998) and "Strategies to Combat Cyber Terrorism and Information Warfare in the 21st Century" (October 28, 1999).

Additionally, in conjunction with the multinational research project on "Counter Terrorism Strategies for the 21st Century: National, Regional and Global Agenda," focusing on providing policy-makers with a comprehensive "best practices" counter terrorism strategy for the future, the Inter-University Center for Cyber Terrorism and Information Warfare has developed a supplementary research effort in this important field of public concern. Utilizing the documentary resources available at the Center, in collaboration with other affiliate structures such as the Inter-University Center for Terrorism Studies and the Inter-University Center for Legal Studies (both based in Washington, D.C.), we assembled relevant materials focusing on cyber threats and responses from a U.S. perspective.

The purpose of this four-volume set on *Cyber Terrorism and Information Warfare* is to provide the most comprehensive resource of U.S. documentation on the subject. While many of these sources are familiar to researchers in the field, many more works and other documentation remain virtually unknown or forgotten. This set is being published in Oceana Publications' Second Series of *Terrorism: Documents of International and Local Control.*

The first volume "Executive Branch" consists essentially of White House and FBI materials. The second volume, titled "Legislative Branch" includes Attorney General, Central Intelligence Agency and Department of Defense documents as well as congressional testimony, and court cases. The third volume on "Legal Foundations" covers issues such as private intrusion response, legal authorities database, major federal legislation, and legal impediments to information sharing. The fourth volume on "Research and Development Roadmaps" incorporates recommendations for protecting and assuring critical national infrastructures. It is expected that other volumes covering additional U.S. and other nations' documentation will be published subsequently.

We wish to acknowledge the financial support of the Smith Richardson Foundation for the research project on "Counter Terrorism Strategies for the 21st Century: National, Regional, and Global Agenda" which thus far has resulted in the publication of this four volume set as well as Volumes 15, 16, 17, and 18 of *Terrorism: Documents of International and Local Control—U.S. Perspectives* published earlier this year. Thanks are also due to the project research staff, Dr. Herbert M. Levine, Mr. James T. Kirkhope, and Ms. Kaija E. Schilde.

Yonah Alexander

and

Michael S. Swetnam

Arlington, Virginia, USA
November 1, 1999

ABOUT THE EDITORS

Yonah Alexander

Professor Yonah Alexander is Director of the Terrorism Studies Program at The George Washington University. Concurrently, he is Director, Inter-University Center for Terrorism Studies, which is affiliated with academic institutions around the world. In addition, Professor Alexander is a Senior Fellow, Center for International Studies, Tel Aviv University and Director, Counter-Terrorism Center, Potomac Institute for Policy Studies.

Educated at Columbia University, the University of Chicago, and the University of Toronto, Professor Alexander taught at The George Washington University; The American University; Columbus School of Law at Catholic University; Tel Aviv University; the City University of New York; and the State University of New York. He has also lectured throughout the world, including Amsterdam; Ankara; Beijing; Berkeley; Berlin; Bogota; Brussels; Buenos Aires; Cairo; Caracas; Geneva; Harvard; Lima; London; Los Alamos; Mexico City; Moscow; New Delhi; Oxford; Paris; Princeton; Rome; Santiago; Seoul; Singapore; Stanford; Stockholm; Sydney; Tokyo; and Warsaw.

His research experience includes such appointments as Research Professor of International Affairs, The George Washington University; Director, "Terrorism, Gray Area and Low Level," U.S. Global Strategy Council; Senior Fellow, The Institute for Advanced Studies in Justice, School of Law, The American University; Research Associate, School of Journalism, Columbia University; Senior Staff Member, Center for Strategic and International Studies, Georgetown University; Director, Institute for Studies in International Terrorism, State University of New York; and Fellow, Institute of Social Behavior Pathology, the University of Chicago.

Dr. Alexander is Founder and Editor-in-Chief of *Terrorism: An Electronic Journal and Knowledge Base* and the *International Journal on Minorities and Group Rights*. He also founded and edited *Terrorism: An International Journal* and *Political Communication and Persuasion: An International Journal*. He has published over sixty books on the subjects of international affairs, terrorism, and psychological warfare. He serves as Editor-in-Chief of a series of books on terrorism for Martinus Nijhoff (The Netherlands) and is co-editor of the *International Library of Terrorism* (Dartmouth Publishing, United Kingdom), and *Terrorism: Documents of International and Local Control* (Oceana Publications, New York). In addition, he is General Editor of *Terrorism: An International Resources File* (UMI, Ann Arbor, Michigan).

Michael S. Swetnam

Mr. Swetnam founded the Potomac Institute for Policy Studies in 1994 and has served as its President and Chairman. The Institute is a not-for-profit organization dedicated to the study of technology and technology policy as it effects government and industry. The Institute has conducted studies and provided support to the National Science Foundation, the Defense Advanced Research Projects Agency, NASA, Private Foundations, and several leading industries.

Mr. Swetnam is currently a member of the Technical Advisory Group to the United States Senate Special Select Committee on Intelligence. In this capacity Mr. Swetnam provides expert advice to the US Senate on the R&D investment strategy of the US Intelligence Community. Mr. Swetnam also served as a Special Consultant to President Bush's Foreign IntelligenceAdvisory Board (PFIAB) from 1990 to 1992 where he provided expert advise on Intelligence Community issues including budget, community architecture, and major programs. He also assisted in authoring the Board's assessment of Intelligence Community support to Desert Storm/Shield.

Prior to forming the Potomac Institute for Policy Studies, Mr. Swetnam worked in private industry for as a Vice President of Engineering at the Pacific-Sierra Research Corporation, Director of Information Processing Systems at GTE, and Manager of Strategic Planning for GTE Government Systems.

Prior to joining GTE, Mr. Swetnam worked for the Director of Central Intelligence as a Program Monitor on the Intelligence Community Staff (1986-1990). Mr. Swetnam was responsible for the development and presentation to Congress of the budget of the National Security Agency and helped develop, monitor, and present to Congress the DOE Intelligence Budget. Mr. Swetnam was also assigned as the IC Staff representative to intergovernmental groups that developed the INF and START treaties. He assisted in presenting these treaties to Congress for ratification. Collateral duties included serving as the host to the DCI's Nuclear Intelligence Panel and Co-Chairman of the S&T Requirements Analysis Working Group.

Mr. Swetnam has served in several public and community positions including as a School Board Member (1970-72), Northern United Kingdom Scout Master (1984-85), Chairman, Term limits Referendum Committee (1992-93), President (1993) of the Montgomery County Corporate Volunteer Council, Montgomery County Corporate Partnership for Managerial Excellence (1993), and the Maryland Business Round-table (1993).

CHAPTER 1

CRITICAL FOUNDATIONS
PROTECTING AMERICA'S INFRASTRUCTURES

The Report of the President's Commission on Critical Infrastructure Protection

October 13, 1997

Contents

Foreword

Executive Summary

Part One: The Case for Action

- Chapter One Acting Now to Protect the Future
- Chapter Two The New Geography
- Chapter Three New Vulnerabilities, Shared Threats, Shared Responsibility
- Chapter Four Findings and Policy

Part Two: A Strategy for Action

- Chapter Five Establishing the Partnership
- Chapter Six Building the Partnership
- Chapter Seven Structuring the Partnership
- Chapter Eight Report on Awareness and Education
- Chapter Nine Leading by Example
- Chapter Ten Legal Initiatives
- Chapter Eleven Research and Development
- Chapter Twelve Implementation Strategy

Onward

Appendices

- Appendix A Sector Summary Reports
- Appendix B Glossary

The President
The White House
Washington, DC 20500

Dear Mr. President:

It is a privilege to forward the report of the President's Commission on Critical Infrastructure Protection, *Critical Foundations*. You asked us to study the critical infrastructures that constitute the life support systems of our nation, determine their vulnerabilities and propose a strategy for protecting them into the future. I believe our report does this.

There is no doubt that our critical infrastructures are the best in the world–largely the result of the tremendous efficiency and global reach made possible by incorporation of our rapidly advancing information and communication technology. In fact, we found all our infrastructures increasingly dependent on information and communications systems that criss-cross the nation and span the globe. That dependence is the source of rising vulnerabilities and, therefore, it is where we concentrated our effort.

We found no evidence of an impending cyber attack which could have a debilitating effect on the nation's critical infrastructures. While we see no electronic disaster around the corner, this is no basis for complacency. We did find widespread capability to exploit infrastructure vulnerabilities. The capability to do harm–particularly through information networks–is real; it is growing at an alarming rate; and we have little defense against it.

Because the infrastructures are mainly privately owned and operated, we concluded that critical infrastructure assurance is a shared responsibility of the public and private sectors. The only sure path to protected infrastructures in the years ahead is through a real partnership between infrastructure owners and operators and the government. Consequently, in addition to our recommendations about improving our government's focus on infrastructure assurance in the Information Age, you will find some recommendations for collaborative public and private organizational arrangements that challenge our conventional way of thinking about government and private sector interaction.

Thank you for the opportunity to serve our nation on this Commission, and for the chance to work with a talented and patriotic group of Commissioners and staff from both government and the private sector.

Respectfully,

Robert T. Marsh
Chairman

President's Commission on Critical Infrastructure Protection

COMMISSIONERS

Robert T. Marsh, *Chairman*

John R. Powers, *Executive Director – Federal Emergency Management Agency*

Merritt E. Adams – *AT&T*

Richard P. Case – *IBM*

Mary J. Culnan – *Georgetown University*

Peter H. Daly – *Department of the Treasury*

John C. Davis – *National Security Agency*

Thomas J. Falvey – *Department of Transportation*

Brenton C. Greene – *Department of Defense*

William J. Harris – *Association of American Railroads*

David A. Jones – *Department of Energy*

William B. Joyce – *Central Intelligence Agency*

David V. Keyes – *Federal Bureau of Investigation*

Stevan D. Mitchell – *Department Justice*

Joseph J. Moorcones – *National Security Agency*

Irwin M. Pikus – *Department of Commerce*

William Paul Rodgers, Jr. – *National Association of Regulatory Utility Commissioners*

Susan V. Simens – *Federal Bureau of Investigation*

Frederick M. Struble – *Federal Reserve Board*

Nancy J. Wong – *Pacific Gas and Electric Company*

EXECUTIVE STAFF

Phillip E. Lacombe, *Staff Director*

James H. Kurtz, COL, USA, *Chief of Staff/Executive Secretary*

Janet B. Abrams, *Director of External Affairs/White House Liaison*

Robert E. Giovagnoni, Col, USAF, *General Counsel*

Adrienne M. Griffen, *Executive Assistant to the Chairman*

Elizabeth (Betsy) Harrison, *Director of Legislative Affairs*

Brian P. Hoey, Lt Col, USAF, *Executive Assistant to the Chairman*

Nelson M. McCouch III, MAJ, USA, *Director of Public Affairs*

Monica Y. McNeil, *Executive Assistant to the Chief of Staff/Assistant Executive Secretary*

Annie N. Nelson, *Director of Administration*

Carla L. Sims, *Director of Public Affairs*

Lawrence P. St. Marie, SMSgt, USAF, *Executive Officer*

Sona A. Virdi, *Executive Assistant to the Staff Director*

PROFESSIONAL STAFF

Elizabeth A. Banker

Gary R. Boyd

Patricia E. Burt

Julie Consilvio

Frederick S. Davidson

L. C. J. Jacobson

Gary P. Kosciusko

Lloyd E. Lutz Jr., Lt Col, USAF

Carol M. Medill

T. Lynette Proctor

Pamela D. Saunders

James J. Stekert

Stephen T. York

SUPPORT STAFF

Bernard R. Robinson, *Deputy Director of Administration*

Bonnie L. Julia, SFC, USA, *NCOIC*

Karen R. Allen, SrA, USAF

Robert W. Boyd, YN2, USN

Joseph A. Broadway, YN1, USN

Patrick Barlow

Eric J. Cline

James E. Crawford, SSG, USA

Debra A. Dawson, SSG, USA

Roada Dickerson, SrA, USAF

Elizabeth S. Ellingboe, SSgt, USAF

Jeffrey G. Estep, SSgt, USAF

Troy L. Joyner, SSG, USA

Peter D. LeNard

Becky Love

Gerald T. Posey, TSgt, USAF

Sandra M. Robinson, SSgt, USAF

Sandra L. Scroggs

Mike Seabron

Sherrie M. Smith, SGT, USA

Sharon S. Strippoli

Shawn R.L. Vincent, Sgt, USAF

Scott A. Ward

Brian W. Young, SrA, USAF

Ed Young

SENIOR CONSULTANTS

William A. Buehring

Mary F. Dunham

Ron E. Fisher

Paul W. Hanley

Peter Gossens

Duane G. Harder

Michael T. Hovey

Joelle Jordan

Ramesh Maraj

Gabe Maznick

Willis J. Ozier

Paul Byron Pattak

James P. Peerenboom

George J. Rothstein

Lee M. Zeichner

And special thanks to the following for their advice and support:

Ed Appel

Frederick L. Frostic

Bill Garber

Seymour Goodman

David Graham

Michael Leonard

Stephen J. Lukasik

Paul H. Richanbach

Kathleen Robertson

Elizabeth Sauer

Paula Scalingi

James Schlesinger

Suzy Tichenor

Larry Welch

FOREWORD

The task given us by the President was daunting. America's critical infrastructures underpin every aspect of our lives. They are the foundations of our prosperity, enablers of our defense, and the vanguard of our future. They empower every element of our society. There is no more urgent priority than assuring the security, continuity, and availability of our critical infrastructures.

After fifteen months of evaluating the infrastructures, assessing their vulnerabilities, and deliberating assurance alternatives, our fundamental conclusion is that we have to think differently about infrastructure protection today and for the future.

We found that the nation is so dependent on our infrastructures that we must view them through a national security lens. They are essential to the nation's security, economic health, and social well being. In short, they are the lifelines on which we as a nation depend.

We also found the collective dependence on the information and communications infrastructure drives us to seek new understanding about the Information Age. Essentially, we recognize a very real and growing cyber dimension associated with infrastructure assurance.

In the cyber dimension there are no boundaries. Our infrastructures are exposed to new vulnerabilities–cyber vulnerabilities–and new threats–cyber threats. And perhaps most difficult of all, the defenses that served us so well in the past offer little protection from the cyber threat. Our infrastructures can now be struck directly by a variety of malicious tools.

Our new thinking must accommodate the cyber dimension./ We must develop a new set of "street smarts" to deal with it, and we must apply them in new ways. One of the most important is recognizing that the owners and operators of our critical infrastructures are now on the front lines of our security effort. They are the ones most vulnerable to cyber attacks. And that vulnerability jeopardizes our national security, global economic competitiveness, and domestic well being.

It is with this in mind that we offer our report.

EXECUTIVE SUMMARY

"Our responsibility is to build the world of tomorrow by embarking on a period of construction–one based on current realities but enduring American values and interests . . ."

President William J. Clinton, "A National Security Strategy for a New Century," May 1997

Introduction

Our national defense, economic prosperity, and quality of life have long depended on the essential services that underpin our society. These critical infrastructures–energy, banking and finance, transportation, vital human services, and telecommunications–must be viewed in a new context in the Information Age. The rapid proliferation and integration of telecommunications and computer systems have connected infrastructures to one another in a complex network of interdependence. This interlinkage has created a new dimension of vulnerability, which, when combined with an emerging constellation of threats, poses unprecedented national risk.

For most of our history, broad oceans, peaceable neighbors and our military power provided all the infrastructure protection we needed. But just as the terrible long-range weapons of the Nuclear Age made us think differently about security in the last half of the 20th Century, the electronic technology of the Information Age challenges us to invent new ways of protecting ourselves now. We must learn to negotiate a new geography, where borders are irrelevant and distances meaningless, where an enemy may be able to harm the vital systems we depend on without confronting our military power. National defense is no longer the exclusive preserve of government, and economic security is no longer just about business. The critical infrastructures are central to our national defense and our economic power, and we must lay the foundations for their future security on a new form of cooperation between government and the private sector.

The Case for Action

A satchel of dynamite and a truckload of fertilizer and diesel fuel are known terrorist tools. Today, the right command sent over a network to a power generating station's control computer could be just as devastating as a backpack full of explosives, and the perpetrator would be more difficult to identify and apprehend.

The rapid growth of a computer-literate population ensures that increasing millions of people around the world possess the skills necessary to conduct such an attack. The wide adoption of common protocols for system interconnection and the availability of "hacker tool" libraries make their task easier.

While the possibility of chemical, biological, and even nuclear weapons falling into the hands of terrorists adds a new and frightening dimension to physical attacks, such weapons are difficult to acquire. In contrast, the resources necessary to conduct a cyber attack have shifted in the past few years from the arcane to the commonplace. A personal computer and a telephone connection to an Internet Service Provider anywhere in the world are enough to cause harm.

Growing complexity and interdependence, especially in the energy and communications infrastructures, create an increased possibility that a rather minor and routine disturbance can cascade into a regional outage. Technical complexity may also permit interdependencies and vulnerabilities to go unrecognized until a major failure occurs.

We know our infrastructures have substantial vulnerabilities to domestic and international threats. Some have been exploited–so far chiefly by insiders. Although we know these new vulnerabilities place our infrastructures at risk, we also recognize that this is a new kind of risk that requires new thinking to develop effective countermeasures. Coping with increasingly cyber-based threats demands a new approach to the relationship between government and the private sector. Because it may be impossible to determine the nature of a threat until after it has materialized, infrastructure owners and operators–most of whom are in the private sector –must focus on protecting themselves against the tools of disruption, while the government helps by collecting and disseminating the latest information about those tools and their employment. This cooperation implies a more intimate level of mutual communication, accommodation, and support than has characterized public-private sector relations in the past.

The Commission has not discovered an immediate threat sufficient to warrant a fear of imminent national crisis. However, we-are convinced that our vulnerabilities are increasing steadily, that the means to exploit those weaknesses are readily available and that the costs associated with an effective attack continue to drop. What is more, the investments required to improve the situation–now still relatively modest–will rise if we procrastinate.

We should attend to our critical foundations before we are confronted with a crisis, not after. Waiting for disaster would prove as expensive as it would be irresponsible.

A Strategy for Action

The Commission recommends several practical measures to realize our vision of a new government-private sector partnership.

The quickest and most effective way to achieve a much higher level of protection from cyber threats is a strategy of cooperation and information sharing based on partnerships among the infrastructure owners and operators and appropriate government agencies.

To facilitate this new relationship between government and industry, new mechanisms will be needed, including sector "clearing houses" to provide the focus for industry cooperation and information sharing; a council of industry CEOs, representatives of state and local government, and Cabinet secretaries to provide policy advice and implementation commitment; a real-time capability for attack warning; and a top-level policy making office in the White House.

Other measures are also required. Infrastructure protection must be ingrained in our culture, beginning with a comprehensive program of education and awareness. This includes both infrastructure stakeholders and the general public, and must extend through all levels of education, both academic and professional.

The federal government must lead the way into the Information Age by example, tightening measures to protect the infrastructures it operates against physical and cyber attack.

The government can also help by streamlining and clarifying elements of the legal structure that have not kept pace with technology. Some laws capable of promoting assurance are not as clear or effective as they could be. Others can operate in ways that may be unfriendly to security concerns. Sorting them out will be an extensive undertaking, involving efforts at local, state, federal, and international levels. We have offered a number of preliminary legal recommendations intended to jump-start this process of reform.

Another area where government must lead is in research and development. Some of the basic technology and tools needed to provide improved infrastructure protection already exist, but need to be widely employed. However, there is a need for additional technology with which to protect our essential systems. We have, therefore, recommended a program of research and development focused on those needed capabilities.

In summary, all of us need to recognize that the cyber revolution brings us into a new age as surely as the industrial revolution did two centuries ago. Now, as then, our continued security requires a reordering of national priorities and new understanding about our respective roles in support of the national goals. The relationships that have stood us in such good stead through the end of the second millennium must give way to new ones better suited to the third.

PART ONE
THE CASE FOR ACTION

CHAPTER ONE
ACTING NOW TO PROTECT THE FUTURE

"We are at the dawn of a new century. Now is the moment to be farsighted as we chart a path into the new millennium."

President William J. Clinton, "A National Security Strategy for a New Century," May 1997

Life is good in America because things work. When we flip the switch, the lights come on. When we turn the tap, clean water flows. When we pick up the phone, our call goes through. We are able to assume that things will work because our *infrastructures* are highly developed and highly effective. By *infrastructure* we mean more than just a collection of individual companies engaged in related activities; we mean a network of independent, mostly privately-owned, manmade systems and processes that function collaboratively and synergistically to produce and distribute a continuous flow of essential goods and services.

Businesses, too, depend on infrastructures. Private companies are able to guarantee on-time performance because our infrastructures permit low cost transport and instantaneous tracking of shipments. Managers take for granted that the goods and services essential to their operations will be there when needed.

Reliable and secure infrastructures are thus the foundation for creating the wealth of our nation and our quality of life as a people. They are fundamental to development and projection of the military power that enables our diplomacy to be effective. They make it possible for us to enjoy our inalienable rights and take advantage of the freedoms on which our nation was founded. Certain of our infrastructures are so vital that their incapacity or destruction would have a debilitating impact on our defense and economic security.

The ***transportation*** infrastructure moves goods and people within and beyond our borders, and makes it possible for the United States to play a leading role in the global economy.

The ***oil and gas production and storage*** infrastructure fuels transportation services, manufacturing operations, and home utilities.

The ***water supply*** infrastructure assures a steady flow of water for agriculture, industry (including various manufacturing processes, power generation, and cooling), business, firefighting, and our homes.

The *emergency services* infrastructure in communities across the country responds to our urgent police, fire, and medical needs, saving lives and preserving property.

The *government services* infrastructure consists of federal, state, and local agencies that provide essential services to the public, promoting the general welfare.

The *banking and finance* infrastructure manages trillions of dollars, from deposit of our individual paychecks to the transfer of huge amounts in support of major global enterprises.

The *electrical power* infrastructure consists of generation, transmission, and distribution systems that are essential to all other infrastructures and every aspect of our economy. Without electricity, our factories would cease production, our televisions would fade to black, and our radios would fall silent (even a battery-powered receiver depends on an electric-powered transmitter). Our street intersections would suddenly be dangerous. Our homes and businesses would go dark. Our computers and our telecommunications would no longer operate.

The *telecommunications* infrastructure has been revolutionized by advances in information technology in the past two decades to form an *information and communications* infrastructure, consisting of the Public Telecommunications Network (PTN), the Internet, and the many millions of computers in home, commercial, academic, and government use. Taking advantage of the speed, efficiency and effectiveness of computers and digital communications, all the critical infrastructures are increasingly connected to networks, particularly the Internet. Thus, they are connected to one another. Networking enables the electronic transfer of funds, the distribution of electrical power, and the control of gas and oil pipeline systems. Networking is essential to a service economy as well as to competitive manufacturing and efficient delivery of raw materials and finished goods. The information and communications infrastructure is basic to responsive emergency services. It is the backbone of our military command and control system. And it is becoming the core of our educational system.

Disruption of any infrastructure is always inconvenient and can be costly and even life threatening. Major disruptions could lead to major losses and affect national security, the economy, and the public good. Mutual dependence and the interconnectedness made possible by the information and communications infrastructure lead to the possibility that our infrastructures may be vulnerable in ways they never have been before. Intentional exploitation of these new vulnerabilities could have severe consequences for our economy, security, and way of life.

Technologies and techniques that have fueled major improvements in the performance of our infrastructures can also be used to disrupt them. The

United States, where close to half of all computer capacity and 60 percent of Internet assets reside, is at once the world's most advanced and most dependent user of information technology. More than any other country, we rely on a set of increasingly accessible and technologically reliable infrastructures, which in turn have a growing collective dependence on domestic and global networks. This provides great opportunity, but it also presents new vulnerabilities that can be exploited. It heightens risk of cascading technological failure, and therefore of cascading disruption in the flow of essential goods and services. Computerized interaction within and among infrastructures has become so complex that it may be possible to do harm in ways we cannot yet conceive.

The threat is real enough. The terrorist bombings of the US World Trade Center, the federal building in Oklahoma City, and the El Khobar quarters in Saudi Arabia have demonstrated all too well the malevolent intent of some parties toward the United States. Skilled computer operators have demonstrated their ability to gain access to networks without authorization. Some do it for the thrill or the notoriety. Some do it for financial gain. Some do it to further a cause. Whatever the motivation, their success in entering networks to alter data, extract financial or proprietary information, or introduce viruses demonstrates that it can be done and gives rise to concerns that, in the future, some party wishing to do serious damage to the United States will do so by the same means.

Real vulnerabilities also exist. Infrastructures have always been subject to local or regional outages resulting from earthquakes, storms, and floods. Their owners and operators, in cooperation with local, state, and federal emergency services, have demonstrated their capacity to restore services efficiently. Physical vulnerabilities to man-made threats, such as arson and bombs, are likewise not new. But physical vulnerabilities take on added significance as new capabilities to exploit them emerge, including chemical, biological, and even nuclear weapons. As weapons of mass destruction proliferate, the likelihood of their use by terrorists increases.

Terrorist attacks have typically been against single targets–individuals, buildings, or institutions. Today, more sophisticated physical attacks may also exploit the emerging vulnerabilities associated with the complexity and interconnectedness of our infrastructures. Bombs–even homemade ones–have always been able to damage a pipeline, electrical power transformer, telecommunications switching station, or microwave relay antenna. In the networked world of today, the effects of such physical attacks could spread far beyond the radius of a bomb blast. Adding to our physical vulnerability is the fact that information readily available on the World Wide Web (WWW) may disclose to a terrorist the best place to set explosive charges for maximum disruptive effects.

Our dependence on the information and communications infrastructure has created new cyber vulnerabilities, which we are only starting to understand. In addition to the disruption of information and communications, we also face the possibility that someone will be able to actually mount an attack against other infrastructures by exploiting their dependence on computers and telecommunications (see Figure 1).

Physical means to exploit physical vulnerabilities probably remain the most worrisome threat to our infrastructures *today*. But almost every group we met voiced concerns about the new cyber vulnerabilities and threats. They emphasized the importance of developing approaches to protecting our infrastructures against cyber threats ***before*** they materialize and produce major system damage.

Figure 1. Interdependencies: New Risks and Vulnerabilities

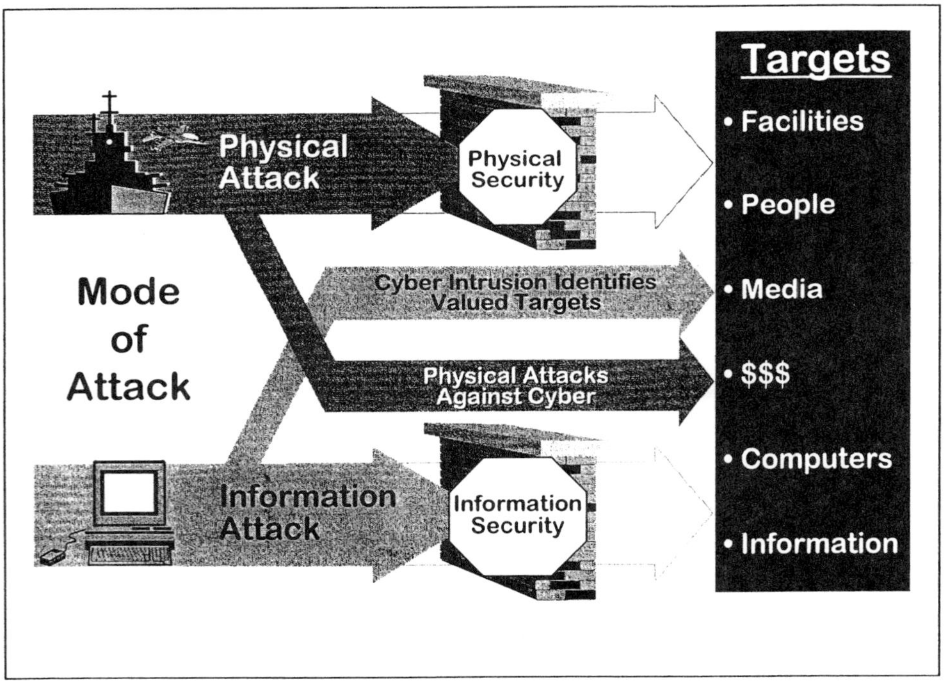

We know our infrastructures have substantial vulnerabilities to domestic and international threats. Some have been exploited–so far chiefly by insiders. Protecting our infrastructures into the 21st Century requires that we develop greater understanding of their vulnerabilities and act decisively to reduce them. It was for just this purpose that President Clinton called into being the President's Commission on Critical Infrastructure Protection in July 1996. In the fifteen months since its creation, the Commission–drawn from the federal government and the private sector–has thoroughly reviewed the vulnerabilities and threats facing our infrastructures, assessed the risks, consulted with thousands of experts, and deliberated at length as to how best to assure our nation's critical foundations in the decades to come. Our analyses, findings, conclusions, and recommendations form the substance of this report.

Our fundamental conclusion is this:

> ***Waiting for disaster is a dangerous strategy. Now is the time to act to protect our future.***

CHAPTER TWO
THE NEW GEOGRAPHY

"As borders open and the flow of information, technology, money, trade, and people across borders increases, the line between domestic and foreign policy continues to blur."

President William J. Clinton, "A National Security Strategy for a New Century," May 1997

Few enemies of the United States have ever had the means to seriously threaten our heartland. Even in the darkest early days of World War II, just after Pearl Harbor, no enemy had the shipping, landing craft, or forces to invade the continental US, or aircraft with the range to reach the mainland and return. For most of our history we've never had to worry much about being attacked at home; broad oceans east and west and peaceable neighbors north and south gave us all the protection we needed.

In the early 1950s, the geography that kept us safe was overcome by Soviet long-range bombers and intercontinental ballistic missiles aimed not only at our military capabilities, but also at the industries and institutions that give our nation its character. We had to learn to think differently about our safety and security. We built backyard bomb shelters, and whole generations of us practiced diving beneath our school desks at the sound of a siren. Our fear of surprise nuclear attack slowly faded as we developed satellites and other early warning capabilities that enabled us to overcome geography and detect a Soviet missile launch in time to launch our own missiles–thus ensuring the credibility of the deterrent policy of Mutual Assured Destruction.

The demise of the Soviet Union, "detargeting" of nuclear missiles, and strategic arms reductions appear to have left America once more relatively invulnerable to physical attack by foreign nations. However, as the threat of a nuclear war has diminished, new technologies have appeared that render physical geography less relevant and our domestic sanctuary less secure. Today, a computer can cause switches or valves to open and close, move funds from one account to another, or convey a military order almost as quickly over thousands of miles as it can from next door, and just as easily from a terrorist hideout as from an office cubicle or military command center. A computer message from Earth can steer a vehicle and point a camera on the surface of Mars. A false or malicious computer message can traverse multiple national borders, leaping from jurisdiction to jurisdiction to avoid identification, complicate lawful pursuit, or escape retribution.

Vulnerability to an adversary using cyber tools was examined during a military exercise[1] conducted in early summer of 1997. The scenario fea-

---

1 Chairman of the Joint Chiefs of Staff Exercise ELIGIBLE RECEIVER 1997.

tured "scripted" attacks on the energy and telecommunications infrastructures (controllers injected incidents into the scenario; military commands and government agencies reacted as though the reported incidents were real). Companies providing electrical power in selected cities were subjected to scripted attack by cyber means, over time, in a way that made the resulting simulated outages appear to be random and unrelated. Concurrently, a "Red Team" used hacker techniques available on the Internet to attempt to penetrate Department of Defense (DoD) computers. With no insider information, and constrained by US law, the team spent three months probing the vulnerabilities of several hundred unclassified computer networks. They were able to penetrate many of these networks, and even gained system administrator level privileges in some.

Simulated cyber attacks on nearby privately owned energy companies and telecommunications service providers and successful penetrations into DoD computers were assessed by controllers as sufficient to have disrupted operations at selected military bases–creating a situation in which our ability to deploy and sustain military forces was degraded. Was this exercise an overstatement of today's vulnerabilities or a glimpse at future forms of terrorism and war? The experience to date, the known vulnerabilities, and the continuing pace of change suggest the latter.

In short, the day may be coming when an enemy can attack us from a distance, using cyber tools, without first confronting our military power and with a good chance of going undetected. The new geography is a borderless cyber geography whose major topographical features are technology and change.

But it is also a global geography. The world's economy is integrated as never before. With rapid movement of capital, labor, goods and services, technology, and above all, information, across frontiers, our businesses have global outlooks, customers, and needs. In this global economy, communications give even small nations equal access to markets. A nation may no longer need to control territory to have access to its resources.

These changes also have a dark side. As a result of global economic integration, made possible in large measure by information technology, operations of US infrastructures extend far beyond our national boundaries, and even beyond our control. As networks extend to new markets and new sources, new points of entry are established, providing conduits of attack to adversaries at home and abroad. International terrorism, narcotics trafficking, and transnational economic crime are also features–undesirable features–of the new geography.

Technology and Change

Fifteen years ago, there were few cell phones or computers and Internet access was limited. The World Wide Web did not exist, nor did today's widely used e-mail systems.

Today, in the United States alone, there are about 180 million computers. Worldwide, there are some 1.3 million local area networks. Computers communicate regionally, nationally, and globally across thousands of wide area networks or through the Internet.

The pace of technological change and our reliance on technology are suggested in Table 1, which compares worldwide populations of 1982 with those of a year ago and those projected to exist in 2002.[2] This table illustrates the growth in the number of potential targets for a cyber attack. It also shows the growth in the number of people having the technical skills necessary to launch such an attack. Of particular significance is the fact that in the past 15 years, the public telecommunications network has become increasingly software driven, remotely managed and maintained through computer networks. The last line of the table shows the population of systems control software specialists who possess the tools and know-how to disrupt or take down the public telecommunications network.

Table 1. Global Technology Trends

Category	15 Years Ago	1996	5 Years Hence
Personal Computers	Thousands	400 million	500 million
Local Area Networks	Thousands	1.3 million	2.5 million
Wide Area Networks	Hundreds	Thousands	Tens of thousands
Viruses	Some	Thousands	Tens of thousands
Internet Devices Accessing the World-Wide Web (WWW)	None	32 million	300 million
Population With Skills for a Cyber Attack	Thousands	17 million	19 million
Telecommunications Systems Control Software Specialists	Few	1.1 million	1.3 million

---

2 Technical population data, programmers and telecommunications, 1982-2025, International Data Corporation, and e-mail and documents from the National Computer Security Center, National Security Agency, July 29, 1997.

Effects of the New Geography on Infrastructures

Profound change within the global marketplace, interdependency, restructuring, and reliance on technology make protection a continuing challenge for business and national leaders. The ever-expanding global information infrastructure underpins the global economy. Both business and government must adjust to a borderless world of unrestricted transactions and communications.

Many major infrastructure industries, particularly telecommunications and electricity, are being affected by deregulation and are restructuring to compete at home and in the global marketplace. Organizations have harnessed information technology to accelerate their delivery of goods and services, tighten the efficiency of their processes, and shed excess inventory and unused reserve capacity. Many businesses are so tightly balanced in their "just-in-time" processes that recovery from even a minor disruption would prove difficult.

In sum, technology and change produce better service at lower cost, new markets and more efficient processes throughout the nation and indeed the world. As a result, we depend more than ever on infrastructure services. But at the same time, market forces result in a diffusion of accountability, a decrease in "end-to-end" or system-wide analysis and responsibility, less research and development investment, and a reduction in reserve capacity. Today's processes are more efficient, but they lack the redundant characteristics that gave their predecessors more resilience.

All of us–government and business, service providers, and service consumers–must pay attention to, and think differently about, a new geography that is global in the physical dimension and without borders in the cyber dimension.

CHAPTER THREE
NEW VULNERABILITIES, SHARED THREATS, SHARED RESPONSIBILITY

"We face no imminent threat, but we do have an enemy–the enemy of our time is inaction."

President William J. Clinton,
State of the Union Address, January 1997

New Vulnerabilities

Each of the infrastructures is vulnerable in varying degrees to natural disasters, component failures, human negligence, and willful human misconduct. The Commission divided its work into five "sectors" based on the common characteristics of the included industries, and found a mix of physical vulnerabilities, many first identified in the 1980s, and newer cyber vulnerabilities. Results of the sector team studies are in Appendix A. Key points are summarized below.

Information and Communications

All critical infrastructures are increasingly dependent on information and communications. The most important impact and vulnerability for this sector is the increasing interdependency of the PTN and the Internet. The Internet depends heavily on the PTN. The PTN, in turn, depends on electrical power for operations and on telephone lines and fiber optic cables that often run along transportation routes. The PTN is increasingly software driven, and remotely managed and maintained through computer networks. Deregulation of the telecommunications industry will markedly increase the number of access points, increasing opportunities for attack.

One well-publicized example of vulnerability associated with our dependence on computers is the "Year 2000" problem, which, if not corrected, has the potential to adversely affect the operations of all our infrastructures. Solving the Year 2000 problem was not part of the Commission's mission, and efforts are under way elsewhere in the federal government and across the country to remedy the problem before computer clocks turn to 00. But within the bounds of our mission, we did observe that resolving the Year 2000 problem requires the complete review and possibly the substantial revision of an affected organization's operational computer programs. Many people will have to be given access to these programs, as well as the authority to modify them and place them in service with less than adequate testing. The vulnerability will be worse if, as we expect, much of the review and modification work is contracted to outside, perhaps even foreign, firms. An adversary with access to a company's operational computer programs could understand aspects of the company's business practices better than the company's own management, which in turn would allow that adversary to design a subtle or comprehensive attack to gather information or reduce system effectiveness.

Energy

Prolonged disruption in the flow of energy would seriously affect every infrastructure.

The significant physical vulnerabilities for electric power are related to substations, generation facilities, and transmission lines. Large oil refineries are also attractive targets. The increase in transportation of oil via pipelines over the last decade provides a huge, attractive, and largely unprotected target array. Oil and gas vulnerabilities include lines at river crossings; interconnects; valves, pumps, and compressors; and natural gas city gates. Large metropolitan areas could be deprived of critical fuel for an extended period by a properly executed attack.

The widespread and increasing use of Supervisory Control and Data Acquisition (SCADA) systems for control of energy systems provides increasing ability to cause serious damage and disruption by cyber means. The exponential growth of information system networks that interconnect the business, administrative, and operational systems contributes to system vulnerability.

Banking and Finance

The principal vulnerabilities of the banking and finance sector are physical in nature. Its payments systems and its securities and commodities exchanges with their clearing and settlement organizations are vital to other parts of the banking and financial system and the economy at large. There are few of them, and in some cases, they are geographically concentrated. To back up its payments systems, the Federal Reserve has three geographically dispersed and "hardened" sites, each capable of completing the full volume of transactions sent over its wire transfer system. Similar back-up and "hardening" of facilities can be found in the other electronic payments and messaging systems, and most exchanges have a variety of contingency arrangements to rechannel trading activities should anyone's facilities become inoperable. In addition, the principal clearing and settlement organizations for the major stock exchanges have back-up sites some distance from the primary sites, as well as cold storage sites for data. These arrangements, together with strong measures to "harden" primary facilities, greatly reduce the overall vulnerabilities of this sector, but there remains risk from any event that disrupts telecommunications service and electric power within the geographic area in which key facilities are concentrated.

Physical Distribution

While the vulnerabilities of the physical distribution sector are still predominantly physical in nature, there are emerging cyber vulnerabilities as the sector increasingly relies on information technology to shorten lead times, route and schedule traffic and more–all on increasingly crowded communications channels. Physically most significant are the bridges over waterways, which are crossed by personal and commercial transportation, railroad tracks, telecommunications cables, and gas and oil pipelines. Vulnerabilities of the information and communications infrastructure also affect every aspect of the transportation industry. The most significant projected vulnerabilities are those associated with the modernization of the National Airspace System (NAS) and the plan to adopt the Global Positioning System (GPS) as the sole basis for radionavigation in the US by 2010.

Vital Human Services

Emergency responders are inadequately trained and equipped to respond to a chemical, biological, or nuclear attack on a civilian target. The 911 system can be overloaded through misuse and mischief, thereby missing life-and-death calls. Response coordination is vulnerable because the allocated radio frequencies used for responder communications are becoming congested and inadequate.

Treated water supplies do not have adequate physical protection to mitigate the threat of chemical or biological contamination, nor is there technology available to allow the detection, identification, measurement, and treatment of highly toxic, waterborne contaminants. Cyber vulnerabilities include the increasing reliance on SCADA systems for control of the flow and pressure of water supplies.

Government services are dependent on mega-databases of a highly confidential nature and containing information on private citizens. The uneven security practices of government agencies allow exploitation through the cyber vulnerabilities of these databases.

Shared Threats

A *threat* is traditionally defined as a *capability* linked to hostile *intent*. Linking capability to intent works well when malefactors are clearly discernible and US intelligence agencies can focus collection efforts to determine what capabilities they possess or are trying to acquire. During the Cold War, for example, weapons with potential to threaten the United States took years to develop, involved huge industrial complexes, and were on frequent display in large military exercises. Today, however, malefactors are no longer necessarily nation-states, and expensive weapons of war are joined by means that are easier to acquire, harder to detect, and have legitimate peacetime applications. The tools designed to access, manipulate, and manage the information or communications components that control critical infrastructures can also be used to do harm. They are inexpensive, readily available, and easy to use.

While poor design, accidents and natural disasters may threaten our infrastructures, we focused primarily on hostile attempts to damage, misuse, or otherwise subvert them. The Commission looked at both physical and cyber threats; however, we concentrated on the fundamentally new security challenges presented by networked information systems. Key points are summarized below.

Physical Threats

Physical threats fall into two general categories. The first includes threats posed by explosives, such as the World Trade Center and Oklahoma City bombings. Also included are a number of less well-known attacks and thwarted attacks on facilities like electric power transformers and utility towers over the past decade. A much more significant aspect of this threat exists in the form of nuclear weapons. Reports from Russia suggest that some so-called "suitcase weapons" are unaccounted for and may have fallen into the hands of terrorists. Federal Bureau of Investigation (FBI) Director Louis Freeh recently testified that while there is no hard evidence to confirm these reports, they are being treated with utmost seriousness.[3] Increasing attention is also being focused on chemical, biological and radiological threats. Chemical agents have already been used by terrorists, in the 1995 Aum Shinrikyo gas attack in Tokyo. In addition, work done for the Commission by a national laboratory found there is a credible threat to the nation's water supply systems from biological and chemical agents.

The second category is electronic weapons designed to attack computer-based systems. Included here are radio-frequency devices that capture computer signals as they emanate from the equipment, and electromagnetic pulse and radio-frequency weapons that are intended to destabilize or destroy sensitive electronic components. We determined that weapons of the latter type are still in exploratory stages.

In examining physical threats, the Commission concentrated on two critical issues:

1) the targeting of key links and nodes whose destruction might ripple through infrastructures or across infrastructures, and

2) coordinated attacks which, in combination, could severely impact the nation's security and economic competitiveness.

Simulation exercises with senior representatives of the infrastructures and government shed some light on potential impacts of such attacks, but

---

3 Testimony of FBI Director Louis J. Freeh before the House Committee on International Relations, October 1, 1997.

much more work is needed to understand the implications of interdependent infrastructures.

Cyber Threats

The Commission focused more on cyber issues than on physical issues, because cyber issues are new and not well understood. We concentrated on understanding the tools required to attack computer systems in order to shut them down or to gain access to steal, destroy, corrupt or manipulate computer data and code. In addition to accidents and negligence, threats to computer systems cover a broad spectrum that ranges from prankish hacking at the low end to organized, synchronized attacks at the high end. But the basic attack tools–computer, modem, telephone, and user-friendly hacker software–are common across the spectrum and widely available.

Potential cyber threats and associated risks range from recreational hackers to terrorists to national teams of information warfare specialists. Repeatedly identified as the most worrisome threat is the *insider*–someone legitimately authorized access to a system or network. Other malefactors may make use of insiders, such as organized crime or a terrorist group suborning a *willing* insider (a disgruntled employee, for example) or making use of an *unwitting* insider (by getting someone authorized network access to insert a disk containing hidden code, for example).

Five examples of new types of attack help illustrate the way commonplace cyber tools can be used to do harm.

A Cyber Attack on the Specific Data Base of an Owner/Operator

In the case of unauthorized entry into a network or system for the purpose of illegal financial transfers, stealing proprietary information, disrupting records, or merely "browsing," owners and operators have a responsibility for prudent and sufficient security systems such as firewalls and passwords and qualified personnel to detect anomalies that indicate a successful entry so that further isolation or deflection measures can be taken to foil the attack.

A Cyber Attack for the Purpose of Gaining Access to a Network

If a particular system or network is discovered through "electronic reconnaissance" to have low security standards and to be interconnected to other networks of interest to the attacker, the attacker will use the most weakly defended pathway for access to the targeted system. This suggests that owners and operators need to consider establishing security standards for those with whom they are connected.

A Cyber Attack for the Purpose of Espionage

Intellectual property is vulnerable to theft in entirely new ways. The threat may come from a witting or unwitting insider, an unscrupulous competitor, or the intelligence service of a foreign power. Competitive advantage may be lost without knowing it was even at risk. This is true in business as well as in government.

A Cyber Attack for the Purpose of Shutting Down Service

Attacks by flooding communication lines have denied 911 service in some communities and shut down e-mail service to major users. Denial-of-service attacks are of concern to all institutions whose business depends on reliable communications. Sharing information about the tools used in these attacks and techniques to deflect or defeat them is therefore of interest to a wide range of public and private institutions.

A Cyber Attack for the Purpose of Introducing Harmful Instructions

An attacker can plant a virus or leave behind a program that will give the attacker critical information, such as passwords that can be used to log in to other networks. A virus may be transmitted within a local area network or passed on to an external net. "Logic Bombs" and "Trojan Horses" are designed, respectively, to destroy software at a preselected time and to enable future access. Given the rate of development of viruses, it is essential that all interconnected users adopt a high level of virus detection.

The Internet

Threats to the Internet are of primary concern because we are becoming increasingly dependent on it for communications–including government and military communications–for commerce, for remote control and monitoring of systems, and for a host of other uses; because our ability to understand its full impact on society seems unable thus far to keep up with its explosive growth; and because it is inherently insecure (see Figure 2).

The Internet was designed in 1968 by the then Advanced Research Projects Agency (ARPA), now the Defense Advanced Research Projects Agency (DARPA), to determine how to build resilient computer networks that could survive physical attacks or malfunctions in portions of the network. The ARPAnet, as it was called, was not designed as a secure network, but depended for security on a small number of users who generally knew and trusted one another.

Figure 2. Increasing Exposure

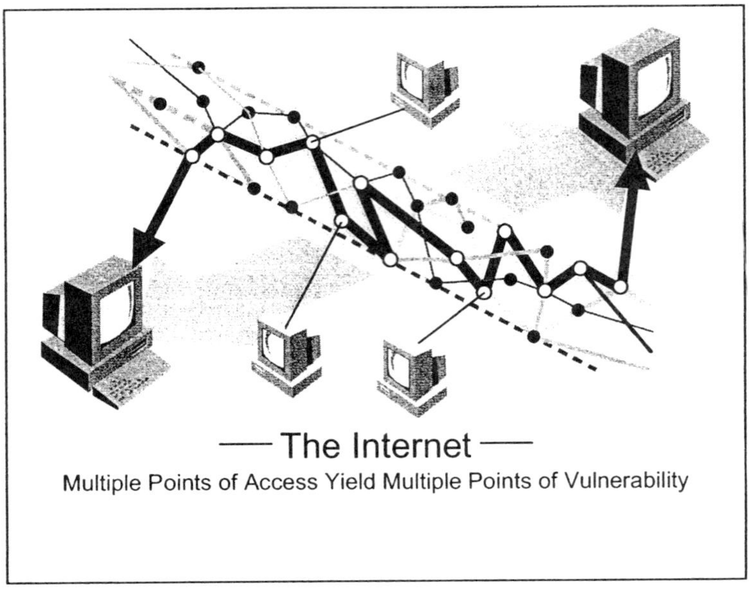

Commercialization of the Internet in the early 1990s, boosted by the WWW, caused incredible growth. Government and the private sector began to seize the advantages of the Internet as an alternative to other unclassified means of communication. The Internet continues to proliferate globally. In general our growing proclivity to network continues to outpace network protection. The price for the efficiency of networking is increased exposure of data and systems to unauthorized and anonymous access. A study done for the Commission by Carnegie-Mellon University's Computer Emergency Response Team (CERT) confirmed that "because the ties between critical infrastructures and the Internet will continue to become stronger and more intricate, the impact of an Internet attack could be devastating."[4] þò46þò

Information Warfare

Even more recent than the evolution of the Internet has been development and open discussion of the concept of Information Warfare (IW). The Gulf War illustrated the importance of infrastructures to national defense–our domination of Iraq's information and communications ensured victory over a well-armed military force with minimum allied losses. Other nations have drawn similar conclusions. Offensive IW, in brief, uses computer intrusion techniques and other capabilities against an adversary's information-based infrastructures. The Commission is aware of little in the way of special equipment required to launch IW attacks on our computer systems;

---

4 CERT report to the Commission, January 1997, p. 3.

the basic attack tools–computer, modem, telephone, and software–are essentially the same as those used by hackers and criminals. And compared to the military forces and weapons that in the past threatened our infrastructures, IW tools are cheap and readily available (see Figure 3).

If the basic cyber attack tools and skills are common across the spectrum, what may distinguish recreational hackers from Information Warriors is ***organization***. Said another way, an IW attack against US infrastructures may be little more than a series of hacker attacks, conducted against carefully chosen and thoroughly reconnoitered targets, synchronized in time, to accomplish specific purposes.

Figure 3. Tools to Do Harm

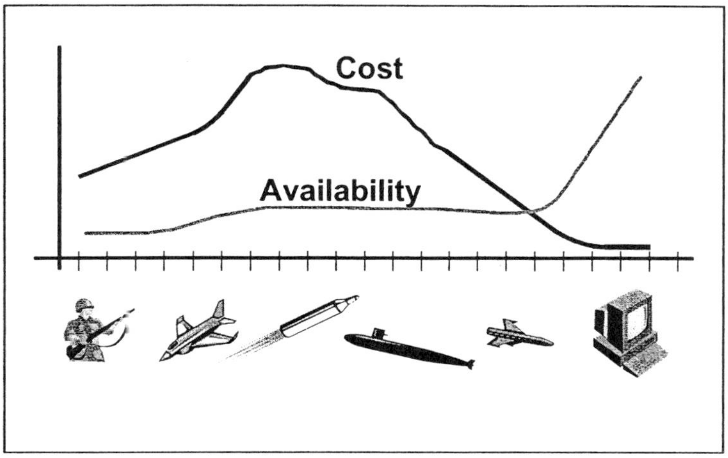

For an adversary willing to take greater risks, cyber attacks could be combined with physical attacks, against facilities or against human targets, in an effort to paralyze or panic large segments of society, damage our capability to respond to incidents (by disabling the 911 system or emergency communications, for example), hamper our ability to deploy conventional military forces, and otherwise limit the freedom of action of our national leadership.

Terrorists frequently choose prominent targets that produce little physical impact beyond the target itself, but widespread psychological impact. For a physical attack on infrastructures, less spectacular targets could be chosen, such as switching stations, communications antennas, pipelines, transformers, pumping stations, and underground cables. Many facilities whose physical damage or destruction would have a disruptive effect on an infrastructure are purposely located in sparsely populated or even unpopulated areas. If they are physically attacked it may take some time to discover the nature of the damage, and in the absence of casualties it may be some time before the attacks are reported. Even when they are reported, each incident is at first a local event, and if several such events occur over a period of weeks or months it may take considerable time before they are recognized as part of a pattern. Recognition that an attack is in progress could be delayed even if physical attacks were to occur simultaneously, if the targets were spread across several jurisdictions and no mass casualties were produced to generate "breaking news" at the national level.

The chances of immediately discovering that a concerted cyber attack is in progress are today even slimmer. Computer intrusions do not announce their presence the way a bomb does. Depending on the skill of the intruder and the technology and training available to their own system administrators, individual companies whose networks are penetrated may or may not detect an intrusion. Intrusions that are discovered may or may not be reported to law enforcement authorities, who may or may not have the resources to investigate them and conclude whether they are the work of an insider, a hacker, a criminal, or someone truly bent on harming the infrastructure. It sometimes takes months, even years, to determine the significance of individual computer attacks. In the highly publicized 1994 Rome Labs case, the main intruder–a London teenager–was caught in the act; but his alleged accomplice and mentor–who turned out to be a Welsh computer specialist only a couple of years older–was not identified and arrested until more than two years later.

In the absence of intrusion detection tools, uniform reporting of incidents as they occur, and some central capability to analyze incidents as they are reported, it is conceivable that an orchestrated attack against US infrastructures could be under way for some time before it is recognized as such and the attacker's motives and objectives can be deduced.

Intelligence Community Challenges

Information Warfare presents significantly new challenges for the intelligence community in identifying and assessing threats to the United States. This is partly because concepts of IW are only now taking shape abroad and because tools and techniques used for IW attack are inexpensive and ubiquitous. It is clear that a number of nation-states are closely following US developments in IW and are themselves exploring IW capabilities. They recognize that modern industrialized states are increasingly dependent on the uninterrupted flow of information. In addition, sub-national groups increasingly rely on advanced information technologies to support their illegal operations, and US intelligence analysts must be on the look-out for indications of interest by these groups in using their technical knowledge to harm the United States by attacking our critical infrastructures.

Recent assessments of foreign IW threats suggest a measured apprehension about the future. While no one is forecasting a sudden and major IW attack on the United States in the next few years, a number of factors support the sense of a growing threat. The US is by no means alone in recognizing and seizing the advantages of the global information and communications infrastructure and thus the increasing likelihood of various forms of international competition in the information arena. It is reasonable to assume that the number of states following our lead will increase. Other states and non-state groups will become increasingly familiar with opportunities for offensive use of computer techniques as they develop their own technology base and necessary cyber defensive capabilities. Finally, computer crime, including that directed against American businesses, will continue to grow in nation-states that do not enforce strong prosecution.

Shared Responsibility

The government and private sector share substantially the same national information infrastructure. Both have been victims of unauthorized computer intrusions, theft, and disruption. In our view, the line separating threats that apply only to the private sector from those associated with traditional national security concerns must give way to a concept of shared threats (see Figure 4). Shared threats demands a shared response, built from increased partnership between government and the owners and operators of our infrastructures.

Figure 4. Threat Spectrum

Factory owners or service providers were not expected in the past to protect themselves from enemy bombers or missiles; that was government's job. In the future, though, the owners and operators may be on the front line, and their networks may be the battlefield. The tools and know-how required to do harm are inexpensive, readily available, and easy to use.

Owners and operators need to protect themselves from the tools and the know-how. Government can help by collecting and disseminating information about all the tools that can do harm. Owners and operators can help by informing government when new tools or techniques are detected. Government has an obligation to collect information about potentially hostile groups and nation-states, and to issue timely warnings alerting owners and operators when new threats are detected.

We must achieve a new understanding of the threats that confront us–an understanding that focuses on the capability to do harm rather than identifying the person, group or nation intent on doing harm. Traditional indicators of developing capability are not present. There are no missile silos to count or railway cars to examine. We must acknowledge that the capacity for harm exists, and act now, as partners, to protect our future.

CHAPTER FOUR
FINDINGS AND POLICY

Analysis of the infrastructures, their vulnerabilities, and shared threats led the Commission to several observations. This chapter sets out those findings and then suggests a policy framework that addresses them. Subsequent chapters deal with specific recommendations resulting from that process.

In some respects our most important finding was the need to think differently about infrastructure protection. The management approach we now use was designed to deal with the Industrial Revolution, then was adjusted to manage successively the stabilization of America after the Civil War, the Depression, World War II, and finally the nuclear stand-off of the Cold War. None of those approaches is particularly applicable to the world as it will look through the lens of information technology in the third millennium.

FINDING:

Information sharing is the most immediate need.

There are many instances in which information is shared between the private sector and government, as in the case of the North American Electric Reliability Council (NERC) and the Presidentially-appointed National Security Telecommunications Advisory Committee (NSTAC). But there are important shortfalls. Increasing the sharing of strategic information within each infrastructure, across different sectors, and between sectors and the government will greatly assist efforts of owners and operators to identify their vulnerabilities and acquire tools needed for protection.

FINDING:

Responsibility is shared among owners and operators and the government.[5]

Owners and operators have always focused on protecting themselves from known threats to their operations, because it is in their interest. The government has always focused on protecting the nation from threats beyond the capabilities of private self-protection. Today, an adversary can bypass our national defense forces to attack directly the infrastructures that underpin our national economic strength. Traditional national secu-

---

5 While sometimes these owners and operators are referred to as the 'private sector," in truth the infrastructures also include publicly-owned and operated activities such as municipal water companies, state and local highway departments, and fire, police, and emergency response agencies.

rity concerns must give way to a concept of shared threats, for which responsibility must be shared between government and infrastructure owners and operators.

FINDING:

> ***Infrastructure protection requires integrated capabilities of diverse federal agencies, and special means for coordinating federal response to ensure these capabilities are melded together effectively.***

The Commission believes that the federal government's job in infrastructure protection includes the traditional defense, law enforcement, intelligence, and other responsibilities as well as the additional effort, resources and processes to respond to the cyber dimension. The structures detailed in our recommendations are designed to expand the reach of existing capabilities, provide a means to coordinate them, and integrate them with the resources of the owners and operators.

FINDING:

> ***The challenge is one of adapting to a changing culture.***

Our culture is changing at an accelerating pace. The Information Age is still unfolding, but it is already clear that it brings with it at least as many adjustments to our way of life as did the Industrial or the Nuclear Age, and that the requirement to adapt will be more urgent. Bold, sweeping measures are required to educate and inform our private sector, public servants, and citizens about the realities of the new environment.

FINDING:

> ***The federal government has important roles in the new infrastructure protection alliance with industry and state and local governments.***

The federal government is in a position to lead by example by adopting best practices, actively managing risk, and improving security planning in its own systems.

FINDING:

> ***The existing legal framework is imperfectly attuned to deal with cyber threats.***

Laws change at a much slower rate than technology. The existing legal framework does not reflect current technology in a number of ways. Legal authorities will need to be modified to allow for greater awareness of information security concerns, to enable response to and recovery from cyber events, to increase deterrence against computer crimes domestically and internationally, and to clarify roles and responsibilities in a world that is increasingly moving away from jurisdictional boundaries.

FINDING:

Research and development are not presently adequate to support infrastructure protection.

New challenges require new resources and new examination of how to protect ourselves. The Commission's proposed research and development (R&D) program identifies specific areas for research to provide the needed technologies.

Toward Recommendations

As we approached making recommendations to assure our critical infrastructures, the Commission adopted a set of principles to guide our decisions.

- *Build on that which exists.* It will be easier and faster to implement, more effective, and more likely to be accepted than creating something new.
- *Depend on voluntary cooperation.* Partnerships between industry and government will be more effective and efficient than legislation or regulation.
- *Start with the owners and operators.* They have a strong economic stake in protecting their assets and maximizing customer satisfaction. They understand the infrastructures and have experience in responding to outages.
- *Practice continuous improvement.* Take action in affordable increments. There is no "magic bullet" solution. Aim not only to protect the infrastructures, but also to enhance them.
- *Coordinate security with maintenance and upgrades.* Security should be incorporated in planned maintenance and scheduled upgrades.
- *Promote government leadership by example.* Government-owned facilities should be among the first to adopt best practices, active risk management, and improved security planning.
- *Minimize changes to government oversight and regulation.* Several of the infrastructures have a long history of government regulation, with a clear legislative mandate and a record of success. We consciously avoided proposing significant changes in regulation.

A Proposed National Policy for Infrastructure Protection

Critical infrastructures underpin the security of our national wealth, our defense capability, the economic prosperity of the people, and, above all, the maintenance of the system of human rights and individual freedoms for which the United States was founded and has stood since 1776. The threat of infrastructure attacks therefore has the potential for strategic damage to the

United States. Accordingly, the assurance of critical infrastructures deserves national attention and leadership by the federal government.

It shall be the policy of the US to assure the availability and continuity of the critical infrastructures on which our economic security, defense, and standard of living depend. The infrastructures will be defended by whatever means necessary, including the full range of business, legal, law enforcement, military, and social tools available.

Further, the US recognizes that assuring infrastructure is not just a government or business responsibility, but is shared by those public and private interests that own and operate the infrastructures and the government agencies responsible for defense, law enforcement, and economic security of the nation.

The interdependent nature of the critical infrastructures and their collective dependence on the information and communications infrastructure have created new assurance challenges that can only be met by a partnership between owners and operators and government at all levels. Only the owners and operators have the knowledge, access, and technology to defend their systems from the growing array of widely available information-based tools. Only the federal government has the legal authority, law enforcement capability, and defense and intelligence resources needed to deter the most sophisticated nation-state and other serious cyber threats. And only the federal government has the intelligence and related capabilities to find the tools that do harm and promulgate information about them throughout the infrastructures.

As a matter of urgency, an Office of National Infrastructure Assurance should be established under the National Security Council (NSC) and given overall program responsibility for infrastructure assurance matters, including policy implementation, strategy development, federal interagency coordination, and liaison with state and local governments and the private sector. Among other responsibilities, this Office will devise and establish mechanisms for the exchange of views and information between the government and the private sector, identify information requirements for infrastructure assurance, and ensure that infrastructure assurance considerations are taken into account in making other government program decisions.

The Office of National Infrastructure Assurance should ensure that a program of public awareness is implemented throughout the country to inform the American public about infrastructure protection. This will include establishment of appropriate curricula in the national education system, from kindergarten through graduate school and including professional training. The Office of National Infrastructure Assurance should also ensure that individual agencies of the federal government implement infrastructure preparedness provisions and update their security plans to include protection against Information Warfare threats.

PART TWO
A STRATEGY FOR ACTION

CHAPTER FIVE
ESTABLISHING THE PARTNERSHIP

Information Sharing–The Indispensable Step

Objective:

Promote a partnership between government and infrastructure owners and operators beginning with increased sharing of information relating to infrastructure threats, vulnerabilities, and interdependencies.

Need for Sharing Information About the Cyber Dimension

The private sector owners and operators of critical infrastructures are engaged in market-based programs offering services at competitive prices. Their risk analysis weighs the cost of physical and cyber disruption against the cost of physical and cyber security. Their willingness to invest in defenses against the cyber tools that may do harm is dependent on their experience with these disruptions and the information they have about them.

While physical security is a mature discipline, our understanding of cyber vulnerabilities and threats is incomplete. Owners and operators do not have sufficient threat and vulnerability information for informed risk management decisions. Some of the information they need may be available from the federal government, particularly from the law enforcement and intelligence communities.

As emphasized earlier, two-way sharing information is indispensable to infrastructure assurance. While infrastructure owners and operators have the fullest appreciation of vulnerabilities, they have access only to their own information or, in some cases, information pertaining to their industry or sector. Consequently, there is no comprehensive body of knowledge available for effective analysis of overall infrastructure vulnerabilities and threats. This is especially true of vulnerabilities created by increased dependence of infrastructures on one another. Current information-sharing mechanisms perform well in matching physical threats to known vulnerabilities, and employing appropriate countermeasures. However, the same cannot be said of the emerging cyber arena.

Overcoming Reluctance

Our contacts with public and private sector stakeholders identified a need to increase the flow of information about cyber threats and vulnerabilities. Many offered a perception that private sector owners and operators share information only when they suffer substantial loss or are convinced of imminent danger to continuity of operations.

Infrastructure representatives expressed reluctance to share information about vulnerabilities because they fear it might be made public, resulting in damage to their reputations, exposing them to liability, or weakening their competitive position. Many also feared that sharing vulnerability information could invite unwanted federal regulation. The degree of reluctance varied according to infrastructure, but was present in each. The latest Computer Security Institute/FBI Computer Crime and Security Survey reinforces these observations, noting that of respondents who experienced an attack during the previous year, only 17 percent reported it to law enforcement authorities.[6]

Owners and operators told us they might have a better idea of actions they should take if the government shared more threat and vulnerability information. Likewise, government representatives told us they could better protect infrastructures if owners and operators would stop withholding information. While it is clear that government and the private sector would benefit from an improved two-way flow of information concerning threats and vulnerabilities, we caution against expecting a sudden revelation. Our classified government briefings and confidential discussions with private sector representatives produced no evidence of some missing piece of information that would make the whole picture suddenly fall into place.

Need for National Analytic Capability

Of course, sharing information isn't enough; we need the analytic tools to examine information about intrusions, crime, and vulnerabilities and *determine what is actually going on in the nation's infrastructures.* Deciding whether a set of cyber and physical events is coincidence, criminal activity, or a coordinated attack is not a trivial problem (see Figure 5). In fact, without a central information repository and analytic capability, it is virtually impossible to make such assessments until after the fact. This is of increasing concern as infrastructure operations become more reliant on information and communications—the very sector about which it is most difficult to make assessments.

---

6 Computer Security Issues & Trends, Vol. III, 1997 Computer Security Institute/FBI Computer Crime and Security Survey.

Figure 5. Coincidence or Attack?

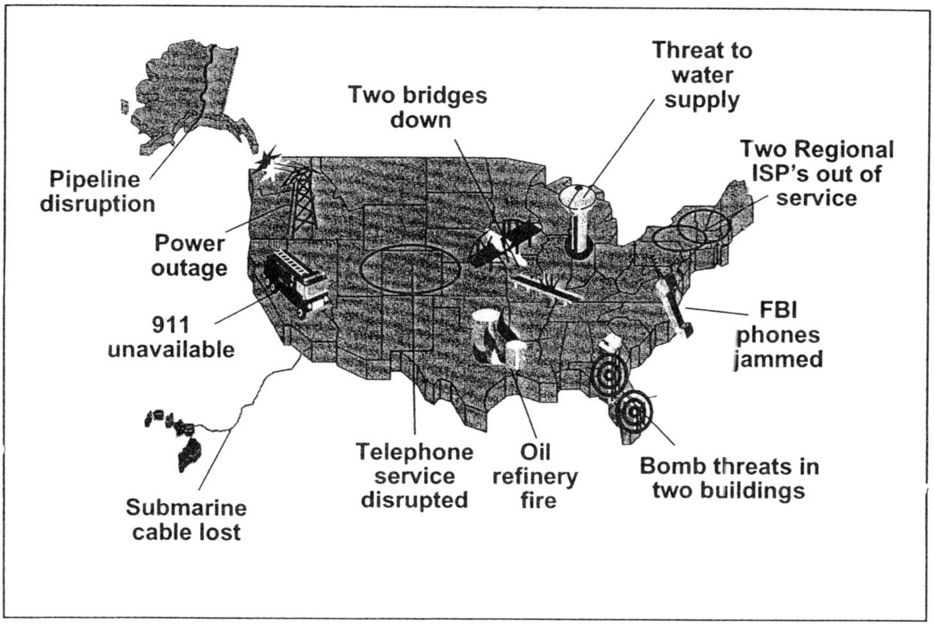

A number of government and private organizations hold and distribute incident reports related to infrastructure protection, but comprehensive analysis of this information is limited. The need for analysis is especially critical to support decision-making about responding to attacks. There is insufficient interagency, federal-to-state and local government, or public/private correlation of data to support crisis action planning in response to a cyber terrorist incident. The need for "a cyber-threat-clearinghouse . . . centralized effort for comprehensive intelligence analysis of cyber issues . . . an industry/government information exchange for threat and vulnerability data" has been documented frequently.[7]

Existing Information Sharing Efforts

Our work did identify highly successful information sharing organizations already at work in other areas. The Centers for Disease Control and the Coordinating Sub-Group on Counter-Terrorism (CSG/CT) in the NSC

7 See, for example, The Future of US Intelligence, report by The Working Group for Intelligence Reform, 1996; NII Risk Assessment: A Nation's Information at Risk, report by the Reliability and Vulnerability Working Group, 1995; and NII: the Federal Role, report of the National Information Infrastructure Security Issues Forum, 1995. More details are contained in an internal Commission paper on Information Sharing.

are useful models for expeditious information sharing to support action planning.

We also found a great deal of information sharing already underway. Trade associations, consortia, and other groups exchange information among their members and, in some cases, directly with government. Many federal, state and local government agencies have existing relationships with infrastructure owners and operators. Within all the infrastructure sectors, at least some portions are subject to regulatory control by government agencies, and information is shared, albeit sometimes within carefully defined constraints.

Several federal agencies provide information to infrastructure owners and operators. The FBI's Awareness of National Security Issues and Response (ANSIR) program gives over 25,000 industry members information that provides threat and vulnerability insights. More narrowly focused programs are the Department of Transportation's terrorist threat notification to the civil aviation industry and the National Security Agency's INFOSEC Vulnerability Assessment Program, which provides information systems-related data to private sector partners. The Comptroller of the Currency operates another system providing advisories on information integrity and security risks to financial institutions.

Information To Be Shared

Common to most of these programs is the narrow range of information collected and shared. In almost every case, they are tightly focused on specific information with no attempt to determine whether the information might also be useful for infrastructure protection purposes. Regulatory information is not generally focused on infrastructure protection. For example, telecommunications carriers report service disruptions of 30 minutes affecting 30,000 or more customers to the Federal Communications Commission (FCC). But that reporting channel would not identify a series of smaller attacks dispersed around the country and designed to slowly weaken public confidence in the system.

Figure 6. Scope of Useful But Unavailable Information

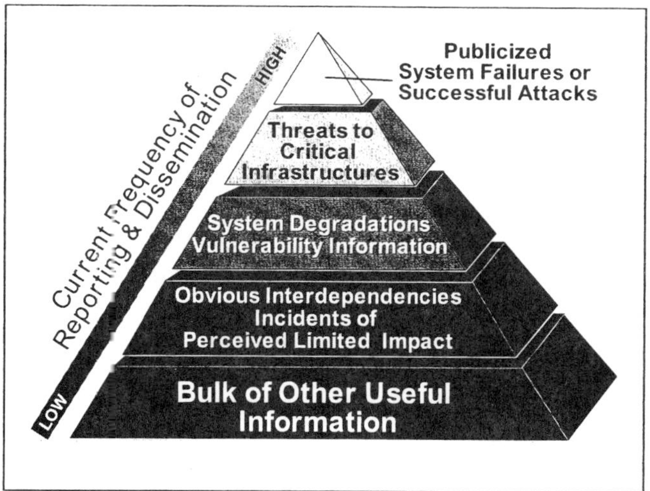

Figure 6 depicts the types of information pertinent to infrastructure assurance and the likelihood that the information is reported to law enforcement. Currently, only information derived from selected threats and difficult-to-ignore, successful attacks is readily shared. This narrow range of reported types of information should be viewed as only the tip of a mountain of data whose compilation would be helpful in infrastructure assurance. Included are such topics as system degradations due to physical acts or cyber-based events; vulnerabilities (hardware failure rates, operator-induced malfunctions, poor maintenance practices, or software flaws); not-so-obvious cyber or physical vulnerabilities resulting from dependence on other infrastructures; incidents of vandalism, malicious mischief, or suspicious activity; and physical or cyber anomalies. Information in government hands, such as criminal statistics and threat data, seldom is scrutinized for revelations about vulnerabilities or interdependencies. The government and infrastructure owners and operators must both push assurance-related data from the bottom towards the top of their respective agendas where it can be more readily analyzed.

Legal Impediments to Information Sharing

We envision the creation of a trusted environment that would allow the government and private sector to share sensitive information openly and voluntarily. Success will depend on the ability to protect as well as disseminate needed information. We propose altering several legal provisions that appear to inhibit protection and thus discourage participation.

Confidential Information

The Freedom of Information Act (FOIA) makes information in the possession of the federal government available to the public on request. Potential participants in an information sharing mechanism may require assurances that their sensitive information will remain confidential if shared with the federal government.

We Recommend:

The proposed Office of National Infrastructure Assurance require appropriate protection of specific private-sector information. This might require, for example, inclusion of a b(3) FOIA exemption in enabling legislation.

Trade Secrets and Proprietary Information

Private sector participants may be reluctant to share sensitive information if appropriate protection mechanisms are not incorporated.

We Recommend:

The proposed Office of National Infrastructure Assurance require appropriate protection of information containing trade secrets or other forms of proprietary information.

Classified Information

Information collected by the government to benefit a threat warning process may require protection in the form of classification.

We Recommend:

The proposed Office of National Infrastructure Assurance consider the need for classification of certain information, or certain bodies of aggregated information, and the impact that classification would have on the dissemination process.

Antitrust

Potential contributors from the private sector are reluctant to share specific threat and vulnerability information because of impediments they perceive to arise from antitrust and unfair business practice laws.

We Recommend:

The Department of Justice (DOJ) offer limited assurances to the private sector that participation in information sharing processes would not run afoul of antitrust laws and consider providing appropriate guidelines to inform participation.

Liability

Information which could prevent harm to a critical infrastructure may arise from participation in a threat and warning capability. Failure to share such information, or to act on such information shared by others, might carry liability consequences for public and private participants.

We Recommend:

The federal government undertake a detailed study of liability issues surrounding participation in an information sharing process.

National Security

Currently, many federal agencies have their own specific guidelines controlling interaction with foreign corporations or corporate entities with significant foreign ownership.

We Recommend:

The NSC study whether the federal government should standardize guidelines for sharing infrastructure assurance information with foreign corporations in light of potential national security risks and benefits.

Appropriate guidelines are needed for sharing information with foreign corporations.

We Recommend:

In the short term, the proposed Office of National Infrastructure Assurance set guidelines for the sharing of infrastructure assurance information with foreign corporations.

State Government Liability and Disclosure

Many of the legal impediments to information sharing identified at the federal level exist at the state level as well. However, diversity among state laws further complicates efforts to maximize participation in information sharing.

We Recommend:

A study group identify legal impediments to information sharing at the state level, propose solutions, and draft model legislation.

Conclusion

We believe information sharing is the critical foundation for an effective partnership to enhance our ability to protect critical infrastructures in the years ahead. Sharing information figures prominently in the additional recommendations we make and the structures we recommend for the public and private sector elsewhere in this report. How then should we build the relationship between private and public sector organizations so they can share, use, and act on information to better protect our critical infrastructures?

CHAPTER SIX
BUILDING THE PARTNERSHIP

Owners and Operators
State and Local Governments

Objective:

Ensure infrastructure owners and operators and state and local governments are sufficiently informed and supported to accomplish their infrastructure protection roles.

Protecting America's infrastructures is neither an entirely public nor entirely private interest. Vulnerabilities pose risks to government, business, and citizen alike. Reducing those risks requires coordinated effort within and between the private and public sectors. The need for infrastructure protection creates a zone of shared responsibility and potential cooperation for industry and government.

Owners and operators have a responsibility to deliver reliable service. While sometimes these owners and operators are referred to as the "private sector," in truth the infrastructures include publicly-owned and operated entities such as municipal water companies, state and local highway departments, and fire, police, and emergency response agencies. Regardless of whether they are primarily accountable to shareholders or taxpayers, owners and operators must take prudent steps to reduce or eliminate their own vulnerabilities–*to protect themselves not so much against a known threat, but against the tools an unknown perpetrator could employ.*

Government has an undeniable role in accomplishing the tasks that government alone can undertake–including law enforcement at local, state and federal levels, and national intelligence, defense and diplomacy.

The Commission found a need for a new partnership between government and owners and operators to assure our critical infrastructures. And we found that the need to share information was a foundation on which we could build that partnership.

Infrastructure assurance is essentially a process of risk management. The process is generally defined to include prevention, mitigation, incident management, and recovery. The many functions associated with infrastructure assurance fit into these four categories.

In considering how these functions are accomplished today, and how they might be in the future, we identified opportunities for enhancing the effectiveness of the owner and operators through increased partnership with the federal government.

National Threats and Public-Private Partnerships

Our approach to partnership for infrastructure assurance was to examine which functions were the responsibility of each partner and the expectations associated with those functions. This led to the specific recommendations contained in this discussion about private sector, and state and local government roles.

A New and Challenging Environment

Infrastructure providers deal with known vulnerabilities and associated risks within their infrastructures. But the rapid introduction of new technologies and interconnected nature of the infrastructures present new challenges. Before interdependencies were as great as they are now, physical attacks and outages were contained. Extensive reliance on computer and telecommunications technologies makes it more difficult for owners and operators to know whether outages result from technical failure or intentional intrusion.

Further complicating the partnership is our dependence on these infrastructures for national defense, economic competitiveness, and quality of life. Realizing this certainly places the role of critical infrastructure owners and operators into new perspective. While they must still respond to normal business pressures–the bottom line, shareholder concerns, and their customers–they must also acknowledge that the government has an increasing interest in infrastructure providers. The critical role of many public utilities exemplifies this situation where health, safety and other public concerns are so dependent on the infrastructure that government interest is unquestioned. Today, the interconnected nature of the infrastructures, the potential for local disruptions to cascade into other infrastructures, and the dependence of national security on those same infrastructures present a clear need to think in new ways.

These facts alone emphasize the need for infrastructure owners and operators and government at all levels to find new ways of working together. These new partnerships must be designed to foster mutual trust and facilitate sharing of the types of information that each partner needs to assure the uninterrupted flow of essential goods and services.

Expectations of Owners and Operators[8]

Owners and operators are the primary players in infrastructure assurance. For all the expected business and operational reasons, they protect their critical systems and facilities based on a perceived set of risks. Better information on emerging threats and vulnerabilities, particularly those stemming from unrecognized or little understood interdependencies, will assist managers in making decisions about investment in security processes, thus improving assurance not only for their own company's operations, but for operation of the infrastructure overall.

The Commission believes it is the responsibility of owners and operators to:

1) Provide and manage facilities delivering services to customers efficiently and effectively.

2) Meet customer expectations for quality and reliability of service.

3) Maintain an effective risk management process adequate to:

- identify vulnerabilities and potential threats that might affect continuity of service;
- prevent and mitigate as many credible threats as economically feasible; and
- maintain emergency response capability to quickly restore service and eventually reconstitute the infrastructure in the event of service interruptions.

4) Give special consideration to the vulnerabilities currently in many information systems.

5) Cooperate within their industry to identify best practices for improving service reliability and security.

6) Report possible criminal activities to law enforcement agencies and cooperate with investigations.

7) Establish a relationship with intelligence and law enforcement to assure that information about warnings and threats is communicated in a timely way and that the industry experience with incidents is available as an input to threat analysis.

---

8 As previously noted, the term "infrastructure owners and operators" includes public agencies or corporations as well as companies and activities in the private sector.

Risk Assessment Best Practices

Conduct a security training program for all employees according to their job responsibilities and access authorizations, integrating this program with existing physical security aspects.

Authenticate the identity of all users of the system, determine the uses of the system for which they are authorized, and restrict access to only the authorized functions and data.

Isolate critical operational control systems from all public and most internal networks, or provide adequate firewalls.

Provide adequate procedural and technical controls to assure data integrity, to detect instances of unauthorized change or deletion, and to recover when necessary.

Authenticate and log the origin of all commands to change the operating conditions of the controlled infrastructure.

Create a CERT, or similar response capability, with the equipment and training needed to investigate suspected intrusions, isolate and recover damaged systems, and restore service to customers.

Provide adequate back-up and recovery capability for the programs and data of any information system that is necessary for normal operations and customer service. To better assure the availability of key control systems, information systems and data, consider redundancy, geographic separation of primary and back-up systems, alternative methods, effective use of encryption, and other relevant security options.

Conduct regular assessments of the vulnerability of information systems using the technical expertise of the National Security Agency (NSA) and others as appropriate to assure that new techniques for attacking systems can be contained by the protective measures currently installed.

Compiled by the President's Commission on Critical Infrastructure Protection

Industry Suppliers

Usually the owners and operators are not the suppliers of the computer hardware and software they use to manage their operations. For significant improvements to be made in the security and integrity of these information systems, suppliers must be involved.

The computing systems industry is highly competitive and normally very responsive to customer needs; however, experience suggests that users may not understand the new vulnerabilities well enough to demand products offering better security. There is recent evidence that major suppliers are giving security and integrity more attention than in the past. We expect this trend to accelerate as owners, operators, and industry associations study their vulnerabilities and demand improved products.

Vulnerabilities Assessments

The more owners and operators understand about their vulnerabilities (see Figure 7), the better able they are to make effective decisions about protecting their operations. A step toward increasing private sector awareness can be taken by increasing federal government participation in the vulnerability assessments conducted by owners and operators.

We Recommend:

The NSA, the Department of Energy (DOE) and DoD:

- continue to perform vulnerability assessments for critical infrastructure owners and operators.
- provide vulnerability assessment training to private sector service providers on a cost-reimbursable basis, e.g. sharing knowledge and expertise of key government centers of excellence.

Figure 7. Vulnerabilities Profile

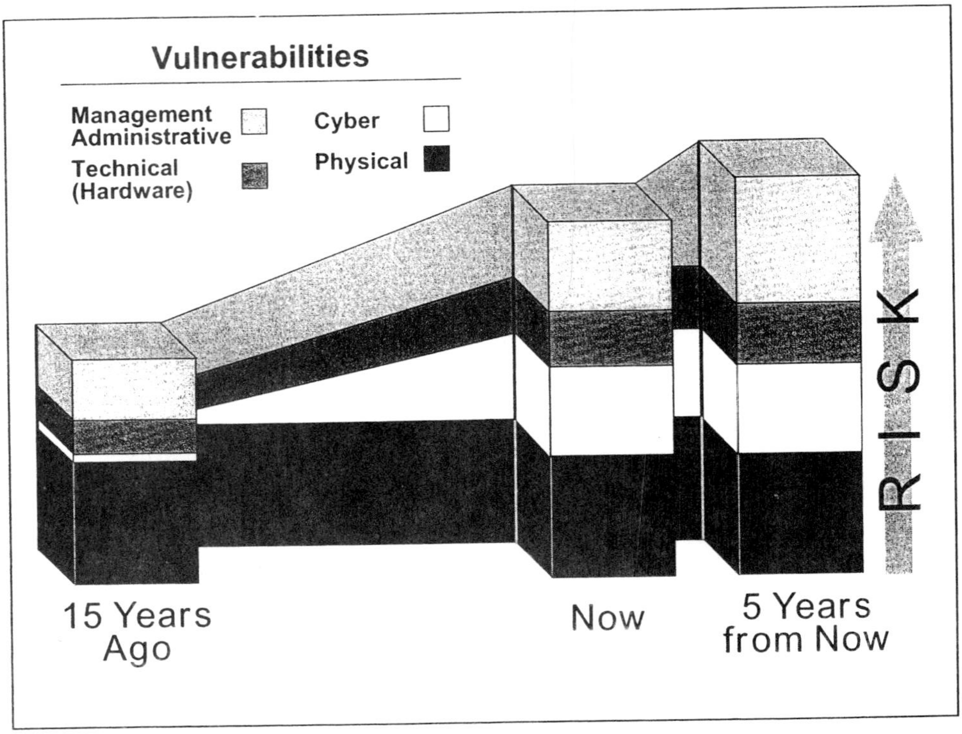

Publicize & Support Application Of Risk Assessment Methodologies

---

We Recommend:

The proposed Office of National Infrastructure Assurance and National Infrastructure Assurance Council encourage private industry to perform periodic risk assessments of critical processes, including information and telecommunications systems. This will enable more informed risk management decisions in the face of rapid, pervasive change.

We Recommend:

The proposed Office of National Infrastructure Assurance encourage the insurance industry to develop its risk methodologies for application to the critical infrastructure industries.

Sensitive Information

In Chapter 5, we described the need to overcome the concern of owners and operators that information they provide to government might not be protected. As the importance of the infrastructures to every aspect of national strength is understood, information that may be useful to an enemy in designing an attack on those infrastructures takes on a new importance.

One example of concern in this area is found in requirements for publication of sensitive information about critical components or the functioning of infrastructures. This information has the potential to serve as a "road map" for a potential infrastructure attack; therefore, its publication may lead to the exploitation of vulnerabilities.

We Recommend:

The President issue an Executive Order requiring that federal agencies accomplish the following before publishing or requiring the publication of information about critical components or functioning of infrastructures.

- bring together the relevant stakeholders to discuss the implications of the requirement.
- identify the purpose for publishing the information and ensure that the information is published in a format that minimizes the likelihood it will be used in ways that are incompatible with infrastructure assurance.
- certify that the positive and negative effects on infrastructure assurance have been fully explored, including that the potential benefits of publication outweigh any identified risks.

Protection of Infrastructure Vulnerability Information

We now must question whether information regarding vulnerabilities–in the aggregate at least–shouldn't be protected in some fashion.

We Recommend:

The US Security Policy Board be tasked to provide a recommendation to the President on criteria for and means of protecting otherwise unclassified private sector information on threats and vulnerabilities to critical infrastructures.

Publication of Infrastructure Assurance Data

The publication of infrastructure assurance-related comparative data within an industry may positively influence performance by motivating increased attention to information security, as well as reliability, without resorting to regulation. This information may prove useful to consumers in light of increased competition and choice between providers of critical infrastructure services. We found such reports of great use in some infrastructures. In electric power, for example, the NERC publishes each month on its WWW site a Performance Honor Roll of companies achieving 95 percent or better reliability in the previous 12 months.

We Recommend:

The Administration direct the FCC's Network Reliability and Interoperability Council to initiate a feasibility study of publishing comparative infrastructure assurance-related data for the telecommunications industry. The study should focus on whether publication is likely to achieve infrastructure assurance objectives, the types of data to collect and publish, whether current data collection efforts are sufficient, and other possible impacts of publication. Similar studies should follow for other infrastructures.

Security Standards

The Commission considers the development of standards to be the responsibility of the infrastructure operators themselves. In our research, we were advised of an initiative of the NERC to apply mandatory reliability standards (which include security) to its members. Currently, the Federal Energy Regulatory Commission (FERC) is dealing with this issue for the whole electric power industry. The activities of these two organizations are headed in directions the Commission applauds and recommends to other infrastructure sectors.

However, we do believe that government should encourage and assist public and private sector standard-setting bodies to broaden their areas of responsibility regarding reliability to include security assurance.

We Recommend:

The National Institute of Standards and Technology (NIST) and NSA work with the proposed Office of National Infrastructure Assurance to offer their expertise and encourage owners and operators of the critical infrastructures to develop and adopt security-related standards. Relevant federal and state regulatory agencies, industry associations and standards groups, and law enforcement and intelligence agencies should also participate in the process of identifying and developing standards.[9] These standards should address not only the technology itself, but also ancillary topics such as tools, policies, procedures, and practices.

---

9 Standard-setting groups include the American National Standards Institute, the Institute of Electrical and Electronic Engineers, and the National Computer Security Association.

Building the Partnership
State and Local Government

State and local governments are integral to the success of the partnership we propose for infrastructure assurance. State and local governments' infrastructure roles cut across the public-private boundaries. They operate infrastructures–certainly emergency services, but also water systems and a host of other vital services. They are also the regulators of many of the infrastructures–particularly those considered to be public utilities. And finally, they are users of the infrastructures–just as dependent on information and communications, energy, and transportation as the federal government.

We met dozens of local officials and held public meetings on infrastructures around the nation. State and local officials consistently expressed their need for federal assistance in key areas relating to infrastructure protection.

High on their agenda is raising public awareness on infrastructure matters, particularly in protecting information networks. They need assistance from the federal government in maintaining competent trained firefighters, policemen, and paramedics prepared to handle infrastructure disasters and threats from chemical, biological, and radiological materials. They need solutions in addressing the crowded spectrum of radio frequencies that emergency services must use to communicate. And they need a forum to share information on infrastructure issues.

Sharing Information

Organizations representing state and local interests have existing relationships with federal government officials. By working through such organizations, we can effectively share information on protecting our critical infrastructures.

We Recommend:

The proposed Office of National Infrastructure Assurance lead the way in making information about infrastructure assurance available to state and local governments through existing organizations such as the National Governors' Association.

Equipping and Training First Responders

The Commission's research found emergency services ill-prepared to deal with chemical and biological attacks. Few "first responders"–firefighters, police, and paramedics–are adequately trained to treat attack victims or equipped with protective gear or supplied with medical treatments, such as atropine.

Legislation initially sponsored by Senators Nunn (D-GA), Lugar (R-IN), and Domenici (R-NM) focused federal resources on providing training, equipment, and information to local first responders. State and local police, fire, and medical officials are requesting an expanded effort in this area, and the Commission agrees that these efforts should be intensified and made more widely available.

We Recommend:

DoD, the Department of Health and Human Services (HHS), and the FBI provide local first responders additional training and equipment for improving the detection, identification and management of chemical, biological, and radiological incidents. Domestic preparedness funding (Nunn-Lugar-Domenici) for these activities should be doubled in FY99.

Spectrum Allocation

Police, firefighters, paramedics, and repair crews must be able to communicate clearly during emergencies. The radio frequencies used for dispatching and communication are congested–making it difficult to use the spectrum effectively.

The FCC has been auctioning segments of the electromagnetic spectrum. As demand rises for commercial bandwidth, spectrum becomes increasingly scarce, placing non-revenue generating public sector users, such as federal, state and local emergency services, under increasing pressure to relinquish relatively under-used portions of their bandwidth allocations.

Addressing this issue, the National Telecommunications Information Administration (NTIA) and the FCC-sponsored Public Service Wireless Advisory Committee (PSWAC) issued a joint recommendation, which the Commission endorses, that the FCC designate inviolate spectrum segments for emergency services–removing them from future auction consideration.

Should circumstances require spectrum reallocation, however, the FCC should ensure compensation of state and local emergency service providers for the costs of replacement equipment, training, and transmission capabilities.

We Recommend:

Expanding NTIA's mission to include representing the interests of state and local governments in addressing access to and use of the electromagnetic spectrum. This advocacy should include efforts to ensure that current needs of these governments are identified and appropriately balanced with commercial and federal sector needs, that interoperability requirements among emergency services–in locales as well as regionally and nationally–are considered, and that adoption of new services and technologies is both facilitated and coordinated across all government levels.

We Recommend:

The FCC and NTIA expeditiously adopt the PSWAC recommendations. In particular, the FCC should:

- allocate an additional 25 MHz of unencumbered spectrum for public safety.
- provide 2.5 MHz in the VHF and UHF bands for interoperability among emergency service providers.
- plan for allocation of an additional 70 MHz for new technology applications in law enforcement and emergency services.
- immediately factor other detailed recommendations of the PSWAC into the spectrum allocation planning process.

These measures will assist state and local governments in meeting their critical infrastructure protection responsibilities, but they are only a first step. We fully recognize that the challenges facing state and local governments go well beyond what can be addressed by the application of such limited means.

Conclusion

In the interconnected, cyber-oriented world of today, the responsibility for infrastructure assurance cannot be divided along traditional lines between government and the private sector or allocated among levels of government. The need to forge a partnership between all players–to achieve joint, integrated, and complementary action–is more acute than ever. With a better understanding of the expectations and roles of the owners and operators, and of state and local governments, comes an appreciation for their increasingly "front line" mission in defending our infrastructures. The federal government should structure itself for its own mission of infrastructure assurance–a mission that now includes facilitating and supporting the efforts of critical infrastructure owners and operators.

CHAPTER SEVEN
STRUCTURING THE PARTNERSHIP

Objective:

Establish national structures that will facilitate effective partnership between the federal government, state and local governments, and infrastructure owners and operators to accomplish national infrastructure assurance policy, planning and programs.

Early in the Commission's deliberations, we recognized the federal government was still organized along Cold War lines. The structures in place had proven very effective at focusing federal attention and resources on physical threats posed by military, terrorist, or criminal entities. Likewise, the relationships between government agencies and infrastructure operators were appropriate to the environment. Except for down-sizing, the structure of the federal government had not changed significantly since the Cold War, and its relationships with infrastructure owners and operators–though less regulatory in nature–had not changed markedly (see Figure 8).

Figure 8. Infrastructure Assurance Today

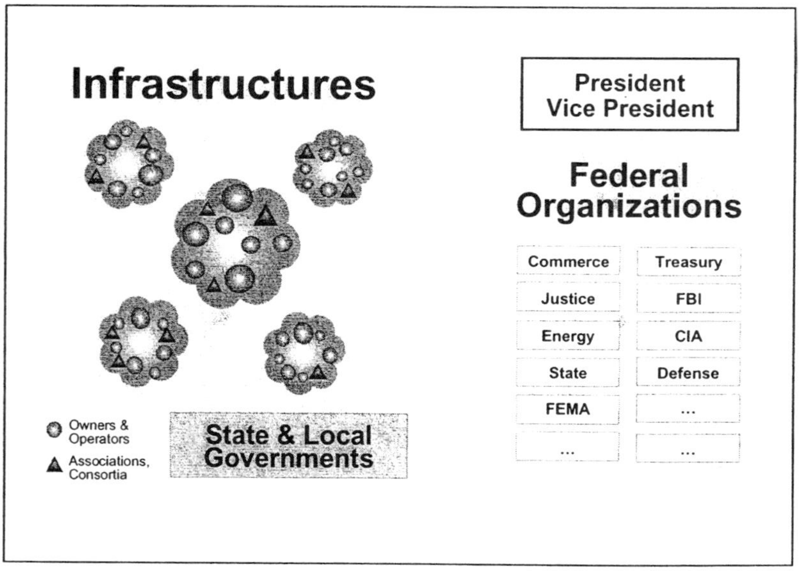

But the federal government today must address the emerging threats to our infrastructures, the new geography discussed earlier in this report, and the requirements of the Information Age. How the government organizes itself is a key factor in the partnership with infrastructure owners and operators that is fundamental to meeting the challenges of the threats we share. Without recommendations that set out clear national organizational structures, the chance for developing a government and industry partnership could elude our grasp.

To address these organizational issues, we examined the functions or actions instrumental to achieving infrastructure assurance and protection at the national level. In each of the five functional areas, the need for partnership and dynamic interaction between the government and infrastructure owners and operators is apparent, as indicated below.

Policy Formulation–The federal government can best assess emerging threats, and the owners and operators can best assess their vulnerabilities. Together they should assess the national risk and determine assurance objectives, strategies, and policy.

Prevention and Mitigation–Owners and operators will have to examine the vulnerabilities of their own systems and networks and put in place the protective measures and practices needed to achieve target levels of assurance. The government can and should support these efforts through R&D, awareness and education, threat assessments, initiatives to facilitate private sector adoption of best practices, and, possibly, through direct financial assistance.

Information Sharing and Analysis–The key products of this functional area are answers to two questions: (1) What unusual is happening among our infrastructures, and (2) what unusual is happening among our adversaries? Owners and operators should take the lead for the former; the federal government (law enforcement and intelligence) for the latter. Analyzing the information provided and synthesizing it into advisories and warnings should be a shared responsibility.

Counteraction (incident management)–The objective of this functional area will be to deter an attack on our critical infrastructures, and, should deterrence fail, to cause the attacker to cease and desist. This area is clearly a federal responsibility, primarily of the law enforcement and defense communities, but there are many important ways in which the owners and operators can and should assist.

Response, Restoration, and Reconstitution (consequence management)–Responding to the basic needs of the populace following a disaster is a responsibility of the states, supported by the federal government. Restoring and reconstituting infrastructures is the responsibility of the owners and operators, supported by their sector. A major restoration and reconstitution effort would require coordinated public and private sector actions.

As we sought to identify what sorts of national structures would best accomplish these functions, we applied the same principles used to guide all of our deliberations.[10]

---

10 These guiding principles are discussed in Chapter 4.

Figure 9. Proposed Roles and Responsibilities

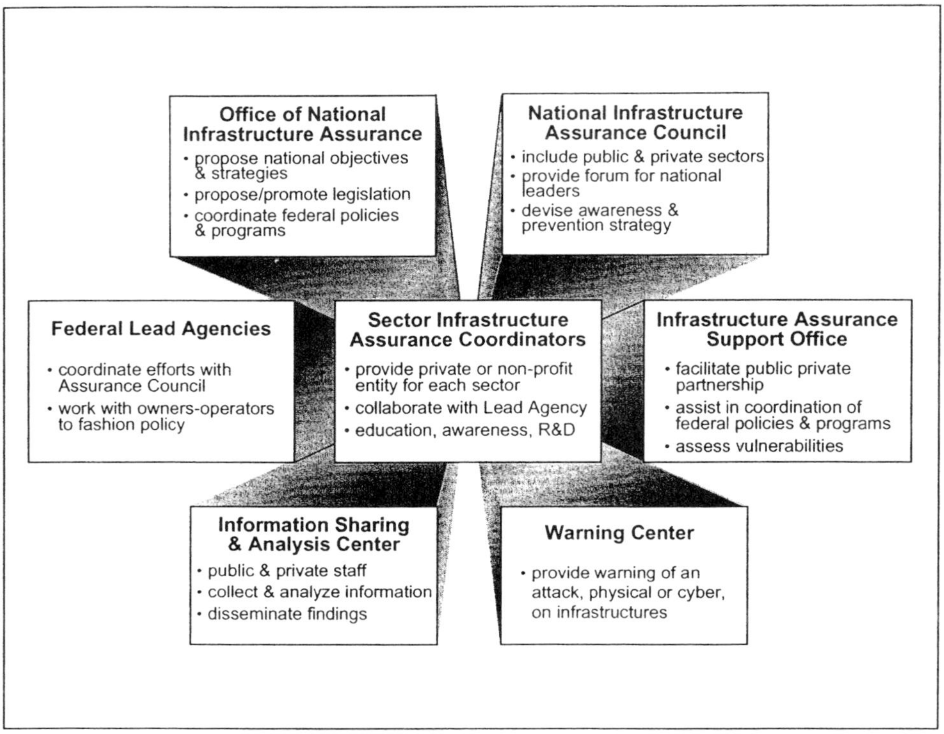

Proposed National Structure for Infrastructure Assurance

The Commission proposes a set of structures and processes within the public and private sectors to facilitate infrastructure assurance functions and complement existing law enforcement, regulatory, and other channels of communication between and among critical infrastructure providers and the government. These new structures and processes will provide trusted and protected channels for sharing public and private infrastructure assurance information, and a means for focusing, enhancing, and generating additional infrastructure assurance efforts throughout the federal government and private sector.

Essentially, we envision the proposed infrastructure assurance structure for the United States as consisting of seven elements (see Figure 9). Each is discussed in detail below.

- An *Office of National Infrastructure Assurance* in the White House to serve as the focal point for infrastructure assurance.

National Infrastructure Assurance Council ("Council")

We Recommend:

The President appoint a high-level council comprised of Chief Executive Officers (CEOs) from throughout the critical infrastructures, senior government officials (Cabinet rank), and representatives of state and local government. The Council would meet regularly to provide a forum for high-level discussion of proposed policies and directions for the nation in this critical area, to encourage and advocate partnership in infrastructure protection, and to make appropriate recommendations to the President.

The Council should provide policy advice to the President. It should meet no less than twice annually, and create whatever sub-structure it needs. A standing executive committee consisting of the Chair, selected Council members, and the Director of the National Office should meet often to manage the Council's work.

Staff support would be provided by the National Infrastructure Support Office. Members of the Council should be permitted to contribute staff and program support from their organizations (both public and private) to assist the Council in its work. Specific functions and duties of the Council should include:

1) Serve as the forum for national debate on infrastructure assurance issues.

2) Promote national objectives and strategies, facilitating discussion among the major stakeholders and government.

3) Review proposals from industry or government for mandatory standards, certifications, and best practices.

4) Provide leadership, advocacy, and support for the education and awareness efforts required to enhance national understanding and support for infrastructure assurance activities. Specifically, the Council should consider advocating, supporting, and encouraging adoption and use of "business" risk assessment tools and methodology.

5) Assist in setting directions for R&D program.

While the National Office and the Council provide avenues for the high level communication needed to develop a partnership in support of infrastructure assurance, the key to success in this arena rests with the existing federal agencies and the infrastructure sectors themselves.

Infrastructure Assurance Support Office ("Support Office")

We Recommend:

The President create a functional office to support infrastructure assurance activities throughout the federal government and the private sector.

The National Office would direct the activities of the Support Office, but it would be located in, and supported by the US Department of Commerce (DOC). The Support Office should be a joint coalition organization, bringing together appropriate national security and non-national security components. Staffing for the new office should reflect this mix among the required 20-30 professional, technical, and support staff including full-time employees, reimbursable details from other federal agencies, and private sector staff obtained under the Intergovernmental Personnel Act or by other means.

Its primary mission would be to support the National Office and the Council. Principal functions for the Support Office would include:

1) Support policy formulation by managing the national risk assessment, providing staff support to the Council and its subcommittees, tracking legislative and regulatory agendas, providing technical assistance to the Sector Infrastructure Assurance Coordinators, consolidating budget requests, drafting the budget proposal, establishing a system for tracking accomplishments, drafting the annual policy, and managing production and distribution.

2) Support prevention and mitigation by assisting the Council and the sectors in consolidating training requirements and developing new programs; by assessing current standards, certifications and best practices; by developing vulnerability assessment instruments (in consultation with the Sector Infrastructure Assurance Coordinators and selected owner and operators) and providing training in their use; and by coordinating the research program.

3) Assist the proposed National Office in the management of the Information Sharing and Analysis Center.

4) Assist the NSC in the preparation of stand-by plans and authorities in coordination with the relevant agencies and private sector entities; and provide technical support to the FBI and Federal Emergency Management Agency (FEMA) for development of policy and plans to manage incidents and consequences.

Federal Lead Agencies ("Lead Agencies")

We Recommend:

The President designate specific federal agencies to take the initiative in bringing together the owners and operators of various infrastructure sectors to create a means for sharing information that is acceptable to all participants. Lead Agencies will not replace the existing relationships, or assume any of the responsibilities of the law enforcement, regulatory or other special function agencies. They will work with sector owners and operators to identify and implement a method of sharing and protecting information.

Many federal and state agencies have interests and responsibilities in the infrastructure sectors. Additionally, each sector is comprised of diverse companies, associations and consortia which may challenge efforts to share information. Assigning leadership responsibility to the highest levels within identified federal agencies creates an opportunity to advocate and generate common purpose among the infrastructure leadership. We anticipate that Lead Agencies will coordinate with the Office of National Infrastructure Assurance to obtain the authorities needed to accomplish the following functions.

1) Establish and maintain channels of communication with all private and public entities having an infrastructure assurance interest in the sector.

2) Facilitate the selection of a Sector Infrastructure Assurance Coordinator (described below).

3) Assist the Sector Coordinator in establishing and operating an effective information sharing program.

4) Provide input to national infrastructure assurance objectives and strategies.

5) Draft new legislation and regulations, as required, and propose the use of federal incentives to facilitate private investment in assurance programs if appropriate.

6) Promote infrastructure assurance education and training, to include advocating use of best practices, within the sector.

7) Assist in developing plans for prevention (long-term reduction of vulnerabilities and short-term defensive actions), mitigation, restoration, and reconstitution.

8) Coordinate, in support of the Federal Response Plan (FRP), as amended, management of the consequences of a successful infrastructure attack and prepare for various contingent attacks through participation in training and exercise programs.

While assigning Lead Agency responsibilities for all critical infrastructures may be novel in some infrastructure areas, in others such a relationship already exists. Clearly, the Departments of Transportation and Energy already perform many of the responsibilities we outline for Lead Agencies. In other infrastructure sectors, telecommunications and information, for example, both DoD and DOC have significant interest and existing relationships in these infrastructure sectors. After much debate, we arrived at a proposal for assigning Lead Agency responsibilities for each infrastructure sector, shown in Table 2.

Table 2. Proposed Lead Agencies

Infrastructures from EO 13010	Commission's Infrastructure Sector	Proposed Lead
Telecommunications	Information & Communications	Joint Department of Defense & Department of Commerce
Electric Power	Electric Energy	Department of Energy
Gas & Oil	Gas/Oil Production & Storage	Department of Energy
Banking & Finance	Banking & Finance	Department of the Treasury
Transportation	All Sub-sectors	Department of Transportation
Water	Water Supply	Environmental Protection Agency
Emergency Services	Emergency Services	Federal Emergency Management Agency
Continuity of Government	Government Services	Office of National Infrastructure Assurance

Perhaps the most challenging responsibility of the proposed Lead Agencies will be facilitating the selection, by the owners and operators, of Sector Infrastructure Assurance Coordinators.

Sector Infrastructure Assurance Coordinators ("Sector Coordinators")

We Recommend:

Each infrastructure sector select or create an entity to facilitate sharing information among providers and with government. These Sector Coordinators will lead the sector in determining, collectively, how best to share the type of information needed for infrastructure protection by the federal government and owners and operators they represent.

Each sector will determine the particular mechanism best able to meet its needs. In some, an association or set of associations may best serve the industry and accomplish the role outlined here. Totally private and voluntary organizations may be selected by some, while others may find an existing regulatory agency more useful in the lead role.

Lead Agencies (described above) will work with infrastructure owners and operators and other government agencies that have industry-specific missions to establish these communication, coordination, and sharing mechanisms. Some sectors already have the kind of industry-wide organization needed. One example of such a partnership is the NERC.

Where a sector has such diverse interests that it cannot settle on a single Sector Coordinator, owners and operators and the Lead Agency may explore innovative solutions, such as a "virtual coordinator" based on existing networked resources.

The functions envisioned for the Sector Coordinators include:

1) Provide the sector with a means to accumulate information, disguise identity of providers, transmit information to the public-private Information Sharing and Analysis Center (described below), receive information from the Center, and disseminate it to the sector's owners and operators.

2) Serve as the focal point within the sector for risk assessment activities; and represent the owners and operators in discussions with other entities of the infrastructure assurance structures as needed.

3) Serve as the clearing house and hub for information sharing within the sector, assist in the analysis of anomalous events, and prepare statistical summaries.

Sector Coordinators will provide the central conduit for the information needed to develop an accurate understanding of what is going on throughout the nation's infrastructures. That is the purpose of the most innovative structure we recommend, a public-private analytic organization.

Information Sharing and Analysis Center ("Center")

We Recommend:

The President propose, and Congress charter, a new organization staffed by federal government employees and infrastructure owner-operator representatives to provide the analyses needed for infrastructure protection. The Center would receive information from all relevant sources, analyze it to determine what is actually happening in the infrastructures, and appropriately inform government and private sector users. Legislative changes will be required to implement this recommendation.

To be effective, the Center must have benefit of the legal initiatives discussed in Chapter 10, including some means to protect sensitive private sector information shared with the government and authority to negotiate non-disclosure agreements with the private sector. It should have direct channels to all interested government agencies to facilitate the flow of information.

Initially, the Center would focus on gathering strategic information about infrastructure threats, vulnerabilities, practices, and resources that will enable effective analyses to better understand the cyber dimension associated with infrastructures. The analysis produced will also allow more effective planning and decision-making about investments required within and outside the government. This information would include technical information of interest to owners and operators needing to better protect their systems from cyber attacks and threat-specific information developed by the government and provided through the Center to the infrastructure owners and operators. The Center would be expected to gather and maintain information about available assurance, protection, and defense resources within both public and private sectors for protection from cyber attack. Additionally, this Center would provide a one-stop/one-call capability for infrastructure assurance with special emphasis on the cyber dimension. When infrastructure owners and operators perceive problems within their information systems, they could call the Center to receive immediate information about available assistance.

The Center will, based on its analysis, issue bulletins, advisories, and other communications that will enable the infrastructure owners and operators to enhance their own levels of protection. It will also provide analysis to the FBI for dissemination to government agencies as required, and to the National Office and Support Office to be included in the policy, planning, R&D, budgeting, and other processes.

The responsibilities envisioned for this Center are:

1) Review reports of unusual occurrences from the owners and operators and the government, and prepare advisories for open release to the infrastructure providers through their Sector Coordinators and the government concerning vulnerabilities, failures, and system deficiencies.

2) Receive intelligence and law enforcement information concerning the development of potentially damaging tools and threats, and prepare advisories.

3) Enable the receipt and validation of anonymous data.

4) Provide technical assistance on a 24-hour basis.

5) Establish an extensive analytical data base accessible by the owners and operators and the government.

The proposed Center should eventually be staffed with between 20 and 40 personnel, about half of whom should be representatives from the infrastructures. Specific cost-sharing details can be negotiated, but to facilitate the speedy implementation of this recommendation, the government should be prepared to deploy the entire "start-up" cadre. The location of this Information Sharing and Analysis Center should be high on the agenda for decision by the Office of National Infrastructure Assurance. We believe an interagency group should investigate creative siting alternatives, especially locating it in the private sector. A number of excellent possibilities are available, among them co-locating with the Carnegie-Mellon University's CERT, another CERT, or a Federally Funded Research and Development Center, or contracting to a private entity (or university).

A significant aspect of the Center would be a government-only cell connected to the FBI's Office of Computer Investigations and Infrastructure Protection (OCIIP), which would serve as the preliminary national warning center for infrastructure attacks and provide the Center with law enforcement, intelligence, and other information needed to ensure the highest quality analysis possible.[11]

Information sharing and analysis will go far toward enabling the infrastructures to better protect themselves and ensuring the government has a more effective picture about what is happening throughout the infrastructures. This will allow us to understand whether diverse events—physical and cyber—are actually coincidental or related actions in an attack on different pieces of our infrastructures.

A Step Toward A National Cyber Warning Capability

We believe the eventual goal in this area is an indications and warning capability that provides immediate, real-time detection of an attempted cyber attack on critical infrastructures. The model for what we have in mind is the air defense and missile warning system. This is a defense system consisting of a monitoring or sensor capability, an analytic capability, and an alerting capability.

Until we are able to field a real-time warning capability, we will need to rely on the proposed Information Sharing and Analysis Center described above and on existing government warning or watch centers. The FBI's

---

11 In July 1996, the Director of the FBI established the Computer Investigations and Infrastructure Threat Assessment Center (CITAC) as a single point of coordination for all criminal, counterintelligence and counterterrorism computer intrusion matters and cases involving threats to critical infrastructures. In August 1997, the Director upgraded the status of this coordination function by creating the OCIIP.

newly-established OCIIP currently has the most potential for this effort and should assume the Warning Center responsibilities. In fact, the FBI has recently established and begun to staff a multi-agency Watch and Threat Analysis Unit within OCIIP. This unit's goal is to use existing criminal, counterintelligence, and counterterrorism authorities to meld information from government sources and cooperating private sector entities to detect cyber threats to critical infrastructures. It will act on that information to provide tactical warning of immediate consequence. OCIIP will use existing mechanisms to issue cyber threat alerts in the same way the FBI now issues terrorist alerts. As new capabilities come on line in the Information Sharing and Analysis Center and with Sector Infrastructure Assurance Coordinators, we expect they will enhance the FBI's alerting mechanisms.

To integrate the capabilities being developed in OCIIP with those proposed elsewhere in this report, the Commission suggests that the OCIIP's function be expanded to include:

1) Operating near real-time secure communications with the proposed Information Sharing and Analysis Center on a 24-hour basis, in addition to the connectivity already being established by the Watch and Threat Analysis Unit with other government watch offices.

2) Integrating anomalous infrastructure events with intelligence and law enforcement information for the purpose of developing indicators that the nation may be "under attack." When such an indication is forthcoming, the FBI would make appropriate notifications and issue warnings, and would alert the Information Sharing and Analysis Center to prepare and disseminate bulletins and threat advisories to infrastructure stakeholders.

We consider development of a warning capability to be of fundamental importance to the future security of our nation. We urge the Director of the FBI to continue to enhance the capabilities of the OCIIP and we reinforce the FBI's requests for the funding needed to establish and maintain capabilities in the cyber arena–beyond those needed for the investigation of criminal, counterintelligence and counterterrorism cases–to include the analytic capacity and the R&D efforts related to threat detection that will enable real warning in the years ahead.

Tables 3 through 7 provide illustrations of how we believe these new structures would interact to accomplish the specific national functions required for infrastructure assurance. We mean them as a guide to the types of relationships, communications, and responsibilities that might develop as the recommendations of the Commission are being implemented.

Table 3. Policy Formulation	
Assess National Risk	Support Office lead, done by contract, reviewed by Council
Integrate Public and Private Sector Perspectives	Council lead, consultation with sectors through Sector Coordinators; Lead Agencies and special function agencies support.
Propose National Objectives and Develop Strategies	Council lead, consultation with sectors through Sector Coordinators, Lead Agencies and special function agencies; proposed to the President through the National Office.
Propose and Promote (New) Legislation	Need identified by all sources, consolidated and analyzed by Support Office, validated by National Office, drafted by Lead Agency or special function agency, reviewed by Council and OMB, submitted to Congress.
Assess and Promote (New) Regulations	Need identified by all sources, consolidated and analyzed by Support Office, drafted by federal or state regulator, reviewed by Council, reviewed by normal regulatory process.
Influence Private Sector Investments	Support Office with contract support identify deficiencies (based on emerging threats) either directly with Sector Coordinators or through the Council; Council recommends to companies through Sector Coordinators.
Prepare, Recommend and Promote Budget	Council, Lead Agencies, special function agencies indicate needs and make recommendations; National Office consolidates with Support Office assistance, package reviewed by Council, Lead Agencies, special function agencies, approved by National Office, submitted to the Office of Management and Budget (OMB).
Manage and Enforce Implementation	Results reviewed by Council.
Shape the International Environment	Subset of CSG/CT prepares, annually, international objectives, meet with Department of State to fashion strategy.
Issue the National Policy	Issued by the President with an endorsement from Chair of the Council.

Table 4. Prevention and Mitigation	
Provide Effective Education and Awareness	Threat analysis provided by Information Center and training requirements identified by Sector Coordinators; consolidated by Council; vendors identified and certified by Council; other education programs coordinated by Support Office; managed by appropriate agencies, such as National Science Foundation (NSF) and Department of Education.
Set Standards, Certifications and Best Practices	Established by Sector Coordinators; forwarded to Lead Agencies if legislation or regulation is desired by companies; Lead Agencies enter into legislative or regulatory process as required, if approved by Council.
Assess Vulnerabilities and Risks of System Components	Council directs Support Office to prepare assessment instrument for each sector requesting one; Sector Coordinators review instrument; assessment vendors identified and certified by Sector Coordinators; owners and operators fund and manage.
Research Advanced Techniques; Develop New Technologies	National Office determines requirements in coordination with Lead Agencies, Council, OSTP, and private sector research organizations; Lead Agencies and/or NSF request funding and manage research as agreed.
Negotiate Funding	Owners and operators identify system upgrades based on risk assessment; Sector Coordinators propose cost share; Council serves as the forum for negotiation with Lead Agencies and representative from National Office.
Acquire the Resources for Protecting Systems	Acquired by owners and operators.
Manage Operations Consistent with Best Practices	Managed by owners and operators; performance reviewed by Sector Coordinators supported by Lead Agencies.

Table 5. Information Sharing and Operational Warning	
Share Information	Owners and operators send information on "unusual events" to Lead Agencies, as required, and to Sector Coordinator information cells (connected to intelligence and law enforcement communities); threshold events and statistics to Information Center (also connected to intelligence and law enforcement).
Analyze Information and Prepare Threat Advisories	Information Center sends general advisories to all or selected participants as needed.
Disseminate Warnings	"Actionable" warning information is relayed by teleconference to the OCIIP for decision, with copy to National Office and CSG/CT duty officer; dissemination directly to Sector Coordinators and owners and operators as per protocol.

Table 6. Counteraction (Incident Management)	
Develop Incident Management Policy and Plan Operations	FBI develops incident management policy and plans; CSG/CT reviews plans; Sector Coordinators develop plans to "close holes and block attacks" using National Office threat information and planning guidance.
Deter, Halt, or Minimize an Attack;	NSC develops deterrence policy. FBI takes lead in any attack, assesses magnitude and requests assistance as required from Defense, Intelligence or other government agencies; lead may be elevated into NSC structure, supported by National Office secretariat.
Implement Defensive Actions	FBI notifies National Office and Sector Coordinators of nature and extent of attack concurrent with standard notifications; Sector Coordinators and FBI consult on recommended provider actions; Sector Coordinators notify owners and operators.
Punish Perpetrators During or After an Attack	FBI takes lead for response to both domestic and international perpetrators unless actions have significant diplomatic implications; in which case, lead is elevated to the NSC.
Control Misinformation and Manage Perceptions	White House stands up public affairs center assisted by law enforcement and intelligence communities, DoD, National Office, and others as needed.
Coordinate Incident and Consequence Management	FBI and FEMA negotiate directly, with National Office participation.

Table 7. Response, Restoration and Reconstitution (Consequence Management)	
Plan for the Response to Consequences	FEMA leads with support of Federal Response Plan agencies and state and local emergency managers.
Manage the Response to Consequences	FEMA leads with support of Federal Response Plan agencies and state and local emergency managers.
Plan for Restoration and Reconstitution	Owners and operators plan for routine disruptions; Sector Coordinators facilitate planning with support of Lead Agencies for major disruptions; planning consolidated by FEMA.
Manage Restoration and Reconstitution	Owners and operators manage routine disruptions; Sector Coordinators work through the Federal Response Plan using Lead Agencies for major disruptions; funding to be determined under provisions of the Stafford Act (PL 93-288, as amended).

Other Federal Responsibilities

Law Enforcement

The basic law enforcement functions are not changed. There is a relatively new class of computer crimes for which classical techniques and training may not be adequate. Federal law enforcement agencies lead the country in developing new capabilities and in conducting training and awareness sessions with state and local agencies.

Intelligence Collection and Analysis

The intelligence community is expected to continue and improve its programs designed to assess the likelihood of an attack from abroad in general and to give specific warning of increasing capabilities or specific hostile intent.

Emergency Planning

Because actions needed to save lives and protect property in the event of a major disruption of infrastructure services are much the same regardless of the cause, we expect that federal emergency planning and response functions will continue as currently constituted. The real key to minimizing losses, however, will be the rapid restoration of the disrupted infrastructures. While private industry has a commendable record of restoring operations after most conventional forms of disruption, in an orchestrated attack there is a potential for damage well in excess of that normally encountered. There may, therefore, be a need to develop plans and capabilities that do not now exist. In Chapter 10, we recommend that FEMA take action to consolidate restoration and reconstitution planning and operations under the auspices of the FRP, using the designated Lead Agencies.

National Defense

Certain threats to our infrastructures may rise to the level of a national defense concern. The magnitude of the threat and required response or the identity of the attacker (from beyond our borders) may shift the lead for a cyber attack from DOJ to DoD. DoD is expected to:

1) plan counteractions to deter, halt, or minimize an attack considering a variety of possible sources and alternative responses, which may include a variety of military options;

2) coordinate selection of specific responses with the National Command Authority; and

3) execute counter-actions as authorized.

In addition, as technology enables increased detection and identification capability for cyber attacks, DoD (including its NSA component) may play an increased role in detecting potential cyber attacks before they enter the nation's domestic communications systems.

International Outreach

In the new geography, protecting our infrastructures at home is not enough. Many aspects of infrastructure operations extend beyond our national borders, and even beyond the control of their owners and operators. The very nature of the cyber dimension renders national borders almost obsolete, and national laws and policies based on those borders of less and less consequence. Initiatives to construct partnerships between and among sectors and infrastructures must of necessity take into account the international character of business. The overall success of our own infrastructure assurance efforts will therefore require substantial international collaboration. The federal government should continue efforts to work with appropriate international bodies to address infrastructure protection concerns and raise the level of international cooperation and coordination on computer intrusion matters. An effective international regime to deter cyber crimes and cyber attacks will be more effective than purely national sanctions. Clarification of the dynamics surrounding a "cyber attack" under international law would also contribute to deterrence. Other issues worthy of international dialogue include the handling of cyber crimes that transcend borders, and legal responsibilities in multi-national infrastructures. Diplomatic efforts can also contribute to the success of our national encryption policy and the development of internationally accepted standards for computer security and information technology.

The United States is not alone in facing the realities of the new geography, but we are definitely in the vanguard of countries which have begun to realize the urgency of the issue. This gives us an opportunity to shape the contours of international cooperation in this universally important area.

Just as the federal government can lead by example in the context of US infrastructure assurance, the US can help create a positive influence on the infrastructure owners and operators–as well as the governments–of the countries that reside with us in the global community.

Conclusion

Managing new risks in the Information Age requires a partnership between industry and government for many purposes, from policy making aimed at preventing a crisis through responding if such a crisis occurs. It also requires adding a cyber dimension to our existing capabilities. Our recommendations in this chapter seek to enable increased partnership with the private sector and, importantly, to increase capabilities within our existing structures (see Figure 10).

Figure 10. Proposed National Structure for Infrastructure Assurance

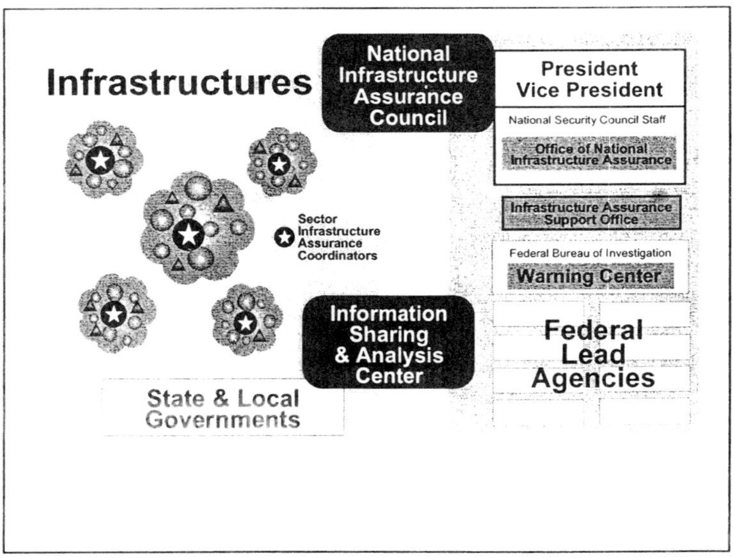

While we strongly endorse a policy of reliance on the private sector for problem-solving, solutions, and technology, we also see a need for a strong government focus on infrastructure protection and a federal framework to implement a national policy on infrastructure protection.

The key to success of the integrated public-private structure we propose will be "buy-in" from all sectors. And the key to their buy-in is heightened awareness of the challenges ahead. The new structures we propose are intended to generate awareness among all participants in infrastructure protection—public and private—and more broadly throughout the nation.

CHAPTER EIGHT
REPORT ON AWARENESS AND EDUCATION

Objective:

Elevate national awareness of infrastructure threat, vulnerability, and interdependency assurance issues through education and other appropriate programs.

The Awareness Challenge

A successful public-private partnership requires a significant level of understanding on the part of the owners and operators of the infrastructures, government, and the public at large. It is clear that one key to a more secure future is broader understanding of the role that information and telecommunications play in our national security and economic competitiveness. That understanding must be supported by an increased knowledge base throughout the nation. We must have new "street smarts" about the Information Age, about computers, and about the communications systems that connect our institutions, homes, and businesses. In short, we need a new awareness throughout the nation.

The National Research Council cited the need for greater sensitivity to information security in a 1991 report:

> *"That today's commercial (computer and software) systems provide only limited safeguards reflects limited awareness among developers, managers, and the general population of the threats, vulnerabilities, and possible safeguards. Most consumers have no real-world understanding of these concepts and cannot choose products wisely or make sound decisions about how to use them. Even when consumers do try to protect their own systems, they may be connected via networks to others with weaker safeguards – like a polluting factory in a densely populated area, one person's laxness in managing a computer system can affect many."*[12]

There are indications that awareness of computer security issues may be increasing, as demonstrated by a recent survey of 10,000 subscribers conducted by *Info-Security News* and Deloitte & Touche. Of the 1,225 responses, 55 percent considered lack of end-user awareness to be a significant barrier to information security. This is an improvement over the 73 percent who provided a similar response two years before, but it still suggests a requirement for greater awareness of the need for special measures to ensure information security.[13]

---

12 Computers at Risk: Safe Computing in the Information Age, National Research Council, National Academy Press, Washington, DC, 1991, pp. 2-3.

13 Info-Security News Industry Survey, May 1997, pp. 20 ff.

An Awareness Program

We have some experience with awareness programs. Forty years ago, wild fires annually destroyed nearly five million acres of land in the United States. Often caused by careless hikers, these fires cost nearly $1 billion per year. The "Smokey Bear" campaign, with its "Only you can prevent forest fires" slogan, saved about $17 billion in its first 30 years.

In the infrastructure protection arena, we need to reach four target audiences: infrastructure owners and operators, corporate infrastructure users, senior governmental officials at the federal and state levels, and the general public.

We Recommend:

The White House sponsor a series of conferences with national leaders in the public and private sectors to define programs to increase the commitment to information security. White House leadership is essential to the success of an awareness program on information security.

We Recommend:

The intelligence and law enforcement communities and the proposed Office of National Infrastructure Assurance expand existing programs of communication with infrastructure owners and operators and senior governmental officials by including periodic briefings on threats and vulnerabilities, recognizing the need to comply with appropriate security considerations.

We Recommend:

The National Academy of Sciences and the National Academy of Engineering establish a Round Table bringing together federal, state, and local officials with industry and academic leaders to develop national strategies for enhancing infrastructure security and to provide continuing support to an awareness program.

We Recommend:

The proposed Office of National Infrastructure Assurance, in coordination with the private sector, spearhead a continuing national awareness campaign, emphasizing infrastructure security.

We Recommend:

The proposed Office of National Infrastructure Assurance establish a program to hold infrastructure assurance simulations involving senior public and private officials. Funded from the proposed R&D budget, the simulations would assess the value of new concepts in improving infrastructure assurance. Reports on the findings of the games would be distributed as a part of the awareness campaign.

We Recommend:

The proposed Office of National Infrastructure Assurance encourage the private sector to develop generally accepted security principles to be used by internal and external audit institutions in their regular operational audit functions in order to sustain awareness in public and private institutions.

We Recommend:

The proposed Office of National Infrastructure Assurance encourage private industry to perform periodic, quantitative risk assessments of their information and telecommunications systems, to enhance awareness of new vulnerabilities. A quantitative risk assessment addresses risk and likelihood of loss in business language and supports cost-benefit analysis for financial risk management.

An Education Program On Computer Ethics

In many families, children are more computer literate than their parents. Lacking experience, the parents seldom offer ethical guidance regarding computer usage. Universities are finding it necessary to establish new protocols for the student population in order to protect privacy and intellectual property. Computer ethics should be introduced as a field of study in all schools, from K-12 through universities.

We Recommend:

The White House convene a conference on the broad issue of computer ethics directed at the K-12 and the general university population, drawing on state and regional leaders who can support the programs in local communities, school systems, and universities.

We Recommend:

The US Department of Education commit at least $5 million per year for five years to assemble and distribute course materials and sponsor appropriate institutions for development of special programs and new course materials for K-12 and university education on the subject of the ethics of computer usage.

A Professional Education Program

There is a significant deficiency in the number of university faculty members equipped to teach information and computer security. Professor Spafford of Purdue University reports that "over the last five years, the four academic institutions teaching information security in computer science programs granted 16 Ph.D.s for security-related research. Of these, three stayed in academia."[14]

In schools of business, students majoring in information systems may learn about computer security. However, the students in other specialties who may become general managers are given little insight into the need to deal with information and communications security even in terms of their study of risk analysis.

The federal government has a number of initiatives under way in information security. DARPA has a research program that is helping in the design of security systems. NSA is developing a continuing workshop in this field. The second will be held in Austin, Texas, in 1998, organized by the University of Texas. These workshops are intended to share information about what is being taught in the field of information and communications security and how these educational programs can be extended to a larger audience.

We Recommend:

The White House convene one or more conferences of academic leaders from engineering, computer science, and business schools to review the status of undergraduate and graduate education in information security and identify changes in the curricula and the resources necessary to initiate needed changes to meet the national demand for professionals in the field.

14 Statement of Dr. Eugene Spafford at "A Briefing on Secure Communications" before the House Committee on Science, February 11, 1997.

We Recommend:

NSF commit $10 million per year for at least five years to university programs on information assurance to support graduate students and faculty in Departments of Computer Science or in Business Schools with a view toward increasing the quality of education, the number of graduates in information and computer security, and the number of faculty members teaching in the field. As a part of such support, authorize the acquisition of advanced equipment when essential to the academic purpose of the program.

A General Education Program

Deficiencies in the training of technicians are reflected in inadequate attention to computer security in day-to-day operations. Education and training are essential to developing the staffs necessary to manage and operate major information systems today. Technicians need a deeper understanding of the systems they manage than they are likely to get if they have only on-the-job training. The rate of growth of the knowledge base makes it necessary to provide for initial training and also refresher training at regular intervals.

There are many commercial institutions in the field of education and training as well as community colleges, university extension programs, professional society programs, and others. While some have good course material, all could benefit from course material developed by the government agencies engaged in work in the field of information assurance.

We Recommend:

NIST, NSA, and the US Department of Education work in collaboration with the private sector to develop programs for education and training of information assurance specialists and for continuing education as technologies change. This effort should also support "training the trainers" to provide an adequate cadre of qualified instructors to teach technicians.

CHAPTER NINE
LEADING BY EXAMPLE

Objective:

Initiate a series of information security management activities and related programs demonstrating government leadership.

Infrastructure assurance is a joint responsibility, but the federal government has an unmistakable duty to lead the effort. Clearly, the federal government must *lead by example* as it reaches out to the private sector and other levels of government. We need to ensure the federal government has the policies and tools required to conduct business in the cyber age. Toward that end, the Commission makes these recommendations.

Improve Government Systems Security

The federal government has not paid sufficient attention to its own computer security needs. While OMB has developed and promulgated guidelines to ensure agencies adopt effective internal computer and network security practices, the effort to identify and replicate best practices throughout the government has fallen short of its target.

We Recommend:

Assigning systems security oversight responsibilities to the proposed Office of National Infrastructure Assurance. This will require legislative changes to restructure those responsibilities from OMB to the new office.

We Recommend:

The Secretary of Commerce and the Secretary of Defense charge NIST and NSA with assisting federal agencies in the implementation of best practices for information security within their individual areas. The process should include a NIST/NSA-facilitated assessment of agency vulnerabilities and security practices with input from the proposed Office of National Infrastructure Assurance.

We Recommend:

The FBI actively recruit college students with appropriate computer-related technical skills to seek employment with the Bureau. The FBI should consider offering part-time employment for skilled college students with regional computer crime squads. This program could produce current benefits as well as future special agent and forensic examiner applicants qualified to investigate cyber crime matters.

We Recommend:

The FBI facilitate hiring and retention of qualified personnel for technical analysis and investigation involving cyber attacks. Three years of service in cyber-related activities could be a condition of employment for those who receive a hiring preference based on computer skills.

Encryption

For electronic commerce to flourish, the information infrastructure must be secure and reliable. Protection of the information our critical infrastructures are increasingly dependent upon is in the national interest and essential to their evolution and full use. A secure information infrastructure requires the following:

- Secure and reliable telecommunications networks.
- Effective means for protecting the information systems attached to those networks.
- Effective means for authenticating communications of trading partners, assuring the integrity of data and non-repudiation of transactions.
- Effective means of protecting data against unauthorized use or disclosure.
- Well-trained users who understand how to protect their systems and data.

Strong encryption is an essential element for the security of the information on which critical infrastructures depend. Establishment of trustworthy key management infrastructures (KMIs) is the only way to enable encryption on a large scale, and must include the development of appropriate standards for interoperability on a global scale. Key recovery is needed to provide business access to data when encryption keys are lost or maliciously misplaced, and court-authorized law enforcement access to the plain text of criminal-related communications and data lawfully seized.

Neither private citizens nor businesses are likely to use the information infrastructure on a routine basis if they lack confidence that their communications and data are safe from modification or unauthorized access. To ensure public confidence in key recovery, stored decryption keys must receive the same sort of legal protections that currently exist for mail, telephone communications, and electronic communications, including e-mail. To fairly balance the competing equities of privacy, electronic commerce, national security and law enforcement, and to ensure public confidence, the following are necessary:

- The public should be free to select an agent to issue digital signatures or to serve as a key recovery agent.
- Law enforcement agencies should have lawful access to the decrypted information when necessary to prevent or detect serious crime. Procedures for judicial review prior to granting government access must be defined in law.
- Individual rights of redress when access is abused should also be defined in law.

We Recommend:

Expediting the several government pilot projects underway or recently announced as a means of testing the technical and policy concepts involved and building public confidence and trust with the KMI key recovery approach. Further, the Administration should promote efforts to plan for the implementation of a KMI that supports lawful key recovery on an international basis. Finally, the federal government should encourage efforts by commercial vendors to develop key recovery concepts and techniques.

Procurement

We Recommend:

An interagency task force identify large pending procurements (such as the new Federal Telecommunications System, FTS 2000) related to infrastructure assurance issues, study whether infrastructure assurance objectives are being considered, determine how they may be adapted, and, based on the lessons learned, propose revisions to the overall procurement process.

Threat Assessments

We Recommend:

The federal government elevate and formalize information threats as a foreign intelligence priority.

NIST Risk Assessment

We Recommend:

NIST and appropriate government agencies continue development of risk assessment methodologies and make these known and available to the private sector, especially owners and operators of infrastructures.

Measuring Performance

The Government Performance and Results Act (GPRA) requires five-year strategic plans and performance measures for major functions and operations of federal agencies to be reviewed by OMB in the budget process. The Information Technology Management Reform Act (ITMRA) requires performance measures related to the use of information technology. The required performance measures do not, however, specifically include information security.

We Recommend:

The Administration direct federal agencies to include assigned infrastructure assurance functions within their GPRA strategic planning and performance measurement framework.

We Recommend:

The Administration and Congress amend the ITMRA to require that agency Chief Information Officers develop performance measures for the security of their information systems and to submit evaluations to OMB as required by the statute.

Certification Programs

The Environmental Protection Agency (EPA), NSA, and others have demonstrated ways of extending the benefits of federal standards or certifications to the private sector. The Commission noted the EPA's ENERGY STAR program as an example of such an effort. A recently-initiated certification partnership between NSA, NIST, and industry is designed to facilitate the evaluation of commercial information assurance products. These low cost, easily administered mechanisms encourage voluntary compliance with federal standards.

We Recommend:

Lead Agencies consider the creation and use of certification programs that are inexpensive to administer and enforce, and that provide incentives for adoption of standards for information security and information technology services and products.

Global Positioning System

The GPS is scheduled to be the sole source of radionavigation for aircraft landing guidance systems by the year 2010. Although cost-efficient, this creates the potential for single-point failure.

We Recommend:

The Secretary of Transportation:

- Fully evaluate actual and potential sources of interference to, and vulnerabilities of, GPS before a final decision is reached to eliminate other radionavigation and aircraft landing guidance systems.
- Sponsor an independent, integrated assessment of risks to civilian users of GPS-based systems, projected through the year 2010.
- Base decisions regarding the proper federal navigation systems mix and the final architecture of the modernized NAS on the results of that assessment.

National Airspace System

The proposed architecture for the modernized NAS appears to have vulnerabilities that should be given full consideration before the final design is approved.

We Recommend:

The Federal Aviation Administration (FAA) act immediately to develop, establish, fund, and implement a comprehensive National Airspace System Security Program to protect the modernized NAS from information-based and other disruptions, intrusions and attack. Program implementation should be guided by the recommendations found in the *Vulnerability Assessment of the FAA National Airspace System Architecture*, prepared for the Commission.

CHAPTER TEN
LEGAL INITIATIVES

Objective:

Sponsor legislation to increase the effectiveness of federal infrastructure assurance and protection efforts.

Infrastructure protection requires the integrated capabilities of diverse federal agencies, and special means for coordinating federal response to ensure that these capabilities are melded effectively together. The first step in defining federal structures to support infrastructure assurance in the Information Age must be to understand how responsibility is assigned today within the legal framework of the federal government.

The interdependence of all the infrastructures and the critical role of the information and communications infrastructure in all aspects of American life create special jurisdictional challenges. These jurisdictional problems are further complicated by the continued growth in cyber attack capabilities across the threat spectrum. The ability to know the origin, purpose and magnitude of an attack is significantly limited today. Consequently, we do not have the sharp and unambiguous jurisdictional cues that guide decisions about response and assignments of responsibility in the more familiar physical arena. We may not know the source of an attack–domestic or foreign. We may not know the identity or motives of the attacker– individual or group, terrorist, criminal, or government. Nor may we know the magnitude of the attack–whether a single system is involved or the attack is perpetuated throughout a network or series of networks. We may not even know if ours is the only nation experiencing the attack.

Given the lack of knowledge available at the initiation of an attack, it is clear that any required federal response will be borne on the Attorney General's authority as the nation's chief law enforcement officer. Elements of the response may require support of the defense, emergency response, intelligence and diplomatic agencies, as well as other agencies within government. There is a clear need to have required decision support, planning capabilities, and response authorities available to the Attorney General and to the White House should the decision reach that level.

The structures we recommend in Chapter 7 recognize that infrastructure assurance is more than a law enforcement, defense, or economic problem. It encompasses the responsibilities of each of these areas, and also of the owners and operators who actually deliver infrastructure services. The federal government must not only integrate the familiar elements of government; it must also lead an effort to enhance the protection capabilities inherent in the infrastructures themselves, and generate the kind of trusted environment that enables a cohesive public-private partnership to accomplish all the functions involved in infrastructure protection.

We recognize also that while responsibilities are widely shared within the government, the current level of technology does not allow the posture of deterrence and forward defense that protects us from foreign military and terrorist threats in the physical dimension. Initially, all cyber attacks will have to be treated as crimes—regardless of where they originated or the purpose of the attack. When investigation provides evidence of foreign government involvement or the magnitude of the attack requires it, other leadership may be assigned. This also will require that the Attorney General have available immediate support from defense, intelligence and elsewhere in the government–especially from those agencies that have special skills and knowledge applicable to the cyber arena.

In making recommendations about increased partnership and better two-way sharing of information, we do not mean to indicate lack of support for existing efforts to build required information centers, watch centers, and command and control facilities. These efforts to enable response to cyber threats—criminal, terrorist or other—must continue. The organizations detailed in our recommendations are designed to expand the reach of existing capabilities, provide a means to coordinate and integrate them with information, knowledge and skills from the infrastructure owners and operators, and generally facilitate their efforts.

In addition to examining these jurisdictional issues, the Commission studied the legal foundations for infrastructure protection, and focused on the need to revisit the current law in light of infrastructure assurance objectives. In so doing, the Commission was able to make recommendations designed to enable the federal government to take a leading role, the private sector to respond, and the government and the private sector to engage in an effective partnership. Some of these recommendations, such as those relating to government model performance and legal impediments to information sharing, are highlighted in other parts of this report. Those recommendations as well as those contained in this chapter will provide a legal foundation for cultural change.

Enabling the Federal Government to Take the Lead

The first set of recommendations revisits existing legal frameworks for federal response to and deterrence of incidents involving the critical infrastructures.

Many areas of federal legislation that enable prevention and mitigation, response, recovery and reconstitution to incidents involving the critical infrastructures were written before the emergence of a recognizable cyber threat. It is not clear whether many of these authorities would apply, or should apply, to a major cyber-related event. Until the dynamics of such an event are better understood, major legislative change is premature. However, the Commission was able to identify key issues and make general recommendations to incorporate infrastructure assurance considerations within these legislative frameworks.

Defense Production Act

The Defense Production Act (DPA) provides authority to assist the reconstitution of critical infrastructures. The Commission reviewed DPA authorities, triggering mechanisms, and current modernization efforts for application to emerging threats, vulnerabilities, and related challenges.

We Recommend:

The Administration and Congress review the DPA in light of infra-structure assurance objectives. Specifically, we suggest:

- Congress consider amending the DPA Declaration of Policy to include a finding that critical infrastructures are essential to national security.
- Lead agencies associated with the critical infrastructures study the energy provision for priorities in contracts as a potential model for reconstituting other critical infrastructures.
- Congress continue funding for the DPA Fund and financial incentives, and make funds available for R&D related to the critical infrastructures.
- The Administration direct federal agencies with authorities pertaining to the critical infrastructures to review DPA authorities and work with industry to use these authorities when needed in response to a critical infrastructure incident.

Stafford Act/Federal Response Plan

The Stafford Act and FRP set parameters for federal response to major disasters as declared by the President. FEMA's authority to prepare for, mitigate, and respond to incidents affecting the operation of the critical infrastructures is unclear under the triggering mechanism currently contained in the statute.

Current FRP capabilities are responsive to infrastructure disruption. The capabilities and expertise to restore and reconstitute the infrastructures reside almost exclusively in the private sector and the main burden for planning and operations falls on the owners and operators of the infrastructure companies themselves.

However, the federal government has a shared responsibility to ensure that these infrastructures are restored rapidly in the event of a major disruption. The federal government should share in the costs of training and exercising, and ensure the availability of critical resources on a yet to be determined cost-sharing basis.

We Recommend:

The proposed Office of National Infrastructure Assurance study the Stafford Act, other authorities, and Federal Response Plan mechanisms for suitability in cyber-induced disasters. The study should address the potential impact of infrastructure failures and the desirability of direct assistance to infrastructure owners and operators.

We Recommend:

FEMA consolidate restoration and reconstitution planning and operations under the auspices of the Federal Response Plan, using the designated Lead Agencies.

Nunn-Lugar-Domenici

The Nunn-Lugar-Domenici legislation focused federal resources on providing training, access to equipment, and information to local first responders. State and local police, fire, and medical officials are requesting an expanded effort in this area. The Commission sees the need for more resources for training and equipment, and possibly an expanded scope to address other infrastructure-related events.

We Recommend:

Congress consider expanding the current Nunn-Lugar-Domenici program to incorporate other critical infrastructure issues, including attacks on infrastructures by means other than weapons of mass destruction, as well as training and information sharing efforts directed at state and local responders.

Adequacy of Criminal Law and Procedure for Infrastructure Assurance—Physical

In addition to the preventive aspects of the DPA, Stafford Act, and Nunn-Lugar-Domenici legislation, deterrence also plays an important preventive role against attacks on critical infrastructures. Deterrence through criminal law should be built not only through federal investigative and prosecutive capabilities, but also state, local, and international response.

Sentencing Guidelines

The Commission concluded there is adequate "legal fortification" from physical attacks. However, we identified several shortfalls relating to deterrence of crimes against critical infrastructures. The Sentencing Guidelines do not adequately address the severity of consequential damages arising from attacks on critical infrastructures–for example, damage resulting from the "downstream" effects of a denial-of-service attack. Consequently, a possibility exists of disproportionately light sentences for some forms of attack on critical infrastructures.

We Recommend:

The US Sentencing Commission expand the Guidelines to include greater flexibility to address actual and consequential damages, including "downstream" damage to property or loss of service resulting from attacks on critical infrastructures.

We Recommend:

The Sentencing Commission consider expanding coverage of its Guidelines to better address consequences of the use of biological and chemical weapons not resulting in death.

Interstate Commerce

The Commission identified two potential deficiencies with respect to purely intrastate attacks against critical infrastructures–even when attacks result in severe damage. In these instances, in order to assume jurisdiction over an investigation or prosecution, the federal government must demonstrate on a case-by-case basis that the incident affects interstate commerce. This is a difficult determination to make at the earliest stages of an investigation, before the scope of an attack is known or its effects are contained.

We Recommend:

Congress consider defining certain critical infrastructures as "instrumentalities of interstate commerce" to enable immediate investigation by federal law enforcement agencies and to subject those convicted to stiffer federal penalties.

Reward/Payment for Information Programs

The Commission reviewed legislation that offers rewards for information leading to the capture of terrorists. Under these legal authorities, Congress authorizes the Attorney General and the Secretary of State to administer rewards and payment-for-information programs. These laws effectively supplement other federal crime legislation to protect critical infrastructures.

We Recommend:

The monetary reward programs for information leading to capture and arrest of criminals be included as a line-item in participating federal agencies' budgets to ensure proper funding and implementation.

Adequacy of Criminal Law and Procedure for Infrastructure Assurance–Cyber

State & Local

Efforts are ongoing in most states to draft effective computer crime legislation. Dealing with juvenile computer crime is an area requiring greater attention. The states and federal government may be able to learn from innovative efforts in this area and consider modification to their laws to address what may be a growing problem.

We Recommend:

DOJ sponsor a comprehensive study aimed at compiling demographics of computer crime, comparing various state approaches to computer crime and discovering effective ways of deterring and responding to computer crime and abuse by juveniles.

Federal

The US Sentencing Commission's revised guidelines for the Computer Fraud and Abuse Act expanded definitions of "harm" and "loss" to include interruptions in service; disruptions or delays in delivery of vital services endangering lives; invasions of privacy; and the cost to the victim of damage assessment, restoration of service and data, and loss of business revenue due to interruption of service.

We Recommend:

The Sentencing Commission consider expanding its broader reformulation of harm and loss (in Guidelines Section 2B1.1, as it applies to violations of the Computer Fraud and Abuse Act and theft of trade secrets) to other forms of electronic crime and crimes relating to information and information technology.

DOJ is currently exploring ways to ease administrative burdens on federal law enforcement officers investigating various forms of computer and high technology crimes that cross federal jurisdictional boundaries. Of specific concern is allowing electronic searches to be conducted across jurisdictional boundaries with the authorization of only one federal judge.

We Recommend:

The Administration endorse and promote efforts currently underway to develop procedural changes to assist law enforcement in the investigation of computer crime, including modification of existing procedures for an effective nationwide trace and search warrant capability.

Congress consider expeditious enactment of such legislation.

International

The US is a leader of efforts to clarify and improve current law enforcement procedures pertaining to computer crime.

We Recommend:

The Administration lead efforts to clarify and improve current procedures for investigating computer crime; work to create a network of international law enforcement agencies and telecommunications carriers to facilitate international investigations of computer crimes; and continue efforts to enhance international cooperation in computer crime investigations.

Legal Impediments to Vulnerability Assessments

Existing laws may create unnecessary legal impediments to the performance of vulnerability assessments on federal computer systems. The Computer Fraud and Abuse Act criminalizes a wide variety of misconduct premised on unauthorized access to government (and private) computer systems. The legislation is silent, however, as to how Red Teams might be *authorized* to attempt penetrations without running afoul of the criminal law. Legislative change does not appear to be required, but agencies should clarify procedures to facilitate sound vulnerability assessment practices.

We Recommend:

Federal agency Chief Information Officers establish procedures for obtaining expedient and valid authorization to allow vulnerability assessments to be performed on government computer systems. This requires a clear designation by agencies regarding who may authorize access to their computer systems for this purpose.

Enabling Private Sector Response

In addition to reviewing federal authorities that could be strengthened or expanded to allow the federal government to more adequately accomplish infrastructure assurance objectives, the Commission also considered potential legal impediments that might prevent owners and operators from taking appropriate action to safeguard portions of critical infrastructure within their control and responsibility. The recommendations contained in this section focus on providing owners and operators greater ability to take protective action.

Private Intrusion Response

Unauthorized intrusions often go undetected; when detected they may not be reported. Currently, computer security specialists and even state-licensed private investigators are gearing up to support private sector needs for computer security services. While their services fulfill some victims' needs for confidentiality and control, potentially valuable information that could be used to assess the scope and nature of the threat is lost. Furthermore, there are no mechanisms in place to ensure the professionalism, qualifications, and methods by which these private investigations are performed.

We Recommend:

Congress consider new ways of facilitating the growth of private sector cyber-security capabilities that encourage increased sharing of information relevant to the scope and nature of the threat.

One approach to this area is nationwide licensing of private security specialists by a professional organization or the government. It might be possible to arrive at a professional licensing scheme that would provide benefits to a number of parties by specifying, for example, qualifications for obtaining a license, levels of insurance required, standards of practice, and conditions to allow for limited information sharing.

Additional prosecutive capabilities may also contribute to the current level of deterrence for computer-related violations. Prosecutive capabilities could be expanded by permitting victims the right to proceed in private civil actions. Civil remedies are currently available at federal and state levels. Improving the international availability of civil remedies is a logical extension of these efforts.

We Recommend:

The President seek to expand the availability of civil remedies for computer-related violations through appropriate multilateral and bilateral agreements.

Privacy Legislation and the Employer-Employee Relationship

"Insiders" provide the most frequent avenue of attack to the nation's critical infrastructures. The federal government guards against insiders' misdeeds through authority to conduct background investigations and periodic reinvestigations. Private employers who operate some of the critical infrastructures do not have the same ability. In many states, private employers do not have access to criminal history information; are prohibited from requesting or using criminal, financial or employment information; and may incur tort liability for revealing unfavorable employment history. These restrictions result from legitimate concerns over privacy, fair employment, rehabilitation, and related questions. We believe security considerations justify limited exemptions from these restrictions.

We Recommend:

The Attorney General convene a group of professionals from law, state and federal legislatures, labor and management organizations, and the privacy community to explore existing laws and recommend measures to balance employers' needs against individual interests in privacy.

We Recommend:

State legislatures consider adopting "consent" as a baseline for allowing employers to request background information from employees and potential employees for sensitive positions within critical infrastructures, subject to fair information practices.

We Recommend:

Congress narrowly expand existing exemptions to the Employee Polygraph Protection Act to include providers of information security services within the scope of its exemptions. This would update the legislation that currently allows polygraphs only to physical security services for certain public services.

CHAPTER ELEVEN
RESEARCH AND DEVELOPMENT

Objective

Increase investment in infrastructure assurance R&D from $250 million to $500 million in FY 99, with incremental increases in investment over a five-year period to $1 billion in FY 04. Target investment in specific areas with high potential to produce needed improvements in infrastructure assurance.

Federal R&D efforts are inadequate for the size of the R&D challenge presented by emerging cyber threats. Only about $250 million per year is being spent on federal infrastructure assurance-related R&D, of which 60 percent–$150 million–is dedicated to information security. There is very little research supporting a national cyber defense. The Commission believes that real-time detection, identification, and response tools are urgently needed. We concluded that market demand is currently insufficient to meet these needs.

R&D for infrastructure protection requires partnership among government, industry, and academia to ensure a successful and focused research and technology development effort.

The President propose an increase in the federal investment in infrastructure assurance research to $500 million in FY99 and incremental increases in annual funding over a five-year period to $1 billion in FY04 for a targeted R&D program focusing on the six R&D areas listed below.

We Recommend:

- *R&D Increases for Information Assurance.* Assurance of vital information is increasingly a key component to the functioning of our interdependent infrastructures. The urgent need to develop new, affordable means of protection is apparent, given the increasing rate of incidents, the expanding list of known vulnerabilities, and the inadequate set of solutions available.
- *R&D Increases for Intrusion Monitoring and Detection.* Reliable automated monitoring and detection systems, timely and effective information collection technologies, and efficient data reduction and analysis tools are needed to identify and characterize structured attacks against infrastructure.
- *R&D Increases for Vulnerability Assessment and Systems Analysis.* Advanced methods and tools for vulnerability assessment and systems analysis are needed to identify critical nodes within infrastructures, examine interdependencies, and help understand the behavior of these

complex systems. Modeling and simulation tools and test beds for studying infrastructure-related problems are essential for understanding the interdependent infrastructures.

- *R&D Increases for Risk Management Decision Support.* Decision support system methodologies and tools are needed to help government and private sector decision-makers effectively prioritize the use of finite resources to reduce risk.
- *R&D Increases for Protection and Mitigation.* Real-time system control, infrastructure hardening, and containment and isolation technologies are needed to protect infrastructure systems against the entire threat spectrum.
- *R&D Increases for Incident Response and Recovery.* A wide range of new technologies and tools are needed for effective planning, response, and recovery from physical and cyber incidents that affect critical infrastructures.

We Recommend:

The National Research Council define, more fully, a national infrastructure assurance research program and lead an effort with departments and agencies already engaged in R&D relevant to each infrastructure.

Assuring Water Quality

Few infrastructures are taken for granted more than our fresh water systems. There is little chance of a threat reducing the quantity of available water sufficiently to endanger the population or cause industrial collapse. But there is risk of malicious attacks over time undermining public confidence. Alternatives for protecting the water supply are few. The most feasible approach we found is a research effort focused on water contamination detection technologies. Effective applications could be developed commercially and implemented at the state and local level.

We Recommend:

The creation of a specific R&D program to provide the scientific knowledge and technology necessary to allow highly toxic chemical and biological agents to be detected, identified, measured and treated in near real-time in the nation's water supply systems. The program should be administered by the EPA.

Provide Early Warning and Response

Real time detection of cyber threats is a special challenge to the R&D community. While this area is included in our recommendation above for additional R&D investment, it is central to the future security of our infrastructures. Some effort is under way, but it requires continued funding and high priority.

Although many industry and government groups are dedicated to ensuring the technical performance of next generation telecommunications networks, there has been no cohesive effort for protecting this infrastructure against the emerging threat of cyber attack. Such effort should include a system of surveillance, assessment, early warning, and response mechanisms to mitigate the potential for cyber threats. Although current methodology for this centralized effort does not exist, several of the basic technical elements required are successfully deployed on a small-scale basis, or in research, and could be integrated into a limited cohesive, national cyber response element.

Conceptually, a successful cyber attack warning and response system would include:

1) A means for near real-time monitoring of the telecommunications infrastructure.

2) The ability to recognize, collect, and profile system anomalies associated with attacks.

3) The capability to trace, re-route, and isolate electronic signals that are determined to be associated with an attack.

We Recommend:

The R&D program include a priority effort to develop such an Early Warning and Response capability.

Chemical and Biological Agent Detectors

Considering the serious and growing threat of a chemical or biological attack, chemical and biological agent detectors and effective protective and clean-up equipment are urgently needed and should be included in R&D efforts.

CHAPTER TWELVE
IMPLEMENTATION STRATEGY

This strategy provides the framework of objectives which will establish the foundations for a longer-term effort to assure our critical infrastructures. It describes major actions leading to fulfillment of each objective, and the expected outcome over the three-year period following a decision by the President to implement the Commission's recommendations. A more detailed implementation plan, with time lines, will be provided during the interagency review of the Commission's recommendations.

Strategic Objectives

Objective 1

Promote a partnership between government and infrastructure owners and operators beginning with increased sharing of information relating to infrastructure threats, vulnerabilities, and interdependencies.

Anticipated Three-Year Outcome

An active program which exchanges information on anomalous activities and suspicious incidents and distributes meaningful integrated analyses of government and private sector data, and threat and warning information, on an almost real-time basis to appropriate decision-markers in both government and private industry.

Action Items

- Develop a planning framework for establishing an Information Sharing and Analysis Center, jointly staffed by government employees and representatives from the critical infrastructures, to receive information from all relevant sources and conduct analyses for dissemination to participants.
- Designate selected federal departments and agencies to assume Lead Agency responsibilities.
- Coordinate with DOJ, other federal agencies, and the private sector to resolve legal impediments to information sharing, including potential antitrust, tort liability, national security, classification, disclosure, and protection of proprietary and trade secret information issues.
- Assist infrastructure stakeholder selection of Sector Infrastructure Assurance Coordinators to facilitate sharing of information among critical infrastructure owners and operators and with the government.
- Develop interagency infrastructure information sharing guidelines.
- Initiate personnel hiring process, identify an appropriate site, and stand up the Information Sharing and Analysis Center.

Objective 2

Ensure infrastructure owners and operators and state and local governments are sufficiently informed and supported to accomplish their infrastructure protection roles.

Anticipated Three-Year Outcome

Infrastructure owners and operators able to make better informed assurance investment decisions; local and state governments better equipped and trained to protect critical infrastructures and respond to untoward events.

Action Items

- Facilitate the efforts of NSA, DOE, and DoD to provide private sector assessments for critical infrastructure owners and operators; facilitate the offer of additional, more encompassing assessments over a range of cyber, physical, and interdependency risks; and provide vulnerability assessment training to private sector service providers on a cost-reimbursable basis.
- Encourage the private sector to develop generally accepted security principles to be used by internal and external audit institutions in their regular operational audit functions.
- Convene a group of professionals from law, state and federal legislatures, labor and management organizations, and the privacy community to examine existing laws in light of infrastructure assurance objectives and recommend measures to balance the legitimate needs of critical infrastructure owners and operators to conduct appropriate employee background investigations with the privacy rights of individual employees.
- Coordinate the continued development of risk assessment technologies, and associated tools, policies, procedures, and practices with appropriate federal agencies; encourage the transfer of these methodologies to the private sector; and encourage private sector performance of periodic quantitative risk assessments.
- Coordinate the development of mechanisms for disseminating information about infrastructure assurance to state and local governments.
- Encourage state legislatures to consider adopting "consent" as a baseline for allowing employers to request background information from employees and potential employees for sensitive positions within critical infrastructures, subject to fair information practices.
- Sponsor federal legislation to narrowly expand existing exemptions to the Employee Polygraph Protection Act to include providers of information security services within the scope of its exemptions.

Objective 3

Establish national structures that will facilitate effective partnership between the federal government, state and local governments, and infrastructure owners and operators to accomplish national infrastructure assurance policy, planning, and programs.

Anticipated Three-Year Outcome

A formal structure that encourages private industry participation in development of a national policy for infrastructure assurance, identifies the capabilities and responsibilities of federal agencies for infrastructure continuity, and facilitates national incident planning, response, mitigation, and restoration activities.

Action Items

- Establish an interagency working group to develop a plan for stand-up of structures that will contribute to the development of a national infrastructure assurance policy, including an Office of National Infrastructure Assurance; a National Infrastructure Assurance Council; an Infrastructure Assurance Support Office; and a Lead Agency to act as the government's focal point for each of the various infrastructure sectors.
- Review the FRP and other applicable documents to assist the FEMA's consolidation of restoration and reconstitution planning relating to cyber infrastructure assurance issues.
- Review results of legislative initiatives and other studies articulating roles and responsibilities of federal agencies for assurance issues; coordinate issues with appropriate entities.
- Coordinate a National Infrastructure Assurance Policy through government and private sector representatives.

Objective 4

Elevate national awareness of infrastructure threat, vulnerability, and interdependency assurance issues through education and other appropriate programs.

Anticipated Three-Year Outcome

A more informed private industry, government and general public who understand critical infrastructures; individuals and institutions who understand the need to protect their own use of information as well as information used by others; general appreciation of the need to develop a broader base of information assurance technical talent; and sharper focus on computer ethics and advanced information security technology in education programs.

Action Items

- Sponsor a series of White House conferences with academic and industry leaders from the public and private sectors to reach consensus on a plan of action that will increase the commitment to information security; emphasize computer ethics for grades K-12 and the general university population; review the status of undergraduate and graduate education relating to infrastructure protection, particularly information security; and define continuing opportunities to meet the national demand for professionals in the field.
- Coordinate with the National Academy of Sciences and the National Academy of Engineering to establish a Round Table in parallel with those in other fields, bringing together federal, state, and local officials with industry and academic leaders to develop national strategies for enhancing infrastructure assurance.
- Obtain NSF funding to support programs of professional education in university computer science departments and business schools.
- Coordinate with intelligence, law enforcement and regulatory agencies to expand programs for CEO briefings relating to infrastructure threats and vulnerabilities.
- Sponsor a feasibility study of publishing comparative infrastructure assurance-related data for certain infrastructures.
- Lead a public service campaign, in coordination with the private sector, to emphasize awareness of the threats and vulnerabilities of infrastructures and methods of improving infrastructure security.

Objective 5

Initiate a series of information security management activities and related programs demonstrating government leadership.

Anticipated Three-Year Outcome

Federal government information and networks are better protected from unauthorized intrusion, disruption, or modification using management procedures recognized as "best practices" and transferable to private industry.

Action Items

- Select a lead agency for assisting federal entities in the implementation of best practices for information security.
- Assign responsibilities for federal computer network security to the proposed Office of National Infrastructure Assurance.

- Encourage law enforcement to initiate new programs to hire and retain qualified personnel for investigative and analytical positions involving cyber issues.
- Fully evaluate threats and vulnerabilities associated with deployment of the GPS prior to elimination of other radionavigation and aircraft landing guidance systems.
- Develop, establish, fund, and implement a comprehensive security program to protect the modernized NAS from information-based and other disruptions, intrusions and attack.
- Resolve issues associated with spectrum allocation for communications among and between emergency service providers.
- Prepare an Executive Order requiring federal agencies to weigh the positive and negative effects on infrastructure assurance before publishing or requiring publication of information about critical components or functioning of infrastructures.
- Facilitate infrastructure assurance simulations within the federal government, and disseminate findings as part of the awareness campaign.

Objective 6

Sponsor legislation to increase the effectiveness of federal infrastructure assurance and protection efforts.

Anticipated Three-Year Outcome

Updated legislation that addresses critical infrastructure issues and enhances law enforcement ability to successfully investigate and prosecute related criminal activities.

Action Items

- Sponsor an interagency task force or other review mechanism to determine applicability of delineating infrastructure assurance objectives in the Information Technology procurement process; the Government Performance and Review Act; the Information Technology Management Reform Act; the Stafford Act; Nunn-Lugar-Domenici; and, the FRP.
- Formalize information threats as a foreign intelligence priority.
- Sponsor legislative activities leading to a finding that certain critical infrastructures are "instrumentalities of interstate commerce."
- Promote broader agency use of programs that provide monetary rewards for information relating to infrastructure attacks.
- Review information required by law to be published to ensure vulnerabilities are not disclosed.

- Coordinate DOJ sponsorship of a study to compile demographics of computer crime offenders, including juvenile offenders.
- Encourage the US Sentencing Commission to consider expanding its broader reformulation of harm and loss (in Guidelines Section 2B1.1, as it applies to violations of the Computer Fraud and Abuse Act and theft of trade secrets) to other forms of electronic crime and crimes relating to information and information technology.
- Endorse efforts currently underway to develop an effective nationwide trace and search warrant capability; and efforts to facilitate international cooperation in computer crime matters.
- Encourage the Sentencing Commission to expand guidelines to include greater flexibility to address actual and consequential damages, including "downstream" damage to property or loss of service resulting from attacks on critical infrastructures and, to better address consequences of the use of biological and chemical weapons not resulting in death.

Objective 7

Increase investment in infrastructure assurance research from $250 million to $500 million in FY99, with incremental increases in investment over a five-year period to $1 billion in FY04. Target investment in specific areas with high potential to produce needed improvements in infrastructure assurance.

Anticipated Three Year Outcome

A focused and accelerated program which delivers usable tools to fill gaps in technology in infrastructure assurance.

Action Items

- Facilitate the establishment of a national focal point for infrastructure assurance R&D efforts and a public/private/academic sector partnership to foster technology advancement and transfer.
- Develop a comprehensive plan to focus R&D on technical solutions to infrastructure assurance issues associated with information security management, intrusion detection, vulnerability assessment and systems analysis, risk management and decision support, protection and mitigation, and incident response and recovery.
- Initiate an R&D program in cooperation with the water system owners and operators to identify vulnerabilities of water supply systems and to evaluate mitigation techniques.

Onward

Originally, we had intended to title this final section of the report as a conclusion. It is anything but conclusion. In fact, it is a beginning. Our entire effort is prologue to a new era of infrastructure assurance.

This is not an exercise in problem solving. It is an attempt to deal with a rapidly changing, technology driven environment in which information and communications technologies add a new dimension of concern. In effect, we are not proposing solutions, but offering a step toward posturing our nation more effectively to deal with a new, still evolving world.

Our nation is in the midst of a tremendous cultural change, which will have a profound effect on our institutions. Accordingly, we are offering first steps toward preparing our critical infrastructures–and our government–to deal with this change. We believe that the only way to assure the future security of the nation is by assuring our critical infrastructures. And doing that will require a vigorous, innovative partnership between our government and the owners and operators of those infrastructures.

We offer these recommendations with a sense of urgency. While we do not believe a debilitating attack is imminent, the threats to our nation and the vulnerabilities in our infrastructures are real.

And the time to act is *now* . . .

Appendix A
Sector Summary Reports

Executive Order 13010 designated as *critical* certain infrastructures whose incapacity or destruction would have a debilitating impact on our defense or economic security. Eight were named: telecommunications; electrical power; gas and oil storage and transportation; banking and finance; transportation; water supply; emergency services (including emergency medical services, police, fire and rescue); and government services. Because some of the eight listed infrastructures lent themselves to similar approaches, the Commission organized into five study teams to address the infrastructure sectors and industries listed below. This appendix provides summaries of the five sector studies, which will be published as separate appendices to the Commission's report.

Sector

Information and Communications—The Public Telecommunications Network (PTN), the Internet, and millions of computers in home, commercial, academic, and government use.

Physical Distribution—The vast interconnected network of highways, rail lines, ports and inland waterways, pipelines, airports and airways, mass transit, trucking companies, and delivery services that facilitate the movement of goods and people.

Energy—The industries that produce and distribute electric power, oil, and natural gas.

Banking and Finance—Banks, non-bank financial service companies, payment systems, investment companies and mutual funds, and securities and commodities exchanges.

Vital Human Services—Water supply systems, emergency services (police, fire, rescue, and emergency medical services) and government services (non-emergency services including Social Security payments, unemployment and disability compensation, and management of vital records).

Information and Communications

Introduction

The US information and communications infrastructure (I&C) sector generates more revenue than most nations produce. Far more than any other nation, the potential of the new technologies has enabled the US to reshape its governmental and commercial processes. We have led the world into the information age, and in so doing have become uniquely dependent on its technologies to keep our economy competitive, our government efficient, and our people safe.

Background

The I&C sector includes the Public Telecommunications Network (PTN), the Internet, and the many millions of computers for home, commercial, academic, and government use. The PTN includes the landline networks of the local and long distance carriers, the cellular networks, and satellite service. Switches automatically establish and disconnect circuits between communicating parties on demand. Prior to the introduction of cellular service in 1983, virtually all switched service was provided by the wireline telephone system. The system's two billion miles of fiber and copper cable remain the backbone of the I&C sector, with the newer cellular and satellite wireless technologies largely serving mobile users as extended gateways to the wireline network. The PTN provides both switched telephone and data services and long term leased point-to-point services.

The Internet is a global network of networks interconnected via routers which use a common set of protocols to provide communications among users. Internet communications are based on connectionless data transport. In other words, the Internet protocol does not establish a circuit between communicating parties during the lifetime of the communication. Instead, each message is divided into small packets of data. Routers forward the packets to other routers closer to their destinations based on address information in the packet headers. To maximize efficient use of the network, the routers may send each packet of a message over a different path to its destination, where the message is reassembled as the packets arrive.

The Internet and the PTN are not mutually exclusive, since significant portions of the Internet, especially its backbone and user access links, rely on PTN facilities. Current trends suggest that the PTN and the Internet will merge in the years ahead; by 2010 many of today's networks will likely be absorbed or replaced by a successor public telecommunications infrastructure capable of providing integrated voice, data, video, private line, and Internet-based services.

The installed base of computers in the US has risen from 5,000 in 1960 to an estimated 180 million today, with over 95 percent of these being personal computers. The remainder includes the majority of the world's supercomoputers and roughly half of the world's minicomputers and workstations. Networking of these machines through the circuits of the PTN and the Internet has grown exponentially over the past 15 years, creating an extended information and communications infrastructure that has changed the way we work and live. This infrastructure has swiftly become essential to every aspect of the nation's business, including national and international commerce, civil government, and military operations.

Threats

The reliability and security of the I&C sector have become matters of critical importance. The primary threats to reliability are natural disasters and system failures. The primary threats to security are deliberate physical and computer, or "cyber," based attacks.

Because they are generally well understood, somewhat predictable, and geographically confined, natural disasters are the most manageable of the threats to I&C reliability. In recent large scale emergencies, telecommunications systems have proven highly resilient. The current policies and organizational arrangements for dealing with natural disasters are working and require no modification at this time.

A second threat to infrastructure reliability, less predictable and potentially farther reaching, is system failure arising from increases in the volume and complexity of interconnection and the introduction of new technologies. The unbundling of local networks mandated by the Telecommunications Act of 1996 has the potential to create millions of new interconnections without any significant increase in the size or redundancy of network plants. Unbundling will be implemented at a time of rapid and large scale change in network technologies. The interaction of complexity and new technologies will almost certainly expand the universe of ways in which system failure can occur, and, unlike natural disasters, there is no assurance that such failures will be localized. Nevertheless, demonstrated system performance, ongoing research, and the ability to modify legislative and technical timetables suggest that the challenge will be successfully managed.

While rapidly increasing complexity has characterized the I&C infrastructure since the breakup of the Bell System and the advent of the Internet, system reliability has remained extraordinarily high. Large scale system failures have occurred very infrequently and have been corrected within hours.

The Federal Communications Commission (FCC) and the telecommunications industry have actively researched reliability issues throughout the 1990s, laying the groundwork for the expected influx of new service providers and technology vendors. Major players in telecommunications have maintained a vested interest in network reliability and can be expected, as in the past, to collectively maintain and improve network performance. Finally, the legislative and technical imperatives underlying the restructuring and can be modified if serious difficulties arise. The current framework of FCC regulation and industry standard setting are self-imposed and are expected to prove capable of accommodating the challenges to reliability posed by complexity and technological advance. This framework can be extended beyond its traditional switched network focus to cover cellular, satellite, cable, and the Internet.

The third and least predictable threat to the infrastructure comes from deliberate attack. Depending on their objectives, attackers may seek to steal, modify, or destroy data stored in information systems or moving over networks, or to degrade the operation of the systems and networks themselves, denying service to their users.

Attackers include national intelligence organizations, information warriors, terrorists, criminals, industrial competitors, hackers, and aggrieved or disloyal insiders. While insiders constitute the single largest known security threat to information and information systems, controlled testing indicates that large numbers of computer based attacks go undetected, and that the unknown component of the threat may exceed the known component by orders of magnitude.

Adversaries can employ a variety of methods against the infrastructure, including traffic analysis, cryptologic attacks, technical security attacks, physical attacks, and cyber attacks. Of these, physical and cyber attacks pose the greatest risk. They have increased rapidly in sophistication and disruptive potential during the 1990s, while the infrastructure's vulnerability has grown. The availability of truck bombs, chemical agents, and biological agents has markedly increased the disruptive potential of physical attacks. At the same time, the vulnerability of the I&C infrastructure to physical attack has increased as service providers have concentrated their operations in fewer facilities.

In the cyber dimension, tools to remotely access, change, or destroy information in vulnerable systems and to control, damage, or shut down the systems themselves have become more sophisticated, easier to use, and more widely available. Department of Defense tests and exercises, together with the rising incidence of documented intrusions and cyber-related losses over recent years, indicate that networked computers are highly vulnerable to these techniques. A broad array of adversaries, including a sizable number of foreign governments, are currently capable of conducting cyber attacks. The Defense Science Board expressed a mainstream view in its November 1996 estimate that limited strategic information warfare capabilities against the US infrastructure will to emerge over the next seven to ten years.

Vulnerabilities

The critical functionality of the PTN—increasingly software driven and remotely managed and maintained—is vulnerable to cyber attack. Deregulation will markedly expand the access points from which to launch an attacks. New entrants will be permitted to interface with the local exchange carrier networks at many different points, including local loops, switches, trunk lines, common channel signaling systems, advanced intelligent network systems, and operating systems. Technical details of the systems are widely available. Open interfaces and common communications protocols will make intrusion easier by standardizing targets and simplify the propagation of attacks from one location in the network to other parts of the architecture.

The introduction of numerous third parties, including foreign companies operating in partnership with US companies or on their own, into every aspect of network operations will alter the trust relationship on which current network architecture is based. The security measures needed to compensate for the loss of trust will take years to develop. During this time, attacks to gain unauthorized access to sensitive data and functions will be easier to accomplish on a widespread basis than at any previous time in the history of telecommunications.

Switching

The susceptibility of the current generation of switching equipment to software based disruption was demonstrated in the collapse of AT&T's long distance service in January 1990. A line of incorrect code caused a cascading failure of 114 electronic switching systems. We believe AT&T's accidental failure could alternatively have been triggered maliciously by relatively small individual actions. Successor generation switching equipment now entering service is likewise potentially vulnerable to remote access, alteration, or control by skilled attackers.

Transport

Another major vulnerability in switched networks is the transport architecture. Transport refers to the transmission facilities used to move traffic between switching and hub offices within a network. Virtually all new fiber optic installations by commercial carriers are currently being configured as Synchronous Optical Networks (SONETs). Most of the elements in SONETs are managed remotely through packet data network connections vulnerable to electronic intrusion. In addition, SONET elements can be remotely attacked through maintenance and testing ports. The first large scale network outage known to be caused by cyber attack was the disruption of a "bulletproof" SONET ring.

Signaling

Common channel signaling (CCS) networks are connectionless data packet networks that carry instructions for call setup, special services, billing, and all other functions involving more than one element across the network. The potential for software-based disruption of common channel signaling was demonstrated in June 1991 when phone service in several cities, including 6.7 million lines in Washington, DC, was disrupted for several hours due to a problem with the network's Signaling System 7 protocol. The problem was ultimately traced to a single mistyped character in the protocol code. Current methods of protecting CCS networks from spurious messages are adequate to detect minor intrusions but are insufficient to protect the network from serious attacks. CCS network elements are also potentially vulnerable to tampering through remote access.

Control

Network operations are controlled by network elements that carry out tasks based on information received via signaling messages or retrieved from network databases. Traditionally, service control for voice telephone service resided in the switches. Implementing new services required physical rewiring of the switching fabric. In recent years, local exchange carriers have been moving service logic to special purpose processing and database systems outside the switches, where it can be upgraded quickly through software changes alone. This control architecture, which permits rapid creation of custom services, is called the advanced intelligent network.

The ability of service logic programs to change the way the network reacts to subscribers' calls makes them a potential source of disruption if they are misprogrammed, corrupted by accident, or accessed and altered by adversaries. Access to service logic of all kinds is set to expand markedly as a 1993 FCC notice providing for access to the advanced intelligent network by third party service providers goes into effect. The FCC ruling states that these service providers must have the ability to incorporate their own service logic and add their own hardware to the network. As the network becomes more open, interfaces to third party providers will provide many new points of entry into the network and its signaling systems, increasing the potential for accidental or deliberate misuse.

Management

Management refers to the tasks associated with running networks on a day-to-day basis, including configuration management and maintenance. These tasks are for the most part automated and carried out from central locations using computer-based operations support systems. Today's high levels of automation and interconnection of network elements make manual management of the network virtually impossible.

Operations support systems are susceptible to a variety of attacks. An attacker can delay, replay, or alter the order in which messages are received, triggering unauthorized management operations. An attacker can alter the contents of management messages, tricking a network node into accepting management parameters that may affect the operations or configuration of the node, interfere with accounting, or disrupt traffic. An attacker can simply prevent exchanges between a managing node and its managed nodes, disrupting network operations.

In the coming years, as subscribers demand greater control over their network services, providers are expected to offer configuration management capabilities unprecedented in today's networks. Misuse of these more powerful capabilities will have the potential to disrupt or halt communications over significant portions of the network.

Network maintenance is increasingly performed through remote access. Remote access allows maintenance personnel to electronically access distant network elements to perform maintenance or management functions. Eliminating the need to physically dispatch repair personnel allows faster response to problems and more efficient use of maintenance staff. The channels used for remote access by authorized maintenance personnel offer potential attack routes for adversaries. Once logged on, an attacker can remove nodes from service and disrupt the network.

Operations support system capabilities have continued to increase in sophistication and in the number of network elements they can control simultaneously. The trend is to reduce the number of operations support systems in the network while expanding their ability to provide a multilevel view of network operations. This has led to the creation of megacenters, which concentrate operations for large segments of the PTN and data communications networks in one location. A megacenter may service central offices extending over a multistate region, giving its operators access to every switch, operations system, and maintenance channel in the central offices served. An adversary with electronic access to a megacenter could target individual circuits, bring down selected services, or disrupt normal operations over large areas.

Another growing vulnerability in network management is the trend by public switched network service providers to manage network elements via the Internet. The Internet was originally built as a vehicle for information sharing in an open and cooperative environment. Security was not a primary design consideration. With its relatively uniform structure and uncomplicated protocols, the Internet offers less resistance than the public switched network to systematic attack. Its growing use in network management offers adversaries the opportunity to attack the PTN by disrupting the Internet. Improved security should be a key priority for the Next Generation Internet.

Findings

Today's level of threat and degree of vulnerability present two risks for national policy to address. The first is the cumulative risk generated by myriad small scale attempts to steal information or money through cyber attack. The vulnerability of individuals and enterprises to cyber theft damages the nation's current and future competitiveness. Losses undermine both the bottom line and public confidence in emerging information technology. For the information and communications infrastructure to realize its full potential as a medium for commerce, government, and military operations, users must have confidence that transactions will be confidential and protected.

The numerous security vulnerabilities in today's I&C infrastructure afford little basis for such confidence today, and the trends are not encouraging. In the meantime, the payoff for successful exploitation is increasing rapidly. With commerce growing exponentially over a medium with minimal protection, criminals and hackers can be expected to develop original and profitable new methods of operation. With larger and larger quantities of imperfectly protected information residing on networked systems, intelligence services and industrial competitors can be expected to find increasingly sophisticated ways to break in. To the extent they succeed, we lose competitiveness. To the extent we are forced to retrench in reaction to losses, we sacrifice opportunity.

The second and more critical risk is that presented by cyber and physical attacks intended to disrupt the US I&C infrastructure and the critical societal functions that depend upon it. With network elements increasingly interconnected and reliant on each other, cyber attacks simultaneously targeting multiple network functions would be highly difficult to defend against, particularly if combined with selected physical destruction of key facilities.

The possibility that such disruption could cascade across a substantial part of the PTN cannot be ruled out. Our experience with very large scale outages is extremely limited, and has dealt with reliability problems rather than deliberate and repeated attack. Network resilience has been asserted, but large scale testing is not feasible. Computer models capable of systematically analyzing security risks associated with large telecommunications networks have not been developed. No one knows how the network would react under coordinated attack. We do know that relatively minor software problems have produced cascading failures in the past. We cannot confidently set an upper limit on the disruptive potential of a planned, large scale campaign.

As the scale and objectives of potential cyber campaigns become more focused, their feasibility and potential for success increases. Achieving selected outages of regional targets, such as financial districts or ports of embarkation for deploying forces, is feasible for a greater number of adversaries than a major disruption of the national infrastructure, particularly if they have access to physical as well as cyber weaponry. Achieving outages of selected equipment, such as high density network elements serving large customer populations, is even more feasible. Noting the large scale outage achieved in a recent cyber attack on a SONET ring, widespread denial of service through remote attack is now a demonstrated capability.

To address the risk posed by the mounting incidence of cyber theft and other small scale attacks, national policy must encourage a cooperative approach to strengthening the security of the infrastructure. To address the risk posed by the vulnerability of the infrastructure to widespread disruption, national policy must ensure that there is an effective national capability to detect and defend against large scale attacks on the I&C infrastructure.

Recommendations

The US has led the world into the information age, and in so doing has become critically dependent on its technologies to conduct national and international commerce, governmental functions, and military operations. The protection of the US information and communications (I&C) infrastructure is a vital national interest.

Six years ago, the National Research Council's report *Computers at Risk* described the growing vulnerability of networked computers and outlined a series of core principles to improve security. Progress in implementing these principles has lagged, while vulnerability and threat have grown significantly. The vast expansion of computer networking, the increasing dependence of the PTN and the Internet on computer-based, remotely-managed control elements, and the increasing levels of interconnectivity and complexity mandated by the Telecommunications Act of 1996 have created new vulnerabilities to I&C reliability and security. Natural disasters, accidents, and system failures pose growing threats to infrastructure reliability, while increasingly powerful methods of physical and cyber attack pose growing threats to infrastructure security. With the I&C infrastructure having become vital to every critical economic, social, and military activity in the nation, effective action to implement effective assurance practices is a matter of great urgency.

Our I&C infrastructure encompasses a wide range of activities extending over vast reaches of physical and virtual space. No entity in government or industry directly controls more than a small fraction of it. The problem of infrastructure security will require shared effort across organizational boundaries. No organization can solve it alone.

Implementing infrastructure protection policies is neither an entirely public nor an entirely private responsibility. The risks are common to government, business, and citizen alike. Reducing those risks will require coordinated effort within and between the private and public sectors. The need for infrastructure protection creates a zone of shared responsibility and cooperation for industry and government. If we are to retain and build upon the competitive edge information technology has given us, we need to work together to substantially improve the trustworthiness of our information systems and networks.

Strengthening Security Through Cooperation Between Industry and Government

To strengthen the security of the information and communications infrastructure, the Commission recommends that the federal government work in cooperation with industry to:

- Strengthen overall public awareness to gain acceptance of and demand for security in information systems.
- Promote the establishment and rapid deployment of generally accepted system security principles, beginning with those concerning password management and imported code execution.
- Promote industry development and implementation of a common incident reporting process.
- Increase accessibility of government threat and vulnerability information, expertise in system security assessment and product evaluation, and operational exercises to assist government and industry risk management decision making.
- Define and maintain metrics for security, along with the current set of reliability metrics, for public telecommunications networks.
- Actively promote network assurance research and development.
- Establish an international framework to support the use of strong cryptography on a global basis.
- Promote the development of effective security enabled commercial information technology and services. Accelerate the development and implementation of usable, affordable tools, methodologies, and practices in information security.
- Support uniform "one call" legislation against the "backhoe threat."

Defending Against Attack

An effective capability to defend the I&C infrastructure against attack in both the cyber and physical dimensions will require new sensing and warning capabilities, an organizational structure capable of dealing with the ambiguities of cyber attack, and new technologies for cyber defense. To ensure that there is an effective national capability to detect and defend against large scale attacks on the information and communications infrastructure, the Commission recommends that the federal government:

- Establish a focal point for national security policy on information infrastructure assurance and a focal point for national operational defense.
- Develop and sustain a robust intelligence collection, analysis, and reporting capability against cyber threats.
- Partner with private industry in developing and implementing indication and warning capabilities.
- Develop technologies needed for defending the nation's infrastructures against cyber attack, including after-action analysis and criminal investigations.

Leadership by Example

To serve as a national model for sound information assurance practices, the federal government should meet or exceed all applicable industry-based best security practices in building, operating, and using its portions of the information and communications infrastructure. Specifically, the Commission recommends that the federal government:

- Implement a common interdepartmental macro-level information systems security policy to standardize procedures and accountability.
- Require participation by all departments and agencies in annual information system vulnerability assessments, online security testing, and operational exercises.
- Establish clear visibility for information system security expenditures in the budgets of departments and agencies to facilitate management.
- Provide appropriate training and professional education in information assurance for all federal system managers, operators, and users, and assist state and local governments in establishing similar programs.

Physical Distribution

Introduction

The physical distribution infrastructure is critical to the national security, economic well being, global competitiveness, and quality of life in the US. The vast, interconnected network of high-ways, railroads, ports and inland waterways, pipelines, airports and airways facilitate the efficient movement of goods and people and provides this nation a distinct competitive advantage in the global economy.

Transportation is a major component of the US economy, representing in 1995 approximately $777 billion, or 11 percent of the Gross Domestic Product (GDP). US commerce depends heavily on the export, import, and domestic movement of raw materials, manufactured goods, foodstuffs, and consumable supplies.

The physical distribution infrastructure includes almost 4 million miles of public roads and highways and more than 360,000 interstate trucking companies, 20 million trucks used for business purposes, and 190 million personal vehicles. It includes more than a hundred thousand miles of track operated by the largest railroads, with 1.2 million operating freight cars and over 18,000 locomotives. It includes airlines that carry more than half a billion passengers a year through 400 airports. It includes almost 6,000 transit entities operating rapid transit rail and bus services. It includes 1,900 seaports and 1,700 inland river terminals on 11,000 miles of inland waterways carrying grain, chemicals, petroleum products, and import and export goods. The physical distribution infrastructure includes more than 1.4 million miles of oil and natural gas pipelines. And it includes delivery services, such as the US Postal Service and many other commercial providers that deliver goods and products on time not only to households, but to manufacturers whose very survival depends on just-in-time delivery of materials and supplies, and to business and even military activities who depend on the rapid delivery of repair parts to keep them in operation.

In this country, transportation is a matter of choice, and of intense competition. Commuters can choose between driving to work or taking mass transit. Travelers can choose to fly, catch a train or bus, or drive the highway. Shippers have their choice among highly competitive, customer focused delivery services and, in the deregulated world of transportation, among trucking firms, railroad companies, barge companies, and deep water shipping companies. Thousands of freight forwarders and consolidators, customs brokers and shipping agents move goods and cargo across the nation and through its ports quickly, cheaply, and effectively.

The US has the world's best transportation and distribution system, which both enables and reflects our having the number one economy in the world. Assuring that this system remains effective is critical to the well being of American citizens and the security of our nation.

Most of our nation's transportation infrastructure is owned by the private sector—railroads and pipelines; the vehicles and equipment operating on our roads, on the water, and in the air; and by state and local governments—our roads, airports, mass transit systems, and ports. The federal government owns the National Airspace System (NAS) operated by the Federal Aviation Administration (FAA), and the locks and dams operated by the US Army Corps of Engineers. The private sector is largely responsible for assuring its own infrastructure and business practices.

Trends

In the past, the business of transportation was conducted with paper–paper contracts and agreements, delivery orders, letters of credit, invoices, manifests, bills of lading, and shipping tags. Today, transportation, like other industries, is becoming increasingly enmeshed in our information-based society with its critical dependence on data and instantaneous communications.

While the transportation system has long been dependent on petroleum fuels, its dependency on other infrastructures continues to increase, for example, on electricity for a variety of essential operations and on telecommunications to facilitate operations, controls, and business transactions.

Demands on the physical distribution infrastructure continue to grow with the population and the economy. However, the ability to expand this infrastructure is limited. Rights of way for new roads, pipelines, railroads, and airports are difficult to obtain and justify. New means must be developed to make the existing system more efficient. Governments and industry have turned to information technology to increase that efficiency. Modernization of the NAS, extensive use and dependence on the Global Positioning System (GPS), and rapidly expanding use of Intelligent Transportation Systems (ITS) all will contribute to a more efficient transportation system.

Electronic commerce and data interchange, which make "just-in-time" delivery the norm rather than the exception, are increasing efficiency and giving companies a competitive edge in the global economy. However, requirements for open access to energy system data, increased dependency on data bases, and placing Supervisory Control And Data Acquisition (SCADA) systems on the public telecommunications network make these systems more vulnerable to unauthorized intrusion. The explosion of telecommunication requirements and intense competition in the communications infrastructure are leading to greater volumes of traffic on existing lines, thereby increasing the potential for "single point failures."

Railroad companies continue to merge, consolidating operations centers and lines, moving more and more traffic onto fewer corridors, and reducing the redundancy of the networks and increasing their vulnerability to physical attack.

Natural gas is being moved by existing pipelines without any agency or organization having a clear picture of the entire system or an understanding of its ability to handle surges in demand, or the tools necessary to evaluate the impact of a system-wide disruption.

The air traffic control system of the FAA is based on decades old technology. The replacement system, while doubtless more efficient, will be more vulnerable unless special security measures are incorporated.

Congestion is common in most metropolitan areas; ITS are being introduced to make more efficient use of existing road systems, but at the same time they will introduce new vulnerabilities. A discussion of the challenges specific to Emergency Services is provided later in this Appendix.

Public Expectations

The American public takes for granted freedom of choice among transportation modes and carriers, and generally wants government intervention limited to matters affecting safety and security. Transportation systems are expected to be reliable and predictable, designed and operated to allow unimpeded flow of goods through ports, across state and international boundaries, with rapid customs and immigration clearance processes and minimal regulatory and bureaucratic impediments.

Infrastructure maintenance and improvement must be adequate to ensure continued foreign investments in the nation's economy. A competitive level playing field within and between modes of transportation is crucial to freedom of choice and an efficient distribution system. Timely delivery of goods and products is essential, so we expect delivery services to be predictable and dependable. Government policies and regulations are expected to foster stability and consistency.

The public reluctantly accepts accidents involving planes, trains, and automobiles. But when the cause is found to be a failure of government oversight, such as substandard aircraft maintenance or a faulty traffic device, the public demands accountability. When a natural disaster affects the physical distribution infrastructure, the public expects rapid restoration. While the public anticipates and tolerates congestion on the nation's roads and highways, government is expected to use effective traffic management systems and techniques to minimize congestion. Gasoline, natural gas, and other energy supplies are expected to be available on demand.

Finally, the public expects a transportation infrastructure ready to respond to national crises, including adequate sea and airlift to move military forces quickly to any trouble spot on the globe.

Federal Role

The US Department of Transportation (DOT) provides national policy, funding, and safety requirements through its operating agencies:

- Office of the Secretary of Transportation (OST)
- Federal Aviation Administration (FAA)
- Federal Highway Administration (FHWA)
- Federal Transit Administration (FTA)
- Federal Railroad Administration (FRA)

- US Coast Guard (USCG)
- Maritime Administration (MARAD)
- Research and Special Programs Administration (RSPA)

DOT works to maintain the integrity of the US transportation infrastructure against terrorist and other criminal acts through a combination of regulations, guidelines, inspections, cooperative agreements, and government investments. Intermodal and interagency intelligence matters and security related actions are coordinated by and with the Office of Intelligence and Security within OST. Security actions are carried out by the DOT operating agencies, commensurate with their respective authorities.

The FAA, Coast Guard, and, to a limited extent, RSPA's Office of Pipeline Safety, are the only DOT agencies with clear statutory authority related to security.

Civil aviation security remains DOT's first priority and primary focus. The FAA has the responsibility and the authority to require contingency measures for air carriers and airports to deal quickly and effectively with immediate threats against civil aviation.

The Coast Guard has authority to respond to threats against cruise vessels and ports in the US and against vessels anywhere in the world carrying US citizens. The Coast Guard can institute regulations to establish and manage security zones around important facilities or operations, and to require certain port facilities and cruise lines to implement security measures.

RSPA regulates the design, construction, testing, operation, and maintenance of natural gas and hazardous liquid pipelines and liquefied natural gas (LNG) facilities; specific security authority exists only for LNG facilities.

Security authority and contingency plans for land transportation, including mass transit, railroads and highways, tunnels and bridges, and for a major portion of the nation's pipeline system, do not exist within DOT. Millions of people use passenger rail daily, and as shown by the 1995 Aum Shinrikyo gas attack in Tokyo and bombings of the subway system in Paris, mass transit remains open and vulnerable to terrorist acts. Millions of miles of pipelines carry natural gas and other hazardous materials throughout the country, and are largely unprotected and vulnerable to sabotage. Railroads carry tons of hazardous materials thorough heavily populated areas with little consideration given to the possible impact of intentional attack. Despite the possible national level political implications of a terrorist attack, protection of railroad, highway, and mass transit facilities remains the responsibility of industry or state and local governments.

Threats and Vulnerabilities

Transportation is inherently vulnerable to a wide range of physical threats. Natural disasters such as floods, earthquakes, landslides, and hurricanes are ever present; when these disasters strike, services are restored through the combined efforts of federal, state and local governments and the affected industry. As for man-made threats, with the exception of civil aviation, few countermeasures are available or appear to protect our transportation systems from physical attack by terrorists or other criminals. In the event of disruption from man-made causes, reconstitution and recovery are the responsibility of the owners and operators of the systems.

While the prospect of physical disruptions has been with the physical distribution infrastructure since its infancy, transportation industries are only beginning to focus on information-based threats or attacks. Many business systems are demonstrably vulnerable; this problem must be addressed by industry. To make intelligent decisions, however, industry leaders need current information on new and emerging threats. This information may be held within other companies in the same industry, in other industries, and within various agencies of the federal government.

Governments and industry have turned to information based systems to increase the efficiency of the public/private transportation system. While these increased efficiencies help keep our industries and companies competitive in the global economy, businesses are now much more vulnerable to electronic penetrations and attack and to disruptions of their supporting infrastructures, particularly telecommunications and electric power.

Conclusions and Findings

Today, information-based attacks cannot cause trains and planes to crash, nor are they likely to cause pipelines to rupture. Tomorrow–perhaps next year, perhaps in ten years–critical transportation systems could be vulnerable to such attacks and crippled unless action is taken now.

Roles, Missions, and Responsibilities

The Department of Transportation has been extremely proactive in counterterrorism efforts, both within the federal government and with the transportation industry. However, based on the Commission's outreach to industry and the federal government, several shortfalls in transportation infrastructure assurance, other than counterterrorism, have been identified:

- No defined roles, mission, and responsibilities for DOT in infrastructure assurance related areas other than counterterrorism.
- Lack of awareness and extremely limited availability of education programs.

- Incomplete or absence of vulnerability assessments of both physical and information-based portions of the transportation infrastructure.
- Limited and untested dissemination of threat information and warnings, and absence of an effective program to share critical information within the industry, and between the industry and the federal government.
- Absence of joint federal government/industry contingency or response plans to respond to an infrastructure threat or attack.
- Absence of security or assurance standards, guidelines, or best practices.

DOT is not well positioned to support the industry in infrastructure assurance efforts. The federal government must be involved with prevention, recovery, and reconstitution efforts within the transportation sector. DOT is neither funded nor staffed to address, with the industry, current or emerging threats to transportation.

Data Collection

Accounting for about 20 percent of all terrorist attacks around the world, transportation systems are a favorite target of terrorism. Better information and data on attacks would assist in development of countermeasures and provide better information for risk management decisions. Some modes of transportation are required to report all safety related incidents and accidents above a certain threshold, while others report through insurance agreements. These data are used to establish programs that can prevent and mitigate incidents and lead to cost effective improvements in safety. The private sector, however, is reluctant to report information-based attacks, fearing public disclosure of vulnerabilities that could be exploited by others and have a negative impact on public confidence in the industry. The physical distribution community is not an active partner in the improvements of data processing and communication systems, and as a result, has become more vulnerable with the extensive adoption of these systems.

Information Sharing and Threat Dissemination

Transportation is essential to the national economy and national security. Transportation is a high visibility terrorist target. Yet no agency or private sector organization is required to have, nor actually has, a program to advise the industry of information-based threats and attacks, nor are intrusions or attacks on the transportation infrastructure generally reported to the federal government.

No tested and effective means exists that facilitates reporting and transfer of information between the government and transportation infrastructure stakeholders on threats and attacks. Information-based threats to the physical distribution system are not addressed by DOT; private sector concern is on a sector-by-sector and company-by-company basis. Established reporting systems, where they do exist within the government and the transportation industry, are "stovepiped," and are not sufficiently shared or coordinated with DOT or with established national indications and warnings processes. Neither the federal government nor the private sector is tasked with identifying, quantifying and tracking information-based threats and attacks, nor is any organization responsible for analyzing and disseminating that data.

The apparent lack of information and sharing about information-based attacks on physical distribution systems limits industry understanding of the extent of the problem and makes it difficult to justify investment in measures to prevent or mitigate the impact of information-based attacks.

Industry representatives who receive information from DOT say they need more on the threat. The DOT is required by statute to notify the civil aviation industry of terrorist threats, but the statute does not require the DOT to notify the remainder of the transportation industry of threats.

Identification of Critical Assets

While industries are aware of critical assets within their companies, the federal government (e.g., DOT) has not identified and does not track those assets critical to the national security. Government/private sector contingency plans generally do not exist for responding to a terrorist threat or attack on the transportation infrastructure. While the transportation industry in general is capable of responding to natural disasters and other similar disruptions to their systems, coordinated plans to evaluate and/or respond to threats of a coordinated series of attacks on the transportation infrastructure have not been developed.

The transportation sector must be aware of and develop a process to protect key assets during heightened threat conditions.

- While the federal government is familiar with some of these assets, industry is in the best position to identify those facilities that require protection during national security events.
- Railroad, airline, highway, port and pipeline operation centers, among other facilities, are critical.
- A coordination process is essential to develop protection and recovery contingency plans.

The National Airspace System (NAS)

The present NAS is relatively immune from intrusions. It is composed of difficult-to-penetrate, dedicated subsystems, with the subsystems having different designs and older, specialized versions of software. However, the modernized NAS will undergo major new developments, including open systems architecture, and will depend on communications technology that permits wide interchanges of information among many of its subsystems.

The NAS would likely become a prime target for terrorism and "rogue" nation states during a national defense emergency.

Because the modernized NAS appears to be particularly at risk from information-based attacks; the FAA and Congress must take firm action to ensure adequate security measures are implemented with the new system.

1) The threat of attacks on the NAS subsystems has been low to date. There have been isolated incidents, including phantom controllers, during and immediately after the 1981 controller strike, and more recently at Roanoke, VA. While the FAA subsystems have not been subject to information warfare in the past, portions of the system still are vulnerable. For example, there have been recent instances of contractor use of FAA communications systems to access and modify software code under development and test.

2) Today's older NAS subsystems have some protective mechanisms built in to trap and remove damaged messages, unauthorized message types, and excessive flight plan filing activity. The subsystems also have many dedicated networks and different versions of proprietary software and communications protocols that make it difficult for intruders. But, their susceptibility will become more severe as the new NAS subsystems are installed.

3) During transition to the future NAS architecture, the level of vulnerability may increase as new systems elements are added to the NAS.

4) The nature of air traffic control operations provides a strong countermeasure that exists in real time within the NAS. Air traffic controllers are constantly observing the traffic under their jurisdiction and pilots are aware of unusual flight circumstances. This controller and pilot detection of system abnormalities will still be important in reducing the impact of any future attacks on the NAS subsystems.

5) The major vulnerabilities inherent in the new architecture are the planned use of new open systems and, using shared communications networks. Use of these new architectures, in conjunction with commercial off-the-shelf (COTS) hardware and software products, will increase the risk of insider and outsider (hacker) access, and the probability of malicious actions that interfere with the operation of NAS subsystems. These actions include data and software corruption, virus and Trojan horse damage. Use of shared networks will significantly increase the FAA system vulnerability to outside attack.

6) Systems with air-ground communications and data links such as the Automatic Dependent Surveillance-Broadcast mode (ADS-B), the Air-

ground Data Link (ADL), and the Wide Area Augmentation System/Local Area Augmentation System (WAAS/LAAS) are susceptible to interference and signal jamming.

7) In the past there was a lack of priority and funding for establishing and conducting a security management program in the FAA, and for implementing information security protection for new automated and open system architectures. The situation is improving in that the top FAA managers forming the Joint Review Council (JRC) have recently identified a need for funding security provisions in the NAS, and have made a preliminary estimate of the funding levels to be included in future FAA budget submissions.

While the FAA has initiated the above effort to address security requirements for the NAS during its upgrade, the security improvements are currently unfunded and need Administration and Congressional support.

Pipelines

Two federal agencies perceive an assurance responsibility for the nation's pipeline system, presenting a unique situation as follows:

- DOT is statutorily responsible for regulating pipeline safety, and in some cases, for security.
- The Department of Energy (DOE) is statutorily responsible for oversight of the nation's energy supply; and as such has in place an effective intelligence and threat dissemination system.

These sometimes overlapping responsibilities must be clarified.

Regarding pipelines, current Environmental Protection Agency (EPA) efforts to publish "worst case scenario" data on the Internet raise serious concerns about the availability of targeting information for both terrorist and nation states. Access to this data should be controlled and made available to the public only on a limited basis. All federal agencies, not just DOT, and the industry, must consider the impact of making potential target information easily and readily available on the Internet and through other anonymous means; at a minimum, access must be controlled on a need-to-know basis.

Global Positioning System (GPS)

The Federal Radionavigation Plan calls for GPS and its augmentations to be this nation's sole radionavigation system by 2010. Current plans, if not modified, could lead to an over reliance on GPS based systems for critical transportation functions. The modernized NAS will depend on GPS and GPS augmentations as its sole navigation and landing systems. Exclusive reliance on any single system creates inherent vulnerabilities; no single system can be guaranteed for 100 percent availability for 100 percent of the time. Possible exclusive reliance on GPS and its augmentations, combined with other complex interdependencies, raises the potential for "single point failure" and "cascading effects."

Recommendations

Agency Roles, Missions and Responsibilities

The Commission is recommending that lead agencies be designated to promote the development of information sharing in respective sectors. Each designated lead agency would take a leadership and coordinating role with the private sector, and also seek appropriate legislation that allows for infrastructure assurance. The DOT, in assuming its responsibilities as sector lead for the Physical Distribution Infrastructure, should consider:

1) Establishing a central office responsible for coordinating intermodal infrastructure assurance as well as terrorism issues, including prevention, mitigation, contingency response, and recovery, and for coordinating with modal and other federal agencies, acting as primary contact points with industry on assurance issues.

2) Developing joint government/industry response and recovery plans with the private sector.

3) Establishing an improved information dissemination and sharing process.

4) Testing the effectiveness of the dissemination process and of established security procedures.

5) Working closer with industry on R&D and education.

6) Requesting for funding and positions to manage these emerging issues and responsibilities.

7) Reviewing of all proposed legislation for adherence with infrastructure assurance policies.

8) Obtaining appropriate executive and legislative authorities necessary to accomplish Lead Agency responsibilities.

Specific funding is necessary within OST and within individual DOT agencies for the following:

1) Providing government security clearances to industry, particularly for CEOs and CIOs.

2) Developing security and infrastructure assurance education programs.

3) Performing cross-cutting research on assurance issues, including GPS, the NAS, and train control systems, and for interagency research with agencies such as NASA and the Department of Defense (DoD).

4) Developing security standards or guidelines, and reporting systems.

5) Using secure telephones (STU III's), and encryption, strengthened firewalls, and other security measures.

6) "Red Teaming" and testing of critical DOT systems and industry systems on a cooperative, selective basis.

7) Conducting DOT sponsored industry symposia and workshops.

Education

Information security programs in the nation's business schools are very limited. Federal sector leads should promote and support development of undergraduate and graduate level programs and courses of instruction, including information security, with concentration in their specific sectors.

Pipelines

The Commission recommends the DOT and the DOE establish a formal process for addressing pipeline assurance issues in partnership with industry, clearly defining the responsibilities of each Department for security related processes, including threat dissemination, coordination, appropriate federal response to threats, possible establishment of threat levels, and developing plans for addressing potential and actual long-term, serious disruptions in the nation's energy supply.

Global Positioning System

The Commission recommends the Secretary of Transportation:

1) Fully evaluate actual and potential sources of interference to, and vulnerabilities of, GPS before a final decision is reached to eliminate all other radionavigation and aircraft landing guidance systems.

2) Sponsor a risk assessment for GPS-based systems used by the civilian sector, projected from now through the year 2010.

3) Base decisions regarding the proper federal navigation systems mix and the final architecture of the NAS on the results of that assessment.

The DOT and FAA must develop a better understanding of interference and other vulnerabilities of GPS before a final decision is reached concerning the status of all other radionavigation and landing guidance systems. A federally sponsored thorough, integrated risk assessment would lay a sound foundation for decisions on future courses of action.

The National Airspace System

The Commission recommends the FAA act immediately to develop, establish, fund, and implement a comprehensive National Airspace System Security Program to protect the modernized NAS from information-based and other disruptions, intrusions and attack. Program implementation should be guided by the recommendations found in the *Vulnerability Assessment of the NAS Architecture*, prepared for the Commission. The Vulnerability Assessment included the following recommendations:

General:

1) The FAA must clearly define responsibility for information security and accountability within its organization. The leadership should be able to make risk decisions that have budget and operational impacts. The FAA has established a NAS Information Security Group (NISG) to coordinate the information security activities of their many organizations, but the group does not yet have decision making authority on the information security (INFOSEC) that will be implemented in the NAS.

2) The FAA should enhance its security protection program with such traditional practices as: implementing and rigorously enforcing a highly visible security policy; planning countermeasures for known open system and COTS weakness; identifying and drawing from emerging infrastructure protection concepts; maintaining a Red Team for independent protection verification; providing an adequate level of electronic security staffing; and establishing a program for security education, training and awareness.

3) The FAA should consider the implementation of full "trusted" hardware and software security capabilities for only the FAA's most vulnerable future subsystems, since the software cost for embedded applications, together with full audit, tracking, and monitoring, may be too great if applied to all subsystems. Relaxation of the full capabilities, such as less rapid revalidation (e.g., a slower fifteen minutes down time) and less constant vigilance of data integrity, should be considered on a case-by-case basis for less critical subsystems, particularly in situations where existing air traffic control recovery procedures exist.

4) The FAA should conduct a comprehensive investment analysis of NAS INFOSEC in order to determine the degree of security protection that is needed.

5) The FAA should program funds for security provisions for the most critical subsystems in the range of two to four percent of subsystem cost, with additional funds as these subsystems become operational. The FAA should refine these percentage estimates and the identification of the subsystems they apply to, through a study of risk mitigation consequences, the degree of penetration testing needed, and INFOSEC life-cycle costs.

Automation: The FAA should provide virus protection, software distribution protection, and access control protection during the design, development, testing, and life cycle support of the future subsystems. Banner or warning screens should be used on all areas accessed by outsiders on computer and communications networks.

Communication:

1) Design the communications networks of the NAS in such a way that interconnections between the FAA administrative network and the NAS operational networks are kept to an absolute minimum and use well managed state-of-the-art protection methods including firewalls.

2) Monitor the use of Internet for backup communications, now in its preliminary planning stage, to avoid intrusions during systems outages.

3) Continue to use dedicated circuits for the most critical NAS assets.

4) Provide comprehensive security protection and maintain a physical separation between the Administrative Data Transmission Network (ADTN) and the Internet, and maintain a physical separation of this network from all critical operational subsystems. The current architectural plans call for multiple ties to the Internet. Computers used for administrative purposes at operational facilities should have no connection to any operational system.

5) Provide backup communications links and standby service contracts to support satellite communications links that fail or are jammed or flooded. The backup links should be capable of the automatic assumption of communications.

6) Ensure that Internet uses from computers that are connected to operational systems are avoided or have strict high-level approval and accountability for access to these systems.

7) Use encryption (never a complete solution for security requirements) on all critical communications links that have National Security implications (future and actual flight plans for Presidential aircraft, and key governmental and military officials).

Navigation and Landing:

1) Establish and maintain a backup navigation and landing system capability, possibly retaining elements of the current navigation and landing systems.

2) Provide full or partial backup for satellite uplinks for the WAAS system.

Surveillance: Provide surveillance backup for the ADS-B system, particularly in high density terminal areas, and in the center of the continental US, where primary radars are scheduled for removal.

Intelligent Transportation System

The Commission recommends that the DOT develop security standards or guidelines for ITS to assist agencies and companies in designing security into ITS systems during development and installation phases.

Crime

The Commission recommends that an intermodal forum, sponsored individually or jointly by industry or the DOT, be established to address the issue of criminal intrusion into unsecured shipping company databases and electronic data interchange, and the potential impact on critical business practices. The forum should be used to bring this issue to the attention of senior government officials and corporate management, assess the scope of the problem, and share best practices.

Anti-terrorism Legislation

The DOT has submitted legislation (H.R. 1720, the "Surface Transportation Safety Act of 1997") designed to protect the passengers and employees of railroad carriers and mass transportation systems and the movement of freight by railroad from terrorist attacks. The Commission recommends this legislation be given strong support from the Administration and Congress.

Energy

Introduction

The security, economic prosperity, and social well being of the US depend on a complex system of interdependent infrastructures. The lifeblood of these interdependent infrastructures is energy, the infrastructure composed of three distinct industries that produce and distribute electric power, oil, and natural gas. Profiles of these three industries are shown in Figures A-1, A-2, and A-3.

In addition to being a key component of the other infrastructures, the energy infrastructure is critical to our economy, with estimated revenues from retail sales of electricity in the US exceeding $200 billion annually, and revenues from oil and gas almost $400 billion. US energy consumption by fuel type is depicted in Figure A-4.

Today our energy infrastructure is the most reliable and robust in the world. While energy shortages and outages have made national and international news, they are rare. The handful of major incidents in modern times, dating back to the 1965 Northeast Blackout, includes the gasoline shortages of 1973 and 1979 and the electric power outages in the western US in 1996. Despite the proven reliability of the US energy infrastructure, however, there are significant challenges to sustaining this robustness and resilience in the near future.

Disparities in prices across the country are partially responsible for the recent restructuring of the electric power industry and natural gas industry (Figure A-5). New information systems for electronic commerce, for data interchange and for improving operational efficiencies are now essential business elements of the energy infrastructure. Electric utility and natural gas companies are merging and consolidating resources, while at the same time new transmission line rights-of-way are almost impossible to obtain because of the "not in my back yard" syndrome or environmental concern. With the advent of natural gas and electricity commodity markets, the number of marketing companies has grown exponentially, from eight in 1992 to more than 250 in 1996, while electric power capacity reserves continue to shrink. As in the telecommunications industry, the customer of the electric power industry faces a complex service industry in which no one company will provide end-to-end service.

The reliability of electricity has become more critical to our nation's competitiveness and standard of living in the Information Age. The use of natural gas to generate electricity is growing rapidly, as it is the current clean fuel of choice. And we are becoming ever more dependent on the supply of foreign oil, recently surpassing the 50 percent level for oil imports.

While the nation's dependence on less expensive foreign oil continues to grow, refineries in this country are being closed and, in light of thin profit margins and environmental constraints, no new refineries are planned in the US. Over the last decade the oil industry has lost 450,000 US jobs to overseas operations, and one major company alone has reported a work-force reduction from 30,000 to 20,000 employees during the early 1990s. Some of these job losses are blamed on federal and state environmental regulations. To minimize costs and increase efficiency, many companies are dramatically expanding their automation and networking systems and are linking their control, administrative, and business information systems. Many companies are also consolidating their computer centers, with one major worldwide company consolidating its operations into a single megacenter.

Most physical threats to the energy infrastructure are well known and documented. As a result of concern about terrorists attacks, the National Security Council, Congress, the Department of Energy (DOE), and the energy industry focused on the physical security of the infrastructure during the 1980s. This activity led to public hearings by Senator John Glenn of Ohio in February, 1989, and was documented in the Office of Technology Assessment report *"Physical Vulnerability of Electric Systems to Natural Disasters and Sabotage."* In response to the government's concerns, the energy industries compiled internal lists of their critical assets and spare components. Processes were established to disseminate threat information, selected security personnel were cleared to receive classified information and forums were established to share information. As a step toward ensuring viability of energy infrastructures, DoD and the Federal Bureau of Investigation (FBI) initiated their Key Asset Protection Programs. Joint private sector and government exercises were conducted for mutual education and to test the emergency response capabilities.

As farsighted and laudable as these efforts were, however, interdependencies within the energy infrastructure and with the other infrastructures were not studied, nor was the energy sector's growing dependence on information systems. Without electric power other critical infrastructures, such as telecommunications and banking and finance cannot function. The transportation infrastructure would cease to operate as it relies almost exclusively on oil products. These linkages reflect the growing interdependencies between the infrastructures.

Some analysts postulate that the 1996 western power outage and the New England and MidWest summer power shortages were not isolated instances, but are indicators of an industry experiencing a weakening in its historically strong assurance program.

Threats

Threats to the US energy system arise from a number of sources including hostile governments, terrorist groups, other organized groups or individuals, disgruntled employees, malicious intruders, complexities, natural disasters, and accidents. More than a thousand reported incidents directed against the US energy system have been documented by the DOE over the last 15 years; some involved outages and significant damage. In recent years, cyber incidents, including deliberate as well as accidental malfunctions, have begun to appear.

Organized attacks on the energy infrastructures in other countries include an Irish Republican Army (IRA) plot to blow up energy and water installations and cause massive disruption across London in the summer of 1996. A police raid in south London found 36 devices; the planned targets included six electrical substations, gas valves and pipelines, and water pumping stations. Six participants were found guilty of conspiracy and were sentenced to 35 years in prison. A more recent event occurred in Texas this April, when a group planted explosive devices on three natural gas holding tanks at a processing plant to divert police attention during a robbery attempt. It was believed the explosions would have released toxic fumes which could have wiped out half of the county.

The most common disrupter of energy supplies is inadvertent damage to buried cables or pipelines, such as is frequently caused by a "back hoe." However, these disruptions are usually localized and have no national level impacts.

Downsizing of the industries, partially in response to restructuring and consolidating pressures, leads to a significant loss of expertise that is difficult to replace. Downsizing also disrupts the traditional compact between employer and employee and creates a potential cadre of disgruntled "insiders." An estimate provided to the Commission by industry security directors was that 75 to 80 percent of the security incidents they experience are caused by persons from within the organization.

Managing consequences of natural disasters and accidents is an inherent part of the energy industries' operational processes. Their mitigation and response efforts and activities have high public visibility, and have resulted in an outstanding response by the industry.

Vulnerabilities

Specific areas of vulnerability addressed by the Commission's Energy team are categorized as:

- Electric Power: power generation (including fuel supply) systems, transmission systems, distribution systems, electric network control and protection systems.
- Oil and Natural Gas: supply, transportation, storage and distribution (pipelines are a joint effort with the Commission's Physical Distribution team).

The Commission's review focused on those elements of the infrastructure in which exploitation of a vulnerability could cause extended regional or national impacts. Nominal impact figures used were 500,000 people/customers affected for at least 12 hours.

Vulnerabilities facing the energy industries include:

- Those created in the operating environment by the rapid proliferation of industry-wide information systems based on open-system architectures, centralized operations, increased communications over public telecommunications networks and remote maintenance;
- Supervisory Control and Data Acquisition (SCADA) systems that are vulnerable because of use of commercial off-the-shelf (COTS) hardware and software, connections to other company networks, and the reliance on dial-back modems that can be bypassed;
- Increased availability of vulnerability information, much of which is mandated by regulatory bodies to facilitate competition, and the tools for exploiting those vulnerabilities;
- Rapid assimilation of advanced technologies with their inherent complexities;
- Consolidation of infrastructure corridors (e.g., communication, electric transmission lines, pipelines, etc.); and

- Previously identified physical vulnerabilities of critical assets that have not been adequately addressed throughout the industry.

Electric Power Vulnerabilities

Of particular concern are the bulk power grid (consisting of generating stations, transmission lines with voltages of 100 kV or higher, plus 150 control centers and associated substations) and the distribution portion of those electric power systems whose interruption could lead to major metropolitan outages. (Note: this report covers the "grid," a North American system comprising the US, Canada, and a small part of Mexico.) On the cyber side, the focus was on the larger networks, including those that interconnect a company's information and operation systems and those that interconnect company systems to each other (Figures A-6 and A-7).

The most significant physical vulnerabilities appear to be related to substations, although certain generation facilities and transmission lines are also inviting targets. There is general agreement that since the industry designs for stability during single and certain double failures, a coordinated attack on multiple targets would be required to cause a significant disruption of service. Furthermore, such an attack would need to hit multiple targets simultaneously or in rapid sequence.

Because of the complexity of the grid, attackers would have difficulty replicating cascading outages such as the two western power outages of July and August 1996. More research is needed to better understand the dynamics of the grid, particularly the phenomenon of voltage collapse, which can lead to a cascading outage.

From the cyber perspective, SCADA systems offer some of the most attractive targets to disgruntled insiders and saboteurs intent on triggering a catastrophic event. With the exponential growth of information system networks that interconnect the business, administrative, and operational systems, significant disruption would result if an intruder were able to access a SCADA system and modify the data used for operational decisions, or modify programs that control critical industry equipment or the data reported to control centers.

Oil and Gas Vulnerabilities

Large refineries (greater than 250,000 barrel capacity) in California, Texas and Louisiana would be attractive targets for physical or cyber attack. The significant increase in the proportion of oil transported via pipelines over the last decade provides a huge, attractive, and largely unprotected target array for saboteurs. Elements of the pipeline system that could be targeted include lines at river crossings, interconnects, valves, pumps, and compressors. Three major pipelines in the country offer the greatest potential for significant impact if attacked successfully. However, on the positive side, over the last five years, many interconnections have been added to natural gas pipelines, making rerouting around a break easier, but this may not always be possible if the line is at capacity.

As in the electric power industry, SCADA systems used in the oil and gas industries are subject to electronic intrusion. If accessed, information could be manipulated or control programs modified. Under certain circumstances, a hammering effect could then be induced in pipelines, possibly leading to breaks. More research is needed to determine the feasibility of such attacks.

Status and Assessment of Current Energy Infrastructure Assurance Programs

The DOE is the lead federal government organization for response to energy emergencies but has limited authority in the infrastructure assurance area. The Federal Energy Regulatory Commission (FERC) oversees wholesale electric and gas rates and service standards, as well as the transmission of electricity and gas in interstate commerce. The North American Electric Reliability Council (NERC) has assumed primary private sector responsibility for the reliability of the bulk power system (that is, the portion of the electric utility system that encompasses the electrical generation resources and transmission system shown in Figure A-1). The Security Committee of the Edison Electric Institute (EEI) provides a forum with a focus toward physical security and law enforcement activities for the security directors of investor owned utilities. The National Petroleum Council (NPC) is an advisory committee of 175 CEOs from the oil and gas industries, and the American Petroleum Institute's (API) and American Gas Association's (AGA) Telecommunications Committees provide forums for telecommunications specialists. The Electric Power Research Institute (EPRI), the Gas Research Institute (GRI), and the Institute of Gas Technology (IGT) are the leading energy technology organizations. The DOE National Laboratories are another source of significant expertise for solutions to the complex technical problems associated with infrastructure assurance.

As a result of the restructuring of the electric power industry, NERC has made significant changes to its organization, including the new requirement of mandatory compliance to its policies and procedures, compliance monitoring, enforcement measures, and increased and broadened membership. However, since NERC is a voluntary organization, enforcement is questionable. Also, a tension exists between different industry groups.

Another notable effort is the Secretary of Energy's Task Force on Electric System Reliability, which was recently established to provide advice on ways to address key institutional, technical, and policy issues associated with maintaining bulk electric system reliability in the new era of a competitive electric industry. An interim report, published in July, focuses on institutional recommendations to enhance overall reliability of the electric power system.

The critical components of the energy infrastructure remain vulnerable to physical attack, and replacement of many of these components involves lead times measured in months. However, most major companies have improved the physical security of their critical sites. From a cost benefit perspective, the companies believe they have taken prudent measures. Many companies' restoration programs are tested all too frequently by nature (hurricanes, earthquakes, tornadoes, fires, and floods) and existing mutual aid agreements have enabled restoration of service in reasonable time, even after the worst disasters.

From the cyber perspective, much needs to be done, and many issues have arisen, of which there is only limited awareness. Even the leading companies have only recently focused on information assurance issues. Despite increasing concern about vulnerabilities, many companies are understaffed in the cyber security area. Where cyber security experts are employed, their main focus appears to be on the business data processing side of the company, with a large share of their effort being expended on virus contamination problems. In most companies, information systems (business, administrative, and operations) are being networked, both internally and externally. Although many industry officials are aware of the significant vulnerabilities introduced by connecting to the Internet, most companies are making such connections. However, in attempts to provide security for information systems, many companies are placing confidence in individual measures, such as firewalls and dial-back modems, to secure their networks. The Commission's studies show that a more systematic approach is needed.

Several proactive information assurance efforts should be noted. The NERC has recently under-taken an initiative to collect information on cyber intrusions. EEI has volunteered to work with other interested groups to further scope the issues and activities in the cyber security area of the electric power infrastructure. EPRI has taken the lead in cyber security for the electric power industry, while DOE has assumed the lead for the federal government. For example, EPRI and DOE have joined forces to assess the security design and development of the information system of the Independent System Operator recently established in California. Also, DOE's Office of Nonproliferation and National Security has outreach programs on cyber security, energy emergencies, and threat assessments. Another notable effort was the three day seminar conducted by IGT (Emergency Response and Critical Infrastructure Protection in the Gas and Electric Industries) in June 1997.

Findings

1) The authorities and responsibilities for energy infrastructure assurance in the federal government need to be clarified.

2) The respective responsibilities of government and private sector for infrastructure assurance are not clearly understood.

3) Improved sharing of threat information and "indications and warning" (I&W) information is needed. Improved sharing of industry experience is needed (e.g., a fully populated cyber intrusion database).

4) More training and awareness in infrastructure assurance is needed, focusing on risk management, vulnerabilities, performance testing, and cyber security.

5) Infrastructure assurance technology advancements could add significantly to the overall protection of industry assets.

6) Adopting uniform physical and cyber security guidelines, standards or best practices would enhance protection.

Recommendations

Energy Infrastructure Assurance Strategy

Historically, the energy infrastructure's strategy has focused on robustness and resilience. The physical vulnerabilities of the pipelines and transmission grid are widely acknowledged and understood, and the philosophy has been to mitigate the natural and man made events that can exploit those vulnerabilities so that service to the customer is not interrupted or, if interrupted, only for the shortest possible time. To assure the energy infrastructure in the future, owners, operators, and the government must work together to develop a strategy focused on the primary objectives of prevention, mitigation and recovery.

Owners and operators can further expand communication channels with the government for sharing information on threats and vulnerabilities to ensure that they are making informed risk management decisions; enhance their research and development, focusing on cyber security and reliability projects; and as major customers of the telecommunications and software industries, make demands for more secure products and services.

Government agencies can contribute to the prevention, mitigation, and recovery of infrastructure losses by assuring that appropriate information sharing paths are established between owners/operators and the government; that existing or new regulations do not adversely impact the protection of the infrastructure; that a level playing field exists for the industry to invest in long term preventive measures; that threat and vulnerability information is provided to assist industry in making informed risk management decisions; and that long-term research and development activities are conducted to enhance assurance.

Implementation of Assurance Strategies

The Commission recommends:

1) Expanded roles and responsibilities for owners and operators, and the government to provide balance for the recommended strategy.

Owners and Operators

- Provide CEO level advisory counsel on infrastructure assurance issues, much as NSTAC provides advice for telecommunications. Representation could come from EEI and NPC.
- Provide threat dissemination and information sharing through associations such as NERC, API, INGAA and AGA.
- Formalize cyber security activities through such organizations as EEI, NERC, EPRI, API, and AGA.
- Fund enhanced infrastructure assurance near-term R&D through such institutions as EPRI and GRI.
- Emphasize education, training, and awareness using such resources as NERC and IGT.
- Provide a forum for development of enhanced physical and cyber security standards/guidelines through such organizations as the Institute of Electrical and Electronics Engineers (IEEE).

Federal Government

- Provide national direction through planning, policy, and legislation to maintain a level playing field for owners and operators investing in infrastructure assurance. Emphasize technology research, training and awareness, emergency response, and information sharing efforts.
- Develop and promulgate a mandated energy infrastructure assurance mission for the Department of Energy to address the responsibilities of the leadership and coordinating role as a federal government lead agency.
- Clarify the respective roles and responsibilities for pipeline security between the Departments of Energy and Transportation (DOT) through a joint effort.
- Provide enforcement/oversight for industry (electric power) reliability standards through FERC.

- Direct and fund the DOE National Laboratories to focus their expertise on infrastructure assurance assessments, response, and energy infrastructure assurance research and development (R&D).
- Expand the existing process for reporting power outages and physical attacks to include cyber attacks, and develop a legislative process to protect sensitive industry data.
- Develop and coordinate an enhanced process for timely, detailed threat information dissemination through the law enforcement and intelligence communities.

State Governments

- Provide assistance in the areas of training and awareness, and assurance exercises.
- Encourage the National Association of Regulatory Utility Commissioners (NARUC) to work through its member state commissions to enhance the protection of public utility infrastructures.

2) Owners, operators, and the government increase funding for R&D in the following technology related areas with security dimensions.

- Cascading effects leading to voltage collapse.
- Online security assessment, including online power flow and transient analysis.
- Transmission and distribution technology, and real time control mechanisms.
- Large scale modeling and evaluation of the power grid and pipeline systems (regional and nationwide).
- Examine solutions to foreign energy supply vulnerability as a cost-benefit approach.

3) Secretary of Energy provide planning, policy, coordination, technical expertise and training/awareness for infrastructure assurance by:

- Developing an energy infrastructure assurance plan in a coordinated government/private sector forum.
- Encouraging development of physical and cyber security standards/best practices within the industries through the various associations (NERC, EEI, EPRI, IEEE, GRI, AGA, and API).
- Developing and enhancing training, education and awareness programs for energy infrastructure assurance practitioners.
- Providing the technical capability for vulnerability assessments available from National Laboratories to conduct reviews of critical infrastructure assets.

- Funding a test bed/pilot program for energy infrastructure assurance that includes the private sector and government.
- Coordinating with the private sector and DoD on research and development of risk management software and techniques, information assurance software and hardware for real time intrusion detection, enhanced authentication and authorization, and vulnerability assessment tools with a focus on SCADA systems.
- Lead an industry/government effort to define the level of threat (e.g., criminal, insider, experienced hacker with intrusion software development capability) to be established as a goal for industry to defend against.

4) The Commission recognizes the importance of the following industry recommendations and recommends the Secretary of Energy work with the industry to:

- Establish standards for a national "one call" program to address third party interruptions (dig-ins).
- Continue joint effort between federal government, EEI, EPRI, NERC, and the oil and gas industries to further develop issues and activities pertinent to cyber security.
- Review government regulations that require excessive reporting and release of what industry considers sensitive information (e.g., FERC Form 715 – Annual Transmission Planning and Evaluation Report).
- Review regulations that may inhibit efforts by utilities to aid one another in emergency response efforts.
- Form a permanently staffed center, jointly supported by government and industry, for sharing threat and vulnerability information from both public and private sector sources.

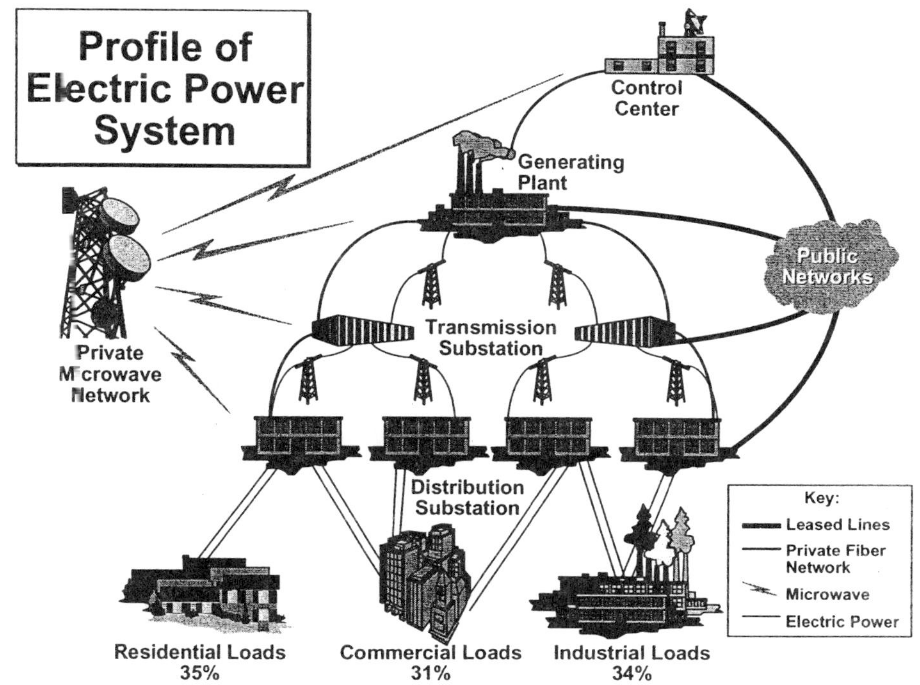

Figure A-1. Profile of Electric Power System

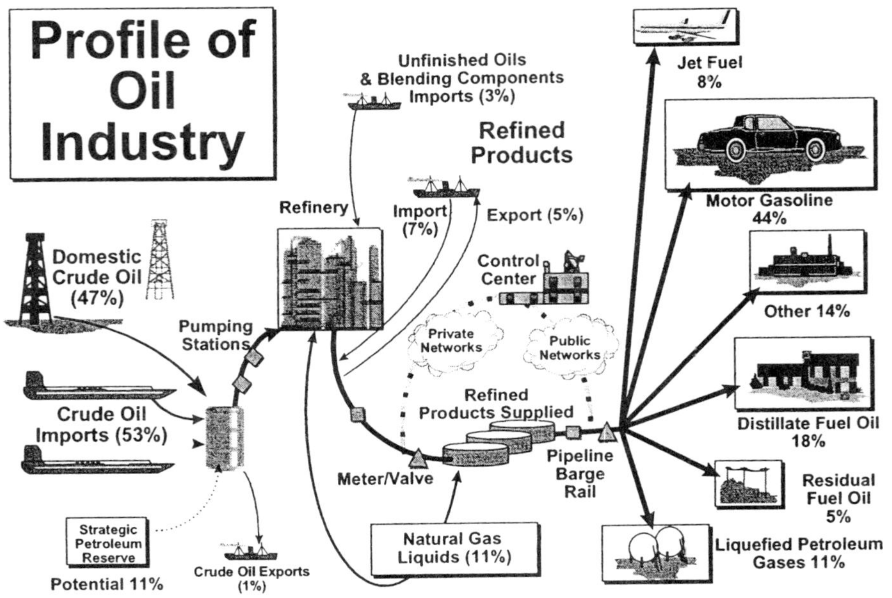

Figure A-2. Profile of Oil Industry

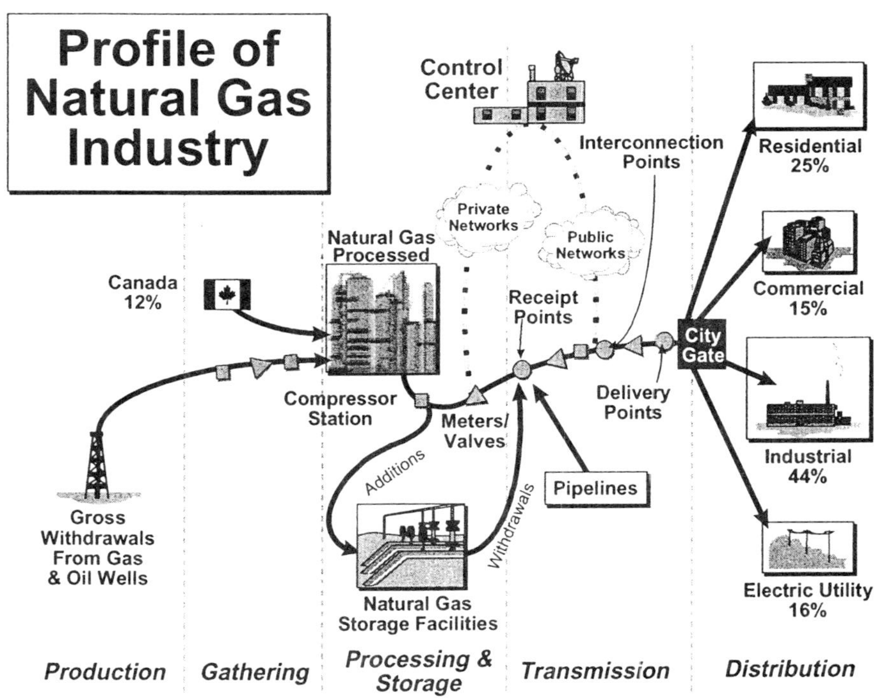

Figure A-3. Profile of Natural Gas Industry

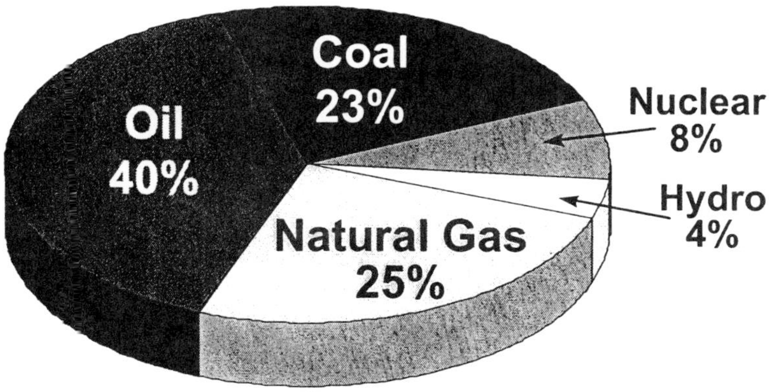

Figure A-4. US Energy Consumption

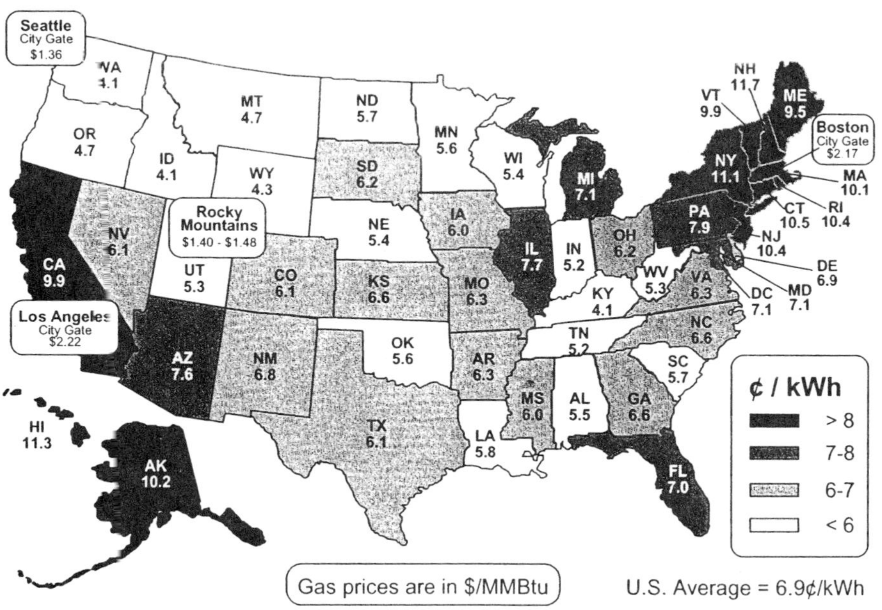

Source: U.S. Department of Energy, Energy Information Administration, DOE/EIA-0562 (96) and DOE/EIA-0348(95)/2 .

Figure A-5. US Electric and Natural Gas Prices

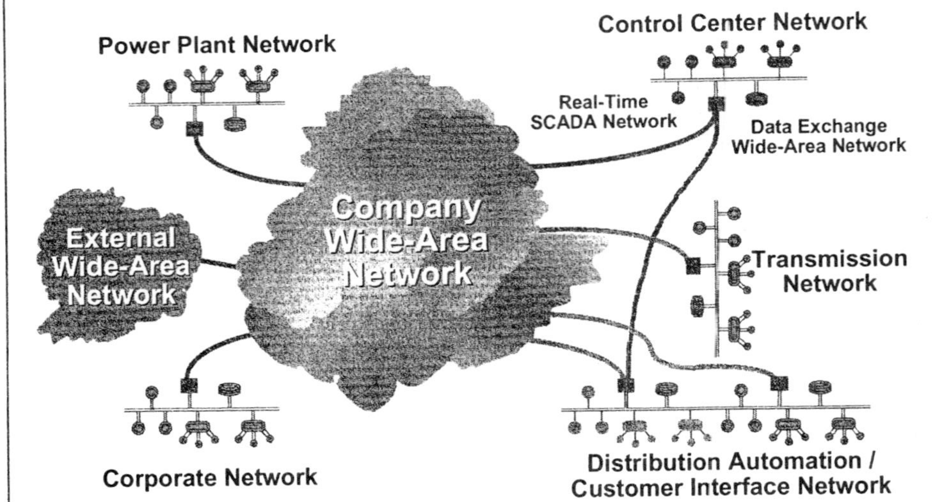

Figure A-6. Integrated Utility Network

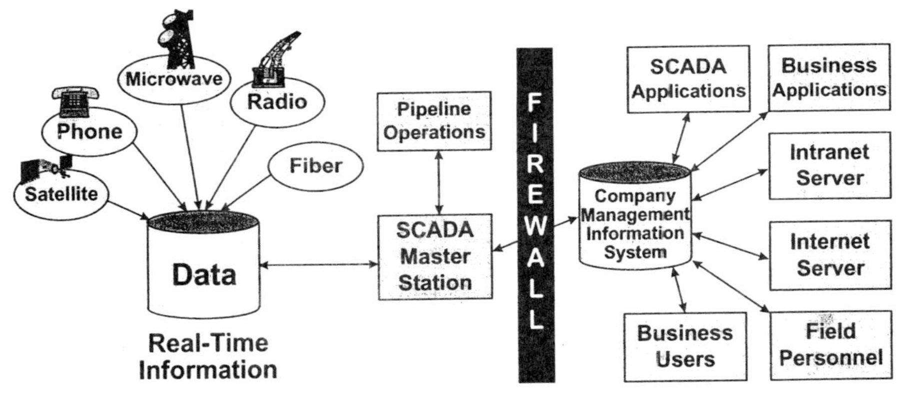

Figure A-7. Oil/Gas Current Internal System Configuration

DESCRIPTION

Volume 1-4. This is an ex-library book and may have the usual library/used-book markings inside.This book has hardback covers. In fair condition, suitable as a study copy. No dust jacket.

Banking and Finance

Introduction

The US financial system is central not only to the functioning of domestic and global commerce, but to the daily lives of virtually all Americans. It represents bank holdings of about $4.5 trillion, a capital market of $7 trillion, investment bank underwriting of $1 trillion, almost $3 trillion in daily payment transactions, and about 10 million jobs.

More than a billion credit cards in circulation in the United Stated account for $500 billion in annual expenditure, or roughly half of all consumer debt. Also, due to the rapid increase in individual retirement accounts of various kinds and the popularity of mutual funds, about half of all households in the United States are investors in the stock market.

The banking and finance infrastructure was defined by the Commission as composed of five principal sectors: banks, financial service companies, payment systems, investment companies, and securities and commodities exchanges. The Commission's banking and finance team conducted a broad-based industry outreach, developed a profile of major participants, geographically mapped industry operations, assessed the level of vulnerability and defense extant within the financial system, and reviewed the analytic structures in prevalent use for such key industry processes as risk analysis and countermeasure investment decision-making.

Our principal finding is that, due to its carefully structured mixture of public oversight and private initiative, the US financial system is among the world's finest. The modern US financial system never has suffered a debilitating catastrophe, and for that reason among others carries an extraordinarily high level of global confidence. Some observers go so far as to characterize it as shock proof.

The Current Situation

The institutions comprising the financial services industry are further ahead than most in employing sophisticated and, in some cases, unique defenses against loss of assets and corruption of core data systems. Consequently, the US financial system is unusually well protected at the national level, and is well prepared to confront a broad range of threats to its operations and integrity.

However, along with other infrastructures studied by the Commission, the banking and financial service industry is undergoing significant structural change. Expansion by banks into previously prohibited business areas such as securities trading; mergers and acquisition activity; heavy and growing engagement in dynamic global financial markets; and the steady move toward electronic commerce combine to present new challenges to the ways vulnerabilities are defined and risks and managed. And, at the operating level, heightened reliance on global information infrastructures and the advanced computing technologies which power them makes the management of those risks more complex.

Deregulation within the telecommunications and electric power industries upon which financial services so heavily rely introduces new factors to the industry's traditional risk management models. Multiple intermediaries have been inserted into what once were end-to-end service systems that–when combined with decreases in reserve capacity margins in these industries resulting from competitive cost pressures–make the operational interdependency among these three gigantic infrastructures even more opaque and complicated.

Risk Management

Managing risk is the principal business of financial institutions. They view protection against physical and cyber threats as a necessary cost of doing business, and they often position security as a competitive advantage and highlight it in advertising designed to attract new customers for such services as remote banking. Security is an integral component of institutional performance and accountability.

Security investments by financial institutions are driven by two primary forces.

- Law and Regulation: Mandatory investments made by the institution for legal or regulatory compliance. Because financial institutions are so heavily regulated, examination processes drive most security investments.
- Risk Management Analyses: Based on internal and external audit findings, industry and technology norms, history and current events, and estimates of future technology or threats, institutions evaluate risk according to probability of occurrence and the likely consequences for the institution. Security investments are made accordingly.

To assist in broadening the industry's recognition of new threats and vulnerabilities that could affect their risk assessments, financial institutions would benefit from better access to reliable current information from government and from across the industry. Reporting is generally compartmentalized by sector; only a few trusted mechanisms now exist for sharing the kinds of information needed to facilitate system-wide risk assessments.

Threats

The major current threats to the overall operation of the financial system are largely physical in nature, consisting either of natural disasters or a direct coordinated attack on the system's more vulnerable points. These are aggravated by the more open availability on the Internet of the kind of information needed to plan such attacks, increasing reliance on global outsourcing of core operations, and the consolidation of bank and other operations centers as a result of merger and acquisition activity.

At the institutional level, however, the most persistent security threat is the insider who might use authorized access to confidential information or operating systems for profit. Financial institutions employ comprehensive and intricate systems of internal controls to counter this threat, but the knowledgeable insider dedicated to corruption is difficult to stop.

There is also the evolving threat of a larger scale cyber attack by a sovereign adversary or organized terrorists with the aim of inflicting serious damage on key elements of the US financial system. The current probability of this threat is estimated to be low but growing, and one of its more troubling features is that its source may be undetectable and the attack itself might be masked as a series of lesser intrusions.

Vulnerabilities

It is important to note some key distinctions in describing financial system vulnerabilities.

First, there is the distinction between vulnerability of the US financial system and opportunities for theft and fraud in individual institutions. Almost all media reporting on vulnerability up to now has risen from single cases of theft. Emblematic of this is the much reported access of Citicorp's electronic money transfer operation by a transnational criminal group in 1994. While this case made dramatic news accounts and was embarrassing to Citicorp, whose ultimate loss amounted to $400,000, it in no way reached the level of a threat to the bank, much less the financial system.

Second, there is the distinction between the financial condition of a single participant in the financial system and the strength of the system as a whole. Recent years have seen some spectacular financial events, such as the Mexican Peso crisis, the failure of Barings Bank due to fraud, and major scandals involving Japanese banks. These shocks were absorbed and managed by appropriate market, regulatory, and central bank actions without lasting harm to the full system.

Based on the sector profiles developed by the Commission, the nation's core payment systems (FedWire, CHIPS, SWIFT) and the organized securities and commodities exchanges seem to present a serious physical vulnerability within the financial system. This is so not because they have failed to take extensive precautionary measures, but rather because there is substantial cross sector dependence on the services they provide, and few if any alternatives available to provide those services in the event of a disabling catastrophe. In contrast, our analysis shows that the other sectors of the financial infrastructure have sufficient diversity to provide for the dispersion of risk among a wide range of alternatives.

As a countermeasure, the FedWire, for example, maintains three hardened operating centers capable of carrying the full volume of its wire transactions. Similarly, the New York Stock Exchange (NYSE), as the nation's most influential exchange, has established extensive system redundancy, alternate power sources, and diverse communication links. Still, the physical concentration of its data processing and operations centers makes more plausible the possibility of an event or series of events that could disable both sites. Even in that event, however, contingency trading arrangements required by the Securities and Exchange Commission (SEC), although never tested, have been described by the SEC as able to restore NYSE operations within several days. Other major exchanges, such as the Chicago Board of Trade and the Chicago Mercantile Exchange, have similar recovery plans, as do lesser securities exchanges.

Public Confidence

Financial institutions are acutely aware that public confidence is their most critical asset. In that respect, the financial service industry shares with government a fundamental dependence on public support for its viability. This linkage is the basis for the important role government has in assuring the safety and soundness of US financial system.

Because of its sensitivity, however, financial institutions generally oppose reporting which goes beyond the existing mandatory regulatory and law enforcement channels. While it is understandable that these institutions wish to avoid costly reporting requirements and potentially damaging disclosures, taking such a position fuels critics who claim that there are large unreported losses — especially related to computer intrusions of various kinds. Any new mechanism for the exchange of information must establish an acceptable climate of trust and control which will encourage participation by financial institutions yet meet emerging governmental national security requirements for more coherent, systemic risk assessments.

Market Forces and Government Action

There is little doubt that ultimately market forces will generate the appropriate level of investment in risk management tools necessary to secure the financial infrastructure into the future. Nevertheless, the financial system regulatory changes under consideration by the Administration and Congress all have in common an important government role in setting ground rules for new forms of competition; providing a level of fundamental indemnity for customers of the system, thereby relieving companies of some risk; and protecting the public interest in the safety, soundness, and fairness of the system as a whole.

Market forces work best when businesses see the investment as a necessary cost of operation, as consistent with their concepts of risk, as providing a competitive advantage, and protecting their brand (Figure A-8 illustrates this process). In this context, one of the problems with sole reliance on market forces for long-term investments in research and development of security tools, for example, is the lack of actuarial data upon which the risks associated with new threats and vulnerabilities might be calculated. Neither can the benefits of the investment be specified. Consequently, the economics of long term prevention measures often works against their development.

For example, for such major preventive measures as the establishment of redundant communications systems for the industry to rely on in the event of a catastrophic telecommunications or electric power failure, or the construction of an alternate trading site for the securities exchanges, payoffs are not easily envisioned. In the absence of measurable risk, the net present value of such investments is minimal if not zero, and justifying present costs to shareholders, investors, and securities analysts by citing general benefits in the absence of clear risk or competitive advantage is not likely to succeed. Therefore, business on its own will not invest in the kind of ultra secure contingency measures usually found in the military or national security arena without either more information about the risk, or some other incentive.

Summary

The current security of the US banking and finance infrastructure is strong. The government, through regulation, plays a central role in assuring the financial system's safety, soundness, and fairness, but the industry itself has over many years developed a diligent culture of security. Both the role of government and industry diligence will continue even as governing statutes and regulations are changed, technology advances, and the industry restructures and competes in the global market.

However, it is important to note that scrutiny of the financial services industry goes beyond government regulation and law enforcement. Because of its centrality to the nation's economy and the daily lives of most Americans, the extraordinarily high value of its assets, and its high global visibility, the industry's operations and behavior are closely observed by securities analysts, investors, major customers, journalists, and the public in general. This provides powerful additional incentive for financial institutions to assure their integrity and take the actions necessary to continue to earn and retain broad public confidence, retain customers, and achieve growth.

Overall, industry risk management efforts concentrate on prevention of loss, with mitigation of loss following in importance. However, contingency planning also is a high priority, as disablement resulting from an attack or natural disaster remains the industry's largest current risk.

Cyber risks on a system-wide scale may emerge from the possibility of unforeseen instability in the telecommunications and electric power industries as they deregulate and disaggregate. On an institutional level, increasing use of electronic banking mechanisms, requiring multiple ports of entry and perhaps an entirely new infrastructure to accommodate the demand for rapid data recall and payment processing, will create new forms of risk to information systems. For example, connections to the Internet for this purpose present a risk of unauthorized access to operating systems if the Internet connection is not effectively partitioned by firewalls and other such tools. Banks and others are approaching this with caution, although the attraction is strong in terms of operational efficiency and expanded market reach.

In the longer term, risk emerges from the maturing of Information Warfare capability among organized adversaries who may wish to attack the US by destabilizing large portions of the financial system, and erode public confidence in it as well as in the capability of the US to defend against such attacks. Both the government and industry would significantly benefit from improved flows of threat and vulnerability information so that precautionary measures can be developed and deployed system-wide at a pace sufficient to provide an effective defense.

Recommendations

Information Sharing

Regulators, law enforcement officials, and industry associations should coalesce to establish a trusted forum for the exchange of relevant threat and vulnerability information so as to facilitate the assessment of risk on a system-wide basis.

Contingency Planning

Regulators and industry associations should sponsor strategic simulations designed to test the adequacy of existing industry recovery plans under a variety of conditions. These should feature the emerging risk factors of growing interdependence complicated by deregulation and global expansion of operations.

Insider Threat

Regulators, private auditors, and the industry should continue to work together to improve examination processes, audit practices, internal controls, and physical security measures to accommodate new kinds of risks and to help deter the insider threat.

Back-up Facilities

National security, law enforcement, and regulators should decide whether the establishment of such security measures as a contingency trading site for major exchanges, contingency data storage centers, and dedicated communications systems, the cost of which probably exceeds the reasonable business risk involved, are appropriate as government-funded national security measures.

Education

Industry associations may wish to take the lead in establishing information security education and awareness programs within academia and in the general public.

CHAPTER 2

THE PRESIDENT'S COMMISSION ON CRITICAL INFRASTRUCTURE PROTECTION

STATEMENT OF ROBERT T. MARSH

Before the Subcommittee on Technology, Terrorism and Government Information

Committee on the Judiciary
United States Senate
November 5, 1997

Introduction

Good afternoon, Mr. Chairman and members of the subcommittee. My name is Tom Marsh and I served as the Chairman of the President's Commission on Critical Infrastructure Protection. On behalf of my fellow Commissioners—several of whom are with me today—I am pleased to be here today to discuss with you the work of the Commission and outline its principal findings and recommendations reflected in our report, *Critical Foundations.*

Before my prepared remarks, I'd like to express my appreciation to you, Mr. Chairman. Your farsighted vision for the nation's future needs and well-being was key to the establishment of the Commission, and also helped spur other government activities in this area, most notably by the FBI and the Defense Department, including the Eligible Receiver exercise you were just briefed on. I can confidently speak for my fellow Commissioners and the nation when I say that we all are deeply indebted for your selfless contribution.

To give you some perspective on the Commission's challenge, imagine, if you will, that. . . .

- the power goes out in the Northwest;
- the 911 is disrupted in a major city because someone has flooded the phone lines with repeat calls;
- two bridges across the Mississippi River are destroyed—bridges that not only carry trucks and trains, but also telephone cables; and
- two Internet service providers in New York city are out of service.

What do we do in such a situation? Who is in charge? Is it merely coincidence? Or a concentrated attack?

These are the types of questions the Commission has been considering —questions to which there are no easy answers.

Questions, we hope, our recommendations will help lay the foundation for answering.

Overview

I appreciate the opportunity to talk about the Commission's work, discuss why protecting our infrastructures is important in light of the new vulnerabilities and threats of the cyber age, present our key findings, and then briefly summarize our recommendations.

I must say right up front: our findings, conclusions, and recommendations are very different from what we anticipated—and different from what our stakeholders anticipated. Many thought this was a problem that government alone could resolve in a few easy steps. But during the past year and a half, we concluded that **protecting our infrastructures is a public-private undertaking that requires a new partnership** and **protecting our infrastructures will take time—and will require long-term efforts and a new way of thinking.**

Mission

The Commission was established by Executive Order 13010 on July 15, 1996. A joint government and private sector endeavor, it was charged to develop a national policy and implementation strategy for protecting our critical infrastructures from physical and cyber threats and assuring their continued operation.

The Infrastructures

The President identified eight infrastructures as our national life support systems. These national infrastructures are vital in that their incapacity or destruction would have a debilitating impact on the defense and economic security of the United States. These are the infrastructures:

- telecommunications
- electric power
- transportation
- oil and gas delivery and storage
- transportation
- water
- emergency services, and
- government services.

Why Attack Infrastructures?

Critical infrastructures have long been lucrative targets for anyone wanting to attack another country. Our nation relies on its infrastructures for national security, public welfare, and its economic strength.

Those who would attack the infrastructures would do so to:

- reduce our ability to act in our own interest,
- erode public confidence in critical services, or
- reduce American economic competitiveness

In the Gulf War, for example, disabling Iraq's infrastructures was one of the keys to our success—a lesson noted with much interest by many countries around the world.

The Commission

The Commission was uniquely tailored for the task. Recognizing that the critical infrastructures are largely owned and operated by the private sector, the Commission structure was a joint public-private undertaking.

The *Commission* was comprised of representatives from both industry and government.

The *Steering Committee* of senior government officials oversaw the work of the Commission and guided us through myriad government concerns.

A *Presidentially-appointed Advisory Committee* of key industry leaders provided the unique perspective of owners and operators of the infrastructures.

The Infrastructure Protection Task Force was established at the same time as the Commission to support infrastructure protection until the Commission's recommendations are enacted.

Guiding Principles

Our approach recognized that most of the infrastructures operate within an existing framework of government policy and regulation. But they are also privately owned competitive enterprises; as such, protection recommendations should not undermine a company's competitive position. We recognized that any solution would have to be viable in the marketplace as well as the public policy arena.

Thus, we adopted the following guiding principles:

First, we knew this could not be a Big Government effort. Government must set the example, but it is the owners and operators who are the key to success. They have a strong economic stake in protecting their assets and maximizing customer satisfaction. They understand the infrastructures and know best how to respond to disruptions.

Second, while we may be undergoing an information revolution, we felt that utilizing the best ideas and processes from current structures and relationships was the proper way to proceed. This means building on existing organizations and relationships as well as fostering voluntary cooperation. Partnership between industry and government will be more effective and efficient than legislation or regulation.

Finally, this is a long-term effort which requires continuous improvement. We must take action in practical increments. There is no "magic bullet" solution. We must aim not only to protect the infrastructures, but also to enhance them.

Outreach

Outreach was a cornerstone of our effort. In fact, our conclusions and recommendations result directly from the conversations and meetings we had with approximately 6,000 individuals from industry, academia, science, technology, the military and government.

We held five public meetings around the country, participated in numerous conferences; hosted simulations, games and focus groups; and sought to increase awareness of this effort through the media and our Web site.

Evolution of Threat

In the past, broad oceans and peacable neighbors provided all the infrastructure protection we needed.

That all changed during the Cold War. Technology became preeminent and geography became irrelevant. Soviet and U.S. nuclear weapons were targeted against each other's power grids, rail networks, and energy industries. But the cost remained high and the ability to carry out such an attack was available to only a few major powers.

Computers and electrons change the picture entirely. Now the capability is widely available at relatively little cost.

This is the "new geography" in which the Commission has focused its efforts—a borderless cyber geography whose major topographical features are technology and change.

Vulnerabilities

We have long understood physical threats and vulnerabilities, but not so in cyber space. The fast pace of technology means we are always running to catch up in the cyber dimension. Thus the Commission's work and our report focus primarily on coping with the cyber threat.

Our foremost concern is the interdependencies presented by the "system of systems" we rely on for the daily operation of our critical infrastructures.

Furthermore, information that describes our vulnerabilities is increasingly accessible. Most of it is unclassified, and much of it is available on the Internet. We had to be careful in compiling this information not to provide a handbook for those who would use it for harmful purposes.

Threat

So, who is the threat? We view the threat as anyone with the capability, technology, and intent to do harm.

While we have not found a "smoking keyboard"—that is, we do not know who has the intent to do harm—we do know that the threat is a function of capability and intent.

We characterize capability as a combination of skills and tools—skills that even most teenagers have, and tools that are readily available.

In short, the opportunity to do harm is expansive and growing.

Threat Spectrum

The bad actors who use these tools range from the recreational hacker—who thrives on the thrill and challenge of breaking into another's computer—to the national security threat of information warriors intent on achieving strategic advantage.

Common to all threats is the insider. We could spend millions on technology to protect our infrastructures, but a well-placed insider or disgruntled employee could render nearly all protection useless.

The New Arsenal

The new arsenal of "weapons of mass disruption" in the cyber world include "Trojan horses," viruses, and e-mail attacks that can be used to alter or steal data. These tools recognize neither borders nor jurisdictions. They can be used anywhere, anytime, by anyone with the capability, technology, and intent to do harm. And they offer the advantage of anonymity.

Responsibilities

We examined the respective roles of the private sector and the federal government in light of this new threat and the potential bad actors.

We concluded that the private sector has a responsibility to protect itself from the local threats, such as individual hackers and criminals. And that the federal government has a larger responsibility to protect our citizens from national security threats. **In short, we found that infrastructure protection is a shared responsibility.**

The private sector is responsible for taking prudent measures to protect itself from commonplace hacker tools. If these tools are also used by the terrorist, then the private sector will also be protecting against cyber terrorist attack and will be playing a significant role in national security.

The federal government is responsible for collecting information about the tools, the perpetrators, and their intent from all sources, including the owners and operators of the infrastructures. The government must share this information with the private sector so that industry can take the necessary protective measures.

Findings

In some respects, our most important finding is that adapting to this challenge requires thinking differently about infrastructure protection. We must look through the lens of information technology as we approach the third millennium.

Specifically, we found that:

- Information sharing is the most immediate need.
- Responsibility is shared among owners and operators and government.
- The federal government has an important role in the new alliance.
- Infrastructures protection requires a focal point.
- We must develop an analysis and warning capability.
- The existing legal framework is imperfectly tuned to deal with cyber threats.
- Research and development efforts are inadequate to support infrastructure protection.

Fundamental Conclusion

We know our infrastructures have substantial vulnerabilities to domestic and international threats. Some have been exploited - so far chiefly by insiders. Protecting our infrastructures into the 21st Century requires that we develop greater understanding of their vulnerabilities and act decisively to reduce them. In the last fifteen months, the Commission has thoroughly reviewed the vulnerabilities and threats facing our infrastructures, assessed the risks, consulted with thousands of experts, and deliberated at length as to how best to assure our nation's critical foundations in the decades to come. Our fundamental conclusion is as follows.

Waiting for disaster is a dangerous strategy. Now is the time to act to protect our future. And this action requires a new partnership to address the risks of protecting our nation's infrastructures.

Recommendations

The Commission's recommendations are the products of much research, discussion, and deliberation. They are founded on shared core principles and they are based on fact. They are aimed at improving coordination and establishing roles for infrastructure protection, fostering partnerships among all stakeholders, and coordinating diverse interests.

Every recommendation was discussed at length in a series of deliberations that addressed all feasible options and the pros and cons of each. All Commissioners accepted the final report as reasonable, balanced, and acceptable for submission to the President.

The Commission's recommendations fall generally into three categories:

- actions the federal government must take;
- actions the owners and operators of the infrastructures must take; and
- actions that require partnership between government and industry.

Information Sharing

During our extensive outreach efforts, we heard time and again that the owners and operators of the infrastructures need more information about cyber threats. They also said that a trusted environment must be built so that they can freely exchange information with each other and with government without fear of regulation, loss of public confidence, incurred liability, or damaged reputation.

The Commission's recommendations lay the foundation for creating a new collaborative environment that includes a two-way exchange of information, not more burdensome regulation.

Our recommendations focus on protecting proprietary information and ensuring anonymity when necessary; reviewing legal impediments to information sharing, such as antitrust provisions and the Freedom of Information Act; and creating information sharing mechanisms both within industry and between industry and government.

Federal Assistance

As to actions the government should take, we recommend specific steps to ensure owners and operators and state and local governments are sufficiently informed and supported to accomplish their infrastructures protection roles, to include:

- Designated federal agencies continuing and expanding the availability of risk assessment services to the private sector, and
- Encouraging industry—and assisting when necessary—to develop risk methodologies.

- The US Security Policy Board should study and recommend how best to protect specific private sector information on threats and vulnerabilities to critical infrastructures.
- The funds appropriated under the Nunn-Lugar-Domenici domestic preparedness program should be doubled to expand and accelerate sharing of capabilities to mitigate the effects of weapons of mass destruction attacks.

Education and awareness

Key to the success of these initiatives is educating our citizens about the emerging threats and vulnerabilities in the cyber dimension. The culture has changed, and our way of thinking about technology and the resulting threats and vulnerabilities must also change.

The Commission's recommendations are aimed at all levels of education, from grammar to graduate school and beyond. They include:

- A series of White House conferences to spur new curricula in computer ethics and intellectual property for elementary and secondary schools.
- A nationwide public awareness campaign, simulations, and Round Table discussions to educate the general public as well as industry and government leaders.
- Partnership between the Department of Education and industry to develop curricula and market demand for properly-trained information security technicians and managers.

Leading by example

Infrastructure assurance is a joint responsibility, but the federal government has an unmistakable duty to lead the effort. Clearly, the federal government must lead by example as it exhorts the private sector and state and local governments to raise the level of security of their systems.

The federal government must pursue the tools, practices, and policies required to conduct business in the cyber age. This includes:

- Improving government information security through developing, implementing, and enforcing best practices and standards—and then conducting certification and measures against those standards.
- Working with industry to expedite efforts for alternative information security and encryption key management pilot programs.
- Elevating and formalizing Information Assurance as a foreign intelligence priority.

- Recruiting and retaining adequate numbers of law enforcement personnel with cyber skills.
- Conducting a thorough risk assessment of the National Aerospace System and the planned sole reliance on the Global Positioning System.

Legal Initiatives

We examined a full range of legal issues relating to protecting the critical infrastructures with three goals in mind:

- increasing the effectiveness of government's protection efforts;
- enhancing the private sector's ability to protect itself; and
- enabling effective public-private partnership where most needed.

We propose the further review of major federal legislation as it relates to the critical infrastructures and the cyber threat.

We have modest recommendations in the area of criminal law and procedure —specifically the Federal Sentencing Guidelines—to take into account the true harm done by attacks on the critical infrastructures.

We call for an expert study group—representing labor, management, government, and privacy interests—to make recommendations for long-term reform in the employer-employee relationship, yet balances security and privacy.

We recommend easing legal impediments to information sharing such as antitrust provisions, federal and private liability, and the Freedom of Information Act.

Research and Development (R&D)

Federal research and development efforts are inadequate to meet the challenge presented by emerging cyber threats. About $250 million is spent each year on infrastructure assurance-related R&D, of which 60 percent—$150 million—is dedicated to information security. There is very little research supporting a national cyber defense. The Commission believes that real-time detection, identification, and response tools are urgently needed, and we concluded that market forces are currently insufficient to meet these needs.

Thus we recommend doubling federal R&D funding for infrastructure protection to $500 million the first year, with 20% increases each year for the next five years. We recommend this funding target:

- Risk management, simulation and modeling, and decision support;

- Contingency planning, incident response, and recovery;
- Information assurance, vulnerability assessment, and system analysis; and
- Early warning and response, monitoring, and threat detection.

Building the Partnership

I need to talk a little about how the federal government and industry can work together to address infrastructure protection concerns. It might be easiest if I first explain a little about our methodology before jumping right into our partnering recommendations.

First, the Commission identified five general functions that are the foundation of infrastructure protection and assurance efforts:

- policy formulation
- prevention and mitigation
- operational warning
- incident management
- consequence management

Next, we fleshed them out to include all the tasks that must be performed to assure our infrastructures.

We knew that a great many people and organizations needed to accomplish these tasks, but we were not sure who or where.

We devised a framework or matrix to help determine who should be responsible for each task.

Along the top of this matrix, roles range from the purely public to the purely private. Along the side, roles range from decentralized to centralized. The top left quadrant, for example, is the role of the federal government—centralized and public. The bottom right quadrant is the place for individual companies—decentralized and private.

Using this framework, we plotted the specific tasks of infrastructure assurance where we thought they should be performed.

The result is a high concentration in the four distinct quadrants, but also a high concentration along the borders. It was the concentrations along the borders that gave us pause, for these are the functions that require a new awareness, a new way of doing business, and a new partnership.

We next looked at how infrastructure assurance is being performed today. There are many players in this game, including the privately-owned infrastructures as well as federal, state, and local governments. There are also a great many existing relationships—such as regulating or enforcing laws—but there are no specific relationships for infrastructure protection and assurance. We focused on bridging this gap.

And this is how we propose to facilitate the public-private partnership—how to bridge the gap—to best protect our infrastructures.

At the policy-making level, we recommend:

- an **Office of National Infrastructure Assurance**—located within the White House—to serve as the federal government's focal point for infrastructure protection;
- a **National Infrastructure Assurance Council** comprised of selected infrastructure CEOs and Cabinet officials to propose policy and advise the President; and
- an **Infrastructure Assurance Support Office** to support both the Council and the National Office.

At the operational level, we recommend:

- **Sector Infrastructure Assurance Coordinators** or clearinghouses as focal points within each infrastructure to share information;
- federal **Lead Agencies** to promote and assist in establishing the sector coordinators;
- an **Information Sharing and Analysis Center** staffed by both private industry and government to receive and share information about infrastructure intrusions to be located in the private sector; and
- a **Warning Center** designed to provide operational warning whenever possible of an attack on the infrastructures, either physical or cyber, located within the FBI.

Conclusion

Just as the risks are shared between the public and the private sectors, so will the solutions be found. Our national and economic security has become a shared responsibility—one that will require a new kind of partnership between government and industry—one which encourages information sharing and one which requires the government to lead by example.

I believe the findings and conclusions of the Commission are based on accurate and reasonable information and analyses. Our recommendations, if implemented, will create the partnerships and structures essential to reducing vulnerabilities in our infrastructures. They will provide the impetus for research and development efforts to increase information security and provide the a cyber defense system. They will increase the nation's ability to prepare, protect, and respond to any threats, strategic or otherwise, directed against our infrastructures, thereby ensuring their continued, effective operation in support of our defense, economic growth, and general well being.

This completes my statement, Mr. Chairman. I will be pleased to answer any questions you or your colleagues may have.

CHAPTER 3

WHITE PAPER

THE CLINTON ADMINISTRATION'S POLICY ON CRITICAL INFRASTRUCTURE PROTECTION: PRESIDENTIAL DECISION DIRECTIVE 63

May 1998

This White Paper explains key elements of the Clinton Administration's policy on critical infrastructure protection. It is intended for dissemination to all interested parties in both the private and public sectors. It will also be used in U.S. Government professional education institutions, such as the National Defense University and the National Foreign Affairs Training Center, for coursework and exercises on interagency practices and procedures. Wide dissemination of this unclassified White Paper is encouraged by all agencies of the U.S. Government.

I. A GROWING POTENTIAL VULNERABILITY

The United States possesses both the world's strongest military and its largest national economy. Those two aspects of our power are mutually reinforcing and dependent. They are also increasingly reliant upon certain critical infrastructures and upon cyber-based information systems.

Critical infrastructures are those physical and cyber-based systems essential to the minimum operations of the economy and government. They include, but are not limited to, telecommunications, energy, banking and finance, transportation, water systems and emergency services, both governmental and private. Many of the nation's critical infrastructures have historically been physically and logically separate systems that had little interdependence. As a result of advances in information technology and the necessity of improved efficiency, however, these infrastructures have become increasingly automated and interlinked. These same advances have created new vulnerabilities to equipment failures, human error, weather and other natural causes, and physical and cyber attacks. Addressing these vulnerabilities will necessarily require flexible, evolutionary approaches that span both the public and private sectors, and protect both domestic and international security.

Because of our military strength, future enemies, whether nations, groups or individuals, may seek to harm us in non-traditional ways including attacks within the United States. Our economy is increasingly reliant upon interdependent and cyber-supported infrastructures and non-traditional attacks on our infrastructure and information systems may be capable of significantly harming both our military power and our economy.

II. PRESIDENT'S INTENT

It has long been the policy of the United States to assure the continuity and viability of critical infrastructures. President Clinton intends that the United States will take all necessary measures to swiftly eliminate any significant vulnerability to both physical and cyber attacks on our critical infrastructures, including especially our cyber systems.

III. A NATIONAL GOAL

No later than the year 2000, the United States shall have achieved an initial operating capability and no later than five years from the day the President signed Presidential Decision Directive 63 the United States shall have achieved and shall maintain the ability to protect our nation's critical infrastructures from intentional acts that would significantly diminish the abilities of:

- the Federal Government to perform essential national security missions and to ensure the general public health and safety;
- state and local governments to maintain order and to deliver minimum essential public services;
- the private sector to ensure the orderly functioning of the economy and the delivery of essential telecommunications, energy, financial and transportation services.

Any interruptions or manipulations of these critical functions must be brief, infrequent, manageable, geographically isolated and minimally detrimental to the welfare of the United States.

IV. A PUBLIC-PRIVATE PARTNERSHIP TO REDUCE VULNERABILITY

Since the targets of attacks on our critical infrastructure would likely include both facilities in the economy and those in the government, the elimination of our potential vulnerability requires a closely coordinated effort of both the public and the private sector. To succeed, this partnership must be genuine, mutual and cooperative. In seeking to meet our national goal to eliminate the vulnerabilities of our critical infrastructure, therefore, the U. S. government should, to the extent feasible, seek to avoid outcomes that increase government regulation or expand unfunded government mandates to the private sector.

For each of the major sectors of our economy that are vulnerable to infrastructure attack, the Federal Government will appoint from a designated Lead Agency a senior officer of that agency as the Sector Liaison Official to work with the private sector. Sector Liaison Officials, after discussions and coordination with private sector entities of their infrastructure sector, will identify a private sector counterpart (Sector Coordinator) to represent their sector.

Together these two individuals and the departments and corporations they represent shall contribute to a sectoral National Infrastructure Assurance Plan by:

- assessing the vulnerabilities of the sector to cyber or physical attacks;
- recommending a plan to eliminate significant vulnerabilities;
- proposing a system for identifying and preventing attempted major attacks;
- developing a plan for alerting, containing and rebuffing an attack in progress and then, in coordination with FEMA as appropriate, rapidly reconstituting minimum essential capabilities in the aftermath of an attack.

During the preparation of the sectoral plans, the National Coordinator (see section VI), in conjunction with the Lead Agency Sector Liaison Officials and a representative from the National Economic Council, shall ensure their overall coordination and the integration of the various sectoral plans, with a particular focus on interdependencies.

V. GUIDELINES

In addressing this potential vulnerability and the means of eliminating it, President Clinton wants those involved to be mindful of the following general principles and concerns.

- We shall consult with, and seek input from, the Congress on approaches and programs to meet the objectives set forth in this directive.
- The protection of our critical infrastructures is necessarily a shared responsibility and partnership between owners, operators and the government. Furthermore, the Federal Government shall encourage international cooperation to help manage this increasingly global problem.
- Frequent assessments shall be made of our critical infrastructures' existing reliability, vulnerability and threat environment because, as technology and the nature of the threats to our critical infrastructures will continue to change rapidly, so must our protective measures and responses be robustly adaptive.

- The incentives that the market provides are the first choice for addressing the problem of critical infrastructure protection; regulation will be used only in the face of a material failure of the market to protect the health, safety or well-being of the American people. In such cases, agencies shall identify and assess available alternatives to direct regulation, including providing economic incentives to encourage the desired behavior, or providing information upon which choices can be made by the private sector. These incentives, along with other actions, shall be designed to help harness the latest technologies, bring about global solutions to international problems, and enable private sector owners and operators to achieve and maintain the maximum feasible security.
- The full authorities, capabilities and resources of the government, including law enforcement, regulation, foreign intelligence and defense preparedness shall be available, as appropriate, to ensure that critical infrastructure protection is achieved and maintained.
- Care must be taken to respect privacy rights. Consumers and operators must have confidence that information will be handled accurately, confidentially and reliably.
- The Federal Government shall, through its research, development and procurement, encourage the introduction of increasingly capable methods of infrastructure protection.
- The Federal Government shall serve as a model to the private sector on how infrastructure assurance is best achieved and shall, to the extent feasible, distribute the results of its endeavors.
- We must focus on preventative measures as well as threat and crisis management. To that end, private sector owners and operators should be encouraged to provide maximum feasible security for the infrastructures they control and to provide the government necessary information to assist them in that task. In order to engage the private sector fully, it is preferred that participation by owners and operators in a national infrastructure protection system be voluntary.
- Close cooperation and coordination with state and local governments and first responders is essential for a robust and flexible infrastructure protection program. All critical infrastructure protection plans and actions shall take into consideration the needs, activities and responsibilities of state and local governments and first responders.

VI. STRUCTURE AND ORGANIZATION

The Federal Government will be organized for the purposes of this endeavor around four components (elaborated in Annex A).

1. **Lead Agencies for Sector Liaison**: For each infrastructure sector that could be a target for significant cyber or physical attacks, there will be a single U.S. Government department which will serve as the lead agency for liaison. Each Lead Agency will designate one individual of Assistant Secretary rank or higher to be the Sector Liaison Official for that area and to cooperate with the private sector representatives (Sector Coordinators) in addressing problems related to critical infrastructure protection and, in particular, in recommending components of the National Infrastructure Assurance Plan. Together, the Lead Agency and the private sector counterparts will develop and implement a Vulnerability Awareness and Education Program for their sector.

2. **Lead Agencies for Special Functions**: There are, in addition, certain functions related to critical infrastructure protection that must be chiefly performed by the Federal Government (national defense, foreign affairs, intelligence, law enforcement). For each of those special functions, there shall be a Lead Agency which will be responsible for coordinating all of the activities of the United States Government in that area. Each lead agency will appoint a senior officer of Assistant Secretary rank or higher to serve as the Functional Coordinator for that function for the Federal Government.

3. **Interagency Coordination**: The Sector Liaison Officials and Functional Coordinators of the Lead Agencies, as well as representatives from other relevant departments and agencies, including the National Economic Council, will meet to coordinate the implementation of this directive under the auspices of a Critical Infrastructure Coordination Group (CICG), chaired by the National Coordinator for Security, Infrastructure Protection and Counter-Terrorism. The National Coordinator will be appointed by and report to the President through the Assistant to the President for National Security Affairs, who shall assure appropriate coordination with the Assistant to the President for Economic Affairs. Agency representatives to the CICG should be at a senior policy level (Assistant Secretary or higher). Where appropriate, the CICG will be assisted by extant policy structures, such as the Security Policy Board, Security Policy Forum and the National Security and Telecommunications and Information System Security Committee.

4. **National Infrastructure Assurance Council**: On the recommendation of the Lead Agencies, the National Economic Council and the National Coordinator, the President will appoint a panel of major infrastructure providers and state and local government officials to serve as the Na-

tional Infrastructure Assurance Council. The President will appoint the Chairman. The National Coordinator will serve as the Council's Executive Director. The National Infrastructure Assurance Council will meet periodically to enhance the partnership of the public and private sectors in protecting our critical infrastructures and will provide reports to the President as appropriate. Senior Federal Government officials will participate in the meetings of the National Infrastructure Assurance Council as appropriate.

VII. PROTECTING FEDERAL GOVERNMENT CRITICAL INFRASTRUCTURES

Every department and agency of the Federal Government shall be responsible for protecting its own critical infrastructure, especially its cyber-based systems. Every department and agency Chief Information Officer (CIO) shall be responsible for information assurance. Every department and agency shall appoint a Chief Infrastructure Assurance Officer (CIAO) who shall be responsible for the protection of all of the other aspects of that department's critical infrastructure. The CIO may be double-hatted as the CIAO at the discretion of the individual department. These officials shall establish procedures for obtaining expedient and valid authorizations to allow vulnerability assessments to be performed on government computer and physical systems. The Department of Justice shall establish legal guidelines for providing for such authorizations.

No later than 180 days from issuance of this directive, every department and agency shall develop a plan for protecting its own critical infrastructure, including but not limited to its cyber-based systems. The National Coordinator shall be responsible for coordinating analyses required by the departments and agencies of inter-governmental dependencies and the mitigation of those dependencies. The Critical Infrastructure Coordination Group (CICG) shall sponsor an expert review process for those plans. No later than two years from today, those plans shall have been implemented and shall be updated every two years. In meeting this schedule, the Federal Government shall present a model to the private sector on how best to protect critical infrastructure.

VIII. TASKS

Within 180 days, the Principals Committee should submit to the President a schedule for completion of a National Infrastructure Assurance Plan with milestones for accomplishing the following subordinate and related tasks.

1. **Vulnerability Analyses**: For each sector of the economy and each sector of the government that might be a target of infrastructure attack intended to significantly damage the United States, there shall be an

initial vulnerability assessment, followed by periodic updates. As appropriate, these assessments shall also include the determination of the minimum essential infrastructure in each sector.

2. **Remedial Plan**: Based upon the vulnerability assessment, there shall be a recommended remedial plan. The plan shall identify timelines for implementation, responsibilities and funding.

3. **Warning**: A national center to warn of significant infrastructure attacks will be established immediately (see Annex A). As soon thereafter as possible, we will put in place an enhanced system for detecting and analyzing such attacks, with maximum possible participation of the private sector.

4. **Response**: A system shall develop a system for responding to a significant infrastructure attack while it is underway, with the goal of isolating and minimizing damage.

5. **Reconstitution**: For varying levels of successful infrastructure attacks, we shall have a system to reconstitute minimum required capabilities rapidly.

6. **Education and Awareness**: There shall be Vulnerability Awareness and Education Programs within both the government and the private sector to sensitize people regarding the importance of security and to train them in security standards, particularly regarding cyber systems.

7. **Research and Development**: Federally-sponsored research and development in support of infrastructure protection shall be coordinated, be subject to multi-year planning, take into account private sector research, and be adequately funded to minimize our vulnerabilities on a rapid but achievable timetable.

8. **Intelligence**: The Intelligence Community shall develop and implement a plan for enhancing collection and analysis of the foreign threat to our national infrastructure, to include but not be limited to the foreign cyber/information warfare threat.

9. **International Cooperation**: There shall be a plan to expand cooperation on critical infrastructure protection with like-minded and friendly nations, international organizations and multinational corporations.

10. **Legislative and Budgetary Requirements**: There shall be an evaluation of the executive branch's legislative authorities and budgetary priorities regarding critical infrastructure, and ameliorative recommendations shall be made to the President as necessary. The evaluations and recommendations, if any, shall be coordinated with the Director of OMB.

The CICG shall also review and schedule the taskings listed in Annex B.

IX. IMPLEMENTATION

In addition to the 180-day report, the National Coordinator, working with the National Economic Council, shall provide an annual report on the implementation of this directive to the President and the heads of departments and agencies, through the Assistant to the President for National Security Affairs. The report should include an updated threat assessment, a status report on achieving the milestones identified for the National Plan and additional policy, legislative and budgetary recommendations. The evaluations and recommendations, if any, shall be coordinated with the Director of OMB. In addition, following the establishment of an initial operating capability in the year 2000, the National Coordinator shall conduct a zero-based review.

ANNEX A: STRUCTURE AND ORGANIZATION

Lead Agencies: Clear accountability within the U.S. Government must be designated for specific sectors and functions. The following assignments of responsibility will apply.

Lead Agencies for Sector Liaison

Commerce	Information and communications
Treasury	Banking and finance
EPA	Water supply
Transportation	Aviation
Highways	(including trucking and intelligent transportation systems)
	Mass transit
	Pipelines
	Rail
	Waterborne commerce
Justice/FBI	Emergency law enforcement services
FEMA	Emergency fire service Continuity of government services
HHS	Public health services, including prevention, surveillance, laboratory services and personal health services
Energy	Electric power Oil and gas production and storage

Lead Agencies for Special Functions

Justice/FBI	Law enforcement and internal security
CIA	Foreign intelligence
State	Foreign affairs
Defense	National defense

In addition, OSTP shall be responsible for coordinating research and development agendas and programs for the government through the National Science and Technology Council. Furthermore, while Commerce is the lead agency for information and communication, the Department of Defense will retain its Executive Agent responsibilities for the National Communications System and support of the President's National Security Telecommunications Advisory Committee.

National Coordinator: The National Coordinator for Security, Infrastructure Protection and Counter-Terrorism shall be responsible for coordinating the implementation of this directive. The National Coordinator will report to the President through the Assistant to the President for National Security Affairs. The National Coordinator will also participate as a full member of Deputies or Principals Committee meetings when they meet to consider infrastructure issues. Although the National Coordinator will not direct Departments and Agencies, he or she will ensure interagency coordination for policy development and implementation, and will review crisis activities concerning infrastructure events with significant foreign involvement. The National Coordinator will provide advice, in the context of the established annual budget process, regarding agency budgets for critical infrastructure protection. The National Coordinator will chair the Critical Infrastructure Coordination Group (CICG), reporting to the Deputies Committee (or, at the call of its chair, the Principals Committee). The Sector Liaison Officials and Special Function Coordinators shall attend the CICG's meetings. Departments and agencies shall each appoint to the CICG a senior official (Assistant Secretary level or higher) who will regularly attend its meetings. The National Security Advisor shall appoint a Senior Director for Infrastructure Protection on the NSC staff.

A National Plan Coordination (NPC) staff will be contributed on a non-reimbursable basis by the departments and agencies, consistent with law. The NPC staff will integrate the various sector plans into a National Infrastructure Assurance Plan and coordinate analyses of the U.S. Government's own dependencies on critical infrastructures. The NPC staff will also help coordinate a national education and awareness program, and legislative and public affairs.

The Defense Department shall continue to serve as Executive Agent for the Commission Transition Office, which will form the basis of the NPC, during the remainder of FY98. Beginning in FY99, the NPC shall be an office of the Commerce Department. The Office of Personnel Management shall provide the necessary assistance in facilitating the NPC's operations. The NPC will terminate at the end of FY01, unless extended by Presidential directive.

Warning and Information Centers

As part of a national warning and information sharing system, the President immediately authorizes the FBI to expand its current organization to a full scale **Nationai Infrastructure Protection Center** (NIPC). This organization shall serve as a national critical infrastructure threat assessment, warning, vulnerability, and law enforcement investigation and response entity. During the initial period of six to twelve months, the President also directs the National Coordinator and the Sector Liaison Officials, working together with the Sector Coordinators, the Special Function Coordinators and representatives from the National Economic Council, as appropriate, to consult with owners and operators of the critical infrastructures to encourage the creation of a private sector sharing and analysis center, as described below.

National Infrastructure Protection Center (NIPC): The NIPC will include FBI, USSS, and other investigators experienced in computer crimes and infrastructure protection, as well as representatives detailed from the Department of Defense, the Intelligence Community and Lead Agencies. It will be linked electronically to the rest of the Federal Government, including other warning and operations centers, as well as any private sector sharing and analysis centers. Its mission will include providing timely warnings of intentional threats, comprehensive analyses and law enforcement investigation and response.

All executive departments and agencies shall cooperate with the NIPC and provide such assistance, information and advice that the NIPC may request, to the extent permitted by law. All executive departments shall also share with the NIPC information about threats and warning of attacks and about actual attacks on critical government and private sector infrastructures, to the extent permitted by law. The NIPC will include elements responsible for warning, analysis, computer investigation, coordinating emergency response, training, outreach and development and application of technical tools. In addition, it will establish its own relations directly with others in the private sector and with any information sharing and analysis entity that the private sector may create, such as the Information Sharing and Analysis Center described below.

The NIPC, in conjunction with the information originating agency, will sanitize law enforcement and intelligence information for inclusion into analyses and reports that it will provide, in appropriate form, to relevant federal, state and local agencies; the relevant owners and operators of critical infrastructures; and to any private sector information sharing and analysis entity. Before disseminating national security or other information that originated from the intelligence community, the NIPC will coordinate fully with the intelligence community through existing procedures. Whether as sanitized or unsanitized reports, the NIPC will issue attack warnings or alerts to increases in threat condition to any private sector information sharing and analysis entity and to the owners and operators. These warnings may also include guidance regarding additional protection measures to be taken by owners and operators. Except in extreme emergencies, the NIPC shall coordinate with the National Coordinator before issuing public warnings of imminent attacks by international terrorists, foreign states or other malevolent foreign powers.

The NIPC will provide a national focal point for gathering information on threats to the infrastructures. Additionally, the NIPC will provide the principal means of facilitating and coordinating the Federal Government's response to an incident, mitigating attacks, investigating threats and monitoring reconstitution efforts. Depending on the nature and level of a foreign threat/attack, protocols established between special function agencies (DOJ/DOD/CIA), and the ultimate decision of the President, the NIPC may be placed in a direct support role to either DOD or the Intelligence Community.

Information Sharing and Analysis Center (ISAC): The National Coordinator, working with Sector Coordinators, Sector Liaison Officials and the National Economic Council, shall consult with owners and operators of the critical infrastructures to strongly encourage the creation of a private sector information sharing and analysis center. The actual design and functions of the center and its relation to the NIPC will be determined by the private sector, in consultation with and with assistance from the Federal Government. Within 180 days of this directive, the National Coordinator, with the assistance of the CICG including the National Economic Council, shall identify possible methods of providing federal assistance to facilitate the startup of an ISAC.

Such a center could serve as the mechanism for gathering, analyzing, appropriately sanitizing and disseminating private sector information to both industry and the NIPC. The center could also gather, analyze and disseminate information from the NIPC for further distribution to the private sector. While crucial to a successful government-industry partnership, this mechanism for sharing important information about vulnerabilities, threats, intrusions and anomalies is not to interfere with direct information exchanges between companies and the government.

As ultimately designed by private sector representatives, the ISAC may emulate particular aspects of such institutions as the Centers for Disease

Control and Prevention that have proved highly effective, particularly its extensive interchanges with the private and non-federal sectors. Under such a model, the ISAC would possess a large degree of technical focus and expertise and non-regulatory and non-law enforcement missions. It would establish baseline statistics and patterns on the various infrastructures, become a clearinghouse for information within and among the various sectors, and provide a library for historical data to be used by the private sector and, as deemed appropriate by the ISAC, by the government. Critical to the success of such an institution would be its timeliness, accessibility, coordination, flexibility, utility and acceptability.

ANNEX B: ADDITIONAL TASKINGS

Studies

The National Coordinator shall commission studies on the following subjects:

- Liability issues arising from participation by private sector companies in the information sharing process.
- Existing legal impediments to information sharing, with an eye to proposals to remove these impediments, including through the drafting of model codes in cooperation with the American Legal Institute.
- The necessity of document and information classification and the impact of such classification on useful dissemination, as well as the methods and information systems by which threat and vulnerability information can be shared securely while avoiding disclosure or unacceptable risk of disclosure to those who will misuse it.
- The improved protection, including secure dissemination and information handling systems, of industry trade secrets and other confidential business data, law enforcement information and evidentiary material, classified national security information, unclassified material disclosing vulnerabilities of privately owned infrastructures and apparently innocuous information that, in the aggregate, it is unwise to disclose.
- The implications of sharing information with foreign entities where such sharing is deemed necessary to the security of United States infrastructures.
- The potential benefit to security standards of mandating, subsidizing, or otherwise assisting in the provision of insurance for selected critical infrastructure providers and requiring insurance tie-ins for foreign critical infrastructure providers hoping to do business with the United States.

Public Outreach

In order to foster a climate of enhanced public sensitivity to the problem of infrastructure protection, the following actions shall be taken:

- The White House, under the oversight of the National Coordinator, together with the relevant Cabinet agencies shall consider a series of conferences: (1)that will bring together national leaders in the public and private sectors to propose programs to increase the commitment to information security; (2)that convoke academic leaders from engineering, computer science, business and law schools to review the status of education in information security and will identify changes in the curricula and resources necessary to meet the national demand for professionals in this field; (3)on the issues around computer ethics as these relate to the K through 12 and general university populations.
- The National Academy of Sciences and the National Academy of Engineering shall consider a round table bringing together federal, state and local officials with industry and academic leaders to develop national strategies for enhancing infrastructure security.
- The intelligence community and law enforcement shall expand existing programs for briefing infrastructure owners and operators and senior government officials.
- The National Coordinator shall (1) establish a program for infrastructure assurance simulations involving senior public and private officials, the reports of which might be distributed as part of an awareness campaign; and (2) in coordination with the private sector, launch a continuing national awareness campaign, emphasizing improving infrastructure security.

Internal Federal Government Actions

In order for the Federal Government to improve its infrastructure security, these immediate steps shall be taken:

- The Department of Commerce, the General Services Administration, and the Department of Defense shall assist federal agencies in the implementation of best practices for information assurance within their individual agencies.
- The National Coordinator shall coordinate a review of existing federal, state and local bodies charged with information assurance tasks, and provide recommendations on how these institutions can cooperate most effectively.
- All federal agencies shall make clear designations regarding who may authorize access to their computer systems.

- The Intelligence Community shall elevate and formalize the priority for enhanced collection and analysis of information on the foreign cyber/information warfare threat to our critical infrastructure.
- The Federal Bureau of Investigation, the Secret Service and other appropriate agencies shall: (1)vigorously recruit undergraduate and graduate students with the relevant computer-related technical skills for full-time employment as well as for part-time work with regional computer crime squads; and (2)facilitate the hiring and retention of qualified personnel for technical analysis and investigation involving cyber attacks.
- The Department of Transportation, in consultation with the Department of Defense, shall undertake a thorough evaluation of the vulnerability of the national transportation infrastructure that relies on the Global Positioning System. This evaluation shall include sponsoring an independent, integrated assessment of risks to civilian users of GPS-based systems, with a view to basing decisions on the ultimate architecture of the modernized NAS on these evaluations.
- The Federal Aviation Administration shall develop and implement a comprehensive National Airspace System Security Program to protect the modernized NAS from information-based and other disruptions and attacks.
- GSA shall identify large procurements (such as the new Federal Telecommunications System, FTS 2000) related to infrastructure assurance, study whether the procurement process reflects the importance of infrastructure protection and propose, if necessary, revisions to the overall procurement process to do so.
- OMB shall direct federal agencies to include assigned infrastructure assurance functions within their Government Performance and Results Act strategic planning and performance measurement framework.
- The NSA, in accordance with its National Manager responsibilities in NSD-42, shall provide assessments encompassing examinations of U.S. Government systems to interception and exploitation; disseminate threat and vulnerability information; establish standards; conduct research and development; and conduct issue security product evaluations.

Assisting the Private Sector

In order to assist the private sector in achieving and maintaining infrastructure security:

- The National Coordinator and the National Infrastructure Assurance Council shall propose and develop ways to encourage private industry to perform periodic risk assessments of critical processes, including information and telecommunications systems.
- The Department of Commerce and the Department of Defense shall work together, in coordination with the private sector, to offer their expertise to private owners and operators of critical infrastructure to develop security-related best practice standards.
- The Department of Justice and Department of the Treasury shall sponsor a comprehensive study compiling demographics of computer crime, comparing state approaches to computer crime and developing ways of deterring and responding to computer crime by juveniles.

CHAPTER 4

PROTECTING AMERICA'S CRITICAL INFRASTRUCTURES: PDD 63

THE WHITE HOUSE
Office of the Press Secretary
May 22, 1998

This Presidential Directive builds on the recommendations of the President's Commission on Critical Infrastructure Protection. In October 1997 the Commission issued its report, calling for a national effort to assure the security of the United States' increasingly vulnerable and interconnected infrastructures, such as telecommunications, banking and finance, energy, transportation, and essential government services.

Presidential Decision Directive 63 is the culmination of an intense, interagency effort to evaluate those recommendations and produce a workable and innovative framework for critical infrastructure protection. The President's policy:

- Sets a goal of a reliable, interconnected, and secure information system infrastructure by the year 2003, and significantly increased security for government systems by the year 2000, by:
    - Immediately establishing a national center to warn of and respond to attacks.
    - Building the capability to protect critical infrastructures from intentional acts by 2003.
- Addresses the cyber and physical infrastructure vulnerabilities of the Federal government by requiring each department and agency to work to reduce its exposure to new threats;
- Requires the Federal government to serve as a model to the rest of the country for how infrastructure protection is to be attained;
- Seeks the voluntary participation of private industry to meet common goals for protecting our critical systems through public-private partnerships;
- Protects privacy rights and seeks to utilize market forces. It is meant to strengthen and protect the nation's economic power, not to stifle it.
- Seeks full participation and input from the Congress.

PDD-63 sets up a new structure to deal with this important challenge:

- a *National Coordinator* whose scope will include not only critical infrastructure but also foreign terrorism and threats of domestic mass destruction (including biological weapons) because attacks on the US may not come labeled in neat jurisdictional boxes;
- The *National Infrastructure Protection Center* (NIPC) at the FBI which will fuse representatives from FBI, DOD, USSS, Energy, Transportation, the Intelligence Community, and the private sector in an unprecedented attempt at information sharing among agencies in collaboration with the private sector. The NIPC will also provide the principal means of facilitating and coordinating the Federal Government's response to an incident, mitigating attacks, investigating threats and monitoring reconstitution efforts;
- An *Information Sharing and Analysis Center* (ISAC) is encouraged to be set up by the private sector, in cooperation with the federal government;
- A *National Infrastructure Assurance Council* drawn from private sector leaders and state/local officials to provide guidance to the policy formulation of a National Plan;
- The *Critical Infrastructure Assurance Office* will provide support to the National Coordinator's work with government agencies and the private sector in developing a national plan. The office will also help coordinate a national education and awareness program, and legislative and public affairs.

For more detailed information on this Presidential Decision Directive, contact the Critical Infrastructure Assurance Office at (703) 696-9395 for copies of the White Paper on Critical Infrastructure Protection.

CHAPTER 5

SUMMARY OF PRESIDENTIAL DECISION DIRECTIVES 62 and 63

THE WHITE HOUSE
Office of the Press Secretary
May 22, 1998

President Clinton today ordered the strengthening of the nation's defenses against emerging unconventional threats to the United States: terrorist acts, use of weapons of mass destruction, assaults on our critical infrastructures and cyber-attacks.

The Combating Terrorism directive (PDD-62) highlights the growing threat of unconventional attacks against the United States. It details a new and more systematic approach to fighting terrorism by bringing a program management approach to U.S. counter-terrorism efforts.

The directive also establishes the office of the National Coordinator for Security, Infrastructure Protection and Counter-Terrorism which will oversee a broad variety of relevant policies and programs including areas such as counter-terrorism, protection of critical infrastructure, preparedness and consequence management for weapons of mass destruction.

The Critical Infrastructure Protection directive (PDD-63) calls for a national effort to assure the security of the increasingly vulnerable and interconnected infrastructures of the United States. Such infrastructures include telecommunications, banking and finance, energy, transportation, and essential government services. The directive requires immediate federal government action including risk assessment and planning to reduce exposure to attack. It stresses the critical importance of cooperation between the government and the private sector by linking designated agencies with private sector representatives.

For more detailed information on Presidential Decision Directive 63, contact the Critical Infrastructure Assurance Office at (703) 696-9395 for copies of the White Paper on Critical Infrastructure Protection.

CHAPTER 6

THE PRESIDENT'S NATIONAL SECURITY TELECOMMUNICATIONS ADVISORY COMMITTEE

INFORMATION INFRASTRUCTURE GROUP REPORT

September 1998

TABLE OF CONTENTS

EXECUTIVE SUMMARY

1.0 INTRODUCTION

2.0 CHARGE

3.0 ACTIVITIES

- 3.1 Transportation Risk Assessment
- 3.2 Electronic Commerce/Cyber Security
  - 3.2.1 Analysis
  - 3.2.2 Conclusions
  - 3.2.3 Recommendations

INFORMATION INFRASTRUCTURE GROUP MEMBERS–ANNEX A

CYBER SECURITY TRAINING AND FORENSICS ISSUE PAPER–ANNEX B

EXECUTIVE SUMMARY

Since the last meeting of the President's National Security Telecommunications Advisory Committee (NSTAC), December 1997, the Information Infrastructure Group (IIG) has concentrated its efforts on issues related to information assurance, infrastructure protection, electronic commerce, and cyber security. The IIG established two subgroups to investigate these topics, the Transportation Information Infrastructure Risk Assessment Subgroup and the Electronic Commerce (EC)/Cyber Security Subgroup.

The Transportation Information Infrastructure Risk Assessment Subgroup conducted a workshop for the transportation industry on telecommunications and information systems dependencies on September 10, 1997. On the basis of findings from that event, the subgroup submitted an interim report to the December 1997 NSTAC XX meeting. The report recommended that more information be gathered, particularly in the area of intermodal transportation, and concluded that broader participation from the transportation industry was desirable. The subgroup anticipates completing the risk assessment in the beginning of the NSTAC XXII cycle.

The EC/Cyber Security Subgroup was established in response to a briefing the Deputy Secretary of Defense gave at the December 1997 NSTAC XX meeting. The subgroup agreed to examine national security and emergency preparedness (NS/EP) implications of EC because both industry and Government incorporate EC into their business practices. After meeting with key officials in industry and Government on security issues related to EC, the subgroup developed an issue paper that focused on one aspect of EC_cyber security training and forensics. That paper centers on the importance of industry and Government cooperation in addressing cyber security. The subgroup is also developing further analyses of EC to be completed in preparation for the NSTAC XXII meeting.

The following recommendations result from deliberations and assessments of the group on cyber security issues.

Recommendation to the President

The President should direct the appropriate departments and agencies to continue working with the NSTAC for the development of policies, procedures, techniques, and tools to facilitate joint industry_Government cooperation on cyber security.

Recommendation to the NSTAC

The NSTAC should continue to work with the presidentially directed departments and agencies in the development of policies, procedures, techniques, and tools to facilitate joint industry-Government cooperation on cyber security.

1.0 INTRODUCTION

The President's National Security Telecommunications Advisory Committee (NSTAC) has directly addressed information assurance and information infrastructure protection issues since 1995. At the January 16, 1995, NSTAC XVII meeting, the Director of the National Security Agency discussed with the NSTAC principals threats to U.S. information systems and the need to improve the security of the Nation's critical infrastructures. The NSTAC chair drafted a letter to the President in March of that year, stating that "[the] integrity of the Nation's information systems, both government and public, are increasingly at risk from intrusion and attack. . . [and] other national infrastructures . . . [such as] finance, air traffic control, power, etc., also depend on reliable and secure information systems and could be at risk."[1] In July 1995, President Clinton replied to the NSTAC letter, stating that he would "welcome NSTAC's continuing effort to work with the Administration to counter threats to our Nation's information and telecommunications systems."[2] The President further asked "the NSTAC's principalsA with input from the full range of NII usersA to provide me with your assessment of national security emergency preparedness requirements for our rapidly evolving information infrastructure."[3]

On May 15, 1995, the NSTAC's Industry Executive Subcommittee (IES) established the Information Assurance Task Force (IATF) to cooperate with the U.S. Government in identifying critical national infrastructures, determining their importance to the national interest, and scheduling several elements for assessment. Working with representatives from the national security community, law enforcement, civil departments and agencies, and the private sector, the task force narrowed an initial list of critical infrastructures to three for study electric power, financial services, and transportation. These three infrastructures were selected on the basis of their strong interdependencies and their growing reliance on telecommunications and information systems for performing key functions.

The IATF developed a risk assessment methodology and subsequently formed three risk assessment subgroups to address the distinct characteristics and concerns of each infrastructure. The first assessment was completed by the Electric Power Risk Assessment Subgroup in September 1996, and its report and recommendations were approved by the principals before NSTAC XIX. The Financial Services Risk Assessment Subgroup completed its risk assessment in 1997, and the NSTAC approved its report and recommendations in advance of NSTAC XX. The Transportation Information Infrastructure Risk Assessment Subgroup conducted a risk assessment workshop in September 1997 and delivered its findings in an interim report to NSTAC XX.

Following NSTAC XIX, the IES restructured its organization, and the IATF's activities were incorporated into the newly formed Information Infrastructure Group (IIG). The IIG's charge was also broadened to include, among other things, a consideration of cyber security and crime issues. That set of issues gained focus at NSTAC XIX, where the U.S. Attorney General initiated a dialogue with the NSTAC principals on the need to develop a more cooperative industry-Government approach to cyber crime. The IIG's current charge is outlined below.

---

1 Letter from Mr. William T. Esrey, Sprint Corporation and Chair of the President's NSTAC, to the President of the United States, March 20, 1995.

2 Letter from the President of the United States to the Chair of NSTAC, July 7, 1995.

3 Ibid.

2.0 CHARGE

The IES charged the IIG to serve as a focal point for NSTAC information assurance and infrastructure protection activities along multiple dimensions:

- continue to assess the implications of transportation information assurance risks;
- examine inter-infrastructure protection and coordination issues among the telecommunications, transportation, electric power, and financial services industries;
- consider the national security and emergency preparedness (NS/EP) implications associated with electronic commerce;
- monitor any Information Systems Security Board (ISSB) related activity; and
- stand ready to assist the electrical power industry in forming an NSTAC-like body.

The major IIG activities were completed by two IIG subgroups the Transportation Information Infrastructure Risk Assessment Subgroup and the Electronic Commerce (EC)/Cyber Security Subgroup.

3.0 ACTIVITIES

At NSTAC XIX, the principals recommended that the President endorse a private sector ISSB initiative. Subsequently, the IES charged the IIG to monitor any ISSB-related activities. On the basis of previous outreach efforts by the NSTAC's National Information Infrastructure Task Force, a private sector group called the Information Security Exploratory Committee (ISEC) further investigated issues regarding the establishment of the ISSB. (The Information Technology Industry Council, a private nonprofit association, serves as the host organization for the ISEC.) Since December 1996, the ISEC has met regularly, and it issued a final report in December 1997. The IIG believed that the NSTAC was misrepresented in the final ISEC report and on two occasions notified the ISEC of the misrepresentation. The IIG plans to continue to work with the ISEC to clarify the issue.

In addition, the IIG monitored the progression of the recommendations from the Electric Power Risk Assessment Report and the Financial Services Risk Assessment Report. The IIG's focus on issues regarding the transportation infrastructure and the NS/EP implications associated with electronic commerce are discussed below.

3.1 Transportation Risk Assessment

On September 10, 1997, the Transportation Information Infrastructure Risk Assessment Subgroup conducted a Transportation Risk Assessment Workshop to gather data about the industry's dependency on information technology and telecommunications. The findings of this workshop were included in an interim report presented during NSTAC XX. The subgroup recommended that further discussions with the industry be conducted to gain information from a wider sample of transportation modes, and that future efforts capture data relating to intermodal transportation trends. The subgroup has contacted the Department of Transportation as a result of Presidential Decision Directive 63 (PDD-63), to jointly ensure the success and relevance of the risk assessment. The Transportation Information Infrastructure Risk Assessment Subgroup has developed a strategy to encourage involvement by key associations to ensure strong participation by all transportation modes so this task can be completed during the next NSTAC cycle.

3.2 Electronic Commerce/Cyber Security

3.2.1 Analysis

On December 11, 1997, the Honorable John Hamre, Deputy Secretary of Defense, briefed the 20th meeting of the NSTAC on steps that the Department of Defense (DOD) is taking to prepare for potential cyber attacks against the Nation. In his remarks, Dr. Hamre expressed the intent of the DOD to move forward with a paperless operation, utilizing EC technologies, and state-of-the-art security tools and strategies while expecting organizations that work with the DOD to adhere to the same level of protection. The IIG established the EC/Cyber Security Subgroup to examine any NS/EP implications associated with the movement to EC as the DOD, other Federal agencies, financial institutions, and a growing number of corporations incorporate EC into their business practices. Toward that end, the subgroup has scoped the issue and has received briefings from key officials in Government and industry on security issues related to EC. The subgroup will develop an issue analysis of the NS/EP implications of EC, which will be presented at NSTAC XXII.

In the course of scoping the EC issue, the subgroup determined that cyber crime concerns might be a long-term impediment to widespread use of emerging EC technologies. The subgroup also acknowledged that those concerns were related to issues raised by the NSTAC and the U.S. Attorney General at NSTAC XIX, where a dialogue was initiated on the need to develop a more cooperative industry-Government approach to cyber security. The subgroup agreed that there were emerging and timely issues related to the investigative community's ability to address evolving cyber security concerns. Officials from the National Infrastructure Protection Center, the Federal Bureau of Investigation Computer Analysis and Response Team, and the Department of Justice Computer Crime and Intellectual Property Section met with members of the subgroup to assist them in determining what timely NS/EP recommendations the IIG can make regarding cyber crime investigation. On the basis of findings from those discussions and previous IIG scoping of the cyber security problem, the subgroup produced an issue paper that presents findings and recommendations on cyber security. The issue paper is attached as Annex B.

3.2.2 *Conclusions*

In its investigation of cyber security issues, the EC/Cyber Security Subgroup concluded that a stronger partnership between industry and Government is required for the following reasons:

- The NS/EP implications of cyber security necessitate developing a partnership with industry, law enforcement, national security communities, and PDD-63 lead departments and agencies.
- The size and complexity of the cyber crime caseload, coupled with the growing sophistication and proliferation of computer-based tools used to attack information systems, challenge traditional policies, procedures, and protocols for investigating crimes and collecting evidence.
- Cyber investigators also face the challenge of staying abreast of industry's continuous introduction of new and innovative technologies.
- These new technologies introduce additional vulnerabilities that are being exploited by intruders.
- Industry possesses knowledge, tools, and expertise that investigators could leverage as they respond to increasingly complex and sophisticated cyber crimes.

A key element in this partnership will be establishing the appropriate levels of trust and understanding to spur cooperation in addressing cyber security issues. This partnership will also need to foster a dialogue that allows representatives from industry and investigative communities to develop a unique understanding of the perspectives and issues faced by their counterparts. The Government and NSTAC Network Security Information Exchanges and the National Coordinating Center for Telecommunications are examples of effective partnerships between the telecommunications industry and Government. The principals of cooperation exercised in these forums could be applied in efforts to address cyber crime.

With its years of experience working with the Government on NS/EP telecommunications, network security, and information assurance issues, NSTAC is positioned to build on this experience in developing a partnership with the law enforcement and national security communities to address cyber security issues.

3.2.3 *Recommendations*

Based on the findings of the EC/Cyber Security Subgroup, the IIG proposes the following recommendations.

3.2.3.1 Recommendation to the President

The President should direct the appropriate departments and agencies to continue working with the NSTAC for the development of policies, procedures, techniques, and tools to facilitate joint industry Government cooperation on cyber security.

3.2.3.2 Recommendation to the NSTAC

The NSTAC should continue to work with the presidentially directed departments and agencies in the development of policies, procedures, techniques, and tools to facilitate joint industry-Government cooperation on cyber security.

ANNEX A
INFORMATION INFRASTRUCTURE GROUP MEMBERS

INFORMATION INFRASTRUCTURE GROUP MEMBERS

CSC	Mr. Guy Copeland
Unisys	Dr. Dan Wiener
AT&T	Dr. Larry Nelson
Boeing	Mr. Bob Steele
COMSAT	Mr. Ernie Wallace
EDS	Mr. Bob Donahue
GTE	Mr. Lowell Thomas
MCI	Mr. Michael McPadden
Nortel	Dr. Jack Edwards
Raytheon	Mr. John Grimes
SAIC	Mr. Bernie Ziegler
Sprint	Dr. Sushil Munshi
U S WEST	Mr. Jon Lofstedt

ACTIVE PARTICIPANTS

CSC	Mr. Richard Swanson
CSC	Ms. Deborah Jacobs
EDS	Mr. Richard Parodi
GTE	Ms. Ernie Gormsen
ITT	Mr. Joe Gancie
Lockheed Martin	Ms. Dena Kisala
NTA	Mr. Bob Burns
Raytheon	Mr. Bob Tolhurst
SAIC	Mr. Bill Deaver
SAIC	Ms. Rosemary Dew
TRW	Mr. Bob Lentz
Unisys	Mr. Fred Tompkins
DOT	LCDR Tim Custer
NASA	Mr. Art Sigust
NTIA	Mr. Bill Belote
SAIC	Mr. Hank Kluepfel

ANNEX B
CYBER SECURITY TRAINING AND FORENSICS ISSUE PAPER

1.0 INTRODUCTION

Since 1990, the President's National Security Telecommunications Advisory Committee (NSTAC) has addressed network security issues affecting the U.S. telecommunications infrastructure. Those issues include vulnerabilities introduced into telecommunications networks by new technologies, the growing number of cyber intrusions as networks become more pervasive and open, and the changing threat environment characterized by linkages between national and economic security. National security and emergency preparedness (NS/EP) telecommunications rely heavily on information systems for transmission and network control. Historically, these information systems have been high-value targets for intrusions. Cyber security the act of protecting sensitive computers and networked systems from misuse, exploitation, and intrusion involves initiating a number of policies and tools to protect the Nation's security interests from a growing cyber threat.

At the 19th meeting of the President's NSTAC, Attorney General Janet Reno initiated a dialogue with the NSTAC principals on the need to develop a more cooperative industry Government approach to cyber security. Because of the Nation's increasing dependence on the information systems that support the telecommunications and other critical infrastructures, the NSTAC views cyber security as an important NS/EP telecommunications issue. Subsequent discussions between industry and Government officials focused on issues of mutual concern. Based on these discussions, the following objectives were identified:

- an investigative community composed of intelligence and criminal investigators and prosecutors at the Federal, State, and local level who are sufficiently trained and resourced to respond to the cyber threats of the 21st century,
- an industry that is aware of the potential vulnerabilities in a networked society and is actively taking steps to address cyber security, and
- a partnership that incorporates a mutually beneficial dialogue between industry and the investigative community.

The NSTAC recognizes that industry and the investigative community need a cooperative dialogue in order to ensure that the investigative community has ready access to better tools, techniques, capabilities, and training. This is particularly important as society transitions to an information-based economy. Specifically, the NSTAC's Electronic Commerce/Cyber Security Subgroup determined that cyber crime concerns might be a long-term impediment to widespread adoption and use of emerging electronic commerce technologies.

2.0 THE CHANGING ENVIRONMENT

2.1 National and Economic Security

Proliferation and sophistication of information technologies, coupled with the growing use of the Internet, have dramatically transformed how our society communicates and conducts business. The use of the Internet is rapidly growing. The number of Americans using the Internet has grown from less than 5 million in 1993 to nearly 62 million in 1997.[4] UUNET estimates that Internet traffic doubles every 100 days.[5] Corporations are looking toward the Internet as a tool for commerce. Many Government agencies, including the Department of Defense (DOD), are beginning to use electronic commerce technologies to streamline business practices. The growth of electronic commerce, the deregulation of the telecommunications industry, and the interconnection of the Public Switched Network (PSN) and the Internet have blurred traditional barriers and opened the door to an "Information Age." This has tremendous security implications for the Nation, as President Clinton stated, "As borders open and the flow of information, technology, money, trade and people across borders increases, the line between domestic and foreign policy continues to blur."[6]

The risks associated with cyber intrusions into critical systems have risen over the past decade. Networked systems supporting our national security community and our national economy have become increasingly reliant on cyber-based information systems. Vital control systems that support the information infrastructure and other critical infrastructures depend upon network connectivity through the use of the PSN.

These factors introduce vulnerabilities that can be exploited by a wide spectrum of perpetrators: disgruntled insiders, recreational and institutional hackers, organized crime, terrorist groups, or hostile nations. The inability to determine the intentions of an intruder attacking or exploiting a networked system makes addressing this threat more difficult. As a result, "by their nature, developments in cyberspace blur the distinction between crime and warfare, thereby blurring the distinction between police responsibilities to protect U.S. interests from criminal acts in cyberspace, and military responsibilities to protect U.S. interests from acts of war in cyber-

---

4 *Morgan Stanley U.S. Investment Research: Internet Retail.* Morgan Stanley, May 1997.

5 Inktomi Corporation White Paper. 1997.

6 President William J. Clinton, "A National Security Strategy for a New Century," May 1997.

space."[7] In February 1998, the DOD faced widespread intrusions into its networks in a case that became known as Solar Sunrise. The case highlights the complicated nature of cyber intrusions, especially with the challenge of determining whether an intrusion is a national security or law enforcement issue. The intrusion occurred as the United States was facing heightened hostilities with Iraq. However, the perpetrators turned out to be two juveniles in California and a juvenile in Israel. Investigators from the Federal Bureau of Investigation (FBI) and DOD were called on to work together to investigate the intrusion.

2.2 The Cyber Security Challenge

Computers can be used by criminals and other nefarious actors in the cyber arena in two broad areas the computer may be used as a tool to commit the offense (e.g., an intrusion into a network system), or it may be incidental to the crime but act as a repository of evidence (e.g., drug traffickers using encrypted computer information to hide evidence). In some cases, the computer may serve both functions. For example, hackers use their computers as tools to intrude into the networked systems of others as well as to store tools and information that could be used as evidence against them (e.g., hacker software, stolen files, and log information).

In fact, hacker intrusion cases represent a troubling percentage of the total cyber crime caseload. In 1997, the FBI investigated 408 cyber intrusion incidents, a 133 percent increase from 1996. In 1996, the Defense Information Systems Agency (DISA) estimated that nearly 250,000 attacks on DOD systems may have taken place in 1995.[8] In cases where the computer contained evidence of a crime rather than being the tool used to commit the crime, the FBI Computer Analysis and Response Team (CART) conducted forensic examinations in almost 2,200 cases. The FBI reports that nearly 70 percent of those cases were instances of white-collar crime; the remaining cases involved organized crime, violent crime, terrorism, and cyber crime investigative programs.

Complicating matters, cyber security initiatives face several constraints. The cyber crime casework has grown exponentially in size and complexity. The Computer Emergency Response Team (CERT) at Carnegie Mellon University's Software Engineering Institute warned that intruders are using sophisticated techniques for analyzing source code vulnerabilities, have

---

7 Richard Hundley and Robert Anderson, "Emerging Challenge: Security and Safety in Cyberspace" *IEEE Technology and Society Magazine*, Vol. 14, No. 4 (Winter 1995-1996) pp. 19-28.

8 United States General Accounting Office, *Information Security: Computer Attacks at Department of Defense Pose Increasing Risks*, GAO-AIMD-96-84, Washington, DC: USGPO, May 1996.

compromised a variety of operating systems, and are using complex denial of service attacks.[9] The General Accounting Office (GAO) noted the threat posed by cyber intrusion is rapidly increasing, and protection mechanisms have failed to keep pace with the threat.[10] Investigators require ready access to the full range of special tools and resources needed to keep pace with the increasing technical sophistication of cyber criminals. Investigative agencies are also faced with the difficult task of training their staff to investigate crime in nontraditional environments. Within these constraints, management must also balance the immediate need to investigate cases against long-term efforts to train employees. These managers, already faced with tight resources and a shortage of investigators to handle a growing caseload, may not be able to spare a key investigator for a prolonged period of time for cyber investigative training.

3.0 FEDERAL INITIATIVES

The Federal Government has recognized the cyber threats to critical networked systems and the potential national security implications. On May 22, 1998, President Clinton announced the signing of Presidential Decision Directive (PDD) 63, which expressed his intent to develop a national system to protect critical infrastructures from these types of attacks.

Federal, State, and local law enforcement agencies have initiated a number of programs to address the investigation and prosecution requirements of this threat. On February 26, 1998, the National Infrastructure Protection Center (NIPC) was established to create an interagency and public-private partnership to combat physical and cyber threats. The NIPC will provide centralized support, planning, and response capability to deal with domestic cyber crime and potential attacks on infrastructures. To provide a unified, comprehensive response to the cyber crime threat, the NIPC will be staffed with cyber investigators from the FBI and other representatives from the Department of Defense, the intelligence community, other Federal departments, state and local law enforcement, and private industry.

The Department of Justice Computer Crime and Intellectual Property Section (CCIPS) represents another strong Federal initiative on cyber crime. The CCIPS attorneys work with other Government officials and industry to develop a global response to cyber attacks. They litigate cyber crime cases,

---

9 CERT Coordination Center, "Increasing Sophistication of Intruder Community Expertise," CERT Summary CS-96.04, World Wide Web, info.cert.org/pub/cert_summaries/,July 23, 1996.

10 United States General Accounting Office, *Information Security: Computer Attacks at Department of Defense Pose Increasing Risks*, GAO-AIMD-96-84, Washington, DC: USGPO, May 1996.

provide litigation support to other prosecutors, and train Federal law enforcement personnel. Other notable initiatives to address cyber crime include efforts within the Department of Treasury and NASA, as well as statewide cyber crime efforts in Florida and New York.

The national security community has undertaken a number of programs to secure its systems from the cyber threat. As the Manager of the Defense Information Infrastructure (DII), the Defense Information Systems Agency (DISA) initiated critical programs to protect against, detect and react to threats to both its information infrastructure and information sources. The DISA has undertaken a number of initiatives to transform the way DOD users move, share, and use information on critical systems such as the Defense Information System Network, the Defense Message System, the Global Command and Control System, and the Global Combat Support System.

Other DOD initiatives on cyber security include steps taken by the Office of the Assistant Secretary of Defense for Command, Control, Communications, and Intelligence (OASD/C3I) to assist Defense Criminal Investigative Organizations (DCIO) and Military Counterintelligence Agencies to respond to the growing number of cyber intrusions and cases involving computer evidence. On June 28, 1996, OASD/C3I established the Defense Computer Forensics Laboratory (DCFL) to develop state-of-the-art cyber forensics tools and provide forensics analysis to DCIOs. In addition, the DOD Computer Training Program (DCITP) was created to develop and provide computer investigation training to cyber investigators within DOD.

4.0 CYBER FORENSICS

Cyber forensics is a discipline that includes computer forensics techniques by law enforcement investigators and the national security community to investigate potential criminal or national security events that involve the use of information systems. Federal, State, and local government agencies of all sizes and jurisdictional responsibilities are increasingly finding critical investigative information stored on computers. These computers range in size from microcomputers to mainframes and in complexity from simple stand-alone desktop computers to complex networked systems with data distributed worldwide. Computers can be huge electronic filing cabinets containing vast amounts of data, instruments of the crime in cases such as hacker intrusions, or containers of contraband such as stolen passwords and user identification. Given the proliferation and sophistication of information technologies, there is an increasing probability that crucial evidence in the form of computers and related items will play a significant role in future criminal and national security investigations. Cyber forensics is the discipline of acquiring, preserving, retrieving, and presenting data for review from this unique evidence.

Cyber forensics relies on detailed, documented, and certified sets of policies and practices, i.e., protocols, that describe the examination processes in technical detail and that are legally defensible. Protocols are discoverable in both criminal and civil trials. Because these examinations will likely become issues before a court, the cyber forensics process must serve two masters. First it must be technically robust to address questions of complete recovery of probative information without altering the original material. It must then meet the legal requirements of acquisition, storage, examination, and distribution in a manner that is entirely consistent with the rules of evidence. In criminal cases, there may be specific limits on the information that may be recovered and examined (privilege and e-mail issues, for example) and who may be party to the examination (grand jury material, for example). It is unlikely that any informal or ad hoc approach to cyber forensic examination will meet these requirements and produce admissible evidence.

The ideal situation places the computer examination process in a laboratory environment; however, practical issues often require that computer forensic experts travel on site in support of investigations. This requirement is typical of investigations of financial crimes in which seizure of computer systems would interfere with legitimate business operations, health care fraud investigations where computer systems are considered part of the health care delivery system, and investigations where privileged information is likely to be encountered. An attack against national security systems presents a similar challenge. Policies and practices must also be in place for these remote efforts to assure that these on-site requirements are met with the same high degree of technical competence and professionalism as would be expected in the laboratory.

The software utilities necessary to conduct these examinations include disk imaging, restoration of erased and deleted files, slack and free space data recovery, text string search, known file filters, and data integrity checks such as CRC and MD-4 and 5. For these software utilities to be acceptable to the court, any special-use (i.e., not COTS) utilities should be certified by an objective third party. Likewise, protocols should receive thorough peer review and represent the state of the art in good laboratory practice.

The FBI reports that the number of criminal cases for which computer evidence was submitted has grown exponentially in the past few years to nearly 2,200 cases in 1997. An even larger number of cases in which computer evidence is recovered exist at the civil or administrative inquiry level throughout Government and the private sector. All cases, criminal, civil, and administrative, deserve to be handled competently and professionally. The volume of material stored on computers, and the litigious nature of many civil and administrative inquiries, combine to require that plans and policies be prepared, in place, and understood before they are needed.

With a growing cyber crime caseload, it is clear that industry, law enforcement, and the national security community need to work together to ensure that all parties involved with investigating cyber-based events have ready access to the needed tools, techniques, and capabilities.

5.0 EDUCATION, TRAINING, AND AWARENESS

Providing the best tools to cyber investigators is only one step in responding to the rising cyber security threat. With the frequency and sophistication of cyber crime rapidly rising, defense against the threat demands thorough training for investigative officers and systems administrators to prevent, detect, and respond to cyber intrusions. Any successful initiative to address the threat must not only provide investigators with the comprehensive training needed to respond to and investigate intrusion incidents but should also make systems administrators aware of their key roles and responsibilities in responding to cyber intrusions.

The electronic nature of this threat is troublesome for investigators because cyber crime challenges traditional policies, procedures, and protocols for investigating crimes and collecting evidence. A cyber investigator must use traditional investigative skills combined with knowledge of network systems to conduct an investigation in an automated environment. Investigators need comprehensive training to understand the operation of an information system and the type of crimes that target information systems or use computers as an instrument of the crime, as well as provide the context of the interconnected environment in which these crimes are committed. This training requires basic knowledge of computer hardware and software technology, network operating systems, Internet and intranet protocols, and intrusion techniques and tools. Law enforcement agencies have recognized the need to train their cyber investigators and have integrated programs to augment their cyber investigation capability; however, the resources allocated for training, while growing, are not sufficient to keep pace with the rapid growth and sophistication of the cyber security threat. This problem is further exacerbated by the fact that many of the "first responders" to a crime scene are likely to be local law enforcement officers or field agents not trained in cyber investigative techniques. For this reason, training programs need to be extended to the full range of investigators.

System administrators are responsible for maintaining information systems and are often the first responders to an electronic intrusion. System administrators need proper training to be able to help in investigating intrusions. NSTAC concluded in the *Report on the NS/EP Implications of Intrusion Detection Technology Research and Development*, "There is a lack of guidance to employees as to how to respond to intrusions and capture the information required to conduct a law enforcement investigation."[11] Training materials and guidelines must be established to make system administrators aware of what information investigators need and what procedures they must follow to collect evidence on an intrusion.

6.0 FINDINGS

In its investigation of cyber security issues, the Electronic Commerce/Cyber Security Subgroup concluded that a stronger partnership between industry and Government is required for the following reasons:

- The NS/EP implications of cyber security necessitate developing a partnership with industry, law enforcement, and national security communities, and PDD-63 lead departments and agencies.
- The size and complexity of the cyber crime caseload coupled with the growing sophistication and proliferation of computer-based tools used to attack information systems challenges traditional policies, procedures, and protocols for investigating crimes and collecting evidence.
- Cyber investigators also face the challenge of staying abreast of industry's continuous introduction of new and innovative technologies.
- These new technologies introduce additional vulnerabilities that are being exploited.
- Industry possesses knowledge, tools, and expertise investigators could leverage as they respond to increasingly complex and sophisticated cyber crimes.

A key element in a dialogue with industry and Government will involve establishing the appropriate levels of trust and understanding that spur cooperation to address cyber security issues. This dialogue will allow representatives from industry and investigative communities to develop a unique understanding of the perspectives and issues faced by their counterparts. The Government and NSTAC Network Security Information Exchanges (NSIE) and the National Coordinating Center for Telecommunications (NCC) are example of effective partnerships between the telecommunications industry and Government. The principles of cooperation exercised in these forums could be applied in efforts to address cyber crime.

With its years of experience of working with the Government on NS/EP telecommunications, network security, and information assurance issues, NSTAC is positioned to build on this experience in developing a partnership with the law enforcement and national security communities to address cyber security issues.

---

11 NSTAC Network Group, *Report on the NS/EP Implications of Intrusion Detection Technology Research and Development*, December 1997.

7.0 RECOMMENDATIONS

Based on the findings of the Electronic Commerce/Cyber Security Subgroup, the following recommendations are proposed.

7.1 Recommendation to the President

The President should direct the appropriate departments and agencies to continue working with the NSTAC for the development of policies, procedures, techniques, and tools to facilitate joint industry Government cooperation on cyber security.

7.2 Recommendation to the NSTAC

The NSTAC should continue to work with the presidentially directed departments and agencies in furtherance of the recommendation above.

CHAPTER 7

STATEMENT BY THE PRESS SECRETARY
Administration Updates Encryption Policy

THE WHITE HOUSE
Office of the Press Secretary
September 16, 1998

The Clinton Administration today announced a series of steps to update its encryption policy in a way that meets the full range of national interests: promotes electronic commerce, supports law enforcement and national security and protects privacy. These steps are a result of several months of intensive dialogue between the government and U.S. industry, the law enforcement community and privacy groups that was called for by the Vice President and supported by members of Congress.

As the Vice President stated in a letter to Senator Daschle, the Administration remains committed to assuring that the nation's law enforcement community will be able to access, under strictly defined legal procedures, the plain text of criminally related communications and stored information. The Administration intends to support FBI's establishment of a technical support center to help build the technical capacity of law enforcement —Federal, State, and local—to stay abreast of advancing communications technology.

The Administration will also strengthen its support for electronic commerce by permitting the export of strong encryption when used to protect sensitive financial, health, medical, and business proprietary information in electronic form. The updated export policy will allow U.S. companies new opportunities to sell encryption products to almost 70 percent of the world's economy, including the European Union, the Caribbean and some Asian and South American countries. These changes in export policy were based on input from industry groups while being protective of national security and law enforcement interests.

The new export guidelines will permit exports to other industries beyond financial institutions, and further streamline exports of key recovery products and other recoverable encryption products. Exports to those end users and destination countries not addressed by today's announcement will continue to be reviewed on a case-by-case basis.

Very strong encryption with any key length (with or without key recovery) will now be permitted for export, under license exception, to several industry sectors. For example, U.S. companies will be able to export very strong encryption for use between their headquarters and their foreign subsidiaries worldwide except the seven terrorist countries (Iran, Iraq, Libya, Syria, Sudan, North Korea and Cuba) to protect their sensitive company proprietary information.

On-line merchants in 45 countries will be able to use robust U.S. encryption products to protect their on-line electronic commerce transactions with their customers over the Internet.

Insurance companies as well as the health and medical sectors in those same 46 countries will be able to purchase and use robust U.S. encryption products to secure health and insurance data among legitimate users such as hospitals, health care professionals, patients, insurers and their customers.

The new guidelines also allow encryption hardware and software products with encryption strength up to 56-bit DES or equivalent to be exported without a license, after a one time technical review, to all users outside the seven terrorist countries. Currently, streamlined exports of DES products are permitted for those companies that have filed key recovery business plans. However, with the new guidelines, key recovery business plans will no longer be required.

The Administration will continue to promote the development of key recovery products by easing regulatory requirements. For the more than 60 companies which have submitted plans to develop and market key recovery encryption products, the six month progress reviews will no longer be required. Once the products are ready for market, they can be exported, with any bit length — without a license — world-wide (except to terrorist nations) after a one-time review. Furthermore, exporters will no longer need to name or submit additional information on a key recovery agent prior to export. These requirements will be removed from the regulations.

Finally, industry has identified other so-called "recoverable" products and techniques that allow for the recovery of plaintext by a system or network administrator and that can also assist law enforcement access, subject to strict procedures. The Administration will permit their export for use within most foreign commercial firms, and their wholly-owned subsidiaries, in large markets, including Western Europe, Japan and Australia, to protect their internal business proprietary communications.

The Administration welcomes a continued dialogue with U.S. industry and intends to review its policy in one year to determine if additional updates may be necessary to continue a balanced approach that protects the public safety and national security, ensures privacy, enables continued technology leadership by U.S. industry and promotes electronic commerce.

THE WHITE HOUSE
Office of the Press Secretary
September 16, 1998

FACT SHEET
Administration Updates Encryption Policy

Exports of 56-bit DES and equivalent products (hardware and software) will be streamlined (under license exception). Requirements for key recovery plans are eliminated.

Exports of unlimited strength encryption products (with or without key recovery) will be streamlined (under license exception) to certain industries. The sectors are:

Subsidiaries of U.S. Firms, worldwide (except seven terrorist nations).

Insurance companies to the same 45 countries recently approved for exports to banks and financial institution exports.

Health and medical organizations (including civilian government health agencies) in the same 45 countries. Does not include biochemical/pharmaceutical manufacturers.

On-line merchants for client-server applications, in the same 45 countries, with the purpose of securing electronic transactions between merchants and their customers. Does not include manufacturers and distributors of items controlled on the U.S. munitions list.

Key Recovery products will continue to be exportable under license exception worldwide (except seven terrorist nations). Review of foreign key recovery agents is eliminated.

Exports of "recoverable" products will be approved to most commercial firms, and their wholly-owned subsidiaries, in a broad range of countries under encryption licensing arrangements. This group of countries covers most major commercial markets including Western Europe, Japan, and Australia. The policy does not include service providers and manufacturers and distributors of items controlled on the U.S. munitions list.

Exports to end users or destinations outside this policy are possible on a case-by-case basis.

Prior to export, products are subject to a one-time product technical review.

CHAPTER 8

PRESS BRIEFING BY THE VICE PRESIDENT, DEPUTY CHIEF OF STAFF JOHN PODESTA, PRINCIPAL ASSOCIATE DEPUTY ATTORNEY GENERAL ROBERT LITT, ASSISTANT DIRECTOR OF THE FBI CAROLYN MORRIS, UNDER SECRETARY OF COMMERCE WILLIAM REINSCH, DEPUTY SECRETARY OF DEFENSE JOHN HAMRE, AND DEPUTY NATIONAL SECURITY ADVISOR JIM STEINBERG

THE WHITE HOUSE
Office of the Press Secretary
September 16, 1998

The Briefing Room
11:57 A.M. EDT

THE VICE PRESIDENT: Good morning. While my colleagues are coming in here, let me acknowledge them. John Podesta is going to take over the podium after I complete my statement, and he is joined by Bob Litt of the Justice Department, Bill Reinsch of the Commerce Department—Under Secretary for the Export Administration—and John Hamre, Deputy Secretary of Defense.

I also want to acknowledge Carolyn Morris of the FBI; Barbara McNamara of the National Security Agency; John Gordon, Deputy Director of the CIA. And you all should know that this process, the results of—the interim results of which I'm announcing here, is a process that has been run principally by John Podesta and Jim Steinberg, Deputy at the National Security Council. And I also want to thank Sally Katzen at the NEC and David Beier on my staff for the work that they and many others have done on this.

Some of you who have followed this issue know that it is probably one of the single, most difficult and complex issues that you can possibly imagine. But we've made progress, and we're here this morning to announce an important new action that will protect our national security and our safety, and advance our economic interests and safeguard our basic rights and values in this new Information Age.

The Information Age has brought us the Internet, an inter-connected global economy and the promise of connecting us all to the same vast world of knowledge. But with that exciting promise comes new challenges. We must make sure that in the Information Age you get information about the rest of the world and not the other way around. We must ensure that new technology does not mean new and sophisticated criminal and terrorist activity which leaves law enforcement outmatched—we can't allow that to happen. And we must ensure that the sensitive financial and business transactions that now cruise along the information superhighway are 100 percent safe in cyberspace.

Balancing these needs is no simple task, to say the least. That is why, in taking the next step toward meeting these complex goals, we worked very closely with members of Congress from both parties, House and Senate; with industry; with our law enforcement community and with our national security community. And as we move forward we want to keep working closely with all who share a stake in this issue—especially law enforcement—to constantly assess and reassess the effectiveness of our actions in this fast changing medium.

Today I'm pleased to announce a new federal policy for the encryption and protection of electronic communication, a policy that dramatically increases privacy and security for families and businesses without endangering out national security.

Beginning today, American companies will be able to use encryption programs of unlimited strength when communicating between most countries. Health, medical, and insurance companies will be able to use far stronger electronic protection for personal records and information. Law enforcement will still have access to criminally-related information under strict and appropriate legal procedures. And we will maintain our full ability to fight terrorism and monitor terrorist activity that poses a grave danger to American citizens.

With this new announcement, we will protect the privacy of average Americans, because privacy is a basic value in the Information Age, indeed in any age. We will give industry the full protection that it needs to enable electronic commerce to grow and to thrive. And we will give law enforcement the ability to fight 21st century crimes with 21st century technology, so our families and businesses are safe, but on-line outlaws are not safe.

In just a moment you will hear more of the details of this new policy, but I want to conclude by saying that this policy does reflect one of the greatest challenges of these new times. And to state it broadly, it's a challenge of how we can harness powerful new technology while protecting our oldest and most cherished values, such as privacy and safety.

I'm grateful to those who have worked so hard to reach this balance. And with today's announcement I believe that all families and businesses have reason to feel safer, more secure and more confident as we approach the 21st century.

And now I'd like to turn things over to White House Deputy Chief of Staff John Podesta.

Question: Mr. Vice President, before you go, can you tell us what you say to Democratic lawmakers who say the President ought to resign?

THE VICE PRESIDENT: I disagree.

Question: How about the release of that tape? What do you think—

THE VICE PRESIDENT: The President is going to have a press conference shortly and I'm sure that you will not miss the opportunity at this national security press conference with the leader of a foreign country to raise all these questions. Question: What about the videotape, should it be released?

Question: It was staged by the White House—you know that, don't you?

MR. PODESTA: Guess what? I'm here to talk about encryption. Okay. I can see the front row leaving here. (Laughter.) As the Vice President noted, Jim Steinberg and I have co-chaired our process in this matter. I volunteered for that duty because of my well-known fascination with The X Files, which most of you know about.

As you know, this is an important and challenging issue that affects many of our interests in our society. And over the past year we've promoted a balanced approach to the issue, working with all segments of our government and working with industry to find a policy that promotes electronic commerce, preserves privacy, protects national security and law enforcement interests, and permits U.S. industry to secure global markets.

Recognizing the importance of moving this issue forward, last March the Vice President asked us to intensify our dialogue with U.S. industry, to bring industry's technical expertise to bear on this issue with the hope of finding more innovative ways that we might assist law enforcement. We appreciate the efforts of Congress, the law enforcement community and particularly the industry groups.

I would note the Computer Systems Policy Project and the Americans for Computer Privacy, who have been in an intensive dialogue with us over the past many months to foster an environment that has allowed us to come up with a policy which we believe has balanced the elements that are necessary in this regard.

I think all the stakeholders in this process, on our side, as well as on private industry's side, now have a greater appreciation of the issues and intend to continue the dialogue, which I think we're most pleased by. Again, I think some of the people here from industry will be available at the stakeout later to take some comment.

Based on the ideas discussed among the various stakeholders, today we're proposing an update to our policies that we've announced in the past. I'm going to serve kind of as M.C. We're going to start off with Bob Litt from the Justice Department and Carol Morris, who I asked to join us, from the FBI, to talk about the law enforcement-FBI concerns. Then we're going to turn to Bill Reinsch from the Commerce Department to talk about export control and electronic commerce. And finally you'll hear from Dr. Hamre from the Defense Department. I might ask Jim also to join us up here.

Before I give up the floor to Bob and Carol, though, I want to stress that encryption policy is an ongoing process. It's one of adaptation; it's an evolutionary process. We intend to continue the dialogue, and over the course of the next year, determine what further updates are necessary as we work with industry to try to, again, come up with a policy that balances national security, law enforcement, and the real needs for privacy and security in electronic commerce.

Thank you. Let me turn it over to Bob.

MR. LITT: Thank you, John. Good afternoon. The Justice Department and the FBI and law enforcement in general is supportive, very supportive of today's announcement on the updating of our export controls on encryption products, particularly with respect to those products that allow law enforcement to obtain lawful access to the plain text of encrypted information.

We have been very encouraged over the last few months by industry's efforts to work with us to develop and market strong encryption products that provide law-abiding citizens with the ability to protect the privacy of their communications and their electronically-stored data, while at the same time maintaining law enforcement's ability to ensure public safety when these products, when they become commercially available, are used in furtherance of serious criminal activity.

Our goal is through whatever means to ensure that when we have the lawful authority to take steps to protect public safety, we have the ability to do so. And we have been working cooperatively with industry for many months to develop approaches that will deal with that.

Carolyn Morris will now talk a little bit about the technical support center that is being proposed.

MS. MORRIS: Thank you very much, Bob.

Good afternoon, ladies and gentlemen. We in federal, state, and local law enforcement, are pleased with the administration's support to establish a technical support center. This center will provide federal, state, and local law enforcement with the resources and the technical capabilities we need to fulfill our investigative responsibilities.

In light of strong, commercially available encryption products that are being proliferated within the United States, and when such products are used in the furtherance of serious criminal activity, this center becomes very, very critical to solving the encryption issues that we need to make cases. As a matter of fact, the FBI has already begun planning activities of this critical technical support center in anticipation of the availability of funds.

The United States federal, local and state law enforcement community looks forward to a cooperative partnership with American industry, the Congress and the administration to ensure that this technical support center becomes a reality in the near future. With this center the American people can be assured that federal, state, and local law enforcement has the necessary resources and tools we need to fulfill our public safety mission. Thank you very much.

UNDER SECRETARY REINSCH: With respect to export controls, the administration is updating its policy in three areas: Our existing policy and some revisions there, an expansion with respect to certain sectors, and an expansion with respect to so-called recoverable products. And let me address each of these separately. In keeping with the administration's reinvention initiatives, I'm going to try to do it in plain language—or plain English, So that those of you that speak the vocabulary of encryption may find it to elementary, but we can go back and do it again in another language, if you want, later on in questions.

With respect to our existing policy, we have for two years ending this December, permitted the export of 56-bit products after an initial one-time review without further review by the government. What we're announcing today is the maintenance of that window permanently. And so 56-bit products will be freed from export controls after a one-time review, in perpetuity, not ending at the end of this year. We are, however, removing the requirement for key recovery plans or key recovery commitments to be provided in return for that change, which was the initial condition that we extracted.

In addition, we are continuing to permit the export of key recovery products—products that contain those features—without restraint worldwide. We are, however, going to simplify significantly our regulations that relate to those exports. In particular, we're going to eliminate the need for six-month progress reports for the plans that have been submitted, and we're going to eliminate the requirement for any prior reporting of key recovery agent information. For those of you that follow the regulations in detail, that means we're going to eliminate Supplement Five of our regulations on these matters.

Now, with respect to sectors, we're making some new innovations in four areas. Some of you may be familiar with the fact that some time ago we announced expanded treatment of encryption products for export to banks and financial institutions. And what we did at that time, briefly, was to permit the export of encryption products of any length, any bit length, with or without key recovery features to banks and financial institutions in a list of 45 countries.

What we are announcing today is, first, that we are adding insurance companies to the definition of financial institutions, so insurance companies will be treated the same way under this policy as banks and other financial institutions are now. In addition, we are providing the same kind of treatment for exports of these encryption products to the health and medical sector operating in the same set of countries. We are excluding from that biochemical and pharmaceutical producers. But the rest of the health and medical sector will be the beneficiary of the same kind of treatment.

In addition, we are providing also this expanded treatment for that country group to on-line merchants that are operating in those countries. That means that for products that are like client-server applications, like SSL, will be able to be exported to those destinations.

All these things will take place under what we call license exception, which means after initial one-time review to determine whether or not your product is, in fact, what you say it is, they can then go without any further review or intervention by the government to those locations. In addition, there is always the option in the export control system of coming in with an application to export these kinds of products to other destinations beyond the ones that I'm talking about right now, and those will be reviewed one by one on their merits.

Finally, with respect to what we have come to refer to as a class of so-called recovery capable or recoverable products, and these are the products that, among others, include what has become known as the doorbell products, which are products that, among other things, will deal with the development of local area or wide area networks and the transmission of e-mail and other data over networks—we are going to permit the export of those products under a presumption of approval and an export licensing arrangement to a list of 42 countries. And within those countries we are going to permit that export to commercial firms only within those countries. And both in that case and in the case of the on-line merchants that I referred to a few minutes ago, we are going to exclude manufacturers or distributors of munitions items, I think for obvious reasons.

We can go into further details later, if you would like. I think for those of you that are interested in the nitty-gritty of all this stuff, BXA intends to post all the details, including the country lists, on its website and we should have that up later today. Thank you.

DEPUTY SECRETARY HAMRE: Good morning. I'm here to speak on behalf of the national security community. I'm joined today by my enormously capable counterparts and colleagues, Deputy Director Barbara McNamara for the National Security Agency; and Deputy Director John Gordon from the Central Intelligence Agency.

The national security establishment strongly supports this step forward. We think this is a very important advance in a crucial area for our security in the future.

We in DOD had four goals when we entered these discussions. First was to strengthen our ability to do electronic commerce. We're the largest company in the world. Every month we write about 10 million paychecks. We write about 800,000 travel vouchers. One of our finance centers disburses $45 million an hour. We are a major, major force in business. And for that reason, we can't be efficient unless we can become fully electronic, and electronic commerce is essential for us. And this is an enormous step forward.

Second, we must have strong encryption and a security structure for that in order to protect ourselves in cyberspace. Many of you know that we have experienced a number of cyber attacks during the last year. This will undoubtedly increase in the future. We need to have strong encryption because we're operating over public networks; 95 percent of all of our communications now go over public infrastructure—public telephone lines, telephone switches, computer systems, et cetera. To protect ourselves in that public environment, we must have encryption and we must have a key recovery system for ourselves.

The third goal that we had was to help protect America's infrastructure. One of the emerging national security challenges of the next decade is to protect this country, the homeland defense of this country, against attack. We must have strong encryption in order to do that, because most of this infrastructure now is being managed through distributed computer-based management systems, and this is an important step forward.

Finally, it is very important that the Department of Defense and our colleagues in the national security establishment have the ability to prosecute our national security interests overseas. Terrorists and rogue nations are increasingly using these tools to communicate with each other and to lay their plans. We must have the ability to deal with that. And so this policy, it's a balanced and structured approach to be able to deal with all four of those problems.

UNDER SECRETARY REINSCH:

I apologize—in listing my changes, I neglected one very important item that I want to go back to, and that is, in the sector area we are also announcing today the ability to export strong encryption of any bit length, with or without key recovery features, to subsidiaries of U.S. companies to all destinations in the world with the exception of the seven terrorist nations.

MR. PODESTA: Okay, I think we're happy to take your questions now. If you could identify whom you're addressing, because there is a variety of expertise. And I would like to introduce one other person, Charlotte Knepper from the NSC staff, who has been instrumental in pulling this all together.

Question: John, this is a question for you. In October '96 and other White House statements on encryption, there has usually been a line also addressing the domestic side, saying that all Americans remain free to use any strength encryption. I didn't notice anything like that in today's announcement. Are there any conditions under which the White House would back domestic restrictions on encryption?

MR. PODESTA: We haven't changed our policy, and the previous statements are certainly intact. We have made a number of policy statements in the past, since this administration came into office, and I think that you should view this as a step forward, building on the policies that we have put before the American public in the past.

Question: John, could I ask you one question about an un-encrypted matter?

MR. PODESTA: Maybe. (Laughter.)

Question: Democrats on the Hill are now saying, and John Kerry is saying that the President's actions absolutely call for some sort of punishment. What are Democrats telling you about what they feel must be done at this point?

MR. PODESTA: Well, I think I'm not going to stand here and take a lot of questions, but I'm going to give special dispensation, as a Catholic, today—which is I'm going to return your phone calls later. But in deference to the people up here I think we'll handle it that way.

But in specific response, I'll take one, which is that I think that we had a number of productive meetings with Democrats on both sides of the Hill yesterday. They view the President as a person who has led on the issues that are important to them, and I think what they want to do is get back to having him speak out and be a leader on the issues of education and the health care bill of rights, on saving Social Security. And I think they pointed at that and wanted to work with us on that.

I think with regard to the question that you posed with regard to Senator Kerry, I think that's a matter that they are debating amongst themselves more than they are debating with the White House. I think it's probably presumptuous for us at this point to offer them assistance or guidance. I mean, the President has said that what he has done was wrong; he's apologized for it; he's asked for forgiveness. He is moving forward. And I think that this debate is going on, on Capitol Hill, but it's largely going on amongst members themselves.

Question: We haven't heard many of them say they want to get back to the work at hand.

MR. STEINBERG: You heard John, and I'm going to leave it there.

Let me just add a word in response, in connection with the domestic controls issue. I think one of the lessons that we've learned from this exercise is that—actually, two lessons—one, that trying to balance the various interests and equities in this is much less of a zero sum gain than I think some began to look at the question. That is, you heard from Dr. Hamre and others that many of the interests involved have common interests in making sure that we have secure and effective means of dealing with communications and stored data.

And so we found, by looking in a very pragmatic way, that there were ways to solve these problems without very, kind of, broad-based solutions. In particular, I think the idea that there's no one-size-fits-all answer to the problems of meeting the various needs informs the decisions that we reached—that there are a variety of different techniques that respond to the different aspects of the industry, the different aspects of the technology. I think that's what made the progress possible today, is that industry, agencies and Congress sat down together, pulled the problem apart, began to look at its different components and began to fashion very pragmatic solutions.

And so I think we came to this discussion with a spirit of not looking for a kind of single or simple solution to the problem but, rather, how do you tackle and meet the various needs. And I think that's what led to this resolve.

Question: Could you talk a little more about the on-line merchants part of it? I mean, what do you have to do to qualify as an on-line merchant? Do you have to register or can anybody sort of set themselves up in business?

UNDER SECRETARY REINSCH: I think the simplest way to respond to that right now is we'll have a definition in the reg that will be very clear as to what the criteria are for qualification. And those definitions have already been dealt with and agreed to, so we should have them up on the web site this afternoon.

Question: A question for Bill Reinsch. How do you handle, then, 128-bit, to which the Department has given export—or has allowed to be exported after going through this review? Will 128 or things above 56-bit, will they require a license or will they still have to go through plans—

UNDER SECRETARY REINSCH: Well, with respect to the subsidiaries, the health sector, the banks, the financial institutions, the insurance companies, the on-line merchants, and the recoverable products as in the universe defined—no. In the case of all but the recoverable products, they will all go on license exception, which means one-time review and then out the door. With respect to recoverable products, they will come in and go out pursuant to an export licensing arrangement, where we'll have to do a little tailoring depending upon the nature of the product. But there is a presumption of approval for the 42 countries that I indicated.

And that's without reference to bit length—128 or more is all covered by that. Now, if you want to export an 128-bit product that is beyond any of those universes, then you would have to come in for an individual license application.

Question: A question for Mr. Litt. With regard to the technical support center, when do you expect that to be in operation?

MR. LITT: I don't think we have a specific timetable yet. Obviously, it would be helpful for us to have it up and operational as soon as possible, but there are planning and budgetary issues that have to be dealt with.

Question: This is probably a question for Under Secretary Reinsch. The export exceptions now are essentially going to U.S. subsidiaries—foreign subsidiaries of U.S. companies. I was wondering, could you be a little more specific—what size company, what kind of company will be allowed to export powerful crypto to its foreign subsidiaries?

UNDER SECRETARY REINSCH: That doesn't make any difference. The universe is determined by the end user, not by the nature of the American company. But it is not—while part of this relates to subsidiaries of U.S. companies, that is correct, we also intend, on a case-by-case basis, to provide for favorable treatment for export of the same kind of thing to strategic partners of U.S. companies—those foreign companies that are engaged in a closer, say, joint venture, that kind of relationship.

Well, I think that's it.

Question: What about foreign companies that have U.S. subsidiaries, like Seaman's or—or Chrysler—can they get this encryption?

UNDER SECRETARY REINSCH: Well, keep in mind, there are multiple universes here. If you're talking about the financial institutions, the banks and the insurance companies, those aren't necessarily American financial institutions. That's for export to any financial institution, and for their use in any of their branches, aside from the terrorist countries. This is true for the health sector; this is true for on-line merchants as well. Those are not restricted to U.S. companies.

Obviously, if we're going to have a requirement for U.S. subs, it relates to U.S. subs, and wouldn't affect the examples you've described. Now, with respect to recoverable products, which actually is one of the areas where the companies you mentioned would probably be looking because they'd be looking to build a network among their various offices, affiliates of subsidiaries, dealers if necessary, worldwide, the recoverable provisions that I described could be exported to those companies within the territorial universe I described—the 42 countries.

Thank you very much.

CHAPTER 9

PRESIDENT'S EXPORT COUNCIL

SUBCOMMITTEE ON ENCRYPTION (PECSENC)

September 18, 1998

The following findings have been adopted by the PECSENC as a reflection of conditions of international competition prior to the U.S. Government's liberalization of encryption export controls announced on September 16, 1998. The liberalization may affect many of these findings, and the findings will be used as a baseline for a review of the effects of the liberalization in future sessions of the PECSENC.

1. The difference between U.S. encryption controls and those of other nations is a serious—but not the only—factor determining success in the computer security market. With or without controls, both U.S. and foreign products are likely to continue to coexist, and other factors are likely to continue to slow deployment of security products. Many foreign companies, for example, especially those influenced by governments, will continue to favor domestic security solutions, and many computer users will not deploy serious security technology until there have been major incidents with losses that can be attributed to lack of encryption.

2. Nonetheless, the adverse impact of controls on U.S. industry is palpable. For many software applications, business customers simply demand security and encryption; it is a checklist item, and its absence is a deal breaker. While simply counting the number of foreign encryption software products in the market is not an accurate measure of the impact of controls, one particularly serious risk is that non-U.S. companies will use their ability to export stronger encryption as "leverage" to dominate particular applications.

This has happened in at least one field—Internet banking—and may occur in other areas of electronic commerce. Brokat, a German company that scarcely existed four years ago, now has 250 employees and offices in several countries including the United States. Brokat's specialty is Internet banking and electronic commerce, but it broke into that business on the strength of being able to offer stronger encryption than German banks could obtain in Netscape or Microsoft browsers. It is now a major player in this niche, with 50% of the European Internet banking market and enough U.S. customers to justify a 20-person U.S. branch office. Meanwhile, encryption constitutes 10% or less of Brokat's revenue, and it has expanded its initial Internet banking offerings to include support for other forms of electronic commerce. Loss of U.S. competitiveness in the electronic commerce software market obviously raises concerns not just about encryption software but other software opportunities. Indeed, it foreshadows a weakening of the U.S. position as a leader in electronic commerce generally.

3. The persistent emphasis in U.S. export control policy over the past two years on key recovery, or "lawful access," has also taken a toll on the credibility of U.S. security products. Key recovery continues to find a market. Business wants to ensure that data are available for corporate purposes, including litigation. Key recovery is seen as an important feature for stored business data (though not for communicated data in transit).

But the use of export controls to drive the key recovery market further than it would go by itself is hurting U.S. industry. Foreign governments and competitors, particularly in Europe, have misinterpreted this U.S. policy, perhaps deliberately. In essence, foreign customers are told often by their governments as well as local security companies that all U.S. encryption products come with a back door allowing the U.S. government to read the contents. In part this is the result of outmoded "Recovery" supplements to U.S. export rules that demand an unrealistic level of U.S. government access to key recovery products. In part it reflects the hostility of many foreign governments toward U.S. key recovery and access policies. It also reflects the fact that some countries will simply never rely on security products that are not home-grown, and misunderstanding U.S. key recovery policies may simply be a handy stick to beat U.S. products with. But it is unfortunate that the U.S. government has provided such a large and easily wielded stick.

4. U.S. controls are driving many U.S. companies into "cooperative arrangements" with foreign encryption suppliers. These cooperative arrangements allow U.S. companies to provide complete security solutions by encouraging their foreign partners to marry foreign-made crypto with U.S. commercial applications. These cooperative arrangements are highly risky under U.S. law, but they are not unlawful per se. Given the stakes, many companies have been prepared to take risks under U.S. law, and it is expected that more will do the same. The result is that U.S. policy has fostered the development of cryptographic software and hardware skills outside the United States. German, Swiss, Canadian, Russian, and Israeli cryptography companies have all benefited form this unintended consequence of U.S. encryption policy.

5. The U.S. government has made efforts to "level the field" of disparate export controls for encryption through negotiations under the Wassenaar Agreement. The U.S. proposal that 56-bit encryption become a new "floor" for encryption exports under Wassenaar, while certainly better than current policy, is likely to be implemented at least a year and perhaps several years too late. In response to the U.S. KMI initiative, which conditionally decontrolled 56-bit encryption in December 1996, other countries also decontrolled 56-bit DES but more or less unconditionally. The countries include

Canada and apparently the United Kingdom. And by 1996, other countries, such as Germany, already were approving the export of 56-bit DES to virtually any country for virtually any purpose. Most recently, the exhaustion of a 56-bit DES key using a machine built for a quarter million dollars has entirely discredited DES as a serious security tool for valuable secrets. Single DES remains a useful tool for assuring privacy against a wide variety of potential adversaries and snoops, but decontrolling 56-bit encryption will not provide a significant boost to the competitiveness of U.S. technology for serious security applications.

6. Process and timing: In 1995, the State Department approved routine license applications for the export of encryption in less than a week on average. This was when the State Department had jurisdiction over encryption and NSA staffed the State Department's office and handled all encryption license applications.

This is no longer the case. The Commerce Department has staffed up heavily in the encryption field, but its processes now include parallel reviews by the FBI and NSA under a 30-day deadline that can be extended further with a simple "no" vote by either agency. For whatever reason, these agencies are now taking the full 30 days—and often 90 days. Against a backdrop of continued export liberalization over the past four years, this degradation in export control performance strikes a jarring note.

The Commerce Department's performance in this area is not necessarily out of line with the performance of other countries. The German government often takes two to three months to approve a license for a new product and six weeks to approve a license for routine shipments. The difference is that German companies know with certainty that a license will be issued at the end of the process; and the German government imposes no key recovery requirement on exporters. Therefore, they can make commitments to deliver products that require a license even before they get the license. In the United States, both the FBI and NSA have at times cast votes intended to roll back existing policies, and they have at a minimum managed to stall licenses that seemed to fit existing policy. A key recovery policy, for example, has been applied sporadically to U.S. multinationals and with some inconsistency to other exports. For this reason, it is not prudent for exporters to assume that a license will be issued or to make commitments on the assumption that the license will be issued - even when existing policy makes it seem likely that a license will eventually be granted. Because an RFP by a foreign company may provide only 30 days for responsive proposals, and the proposals often must include an assurance that an export license will be obtained, some U.S. companies lose bidding opportunities simply because the U.S. government does not process licenses quickly enough.

In other respects, of course, Commerce Department practice is a large improvement over State's performance. This is particularly true for controversial licenses, on which Commerce typically forces a decision over a course of months. In contrast, State Department licenses could be held up for months without any explanation and there were no deadlines for resolving interagency disputes. Nonetheless, it seems clear that the Commerce Department and the other participants in the encryption licensing process should adopt additional procedures to speed the granting of relatively non-controversial licenses.

CHAPTER 10

TERRORISM, CYBER SECURITY AND THE BILL OF RIGHTS

REMARKS BY RICHARD A. CLARKE
NATIONAL COORDINATOR FOR SECURITY, INFRASTRUCTURE PROTECTION, AND COUNTER-TERRORISM

American Bar Association Committee on Law and National Security

November 12, 1998

Thank you for this opportunity to address you, but I must admit that I will understand if you are a little disappointed that this is what the NSC counter-terrorism and cyber security guy looks like.

After all, I too have seen the recent movies in which in the middle of some terrorist crisis that threatens America, well, threatens at least New York City, an attractive young woman shows up, whips out a badge, and proudly announces to the relief of all: "I'm in charge, I'm from the NSC!"

I don't do that in, as they say, real life. Nor, I must admit, do I really know whether you cut the red wire or the green wire when you find a nuclear bomb ticking on First Avenue.

Popular Misconceptions about Responding to Terrorism

Beyond propagating misleading notions about the NSC, there are more troublesome misconceptions being spread by such movies and other media concerning the way in which America would, or should, respond to terrorism:

- terrorism from a rent-a-truck,
- terrorism from a chem-bio lab, or
- terrorism on the Internet.

There seems to be a growing and unfounded conventional wisdom that the response to terrorism and other threats to civil order and national security must be the erosion of rights.

I, for one, would like to dissent.

Learning From History

It is certainly true that this country and other democracies have in the past responded badly to perceived internal threats:

- suspending *habeas corpus* during the Civil War and regretting it after the fact in *Ex Parte* Mulligan;
- interning Japanese Americans during World War II and taking three decades for the Court to rule error in Korematsu;
- witch hunting in the McCarthyite Red Scare of the Cold War until Margaret Chase Smith, Edward R. Murrow and others of courage initiated the backlash.

Given George Santayana's choice of reliving the lessons of history or learning from them, I believe we have chosen the latter. When the Murrah Federal Building in Oklahoma City was attacked and its occupants murdered by terrorists, President Clinton warned against unproven assumptions about who may have done it, or what their ethnicity may have been.

He was right: Right to warn about collective guilt and right to warn about unjustified stereotyping.

When TWA 800 exploded over the waters off Long island and there was rush to assume a Middle Eastern terrorist group had attacked us, the President and the FBI warned against rushing to conclusions. After the most exhaustive investigation in history, we learned that the aircraft exploded because of a system malfunction.

We did, however, increase airport security after the crash. Members of some ethnic groups reported that they were harassed as a result of those new measures. If we had not learned from our history, we might have ignored those complaints. Instead, we in the White House, the Department of Transportation, and the FAA have worked with the carriers to insure that ethnicity is not a reason for special attention from airline security staff.

Those airline security procedures do have a price: in the cost of an airline ticket, in the amount of time you need to get to your flight. Terrorists can cause us to resort to costly measures, but they must not cost us our rights.

The reason that many terrorist groups abhor the United States is that we acknowledge and protect the rights inherent in every human. The terrorists represent those that seek to deny those rights. They seek to eliminate societies that not only protect but advocate those rights.

It is, therefore, not only trite but true that by infringing on rights in response to terrorism we do what the terrorists want: to eliminate a society that cherishes and practices civil liberties for all its people.

I would also note that the history of fighting terrorism shows that societies that respond by infringing on civil liberties enlarge the pool of support for terrorists, instead of isolating and defeating them. Draconian measures are a temporary palliative and a long term corrosive.

But pragmatism is not the fundamental consideration here: the United States government does not legally and can not morally have the authority to remove the Bill of Rights to respond to terrorism.

Terrorism, Cyber Security, and Rights

Whether the terrorists' weapon is a truck bomb or a logic bomb, we must not react by throwing out the Bill of Rights. Because of the potential of destruction and dislocation that can be caused by cyber terrorism and information warfare, we may as a society chose to institute systems that are financially costly to protect our vital computer controlled systems.

Most of what we do as a nation—our banking and finance, our electrical power, our telecommunications, our defense, our transportation systems —depend upon computers and computer controlled systems. By definition such systems are potentially vulnerable to destruction through the new cracker tools and techniques.

Thus, companies and governmental agencies are spending ever increasing amounts of time and money seeking to achieve "information assurance" or "cyber security." The costs of such efforts are passed along to the consumer and taxpayer.

Actually, the costs of not having security upgrades on computers are also passed along: in the considerable dollars it takes to restore systems crashed by crackers, in the cost of fraud when someone assumes your identify using information stored on computers. The costs can even be far greater: if a city is thrown into darkness, or a train derails, or America looses a war.

Thus, there is and will continue to be an increasing focus on systems and procedures to safeguard our computers, the information that is on them, and the functions they control. In addition to being expensive, these measures may also result in inconvenience to the network computer user: you may have to change your password from time to time, to access some systems you may have to have a smart card, in some cases you may no longer be able to dial in from your lap top at home, to access some parts of the network you may need authorization from the systems administrator.

The President has provided me clear direction that in designing systems to safe guard government computers and in working with industries to help them protect their own systems, we must do nothing that erodes civil liberties or privacy rights.

Indeed by seeking to secure government computers and urging private sector companies to protect theirs, we seek to increase privacy, to protect us all from both Big Brother and unauthorized Little Brothers who may crack into what we assume are privileged and protected files that may have information about us.

So let me once, and I hope for all time, drive a stake into certain myths that are about:

- The Administration is not proposing to expand the role of the military to defend non-military computer systems.
- The Administration is not proposing to have the military or intelligence agencies monitoring privately owned computer systems.
- We do not support Government licensing of computer personnel in the private sector, nor are we proposing any expansion in the use of polygraphing as a condition of employment.

We will develop and install software and hardware on the Federal Government's own computers to prevent unauthorized intrusion:

- so that someone can not crack into Social Security, or the VA, or the IRS and send themselves checks;
- so that crackers do not misroute airplanes;
- so that our weapon systems work when we need them to defend America or its allies.

If we fail to act quickly enough, or if we under resource this effort, or if we choose paths which do not work, and there is then an Electronic Pearl Harbor as some have suggested could happen, then there will be enormous pressure for protective measures on computer networks at any cost. We must avoid such an outcome.

We must and are working with critical sectors of the economy, industries such as electric power and telecommunications, to insure that they have all of the information they need to understand the threat to them from information warfare. We will share our ideas about Best Practices. We will eschew new regulation. We may share new software and hardware that results from our R&D on methods to prevent unauthorized cyber-trespass and intrusions onto computer systems. We expect that the flow will be two way, that the Government will have as much to learn from industry as they from us.

We will do all of this so that:

- no group of terrorists can seek to coerce us to change our policies by denying telephone service in a region;

- no criminal cartel can extort us to free prisoners or give in to some other demand by destroying a city's 911 system or throwing financial markets into confusion;
- no nation can prevent us from using our defense capabilities by a cyber attack on our rails, our electrical power grids, our airports.

All of this will be done with the Congress, with industry groups, with interest groups from the ABA and the ACLU through to any concerned group we can find beginning with the letter Z. It will be an open process.

It is my hope that we, the Administration, will have some specific ideas to share at the time of the submission of the President's FY2000 budget in January. We look forward to a continuing dialogue.

Let me leave you with one thought: It is my firm belief that we can safeguard this country and its computer controlled systems without creating either Big Brother or his cybernetic evil twin.

Thank you.

CHAPTER 11

PRESIDENT CLINTON AND VICE PRESIDENT GORE: KEEPING AMERICA SECURE FOR THE 21ST CENTURY

January 22, 1999

"We will meet the emerging security challenges, as we have met such challenges in the past, by remaining confident not only in our ideals, but in our people. We can prevail over terrorism by drawing on the very best of our free society. With strength and clear purpose, we will rise to meet all those who would intimidate us. Let us come together to protect the peace and build the better world we seek."

President Bill Clinton
January 22, 1999

Today, President Clinton delivered an address at the National Academy of Sciences where he announced major new initiatives to strengthen America's defenses against the emerging threats posed by biological and chemical weapons, and attacks to our critical infrastructure, computer systems, and networks.

Safeguarding Our Citizens From The Threat Of Deadly Weapons. President Clinton has made defending the United States against chemical and biological weapons a top priority. The President has safeguarded Americans from this emerging threat through diplomatic and military means abroad and through increased preparedness at home. The President's fiscal year 2000 budget includes requests for:

- *An additional $40 million in research and development funding for the Department of Health and Human Services,* with the bulk of that funding going to research on new vaccines, including smallpox and anthrax, for eventual use in the national medical stockpile;
- *A 23 percent increase in funding for improvements in the public health surveillance system,* including, increased lab capacities, strengthened epidemiological capabilities for state and local health departments and more resources for communications and information technology. This funding will also allow the Centers for Disease Control to create a network of regional labs to provide rapid analysis of select biological agents;
- *Additional funding for Metropolitan Medical Response Systems,* medical teams who respond to biological and chemical weapons emergencies. This funding will support 25 new response teams.

Cyber-Security For the 21st Century. The President is calling for an investment of $1.46 billion in fiscal year 2000 to defend our critical infrastructure, including, power-generation systems, banking, transportation and emergency services and telecommunications. This investment will help secure computer systems and networks that are potentially vulnerable to computer attack. The President's proposal calls for:

- *Research and Development funding to safeguard key computer systems,* with a focus on developing tools that can identify potentially threatening activities within computer networks or computer code installed to disrupt a computer system;
- *Intrusion Detection Systems* modeled on one initiated at the Department of Defense that will ensure that when one computer system is attacked, others in the network will be instantly informed of the source of the intrusion, the mode of attack, and the recommended ways to stop it;
- *Information Sharing and Analysis Centers (ISAC's).* The Clinton Administration will work with private industry to establish ISAC's to foster the development of uniform practices and standards to protect against computer attack, encourage the sharing of vulnerability analysis, and provide outreach and training programs;
- *Recruiting a Cyber Corps to respond to computer crises.* Under the President's plan, government agencies will be encouraged to recruit expert level computer security workers to respond to future computer crises. This program will use existing scholarship and financial assistance programs and examine new scholarship programs to retain, retain, and recruit computer science students.

CHAPTER 12

FUNDING FOR DOMESTIC PREPAREDNESS AND CRITICAL INFRASTRUCTURE PROTECTION

THE WHITE HOUSE
Office of the Press Secretary
January 22, 1999

FACT SHEET

Funding for Domestic Preparedness and Critical Infrastructure Protection

The President's Fiscal Year 2000 budget includes requests for $2.849 billion for critical infrastructure protection, computer security, and domestic preparedness against a weapons of mass destruction attack.

The budget request also proposes $7.162 billion for conventional counterterrorism security programs.

Domestic Preparedness against Weapons of Mass Destruction

In May 1999 the President proposed adding $300 million for a new weapons of mass destruction domestic preparedness program. As a result, the 1999 enacted level was $1.281 billion. The President's FY 2000 funding request for countering the threat of terrorist use of weapons of mass destruction continues and expands the program to $1.385 billion.

The FY 2000 request would include increases of $30 million above the previous level for research into new vaccines and medicines, an additional $15 million to fund Public Health Surveillance to detect an attack, and an additional $13 million to create new metropolitan medical response teams.

Highlights of the FY 2000 budget include:

$52 million to continue procurement of a national stockpile of specialized medicines to protect the civilian population.

$611 million for training and equipping emergency personnel in U.S. cities, planning and exercising for weapons of mass destruction contingencies and strengthening public health infrastructure.

$206 million to protect U.S. government facilities.

$381 for research and development, including pathogen genome sequencing, vaccines, new therapies, detection and diagnosis, decontamination, and disposition of nuclear material.

Critical Infrastructure Protection and Computer Security

The President's FY 2000 request includes $1.464 billion for protection of critical infrastructure and computer security. This represents a 40% increase in the two budget years since the President created the Critical Infrastructure Protection Commission.

The highlights of this program include:

Critical Infrastructure Applied Research Initiative ($500 million).

Intrusion and Detection Systems: In addition to ongoing Department of Defense funding, $2 million will be spent to design and evaluate a similar system for other Federal agencies.

Information Sharing and Analysis Centers (ISACs): As part of the public-private partnership, we will provide $8 million to support the initial establishment of ISACs.

Cyber Corps: This program will address the shortage of highly skilled computer science expertise in the government and enable agencies to recruit a cadre of experts to respond to attacks on computer networks. It will use existing personnel flexibilities, scholarship and financial assistance programs, and $3 million to examine new scholarship programs to retrain, retain and recruit computer science students.

Counter-terrorism Security

In addition to the programs above, the President's FY 2000 budget request for all anti-terrorism and counter-terrorism programs is $8.547 billion, a 12% increase over the FY 1999 enacted level and an 18% increase over FY1998.

The President also requested a supplemental appropriation in FY 1999 of $2.064 billion after the Africa bombings. This includes $1.4 billion to provide additional security measures to diplomatic and consular facilities and rebuild the two embassies destroyed in Dar es Salaam and Nairobi.

CHAPTER 13

KEEPING AMERICA SECURE FOR THE 21ST CENTURY: COMPUTER SECURITY AND CRITICAL INFRASTRUCTURE

THE WHITE HOUSE
Office of the Press Secretary
January 22, 1999

FACT SHEET

Today, President Clinton will announce new initiatives to defend the nation's computer systems and critical infrastructure from cyber-terrorism. The most critical sectors of our economy—power-generation, telecommunications, banking, transportation and emergency services—are potentially vulnerable to disruptions from computer attack.

President Clinton will propose in the Fiscal Year 2000 budget spending $1.46 billion to defend against this emerging threat. This represents an increase of $400 million from the FY 1999 budget proposal. It will include funding for the following initiatives:

Critical Infrastructure Applied Research Initiative: The President's budget proposal includes $500 million for research and development efforts. A portion of these funds will be spent on new initiatives to improve information assurance by safeguarding networks. Funds will also be dedicated to developing tools that can identify anomalous activities and "Trojan Horses" (malicious codes installed by unauthorized users).

Computer Intrusion Detection Networks: The Defense Department has already begun to install intrusion detection systems and create a network to warn key computers of an attack. Under the President's initiative, a similar system for other Federal agencies will be evaluated and designed. These networks will ensure that when one computer system is attacked, others in the network will be instantly warned of the intrusion, informed of the mode of attack used, and provided with methods to stop it.

Information Sharing and Analysis Centers (ISACs): As part of the public-private partnership, we will support the initial establishment of ISACs to foster private sector development of best practices and standards for computer security, to encourage the sharing of vulnerability analysis, and to provide outreach and training programs. These ISACs will enable the Federal government to provide private industry with threat information without compromising privacy, civil liberties or proprietary data.

Cyber Corps: This program will address the shortage of highly skilled computer science expertise in the government and enable agencies to recruit a cadre of experts to respond to attacks on computer networks. It will use existing personnel flexibilities, scholarship and financial assistance programs, and examine new scholarship programs to retrain, retain and recruit computer science students.

CHAPTER 14

REMARKS BY THE PRESIDENT ON KEEPING AMERICA SECURE FOR THE 21ST CENTURY

National Academy of Sciences
Washington, D.C.

THE WHITE HOUSE
Office of the Press Secretary
January 22, 1999

THE PRESIDENT: Thank you very much. Jamie, Dr. Lederberg, I'd like to thank you for your service in this and so many other ways. I would like to thank Sandy Berger for many things, including indulging my nagging on this subject for the better part of six years now.

I was so relieved that Dr. Lederberg not very long ago—well, last year—brought a distinguished panel of experts together to discuss this bioterrorism threat, because I then had experts to cite on my concern and nobody thought I was just reading too many novels late at night. (Laughter.)

Madame Attorney General, Secretary Shalala, Secretary Richardson, Director Witt, Deputy Secretary Hamre, Commandant of the Coast Guard and our other military leaders who are here, Mr. Clarke, ladies and gentlemen. I'm delighted to be here to discuss this subject. With some trepidation, Sandy Berger noted that Dr. Lederberg won a Nobel Prize at 33, and I was governor—you can infer from that that I was not very good at chemistry and biology. (Laughter.)

But any democracy is imbued with the responsibility of ordinary citizens who do not have extraordinary expertise to meet the challenges of each new age. And that is what we are all trying to do. Our country has always met the challenges of those who would do us harm. At the heart of our national defense I have always believed is our attempt to live by our values—democracy, freedom, equal opportunity. We are working hard to fulfill these values at home. And we are working with nations around the world to advance them, to build a new era of interdependence where nations work together—not simply for peace and security, but also for better schools and health care, broader prosperity, a cleaner environment and a greater involvement by citizens everywhere in shaping their own future.

In the struggle to defend our people and values and to advance them wherever possible, we confront threats both old and new—open borders and revolutions in technology have spread the message and the gifts of freedom but have also given new opportunities to freedom's enemies.

Scientific advances have opened the possibility of longer, better lives. They have also given the enemies of freedom new opportunities.

Last August, at Andrews Air Force Base, I grieved with the families of the brave Americans who lost their lives at our embassy in Kenya. They were in Africa to promote the values America shares with friends of freedom everywhere—and for that they were murdered by terrorists. So, too, were men and women in Oklahoma City, at the World Trade Center, Khobar Towers, on Pan Am 103.

The United States has mounted an aggressive response to terrorism—tightening security for our diplomats, our troops, our air travelers, improving our ability to track terrorist activity, enhancing cooperation with other countries, strengthening sanctions on nations that support terrorists.

Since 1993, we have tripled funding for FBI anti-terrorist efforts. Our agents and prosecutors, with excellent support from our intelligence agencies, have done extraordinary work in tracking down perpetrators of terrorist acts and bringing them to justice. And as our air strikes against Afghanistan—or against the terrorist camps in Afghanistan—last summer showed, we are prepared to use military force against terrorists who harm our citizens. But all of you know the fight against terrorism is far from over. And now, terrorists seek new tools of destruction.

Last May, at the Naval Academy commencement, I said terrorist and outlaw states are extending the world's fields of battle, from physical space to cyberspace, from our earth's vast bodies of water to the complex workings of our own human bodies. The enemies of peace realize they cannot defeat us with traditional military means. So they are working on two new forms of assault, which you've heard about today: cyber attacks on our critical computer systems, and attacks with weapons of mass destruction—chemical, biological, potentially even nuclear weapons. We must be ready—ready if our adversaries try to use computers to disable power grids, banking, communications and transportation networks, police, fire and health services—or military assets.

More and more, these critical systems are driven by, and linked together with, computers, making them more vulnerable to disruption. Last spring, we saw the enormous impact of a single failed electronic link, when a satellite malfunctioned—disabled pagers, ATMs, credit card systems and television networks all around the world. And we already are seeing the first wave of deliberate cyber attacks—hackers break into government and business computers, stealing and destroying information, raiding bank accounts, running up credit card charges, extorting money by threats to unleash computer viruses.

The potential for harm is clear. Earlier this month, an ice storm in this area crippled power systems, plunging whole communities into darkness and disrupting daily lives. We have to be ready for adversaries to launch attacks that could paralyze utilities and services across entire regions. We must be ready if adversaries seek to attack with weapons of mass destruction, as well. Armed with these weapons, which can be compact and inexpensive, a small band of terrorists could inflict tremendous harm.

Four years ago, though, the world received a wake-up call when a group unleashed a deadly chemical weapon, nerve gas, in the Tokyo subway. We have to be ready for the possibility that such a group will obtain biological weapons. We have to be ready to detect and address a biological attack promptly, before the disease spreads. If we prepare to defend against these emerging threats we will show terrorists that assaults on America will accomplish nothing but their own downfall.

Let me say first what we have done so far to meet this challenge. We've been working to create and strengthen the agreement to keep nations from acquiring weapons of mass destruction, because this can help keep these weapons away from terrorists, as well. We're working to ensure the effective implementation of the Chemical Weapons Convention; to obtain an accord that will strengthen compliance with the biological weapons convention; to end production of nuclear weapons material. We must ratify the Comprehensive Test Ban Treaty to end nuclear tests once and for all.

As I proposed Tuesday in the State of the Union Address, we should substantially increase our efforts to help Russia and other former Soviet nations prevent weapons material and knowledge from falling into the hands of terrorists and outlaw states. In no small measure we should do this by continuing to expand our cooperative work with the thousands of Russian scientists who can be used to advance the causes of world peace and health and well-being, but who if they are not paid, remain a fertile field for the designs of terrorists.

But we cannot rely solely on our efforts to keep weapons from spreading. We have to be ready to act if they do spread. Last year, I obtained from Congress a 39 percent budget increase for chemical and biological weapons preparedness. This is helping to accelerate our ongoing effort to train and equip fire, police and public health personnel all across our country to deal with chemical and biological emergencies. It is helping us to ready armed forces and National Guard units in every region to meet this challenge; and to improve our capacity to detect an outbreak of disease and save lives; to create the first ever civilian stockpile of medicines to treat people exposed to biological and chemical hazards; to increase research and development on new medicines and vaccines to deal with new threats.

Our commitment to give local communities the necessary tools already goes beyond paper and plans. For example, parked just outside this building is a newly designed truck we have provided to the Arlington,

Virginia, Fire Department. It can rapidly assist and prevent harm to people exposed to chemical and biological dangers.

But our commitment on the cyber front has been strong, as well. We've created special offices within the FBI and the Commerce Department to protect critical systems against cyber attack. We're building partnerships with the private sector to find and reduce vulnerabilities; to improve warning systems; to rapidly recover if attacks occur. We have an outstanding public servant in Richard Clarke, who is coordinating all these efforts across our government.

Today, I want to announce the new initiatives we will take, to take us to the next level in preparing for these emerging threats. In my budget, I will ask Congress for $10 billion to address terrorism and terrorist-emerging tools. This will include nearly $1.4 billion to protect citizens against chemical and biological terror—more than double what we spent on such programs only two years ago.

We will speed and broaden our efforts, creating new local emergency medical teams, employing in the field portable detection units the size of a shoe box to rapidly identify hazards; tying regional laboratories together for prompt analysis of biological threats. We will greatly accelerate research and development, centered in the Department of Health and Human Services, for new vaccines, medicines and diagnostic tools.

I should say here that I know everybody in this crowd understands this, but everyone in America must understand this: the government has got to fund this. There is no market for the kinds of things we need to develop; and if we are successful, there never will be a market for them. But we have got to do our best to develop them. These cutting-edge efforts will address not only the threat of weapons of mass destruction, but also the equally serious danger of emerging infectious diseases. So we will benefit even if we are successful in avoiding these attacks.

The budget proposal will also include $1.46 billion to protect critical systems from cyber and other attacks. That's 40 percent more than we were spending two years ago. Among other things, it will help to fund four new initiatives. First, an intensive research effort to detect intruders trying to break into critical computer systems. Second, crime —excuse me—detection networks, first for our Defense Department, and later for other key agencies so when one critical computer system is invaded, others will be alerted instantly. And we will urge the private sector to create similar structures.

Third, the creation of information centers in the private sector so that our industries can work together and with government to address cyber threats. Finally, we'll ask for funding to bolster the government's ranks of highly skilled computer experts—people capable of preventing and responding to computer crises.

To implement this proposal, the Cyber Corps program, we will encourage federal agencies to train and retrain computer specialists, as well as recruiting gifted young people out of college.

In all our battles, we will be aggressive. At the same time I want you to know that we will remain committed to uphold privacy rights and other constitutional protections, as well as the proprietary rights of American businesses. It is essential that we do not undermine liberty in the name of liberty. We can prevail over terrorism by drawing on the very best in our free society—the skill and courage of our troops, the genius of our scientists and engineers, the strength of our factory workers, the determination and talents of our public servants, the vision of leaders in every vital sector.

I have tried as hard as I can to create the right frame of mind in America for dealing with this. For too long the problem has been that not enough has been done to recognize the threat and deal with it. And we in government, frankly, weren't as well organized as we should have been for too long. I do not want the pendulum to swing the other way now, and for the people to believe that every incident they read about in a novel or every incident they see in a thrilling movie is about to happen to them within the next 24 hours.

What we are seeing here, as any military person in the audience can tell you, is nothing more than a repetition of weapons systems that goes back to the beginning of time. An offensive weapons system is developed, and it takes time to develop the defense. And then another offensive weapon is developed that overcomes that defense, and then another defense is built up—as surely as castles and moats held off people with spears and bows and arrows and riding horses, and the catapult was developed to overcome the castle and the moat.

But because of the speed with which change is occurring in our society—in computing technology, and particularly in the biological sciences—we have got to do everything we can to make sure that we close the gap between offense and defense to nothing, if possible. That is the challenge here.

We are doing everything we can, in ways that I can and in ways that I cannot discuss, to try to stop people who would misuse chemical and biological capacity from getting that capacity. This is not a cause for panic — it is a cause for serious, deliberate, disciplined, long-term concern. And I am absolutely convinced that if we maintain our clear purpose and our strength of will, we will prevail here. And thanks to so many of you in this audience, and your colleagues throughout the United States, and like-minded people throughout the world, we have better than a good chance of success. But we must be deliberate, and we must be aggressive.

Thank you very much. (Applause.)

CHAPTER 15

PRESS BRIEFING BY ATTORNEY GENERAL JANET RENO, SECRETARY OF HHS DONNA SHALALA, AND RICHARD CLARKE, PRESIDENT'S NATIONAL COORDINATOR FOR SECURITY, INFRASTRUCTURE AND COUNTERTERRORISM

THE WHITE HOUSE
Office of the Press Secretary
January 22, 1999

The Briefing Room
11:45 A.M. EST

MR. LEAVY: As you know, the President announced an expansion of the administration's effort to combat the emerging threats of biological, chemical and cyber terrorism. Today, just to go over the details of that and to answer your questions are the President's National Coordinator for Security, Infrastructure, and Counterterrorism, Dick Clarke; Attorney General Reno; and Secretary of Health and Human Services Shalala.

Dick.

MR. CLARKE: The message that we want to get across today is not that we know of an imminent attack—we do not know of any imminent attack being planned on that United States using chemical or biological weapons, or using cyber attack techniques. But we do want to raise consciousness, in the American people, in the scientific community, in the corporate community, and in the Congress, that such attacks are growing increasingly likely.

And as the President said, we need to be ahead of the power curve; we need to be prepared to defend ourselves against those attacks, and in so doing, perhaps to prevent them; at least to be able to mitigate their effects.

The President's announcement today puts our money where our policy is. It's a proposal to spend next year $2.8 billion defending against these two types of threats. That's an increase of 40 percent in federal expenditures over two years ago for cyber defense, to defend America's cyberspace. It is a doubling of the funds over two years for chemical and biological weapons defense.

There are several initiatives within that program. There's $500 million on the critical infrastructure applied research initiative. There's $400 million to research defenses against chemical and biological weapons. There are plans for two networks to defend DOD and federal computer systems. There is a cyber core, a core of federal employees highly trained and skilled who would operate federal government computers, defend them against attack and be able to respond to an attack. There are funds to assist industry groups, which would come together to form their own computer defense facilities. And there is a reinvigoration of the public health surveillance system, to allow us to detect chemical and biological attacks when they occur.

Here to explain the background to all of this are two Cabinet members who have been personally vigorous in urging the President to increase funding, to increase attention on these programs. First, the Attorney General.

ATTORNEY GENERAL RENO: Thank you, Mr. Clarke.

As the President said this morning, our nation has benefitted greatly from cyber technology. It makes America work better than ever before. Intricate networks, power grids and computer systems make up what we call our national infrastructure. If that infrastructure is attacked, we all suffer. That is why we must, and we are, doing everything we can to protect it.

At the Department of Justice we have taken steps to protect it. At the Department of Justice, we have taken steps to protect out critical infrastructure. Since 1998 we have launched the National Infrastructure Protection Center. The Center's mission is to detect, prevent and respond to cyber attacks on our nation's critical infrastructure, and to oversee investigations in this field. It is a true interagency, public-private partnership. Our goal is to have federal agencies working together with state and local officials, working together with the private sector, to exchange information to provide for early detection, to understand the technical issues involved, and to be prepared to prevent it whenever possible.

To what the President has announced today builds on this effort and solidifies this administration's commitment to the protection of our infrastructure, which is so vital to this nation.

And now I'd like to recognize Donna Shalala.

SECRETARY SHALALA: Thank you, Janet. Thank you very much.

We're all here talking about a kind of scenario that we hope that our citizens never have to confront. But the point is to be prepared. And my job—and this is the first time in American history in which the public health system has been integrated directly into the national security system—is to be able to provide tracking and treatment for victims.

And one way of thinking about this is if there is an outbreak—Peggy Hamburg's (phonetic) here, who used to run the New York City Health Department—if there's an outbreak on the New York City subway system, our ability to deal with that depends on the strength of the health system in

New York City—their ability to track and to manage whatever the outbreak is and individual patients as they go to different parts of the health care system.

The tracking system becomes very important, the surveillance system, because, frankly, if there is an outbreak in any major metropolitan area, people spread all over that metropolitan area. And it's our ability to manage the consequences of the outbreak through a health system—not just the public health system, but also the private health system. That kind of surveillance system, that kind of tracking system at the national level, working with local officials is what we've been building here, starting, actually, when we started in the administration, but with a lot more energy over the last two years because of the President's keen interest and the investment of substantial resources.

This budget that the President is proposing is $230 million for fiscal year 2000. That's up from $158 million this year, and it has four components for us—strengthening disease surveillance and the public health network. We're working with the states, of course, to improve their disease surveillance. The medical and the public health system and the medical systems here so that they work together. The surveillance systems are, in general, in this country run by the public health side; the delivery systems, treating the victims, run by the medical people. And we're merging the two of them for these purposes.

The pharmaceutical stockpiles—we have to be assured that the volume and the range of needed pharmaceuticals can be made available quickly. These are not just vaccines we're talking about, we're also talking about antibiotics and about vaccinating large numbers of people within a relatively short period of time as part of the response; and about the research that's needed to do that as well as the stockpiling facilities.

And finally, the research components. We need to better understand the disease agents of bioterrorism. We also need not only vaccines, but the treatment protocols. We have a number of them; now we're upgrading them. This is a continuous improvement strategy.

What the President announced today is significant and the integration, I can't emphasize enough, of the health leaders in this country—from me to the local leaders to the state leaders—into this, which is essentially a systems approach, is very significant.

Question: What is this story that Saddam is trying to perfect some sort of agent that won't attack Arabs, but attacks Westerners?

SECRETARY SHALALA: That actually would make it pretty good for me and Helen. (Laughter.) And, Sam, you're in big trouble. (Laughter.)

I think the point that we're making is whatever it is, this country intends to be ready and to have a system in place to respond.

Question: Have you heard that? Is that just an old wives' tale or is there actually —

MR. CLARKE: I think that's an old wives' tale.

SECRETARY SHALALA: I think it's actually an old Arab's tale.

Question: Madam Secretary, is there any city in America that can identify and cope with a biological weapons attack today?

SECRETARY SHALALA: There are cities who have been putting systems in place for some period of time, and that includes New York, where Peggy Hamburg, Dr. Hamburg, provided the leadership. She's now the Assistant Secretary of the Department in charge of the government's health pieces of this bioterrorism network. And that's a significance, I think, of her appointment, too, because she's put it together at the local level. So there are cities.

The point is we want every place in the United States to be prepared. Remember, we have disaster systems in this country. So it's not a new subject matter to us. It's the introduction of bioterrorism and the fact that if there is an outbreak some place, calling in the Marines is not the solution. It is making sure that you have a public health system, a surveillance and a treatment system, as well as whatever the judicial and the military responses are that are appropriate.

Question: Madam Secretary, how do you respond to critics who say that by its very nature you cannot predict a terrorist attack and that this is just sort of throwing money away, and it's a political—to make people feel better?

SECRETARY SHALALA: You also can't predict natural disasters, as we have learned very well in this country in the last 48 hours, for example. So the point is to have a system in place for surveillance and treatment, so that what we're trying to do is to anticipate. And that's what good leadership is about.

Question: Madam Secretary, do we have any defense against ICBM missiles that would distribute this kind of —

SECRETARY SHALALA: I think you've got the wrong—(laughter.)

Question: Whoever it is, do we have any defense against ICBM missiles that might deliver this kind of thing?

MR. CLARKE: Let me just say we have no intelligence that indicates there is an ICBM-based threat carrying chemical or biological weapons.

Question: What about the threat to rain missiles on Los Angeles?

MR. CLARKE: I'll try again. We have no intelligence that indicates there is a current ICBM threat against the United States with chemical or biological warheads. One of the reasons that the President is looking at a limited ballistic missile defense system, and one of the reasons behind this chemical and biological defense preparation initiative, is our concern that that kind of threat could emerge in the future. That's one of the reasons we're doing all of this.

But as I said at the top, we don't know of any such threat today.

Question: Attorney General, is there a civil liberties issue here? The ACLU seems to think that in setting up a national program and coordination, there may be.

ATTORNEY GENERAL RENO: What I have instructed the Center to do—the National Infrastructure Protection Center—and the FBI, is to work with our lawyers. And we have recruited and trained lawyers who have the technical information and expertise together with the legal expertise to ensure that what we do complies with the Constitution, complies with privacy rights. And we believe firmly that we can continue to meet our obligations in law enforcement in this era of new technology while at the same time complying with the Constitution in every way.

Question: That hasn't been worked out yet—in other words, they're looking at it?

ATTORNEY GENERAL RENO: No, they're working on it daily, as we address issues of search and seizure, as we address issues in computer crime. We are addressing those issues daily. And so far to my satisfaction at least, based on everything I know, we've have been addressing those in accordance with the Constitution.

Question: Attorney General Reno, the President today described the first wave of cyber terrorism from attacks by hackers at this point—computer systems from universities to financial institutions. Can you tell us how big of a threat that is?

ATTORNEY GENERAL RENO: I'd like to address—there are many young people who are far better skilled in cyber tools than I am. They know a lot more about computers than I do. And some of them just don't think it's wrong. They have become very able and adept on computers and they like to challenge themselves and see what they can do. One of the things—one of the points that has been made to me is we've got to let everyone know that it is wrong to invade another person's computer. It is wrong to invade and upset and confuse the data base or disturb the data base. And I think this is one of the responsibilities that we all face and that our public schools face in terms of preparing ourselves for the cyber age.

Question: But aside from it being wrong, how much of a threat is it?

ATTORNEY GENERAL RENO: I think it's a problem that we deal with and we trace and we, again, take appropriate action as based on prosecutions that have been successful to date.

Question: Attorney General, the President said in the New York Times interview that this is the kind of threat that keeps him awake at night. You, too? Or do you—I mean, do you worry about this?

ATTORNEY GENERAL RENO: After six years in this job, I get a good night's sleep. (Laughter.)

MR. CLARKE: Let me expand a moment on the previous question about the hackers being a threat. We're not talking about a few teenagers violating the law and getting into a computer system and having some fun—even though it's illegal and it's serious and we have to deal with it. That's not the threat the President's talking about.

The President's talking about something called information warfare, where a nation or a terrorist group or a criminal cartel could do a systematic national intrusion into the computer systems that control the electric power grids, the telephone networks, the banking and finance system, the transportation nodes, and effectively shut the nation off. In other words, just as in World War II, nations flew bombers over each other's countries to try to destroy infrastructure by dropping bombs. What we're concerned about is in the future nations will have that same capability to destroy each other's infrastructure, not by bombs, but by cyber attack. Now, we can prevent that if we have cyber defenses.

Question: Mr. Clarke and Ms. Reno, Los Angeles is one of the cities that have received both chemical and biological training. And yet in the anthrax hoaxes that we've seen, citizens of Los Angeles have been held for four hours and even longer while people look for anthrax, which is not contagious. What's going on here? Are people—are the trainees, the people who have been trained, not getting the message, or—and are civil liberties being violated by these unnecessary detentions?

ATTORNEY GENERAL RENO: What we are trying to do in the effort that is now underway, which Congress has approved and the administration is pursuing, is we have proposed the development of a national domestic preparedness office. We are working, if that is accepted, we are working with the Department of Health and Human Services, the Department of Defense, FEMA, the specialists, the experts to number one, identify the best means of detection; number two, to work in conjunction with the public health and the medical community in determining what is the appropriate response; and working with lawyers at the Department of Justice to determine what are the implications with respect to privacy and to civil rights.

Question: Attorney General Reno, Mr. Clarke, whenever the government starts sharing information about threats with big companies and not with the general public, that raises certain concerns. Are those fair concerns, and how do you plan to address them?

ATTORNEY GENERAL RENO: This is one of the issues that we are working with state and local authorities, as well as with the private sector, with emergency managers at the state and local level, to make sure that we get out the information that is appropriate and relevant in the best way possible. There are a number of issues involved—how do we disseminate classified information—and we're working through those issues; how do we disseminate information so that it does not unduly alarm people. I think in whatever role you're in, whether it's in forecasting tornadoes or in these circumstances, you've got to balance fully informing the public with not frightening them into inappropriate action.

Question: Mr. Clarke, you had said that there is no apparent threat of ICBM. Was there any apparent threat before the Iraqis starting showering missiles on Tel Aviv?

MR. CLARKE: What I said was we have no indication that there are chemical or biological weapons on warheads that can reach the United States today. In the case of Iraq in 1991, where I happen to have some background, we had a very good idea that they had Scud missiles and that they might attack Israel. That was part of the assumption that went into the planning of the war. We thought we knew how many they were; we thought we knew where they were. Similarly, today, we have a very good idea we knew where they were. Similarly, today, we have a very good idea about where ICBMs are and how many ICBMs there are that can reach the United States.

Question: But what defense do we have against these?

MR. CLARKE: As I said, we do not now have a defense against ICBMs. The President has proposed that we look at a limited ABM defense system. That's one of the things that the Secretary of State, who leaves tomorrow for Moscow, will be discussing with the Russians next week.

Thank you.

CHAPTER 16

INFORMATION TECHNOLOGY FOR THE TWENTY-FIRST CENTURY: A BOLD INVESTMENT IN AMERICA'S FUTURE

THE WHITE HOUSE
Office of the Vice President
January 24, 1999

As part of their FY2000 budget, President Clinton and Vice President Gore are proposing a $366 million, 28 percent increase in the government's investment in information technology research.

This initiative, known as IT2 (Information Technology for the Twenty-First Century), will support three kinds of activities:

- Long-term information technology research that will lead to fundamental advances in computing and communications, in the same way that government investment beginning in the 1960's led to today's Internet;
- Advanced computing for science, engineering and the Nation that will lead to breakthroughs such as reducing the time required to develop life-saving drugs; designing cleaner, more efficient engines; and more accurately predicting tornadoes; and
- Research on the economic and social implications of the information Revolution, and efforts to help train additional IT workers at our universities.

The potential benefits of IT2 are compelling:

- The results of past government research (e.g. the Internet, the first graphical Web browser, advanced microprocessors) have helped strengthen American leadership in the IT industry, which now accounts for 1/3 of U.S. economic growth and employs 7.4 million Americans at wages that are more than 60 percent higher than the private sector average. All sectors of the U.S. economy are using IT to compete and win in global markets, and business-to-business electronic commerce in the U.S. alone is projected to grow to $ 1.3 trillion by 2003.
- Information technology is changing the way we live, work, learn, and communicate with each other. Advances in IT can improve the way we educate our children, allow people with disabilities to lead more independent lives, and improve the quality of health care for rural Americans through telemedicine. U.S. leadership in IT is also essential for our national security.

- Information technology will also lead to a "golden age" of science and engineering. Advances in supercomputers, simulation and networks are creating a new window into the natural world—making IT as valuable as theory and experimentation as a tool for scientific discovery. With computers can make trillions of calculations in a second, scientists and engineers will be able to better predict the impact of climate change, design more efficient and cleaner energy systems, and gain new insights into the fundamental nature of matter.

The initiative builds on previous and current programs in computing and communications, including the High-Performance Computing and Communications program (authorized by legislation introduced by then-Senator Gore), and the Next Generation Internet, authorized by the Congress in 1998. It responds to recommendations made by a private sector advisory committee requested by the Congress (the President's Information Technology Advisory Committee), which concluded that the government was underinvesting in long-term IT research relative to its importance to the Nation. This committee, which is comprised of leaders from industry and academia, concluded that the private sector was unlikely to invest in the long-term, fundamental IT research needed to sustain the information Revolution. The initiative also reflects a strong belief in the research community about the potential of IT to accelerate the pace of discovery in all science and engineering disciplines.

The agencies that will be involved in IT2 include the National Science Foundation, the Department of Defense (including DARPA), the department of Energy, NASA, the National Institutes of health, and the National Oceanic and Atmospheric Administration. Roughly 60 percent of the funding will go to support university-based research, which will also help meet the growing demand for workers with advanced IT skills.

Some of the potential breakthroughs that may be possible as a result of IT2 include:

- Computers that can speak, listen and understand human language, are much easier to use, and accurately translate between languages in real-time;
- "Intelligent agents" that can roam the Internet on our behalf, retrieving and summarizing the information we are looking for in an vast ocean of data;
- A wide range of scientific and technological discoveries made possible by simulations running on supercomputers, accessible to researchers all over the country;

- Networks than can grow to connect not only tens of millions of computers, but hundreds of billions of devices;
- Computers that are thousands of times faster than today's supercomputers, or are based on fundamentally different technology, such as biological or quantum computing; and
- New ways of developing complex software that is more reliable, easier to maintain, and more dependable for running the phone system, the electric power grid, financial markets, and other core elements of our infrastructure.

VICE PRESIDENT GORE: LEADING AMERICA IN THE INFORMATION AGE

For decades, Al Gore has worked to lead the America in the Information Age by offering an exciting vision of the potential of information technology, strengthening America's technological leadership, breaking down the barriers to private sector investment and job creation, putting the future at the fingertips of our children, and ensuring that all Americans have an opportunity to make the most of their lives in the Information Age.

- More than 20 years ago, as a member of Congress, he first popularized the term "information superhighway."
- In 1984, he introduced legislation to promote the development and distribution of high-quality, interactive educational software.
- In 1989, he introduced legislation to authorize a coordinated Federal research program in high-performance computing and communications. He re-introduced this legislation in 1991, which was signed into law later that year as the High-Performance Computing Act. This legislation expanded investment in research networks which led to today's Internet.
- In 1993, as Vice President, he unveiled the Administration's National Information Infrastructure agenda. This agenda called for increased competition in the telecommunications market, more dissemination of government information on the Internet, greater allocation of spectrum for new wireless industries, enhanced privacy protection, and pilot projects of non-profit applications of the Internet and information technology.
- In 1994, he set a national goal of connecting every classroom and library in the United States to the Internet. Two years later, the Telecommunications Act of 1996 provided deep discounts to make Internet access affordable for schools and library. With Al Gore's leadership, the Administration increased its investment in educational technology (com-

puters, software, teacher training) from $23 million in 1993 to over $700 million in 1998.

- Also in 1994, he set forth his vision of a "Global Information Infrastructure" at the International Telecommunications Union. His principles are adopted at the G-7, the ITU, the Summit of the Americas, and APEC.
- In 1996, he called for the creation of the Next Generation Internet, which is 1,000 times faster than today's Internet. Legislation authorizing the NGI is passed by the Congress in 1998.
- In 1998, he called for an "Electronic Bill of Rights" to protect personal privacy. Later that year, the Congress passed legislation that the Vice President had championed on identity theft and children's privacy. The Vice President's leadership was also critical in advancing policies to promote electronic commerce, which is expected to grow to $1.3 trillion by the year 2003 in the U.S. alone.
- Today, the Vice President is announcing a new $366 million increase in long-term information technology research, which will strengthen America's position as the global leader in computing and communications, help create the industries and the high-tech jobs of the 21st century, and accelerate the pace of scientific discovery with high-end computers and simulation.

EXTENDING THE RESEARCH TAX CREDIT: HIGH-TECH JOBS FOR AMERICA

Today, Vice President Gore announced that the Administration will propose extending the Research and Experimentation tax credit for one year, at a cost of $2.4 billion: The R&E credit is currently scheduled to expire on June 1999; the Administration's proposal would extend it to June 30, 2000.

The R&E credit helps stimulate additional private sector investment in research and development: This incentive provides a 20 percent tax credit based on the increase in a firm's research and development.

The R&E credit encourages technological advancement, leads to higher productivity, and helps generate new American jobs:

- Entirely new industries are created as a result of technological innovation. The credit is helpful for R&D-intensive industries such as the information, communications, and electronics sector. Jobs in the high-tech sector and occupations that use information technology pay 60 - 70 percent more than average private sector wages. The credit also benefits major industries such as chemicals, machinery, and motor vehicles.

- Most research and development dollars are directly invested in the salaries of U.S. employees. The credit is only available for research conducted in the United States.
- Economists estimate that half of U.S. productivity growth stems from technological advances.

Private sector R&D investment also leads to a better quality of life for all Americans

- R&D also leads to innovations such as:
  - New life-saving drugs;
  - A more advanced telecommunications infrastructure, capable of transmitting voice, video and high-speed data; and
  - Cleaner and more environmentally-friendly sources of energy.

CHAPTER 17

PRESIDENT'S INFORMATION TECHNOLOGY ADVISORY COMMITTEE REPORT TO THE PRESIDENT

Information Technology Research: Investing in Our Future

February 1999

Table of Contents

Letter of Transmittal

Members of the Advisory Committee

Acknowledgments

Executive Summary

- Findings and Recommendation
- Priorities for Research
- The Government's Essential Role
- Conclusion

Rationale for Government Support of Long Term, Fundamental Research

1. Information Technology: Transforming our Society
   - 1.1 Transforming the Way We Communicate
   - 1.2 Transforming the Way We Deal With Information
   - 1.3 Transforming the Way We Learn
   - 1.4 Transforming the Practice of Health Care
   - 1.5 Transforming the Nature of Commerce
   - 1.6 Transforming the Nature of Work
   - 1.7 Transforming How We Design and Build Things
   - 1.8 Transforming How We Conduct Research
   - 1.9 Transforming Our Understanding of the Environment
   - 1.10 Transforming Government

Table of Transformations

2. Setting Federal Research Priorities: Findings and Recommendations
   - 2.1 Findings
   - 2.2 Recommendations for Research

3. Technical Research Priorities

3.1 Software Research

3.1.1 Findings

3.1.2 Recommendations

3.2 Scalable Information Infrastructure

3.2.1 Findings

3.2.2 Recommendations

3.3 High-End Computing

3.3.1 Findings

3.3.2 Recommendations

4. Socioeconomic Research and Policy Priorities

4.1 Findings

4.2 Recommendations

5. Creating an Effective Management Structure for Federal Information Technology R&D

5.1 Agency Roles

5.2 Policy and Coordination of Information Technology R&D

5.3 Support and Implementation

5.4 Annual Review

5.5 Funding

Can the Proposed Increases be Invested Effectively?

Conclusion

February 24, 1999

The President of the United States
The White House

Dear Mr. President:

We are pleased to present our final report, "Information Technology Research: Investing in Our Future," on future directions for Federal support of research and development for information technology. This report adds detail to the findings and recommendations in our interim report dated August 1998, and strengthens our previous recommendations regarding the importance of social and economic research on the impacts of information technology to inform key policy decisions.

PITAC members are strongly encouraged by and enthusiastically supportive of the Administration's Information Technology for the Twenty-First Century (IT2) initiative. This initiative is a vital first step in increasing funding for long-term, high-risk information technology research and development.

Increased Federal support is critical to meeting the challenge of capturing the opportunities available from information technology in the 21st Century through appropriate research and development. The economic and strategic importance of information technology and the unique role of the Federal Government in sponsoring information technology research make it necessary to increase Federal support over a period of years to ensure our Nation's future well-being.

We hope that our recommendations will be helpful as you consider the priorities for Federal investments. Thank you for your consistent support of our activities over the past year. We look forward to discussing this report with you, with members of your Administration and with members of Congress.

Respectfully,

Bill Joy
Co-Chairman

Ken Kennedy
Co-Chairman

President's Information Technology Advisory Committee

Co-Chairs

Bill Joy
Founder and Chief Scientist
Sun Microsystems

Ken Kennedy, Ph.D.
Ann and John Doerr Professor of Computational Engineering
Rice University

Members

Eric A. Benhamou
CEO and Chairman
3Com Corporation

Vinton Cerf, Ph.D.
Senior Vice President for Internet Architecture and Technology
MCI WorldCom

Ching-chih Chen, Ph.D.
Professor, Graduate School of Library and Information Science
Simmons College

David M. Cooper, Ph.D.
Associate Director of Computation
Lawrence Livermore National Laboratory

Steven D. Dorfman
Vice Chairman
Hughes Electronics Corporation

David W. Dorman
President and Chief Executive Officer
PointCast, Inc.

Robert H. Ewald
Former President and COO
Cray Research, Inc.

David J. Farber
Alfred Fitler Moore Professor of Telecommunication Systems
University of Pennsylvania

Sherrilynne S. Fuller, Ph.D.
Head, Division of Biomedical Informatics, School of Medicine and Director, Health Sciences Libraries and Information Center
University of Washington

Hector Garcia-Molina, Ph.D.
Leonard Bosack and Sandra Lerner Professor, Departments of Computer Science and Electrical Engineering
Stanford University

Susan L. Graham, Ph.D.
Chancellor's Professor of Computer Science, Department of Electrical Engineering and Computer Science
University of California, Berkeley

James N. Gray, Ph.D.
Senior Researcher, Scalable Servers Research Group
Microsoft Research

W. Daniel Hillis, Ph.D.
Vice President and Disney Fellow
Walt Disney Imagineering Research and Development, Inc.

Robert E. Kahn, Ph.D.
President
Corporation for National Research Initiatives

John P. Miller, Ph.D.
Director, Center for Computational Biology, and Professor of Biology
Montana State University

David C. Nagel, Ph.D.
Chief Technology Officer, AT&T and President AT&T Labs

Raj Reddy, Ph.D.
Dean, School of Computer Science, and Herbert A. Simon University Professor of Computer Science and Robotics
Carnegie Mellon University

Edward H. Shortliffe, M.D., Ph.D.
Associate Dean for Information Resources and Technology, Professor of Computer Science, and Professor of Medicine
Stanford University School of Medicine

Larry Smarr, Ph.D.
Director, National Computational Science Alliance, and Professor of Physics and Astronomy
University of Illinois at Urbana-Champaign

Joe F. Thompson, Ph.D.
William L. Giles Distinguished Professor of Aerospace Engineering
Mississippi State University

Leslie Vadasz
Senior Vice President and Director of Corporate Business Development
Intel Corporation

Andrew J. Viterbi, Ph.D.
Vice Chairman, Board of Directors
QUALCOMM Incorporated

Steven J. Wallach
Advisor
CenterPoint Ventures

Irving Wladawsky-Berger, Ph.D.
General Manager, Internet Division
IBM Corporation

ACKNOWLEDGEMENTS

The following members of the community were especially helpful in working with PITAC as panel members or advisors after the interim report was released in August, 1998.

Victor Basili

George Campbell

Jack Dongarra

Jim Flanagan

Donna Hoffman

Jeff Johnson

Brewster Kahle

Ed Lazowska

C. Dianne Martin

Reagan Moore

Donald Norman

Craig Partridge

Dave Patterson

Dan Reed

Michael Skibo

George Spix

Tom Sterling

Rick Stevens

Michael S. Teitelbaum

Mary Vernon

Stephen Wolff

Alexander Wong

William Wulf

PITAC would like to thank the following people for their contribution in writing white papers for PITAC's consideration for the interim report.

Kathie Blankenship

Scott O. Bradner

Michael L. Brodie

Michael Hawley

Rob Kling

Clifford Lynch

Daniel R. Masys

PITAC would also like to acknowledge the contributions of the following people from the National Coordination Office for Computing, Information and Communications, without whom we would not have been able to produce this report:

Kay Howell

Sally Howe

Yolanda L. Comedy

Robert I. Winner

PITAC would also like to thank the numerous Federal employees who gave advice during the writing of this report, in particular, Henry Kelly, Tom Kalil and Lori Perine.

And finally, we are grateful to the entire staff at the National Coordination Office. Our meetings went smoothly because of their careful preparation.

Executive Summary

Information Technology will be one of the key factors driving progress in the 21st century–it will transform the way we live, learn, work, and play. Advances in computing and communications technology will create a new infrastructure for business, scientific research, and social interaction. This expanding infrastructure will provide us with new tools for communicating throughout the world and for acquiring knowledge and insight from information. Information technology will help us understand how we affect the natural environment and how best to protect it. It will provide a vehicle for economic growth. Information technology will make the workplace more rewarding, improve the quality of health care, and make government more responsive and accessible to the needs of our citizens.

Vigorous information technology research and development (R&D) is essential for achieving America's 21st century aspirations. The technical advances that led to today's information tools, such as electronic computers and the Internet, began with Federal Government support of research in partnership with industry and universities. These innovations depended on patient investment in fundamental and applied research.

We have had a spectacular return on that Federal research investment. Businesses that produce computers, semiconductors, software, and communications equipment have accounted for a third of the total growth in U.S. economic production since 1992, creating millions of high-paying new jobs. Government-sponsored University research programs have supported graduate education for many of the leaders and innovators in the field. As we approach the 21st century, the opportunities for innovation in information technology are larger than they have ever been–and more important. We have an essential national interest in ensuring a continued flow of good new ideas and trained professionals in information technology.

After careful review of the Federal programs this Committee has concluded that Federal support for research in information technology is seriously inadequate. Research programs intended to maintain the flow of new ideas in information technology and to train the next generation of researchers are funding only a small fraction of the research that is needed, turning away large numbers of excellent proposals. Compounding this problem, Federal agency managers are faced with insufficient resources to meet all research needs and have naturally favored research supporting the short-term goals of their missions over long-term high-risk investigations. While this is undoubtedly the correct local decision for each agency, the sum of such decisions threatens the long-term welfare of the nation.

The Nation needs significant new research on computing and communication systems. This research will help sustain the economic boom in information technology, address important societal problems such as education and crisis management, and protect us from catastrophic failures of the complex systems that now underpin our transportation, defense, business, finance, and health-care infrastructures. If the results are to be available when needed, we must act now to reinvigorate the long-term IT research endeavor and to revitalize the computing infrastructure at university campuses and other civilian research facilities, which are rapidly falling behind the state of the art. If we do not take these steps, the flow of ideas that have fueled the information revolution over the past decades may slow to a trickle in the next.

To address these problems, the Committee estimated in its Interim report in August 1998 that the Federal government should increase its support for information technology research by a billion dollars per year by FY 2004. Since that time the Committee has sought comments from the community regarding its preliminary findings and recommendations, and convened several panels to review those recommendations. This effort produced a more detailed model for the costs of the research programs and other activities needed to address the problems identified in our report. As a result of these activities, the Committee has further refined the findings and recommendations presented in its Interim Report, and adjusted its funding recommendation accordingly. The Committee now recommends that the Federal government increase annual funding for information technology R&D over the five-year period from FY2000 to FY2004, as follows:

Recommending Funding Increases for Information Technology R&D ($ millions)

Area	FY2000	FY 2001	FY 2002	FY 2003	FY 2004
Software	112	268	276	472	540
Scalable Information Infrastructure	60	120	180	240	300
High End Research	180	205	240	270	300
High End Acquisitions	90	100	110	120	130
Socioeconomic	30	40	70	90	100
Total	472	733	976	1192	1370

These increases are in addition to the programs in existence in FY99. The full report provides additional details on these budget recommendations, including discussions of the method used to produce them.

Although there are unmet needs across the entire spectrum of research activities, priority for increased funding should be on long-term, high-risk investigations. In addition to increases for research itself, the Federal government must also ensure that the research community is equipped with the state-of-the-art facilities needed to carry out advanced projects. Finally, Federal budgets must continue to ensure that advances in information technology work to benefit all Americans and that all Americans have the education and training needed to prosper in a world that will increasingly depend on information technology.

To be successful, the expanded Federal research program we propose must be effectively managed. Current cross-agency coordination mechanisms are working well, but they suffer from the lack of well-defined responsibilities for ensuring that key areas are not overlooked. There has been no agency with the primary responsibility for ensuring that long-term, high-risk research is protected from the pressures that arise in mission agencies. Ideally there should be an agency charged with leading the organization of a fundamental information technology research program appropriate for the 21st century.

The Administration's proposed Federal budget for FY 2000 demonstrates a commitment to sustained growth in information technology research through its initiative, *Information Technology for the Twenty-First Century* (*IT*^{2}). This commitment is an important first step in what must be a continuing effort on the part of the Federal government to increase research dollars and to create a new management system designed to foster innovative research. But the effort cannot stop here. Further increases and continued oversight are needed to remedy the shortfall in long-term research investments that has accrued.

Findings and Recommendation

Federal information technology R&D investment is inadequate. Measured in constant (non-inflated) dollars, support in most critical areas has been flat or declining for nearly a decade, while the importance of information technology to our economy has increased dramatically. As a result, the Nation is gravely underinvesting in the long-term, high-risk research that can replenish the reservoir of ideas that will lead to innovations in information technology in generations to come.

Federal information technology R&D is too heavily focused on near-term problems. Much of the Federal investment in information technology R&D is being funded by mission agencies. In the face of the enormous increases in information technology problems to be addressed, funding agencies have had to prioritize their investments. Inevitably, priority has been given to short-term, mission-oriented goals over long-term research. This reflects the situation in the private sector as well. As a result, investment in long-term, high-risk research has been curtailed. This trend threatens to interrupt the flow of ideas that has driven the information economy in this decade and threatens efforts to solve nationally important problems.

Recommendation: Create a strategic initiative in long-term information technology R&D. To address these problems the Committee recommends that the President create a strategic initiative to support long-term research in fundamental issues in computing, information, and communications. The initiative should increase the total funding base by $1.37 billion per year by FY 2004. The Federal funding agencies should use the resulting budget increases to encourage research that is visionary and high-risk. To do this, they will need to diversify modes of research support and increase the duration of projects. The goal should be to recapture in the universities and research labs much of the excitement that existed at top-rated departments in the past.

Priorities for Research

Four areas of the overall research agenda particularly need attention and must be a major part of a strategic initiative in long-term research and development:

Software—The demand for software has grown far faster than our ability to produce it. Furthermore, the Nation needs software that is far more usable, reliable, and powerful than what is being produced today. We have become dangerously dependent on large software systems whose behavior is not well understood and which often fail in unpredicted ways. Therefore, increases in research on software should be given a high priority. Special emphasis should be placed on developing software for managing large amounts of information, for making computers easier to use, for making software easier to create and maintain, and for improving the ways humans interact with computers. Specifically, the Federal program should:

- Fund fundamental research in software development methods and component technologies.
- Support fundamental research in human-computer interfaces and interaction.
- Support fundamental research in capturing, managing, analyzing, and explaining information and in making it available for its myriad of uses.
- Make software research a substantive component of every major IT research initiative.

Scalable Information Infrastructure – Our Nation's dependence on the Internet is increasing. While this is an exciting development, the Internet is growing well beyond the intent of its original designers and our ability to extend its use has created enormous challenges. As the size, capability, and complexity of the Internet grows, it is imperative that we do the necessary research to learn how to build and use large, complex, highly-reliable, and secure systems. It is therefore important that the Federal government:

- Fund research on understanding the behavior of the global-scale network and its associated information infrastructure. This should include collecting and analyzing performance data and modeling and simulating network behavior.

- Support research on the physics of the network, including optical and wireless technologies including satellites, cable, and bandwidth issues.
- Support research to anticipate and plan for scaling the Internet.
- Support research on middleware that enables large-scale systems.
- Support research on large-scale applications and the scalable services they require.
- Fund a balanced set of testbeds and research infrastructure that serve the needs of networking research, research in enabling information technologies and advanced applications.

High-End Computing – Extremely fast computing systems, with both rapid calculation and rapid data movement, are essential to provide accurate weather and climate forecasting, to support advanced manufacturing design, to design new pharmaceuticals, to conduct scientific research in a variety of different areas, and to support critical national interests. Although they achieve remarkable performance in some cases, the current scalable, parallel, high-end computing systems are not well suited to many nationally important, strategic applications. To ensure that U.S. scientists continue to have access to computers of the highest possible power, funding should be focused on innovative architectures, hardware technologies, and software strategies that overcome the limitations of today's systems. Without major increases in funding in these areas, the realizable performance of new machines will fall far short of their potential. We specifically recommend that the Federal program should:

- Fund research into innovative computing technologies and architectures.
- Fund R&D on software for improving the performance of high-end computing.
- Drive high-end computing research by trying to attain a sustained petaops/petaflops on real applications by 2010 through a balance of software and hardware strategies.
- Fund the acquisition of the most powerful high-end computing systems to support science and engineering research.
- Expand the National Science and Technology Council (NSTC) Computing Information and Communications (CIC) High-End Computing and Computation (HECC) Working Group's coordination process to include all major elements of the government's investment in high-end computing.

Socioeconomic Impact – Information technology will significantly improve the flow of information to all people and institutions in the Nation, and could be a powerful tool for democratization. Our National well-being depends on understanding the potential social and economic benefits of on-going advances in information technology. However, problems are arising from the increasing pace of information technology-based transformations. To realize the promise of the new technologies, we must invest in research to identify, understand, anticipate, and address these problems. We must develop concrete objectives and comprehensive metrics through which to assess the ongoing transformations brought about by the integration of information technology into our lives. We must conduct careful research on the impact of the transformations against the objectives, and develop appropriate policies to deal with the knowledge we gain from the research. The Federal government should:

- Expand Federal initiatives and government/university/industry partnerships to increase information technology literacy, access and research capabilities.
- Expand Federal research into policy issues arising from information technology.
- Fund information technology research on socioeconomic issues.
- Expand the participation of underrepresented minorities and women in computer and information technology careers.
- Create programs to remove the barriers to high bandwidth connectivity posed by geographic location, size, and ethnic history of research, educational institutions, and communities.
- Accelerate and expand education in information technology at all levels–K-12, higher education, and lifelong learning.
- Strengthen the use of information technology in education.

Management and Implementation of Federal Information Technology Research – Building a Federal IT program suited to the needs of the Nation in the 21st century will require new management strategies, new modes of research support, and new implementation strategies. This new approach is demanded by the reality of Federal budget constraints, the need for more long-term cross-disciplinary team research, and the need to maintain a small, efficient, and coordinated research management process. It is essential that the Federal systems responsible for managing and implementing the new IT program be positioned to review the entire information technology research budget, to restore the balance between fundamental and applied research, to encourage long-term and high-risk collaborative research projects, and to employ a systematic review by participating Federal agencies and the private sector.

To achieve these goals, we recommend that the existing Federal information technology management and implementation structure be enhanced as follows:

- Strongly encourage NSF to assume a leadership role in basic information technology research. Provide NSF the necessary resources to play this role.
- Designate a Senior Policy Official for information technology R&D.
- Extend the HPCC program coordination model to all major Federal information technology R&D activities.
- Diversify the modes of research support to include more projects of broader scope and longer duration, placing a renewed emphasis on research carried out in teams.
- Fund collaborations with applications to drive information technology research, but take measures to ensure that research remains a primary goal.
- Fund centers for Expeditions into the 21st Century.
- Establish a program of Enabling Technology Centers that will drive research by examination of critical applications areas.
- Establish an annual review of research objectives and funding modes.

The Government's Essential Role

While the importance of information technology to the future of the economy and the government is clear, it may not be immediately obvious that government investment is needed to ensure continued progress. The PITAC members from industry were unanimous in their opinion that it is not feasible for the private sector to assume responsibility for long term, high-risk research, in spite of the success of the information technology industry. Their opinion is found in the attached sidebar.

We believe that the Federal Government must retain and expand its role in leading long-term fundamental research in information technology. Advanced Government services and national security depend on it. The benefits to our Nation and society will be huge. A loss of international leadership in information technology would be economically devastating.

We cannot rely on industry to fund the needed research because they necessarily focus, in view of economic realities, on the short term. Industry cannot and will not invest in solving problems of importance to society as a whole unless such investments make sense from a business perspective.

The rationale for funding long-term information technology research goes far beyond economic benefit and national security needs. Enormous societal gains can be reaped from advances in information technology. Only through research on a scale substantially greater than is being carried out today, can we build an infrastructure that will be available, affordable and usable by all citizens–one that can support the compelling "transformations" discussed in the next chapter of this report.

Federal investment in information technology directly supports the education and preparation of our young people for careers in information technology research, and the training of workers who need to upgrade their skills to keep pace with a changing marketplace. Trained people are not just a byproduct but rather a major product of publicly supported research. These trained professionals are critical national infrastructure, and will create and develop new ideas, from a talent pool for existing business, and launch new companies.

The benefits that the transformations described in the next chapter of this report can have for our Nation's future are extraordinary. A networked society can reach out to all its citizens, and can bring our Nation closer together and address many societal issues. While it cannot resolve all these issues, information technology can give us leverage toward their solution.

But the realization of the positive transformations we will describe in the next chapter of this report is not guaranteed. The realization of each transformation will depend on the results of aggressive, well-managed Federal research programs. Long-term Federal investment in information technology research is necessary to incubate ideas to the point of clear commercial viability, and to develop methods of measuring and tracking our progress toward realizing our positive vision.

Conclusion

Information technology research is essential for the continued growth of the economy and for the solution of some of the most critical problems facing the Nation. Unless steps are taken now to reinvigorate Federal research in this critical area, we are very likely to see a significant reduction in the rate of progress over the coming decades. The cost to the Nation of such a reduction will be significantly greater than the investment needed to address the problem.

PITAC Industry Member's Rationale for Government Support of Long Term, Fundamental Research

Our PITAC report shows that as information technology has grown to be an extremely successful segment of the economy, and more important to the future of the economy and the government, it has outpaced the research budgets available to support it. At a time when we need more fundamental understanding of information technology, and more long term, high-risk projects, tighter overall budgets and near-term market pressures are squeezing research budgets.

Should we expect information technology research to be funded primarily by commercial interests? To understand this question it helps to review the history of Federal funding in this area.

Since World War II, the Federal government has funded advanced information technology research to meet its own requirements, which have ranged from critical national-defense applications to weather forecasting and medical sciences. Federal funding has seeded high-risk research and yielded an impressive list of billion-dollar industries (such as the Internet, high performance computers, RAID disks, multiprocessors, local area networks, graphic displays). Federally funded university research has trained most of our leading IT researchers. Information technology industries provide hundreds of thousands of jobs and much of the nation's recent economic growth. The Federal investment to date has had tremendous benefits for our Government, our Nation and our economy.

The information technology industry expends the bulk of its resources, both financial and human, on rapidly bringing products to market. The U.S. information technology industry has created an awesome and continuous growth of capabilities based on the most intensely competitive marketplace the world has ever seen. Nearly every available person and dollar in this industry is focused on bringing the next version or the next product to market. Delivery product cycles are as short as every three to six months. The company that fails here misses the next short-term cycle and will not be successful.

The information technology sector has continued to invest extensively in research and development, and has more than doubled its annual R&D investment over the past 10 years, to reach the current level of about $30B. For information technology companies, the fraction of the total budget devoted to R&D is roughly twice the U.S. industry average. Over 90 percent of information technology R&D expenditures are allocated to product development, with the major portion of the remaining expenditures going toward near term, applied research.

Data from the top 50 information technology companies between 1990 and 1997 show that while both overall revenues and overall R&D spending grew, the portion of revenue invested in R&D declined. The ratio of R&D expense to overall revenue declined from about 9 percent to about 7 percent at the end of the period. The declining percentage of investment in R&D reflects the intense pace of the information technology marketplace, and the equally intense competitive pressures on prices and profit margins. Today, industry's efforts are mostly focused on projects that will generate near-term revenues. The reasons for this short-term focus can be found in the changes that have taken place in the information technology industry through the 1990s.

The decade of the 90s has been characterized by a major change in the nature of information technology industry sales. There has been a shift from complex products with large margins to more commodity-like products, such as PCs, with lower margins. Services is the fastest growing segment in the industry, and profit margins in the people-intensive services segment are often lower than those in software and hardware products. Such price and resulting profit declines have required widespread cost control programs. These programs impacted both the expenditures and the nature of the R&D programs.

Another inhibitor to long-term research spending is intense global competition, which puts great pressure on prices, profits, and expenses. Such pressures escalate as new companies are formed at a rapid pace and try to dislodge each other and existing companies, often by competing on price. Further, the incentives associated with long term research, such as prospects for higher P/E ratios, opportunities to receive long term R&D tax credits or to write-off the investments of "in-progress R&D" are either non-existent, or weak and getting weaker.

The fundamental research investments in university research by the Federal government have served to train the majority of our information technology professionals. In recent years, however, the shortage of workers with adequate skills makes it difficult for companies to grow both near and long-term research, even if budgets allowed. This alone suggests the critical need for additional government support of university research.

Past Government investments in high-risk research have helped fuel the intense pace of the information technology marketplace. The U.S. has the most energetic, viable and productive technology transfer mechanism in the world. Ideas freely flow from universities and national labs to existing and new companies. In 1998, over $12 billion dollars were invested by venture funds in new companies. The basic feedstock for these investments has been Government support of basic information technology research. If this feedstock is allowed to deplete, this economic growth engine could slow or disappear.

The United States must not only continue, but also substantially increase, long-term fundamental information technology research programs. Once innovative research ideas have been explored, American companies are well positioned to seize the viable ideas, commercialize and distribute them. The government must increase its investment in the pipeline that generates these ideas and the researchers to work on them. This will create a better society and bring even greater prosperity to our nation.

[1] *Endless Frontier, Limited Resources: U.S. R&D Policy for Competitiveness*, Council on Competitiveness, Washington, D.C., 1996: 110.

[2] *Ibid*, 111.

[3] Data from Compustat, an S&P subscription database. Data was compiled by Compustat from SEC 10-K and 10Q report filings, 1990-1997, then summarized by IBM.

[4] "Venture Capital Investments Hit Another Record High Despite Stock Market Turmoil." *Money Tree Highlights/Q3: 1998.* Price Waterhouse Coopers. http://209.67.194.61/reportq398_highlights.asp

Industry Members of PITAC

Bill Joy, PITAC Co-Chair
Sun Microsystems

Eric Benhamou
3Com Corporation

Vinton Cerf
MCI Communications

David W. Dorman
PointCast, Inc.

Steve D. Dorfman
Hughes Electronics Corp.

Robert Ewald
Former Cray Research, Inc.

James N. Gray
Microsoft Research

W. Daniel Hillis
Walt Disney Imagineering

David C. Nagel
AT&T Labs

Leslie Vadasz
Intel Corporation

Andrew J. Viterbi
QUALCOMM

Steven J. Wallach
CenterPoint Ventures

Irving Wladawsky-Berger
IBM Corporation

1. Information Technology Transforming Our Society

Information technologies are an integral part of people's lives, businesses, and society. Advances in microprocessors, memories, storage, software, and communication technologies make it possible to build computers and computing devices that are increasingly affordable, as well as to enable the development of increasingly powerful systems at reasonable costs. The wide acceptance of Internet standards and technologies is helping us build global computer networks capable of connecting everything and reaching everyone.

Since ancient times, networks have offered opportunities for growth and innovation and have supplied structure to our economic and social systems. From the roads and aqueducts of the Roman Empire, to nineteenth century continental railroad systems, to the telecommunications, broadcast, and satellite networks of the twentieth century, networked capabilities have allowed us to overcome barriers of time and space, and to access and open new frontiers for human interaction and ingenuity.

The free flow of information is essential to a democratic society. Advances in information technology have the potential to significantly enhance this flow of information, and thus strengthen the institutions of our society, from financial markets to government agencies. The flow of information must, however, not only be "free," but "fair." Financial markets, for example, have learned that they must guard against abuses, such as insider trading. Businesses and healthcare institutions must guard against the misuse of personal information put in their trust by their customers. As we have the opportunity to use information technology to strengthen our societal institutions, we must understand the potential pitfalls, and the safeguards we must put in place to achieve both a free and fair flow of information.

As we approach the new millennium, it is clear that the "information infrastructure"–the inter-connected networks of computers, devices, and software–may have a greater impact on worldwide social and economic structures than all networks that have preceded them. The advances in computing and communications technologies of the last decade have already transformed our society in many ways. These advances have transformed the ways in which we view ourselves, our relationships with each other and with other communities, and the ways in which we obtain services, ranging from entertainment and commerce to education and health care. Even so, we have only just begun to grasp the opportunities and experience the transformations that will occur as these technologies mature.

Major technical advances are needed to build a smoothly functioning information infrastructure that links together all people, institutions, and relevant devices (e.g., cars, gas meters, home thermostats, air conditioners) in our Nation and beyond. Only vigorous information technology research and development programs will enable us to achieve our objectives.

But, hard as the technical challenges might be, we must keep in mind the great socioeconomic issues ahead of us on the road to becoming a fully networked society. Thus, we complement the call for research to support the required technical advances with a call for research programs to help us understand and enhance the positive effects of information technology on our economy, society, culture and political system.

The information revolution puts a premium on basic knowledge, not just information technology literacy, but basic skills in reading, writing, communications, and teamwork. Education and training have become lifelong pursuits for our workforce, as new jobs requiring new skills are created, and older jobs and skills become obsolete. The Nation must ensure that access to the benefits of the information infrastructure are available to everyone in our Nation: to those living in small towns and rural areas as well as in big cities, to those living in poor inner city neighborhoods and tribal reservations, as well as in well-to-do suburbs and those who face daily challenges from disabilities. We should use information technology to bridge the gaps in our society, not to create new ones.

A significant portion of our national progress in computing and communications over the past decade has been leveraged from the Federal research programs established by the High Performance Computing Act of 1991 (P.L. 102-194). These programs comprised the High Performance Computing and Communications (HPCC) initiative, which was responsible for moving the U.S. into an era of teraflop computers and gigabit networks. A major focus of the HPCC initiative was a set of Grand Challenges, difficult scientific and engineering problems whose solution were advanced by applying high performance computing and communications technologies and resources. Our recommendations build upon the solid foundation of the HPCC program and expand the HPCC vision to meet the challenges facing us in the 21st century.

To ensure a rapid, smooth, and extendible transition into the 21st century, the President's Information Technology Advisory Committee has identified ten critical "National Challenge Transformations." These information technology transformations will affect how we communicate, how we store and access information, how we become healthier and receive proper medical care, how we learn, how we conduct business, how we work, how we design and build things, how we conduct research, how we sustain a livable environment, and how we manage our government in the next millennium. Exploring these dynamic transformations enables us to identify common information technology challenges critical to our Nation's future and provides a framework for our recommendations for Federal research investments.

1.1 Transforming the Way We Communicate

Vision: At least one billion people worldwide can access the Internet simultaneously and engage in real-time electronic meetings, download the daily news, conduct secure financial transactions, or talk to friends and relatives around the world. This can be done regardless of the language the participants speak, since language translation can be done instantaneously, and regardless of physical limitations of the individual, because devices can accept and provide input and output in many different ways.

The Internet lies at the heart of our communications revolution. But, the current Internet must be expanded in scale to accommodate anticipated growth in use and demands for reliability comparable to that of the modern telephone system. New and improved modes of human interaction with computers must be developed to enrich and simplify the way we communicate. We must understand the behavior of extremely large and complex systems and address the potential fragility of large numbers of autonomously interacting software systems. Global networking raises a host of international issues and even poses questions about the nature of national boundaries as information flows across them invisibly and multinational corporations use worldwide networks to pursue their global interests. Perhaps the biggest challenge of all is to understand how human beings can best take advantage of the new electronic communication possibilities, both individually and in groups.

1.2 Transforming the Way We Deal With Information

Vision: An individual can access, query, or print any book, magazine, newspaper, video, data item, or reference document in any language by simply clicking a mouse, touching a computer screen, talking to a computer, or blinking an eye. Individuals can easily select among modes of presentation: data, text, images, or audio. Information can be referenced and derivations can be incorporated in many new ways, adding value and revealing insights through networked and software-enabled tools. Entertainment can be richer and more personalized, enabling individuals to access music, videos or live events that appeal to them.

This transformation requires significant improvements in data access methods, including high performance information systems and tools to help individuals locate information and present, integrate, and transform the information in meaningful ways. Systems will require interfaces accessible both to experts and novice or infrequent users regardless of physical ability, education, or culture. Multi-modal human-computer interaction technologies are needed including speech, touch, and gesture recognition and synthesis. There are research requirements for topics ranging from network reliability and bandwidth, to scalable software support, database structure and retrieval algorithms, high-performance computing, and robust, reliable, secure ways to deliver and to protect critical information. Providing high-quality entertainment over the Internet requires considerable more bandwidth, advances in audio and video streaming, and better tools to attract good content creators. Challenging issues regarding dissemination of information in electronic form including copyright, intellectual property rights, and the development of realistic business models remain important policy and research topics.

1.3 Transforming the Way We Learn

Vision: Any individual can participate in on-line education programs regardless of geographic location, age, physical limitation, or personal schedule. Everyone can access repositories of educational materials, easily recalling past lessons, updating skills, or selecting from among different teaching methods in order to discover the most effective ways of learning. Educational programs can be customized to each individual's needs, so that the information revolution reaches everyone and personal digital libraries provide a mechanism for managing ones accumulated knowledge resources. Learning involves all our senses, to help focus each student's attention and better communicate educational material.

Information technology is already changing how we teach, learn, and conduct research, but important research challenges in the field of education remain. In addition to research to meet the scalability and reliability requirements for information infrastructure, improvements are needed in the software technologies to enable development of educational materials quickly and easily and to support their modification and maintenance. We know too little about the best ways to use computing and communications technology for effective teaching and learning, in particular, how to effectively use multimedia capabilities to create a richer, and more appealing learning experience. We need to better understand what aspects of learning can be effectively facilitated by technology and which aspects require traditional classroom interactions with the accompanying social and interactive contexts. We also need to determine the best ways to teach our citizens the powers and limitations of the new technologies and how to use these technologies effectively in their personal and professional lives.

1.4 Transforming the Practice of Health Care

Vision: Telemedicine applications are commonplace. Specialists use videoconferencing and telesensing methods to interview and even to examine patients who may be hundreds of miles away. Computer-aided surgery with Internet-based video is used to demonstrate surgical procedures to others. Powerful high-end systems provide expert advice based on sophisticated analysis of huge amounts of medical information. Patients are empowered in making decisions about their own care through new models of interaction with their physicians and ever-increasing access to biomedical information via digital medical libraries and the Internet. New communications and monitoring technologies support treatment of patients comfortably from their own homes.

Future requirements for electronic medical records and health-system intranets will lead to increased reliance on the national infrastructure for communications, data sharing, and direct provision of care at a distance. Privacy and knowledge repositories are important research topics. Research in user interfaces is needed to understand how to make telemedicine applications not only efficient, but also satisfy the more human needs of both patient and physician. Robotics and remote visualization methods, supported by high-reliability, low-latency communications, are needed to support applications such as telepresence surgery.

1.5 Transforming the Nature of Commerce

Vision: Any company can be easily reached by its customers, regardless of location. It can receive immediate customer feedback, and rapidly adjust marketing strategies, prices, or product inventories based on that feedback. Consumers can shop for the best products, services, and prices from the convenience of their hotel room, home, or office. Electronic purchases can be made securely and with total privacy, providing suppliers and retailers with immediate access to cash generated by sales and consumers with automated statements detailing spending and purchases that allow for improved personal financial management.

Electronic communication is already dramatically changing how commercial transactions between companies are conducted, how digitally based goods and services are distributed, and how retail sales are made. Companies are using information technology to get closer to their customers and suppliers. Technology is also helping to reduce paper work and purchasing costs by streamlining the acquisition process and allowing companies to more efficiently find the best suppliers. Privacy and security are critical research topics if electronic commerce is ultimately to earn the full confidence of consumers. The reliability of the communication networks, computers, and business applications are vital to the success of U.S. companies. As the marketplace becomes increasingly global, understanding electronic commerce in the context of international trade relations will also become increasingly important.

1.6 Transforming the Nature of Work

Vision: The workplace is no longer confined to a specific geographic location, as workers can easily access their tasks and colleagues from alternate locations or while en route. Workers can do their jobs without regard to physical proximity to major metropolitan areas. They can choose where they live based on nearness to family or lifestyle preference rather than job market opportunities. A highly flexible workplace is able to accommodate each individual's needs, from working parents to workers with disabilities.

By some projections, as many as 15 million U.S. workers will become telecommuters over the next decade. This should enhance productivity and organizational flexibility as well as provide environmental benefits. To support large numbers of workers in non-traditional office settings, including the rapidly growing number of home businesses, we will need high-speed networking capability, equally available to many workers regardless of location or disability. Software technologies that allow work teams to collaborate effectively will be needed, and the privacy and reliability of the information infrastructure that permits this collaboration will be critical to success. The social and economic implications of telecommuting need to be studied. Computing and communications are also dramatically altering the skill base that workers need to perform their jobs. We need to determine how both employers, employees, and the self-employed can respond effectively to these changes.

1.7 Transforming How We Design and Build Things

Vision: Products and structures, from the highly complex, such as automobiles and buildings, to those used in everyday life, such as consumer appliances and fashion, can be designed with computer simulations that accurately represent the physical properties of the systems being built. Designers, manufacturers/builders, suppliers, and end-users participate in the design process, providing one another immediate feedback. Multiple designs and manufacturing processes can be rapidly explored yielding safer products, higher quality, and lower costs.

Global competition continues to press United States manufacturers to attract new customers and retain current customers by increasing productivity, reducing cost, improving quality, maintaining maximum flexibility, and reducing design cycle time. Information technology has revolutionized the entire product development design cycle and will continue to do so. High-end computing technologies are needed for concept design, simulation, analysis with interactive control and computation steering, the mining of archived data, and the rendering of data for display and analysis. There is a critical need to link engineering development processes with business processes like planning, purchasing, scheduling, investment, and cost management. There is also a need for networked computers that allow simultaneous modification of a standard product to meet customers' needs. As computer prices continue to decrease, these design, engineering and manufacturing capabilities can be applied to many more products.

1.8 Transforming How We Conduct Research

Vision: Research is conducted in virtual laboratories in which scientists and engineers can routinely perform their work without regard to physical location–interacting with colleagues, accessing instrumentation, sharing data and computational resources, and accessing information in digital libraries. All scientific and technical journals are available on-line, allowing readers to download equations and databases and manipulate variables to interactively explore the published research.

High-speed computers and networks are enabling scientific discovery across a broad spectrum–from mapping the human brain to modeling climatic change. Research problems are becoming more complex and interdisciplinary in nature. As a result, researchers are finding innovative ways to collaborate with their colleagues across the globe. Key research technologies include high-end computing to allow higher fidelity models of complex physical phenomena, advances in collaborative environments, visualization of complex datasets, data mining techniques and management of very large datasets and databases.

1.9 Transforming Our Understanding of the Environment

Vision: Information technology can help us to improve a variety of problems from water and air quality to controlling the effects of toxic material. For example, reliable climate models permit us to determine the rate and regional distribution of climate change to support accurate projections by sector and region. Sophisticated models accurately predict the response of ecosystems to changes in temperature, water availability, and atmospheric composition. Fully integrated models allow scientists and policy makers to consider information on climate trends, population trends, resource utilization, and the value of natural and economic resources when making decisions regarding technically feasible and cost-effective options to reduce environmental impacts or adapt to climate change.

To better support national and international energy and environmental policy, the United States requires an unprecedented acceleration and extension of research into climate modeling in order to improve the accuracy of local and regional forecasting. Progress in this area depends on improvements in computational methods. This will require orders of magnitude increases in computing capability to deal with the immense size of these problems in both time and space. We also need other advanced information technologies such as improved numerical methods and algorithms, tools for data storage, management, analysis and visualization, software development and testing techniques, and advanced networks for distributed computing.

1.10 Transforming Government

Vision: Government services and information are easily accessible to citizens, regardless of their physical location, level of computer literacy, or physical capacity. Intelligent systems guide citizens by providing a one-stop shopping experience for locating requested information. Documents and forms can be accessed, completed, and submitted electronically. Automated business processes allow nearly instantaneous response to citizens' requests. In times of natural emergencies, emergency crews have instant access to three-dimensional building models, risk analysis and assessment, high-resolution local weather predictions, stress analyses of damaged structures, rapid evacuation planning tools, and emergency agency coordination.

There is a huge potential to make all government institutions both more efficient and more responsive through information technologies. Technical challenges include significant improvements in systems and methods for accessing data, including high performance data storage and tools to locate and present information. Robust, reliable, and secure networks and software to deliver and protect critical information are important research topics. It is imperative that improvements in government be available to all citizens, so we must understand and surmount barriers to access.

Transforming the Way We...	Challenges	Benefits
1. Communicate	- Scaling for growth and reliability a la the telephone system. - Improving human interaction with computers. -Fragility of systems. -Global networking issues. - Finding best use of newcommunication possibilities, 1-on-1 and in groups.	-One billion users can access the Internet simultaneously, regardless of language and physical limitations.
2. Deal with Information	-Improving data access methods. -Multi-modal human-computer interaction technologies. -Reliability and bandwidth, better audio and video streaming. -Scalable software support. -High performance computing. -Delivering and protecting critical information. -Policy for electronic dissemination of information.	-Everyone can access, query, and print any book, magazine, newspaper, video, data item, or reference document, regardless of language, using mouse, touch screen, speech, or eye blink. -Value is added to information through networked and software-enabled tools.

Transforming the Way We...	Challenges	Benefits
3. Learn	-Scalability and reliability of the information infrastructure. -Improving software technologies for development of education materials and support of their modifications and maintenance. -Determining the best use of computing and communication technology for effective teaching and learning. -Learning which traditional teaching methods to leave alone. -Learning how to teach citizens best use of these new technologies.	-Regardless of location, are, handicaps, or schedule, anyone can participate in on-line education programs. -Everyone can access educational materials to discover the best learning style for them. -Customized educational programs exist for everyone, so no one is left behind.
4. Conduct Commerce	-Having sufficient privacy and security to ensure consumer confidence. -Reliability of communication networks, computers, and business applications needs to be high.	-Customers can reach any company regardless of location. -Immediate feedback facilitates fast adjustment of marketing strategies and inventories. -Consumers shop at their convenience. -Companies can immediately access funds from sales. -Consumers have automated statements permitting improved financial management.
5.Work	-Developing high-speed networking for all, regardless of location or handicap. -Developing software to allow effective collaboration. -Ensuring privacy and reliability of the information infrastructure. -Determining how employers, employees, and the self-employed can respond to changes.	-Workers have access to jobs regardless of proximity to population centers. -Workers can live where they want, not needing to be near jobs. -Workplace can better accommodate individual needs.

Transforming the Way We. . .	Challenges	Benefits
6. Practice Health Care	-Ensuring privacy of information repositories. -Developing robotics and remote visualization methods to support applications such as telepresent surgery.	-Doctors use teleconferencing and telesensing to interview and examine patients. -Surgical procedures can be demonstrated with Internet-based video. -High-end systems provide expert advice. -Patients access biomedical information
7. Design and Build Things	-High-end computing technologies are needed for concept design, simulation, analysis with interactive control and computation steering, mining archived data, and rendering of data. -Need bi-directional engineering development processes linked with business processes.	-Complex designs done via computer simulations. -All parties, including end users, participate in the process. -Safer products, higher quality, lower costs.
8. Conduct Research	-Research problems have become more complex and interdisciplinary. -Researchers need to find innovative ways to collaborate.	-Research is conducted in virtual laboratories, interacting, accessing instrumentation, and sharing data and other resources, all regardless of physical location. -All journals are available on-line.
9. Deal with the Environment	-To accelerate and extend climate modeling research to improve forecasting. -Increasing computing capability by orders of magnitude. -Develop improved numerical methods and algorithms, tools for data storage, management, analysis and visualization, software development and testing, and advanced networks for distributed computing.	-Reliable climate models. -Ecosystem models accurately predict responses to changes in conditions. -Fully integrated models facilitate decision making by scientists and policy makers.

Transforming the Way We...	Challenges	Benefits
10. Conduct Government	-Develop significant improvements in data access: high performance file systems and tools. -Develop reliable, secure networks and software to deliver and protect critical data.	-Government services and information are available to all regardless of location, computer literacy, etc. -One-stop shopping for locating information -Automated business processes accelerate responsiveness. -Enhanced responsiveness to natural disasters.

2. Setting Federal Research Priorities: Findings and Recommendations

Government-funded information technology research has produced enormous innovation. The results have been made readily available to industry. Testbed activities involving academia, government research facilities, and industry have served as powerful engines for technology transfer into the private sector, for the benefit of industry, government, and the Nation.

Today we enjoy the economic, strategic, and societal benefits of well-placed investments in long-term, wide-ranging information technology research begun during the Eisenhower and Kennedy Administrations. These modest investments have yielded massive economic benefits to the Nation. The empirical evidence is unequivocal: today's information technology sector vastly outstrips the current growth of all other sectors of the economy. The Federal Reserve reports that during the past five years production in computers, semiconductors, and communications equipment quadrupled at a time when total industrial production grew by 28 percent. These three industries account for one-third of the total real growth in production since 1992.

No other sector contributes nearly as much to the growth of our economy. The businesses spawned by these technologies employ millions of Americans in manufacturing and information processing jobs that pay wages well above the national average.

History suggests that, to be successful, Federal research investment must be sustained and flexible. Federal policies must support, encourage, and help coordinate long-range technological development. Federal R&D programs must be well designed and must not subsidize activities best left to the private sector. Only in this way can the Federal investment spur those critical areas of technology that either industry neglects or the Government overlooks in the normal course of business because they cut across Federal agency missions.

2.1 Findings

Finding: Total Federal information technology R&D investment is inadequate.

The non-inflationary growth in high-technology businesses fuels U.S. world leadership and creates whole new industries with hundreds of thousands of high-paying jobs. Furthermore, information technology's importance goes far beyond its economic benefits: it is fundamental to the solutions of many nationally important problems. But the Federal R&D investment has been compromised by a shift toward applied R&D. The Committee further finds that the amount of Federal research investment in information technology has not kept pace with information technology's growing economic, strategic, and societal importance to the Nation.

The number of proposals in computer science and information technology competitions provides evidence of the shortage of research funding. For example, the recent competition in Knowledge and Distributed Intelligence sponsored by NSF drew over 1100 letters of intent and over 850 full proposals even though it had been announced that at most seventy-five projects could be funded. The response to the ASCI level 2 centers competition was similar. The reason for the extraordinary number of responses is that funding for information technology is extremely tight. Researchers are forced to participate in nearly every competition for which they might qualify. The end result is that researchers are spending increasing percentages of their time in proposal writing to the detriment of research itself and yet many good ideas are still not being funded.

This trend is not just bad for information technology researchers, it is bad for the Nation. Our ability to produce reliable software, build an information network on which the Nation can run, produce the high-end computing systems needed for advanced science, engineering, and defense tasks, and understand the socioeconomic effects of these revolutionary technologies is threatened by inadequate investments in research and development. By neglecting research, we do more than deplete the stock of fundamental knowledge–we endanger the long-term effectiveness of the entire R&D system and threaten U.S. leadership in the emerging 21st-century information economy.

Finding: Federal information technology R&D is excessively focused on near-term problems.

The NSF defines basic research as the study of the "fundamental aspects of phenomena and of observable facts without specific application toward processes or products." In contrast, "applied research" is aimed at determining the means to meet specific needs, and "development" is defined as the systematic use of knowledge to produce useful materials, devices, or methods. These definitions are widely used in Federal policy and budget accounting, but they accurately describe only part of the nature of research and development

(R&D). R&D is a complex non-linear interaction between concepts and theories, data and experiments, and new products and processes. Basic research is a critically important part of this interwoven system.

During the past decade both industry and Government have altered the balance between basic research and the later stages of technology development and commercialization. At the same time, major corporations have cut back on basic research expenditures, shifting staff from centralized laboratories to operating divisions where applied work is closely tied to commercial products and processes. In both the public and private sectors, the interacting reasons are 1) downward budget pressures, 2) increased focus on mission, and 3) the inefficiency of transitioning long-term research to near-term product.

Although total Federal spending for R&D has remained steady, there has been a marked shift toward support for applied R&D. One example is that DARPA, which funded much of the innovative research in the 1980's, revised its priorities in the 1990's so that all information technology funding was judged in terms of its benefit to the warfighter. In the process of making this change it decreased the time horizon for potential technology transfer. In the early 1990's, DARPA split the Information Systems Office off from its Information Technology Office to address specific military systems. Nevertheless, total funding for basic research in the DARPA Information Technology Office is less than $20 million out of a total office budget of more than $200 million, an inadequate investment in our judgment. DARPA's new leadership plans to reverse these changes, but recent history shows that Government research managers correctly favor their agencies' missions when budgetary pressures grow and they have to choose between long-term research and short-term mission needs.

Most R&D investment restructuring in the early 1990's was essential for industry and the Government to maintain a competitive footing in the global marketplace and to maintain readiness against our present and future adversaries. However, this restructuring came at a high price: a serious decline in basic research activities. Research in computer science is a good example. In 1995, by the Federal Government's own measure,[1] more than two out of every three Federal dollars spent on research in computer science was for applied work.

The Committee finds that the Federal agenda for information technology R&D has moved too far in the direction of near-term applications development, at the expense of long-term, high risk, fundamental investigations in the underlying issues confronting the field. Currently, too high a percent-

---

1 National Science Board, *Science and Engineering Indicators*– 1996. Washington, DC:U.S. Government Printing Office, 1996.

age of research funding comes from mission-oriented agencies whose main goal is not so much to advance the knowledge-base in information technology as it is to solve the immediate problems confronting those agencies. Meanwhile, information technology has begun moving much faster than other fields. In order to intercept tomorrow's challenges we must set long-range, stretch goals rather than focusing on incremental advances. Within information technology, the scalable information infrastructure area is moving extremely fast, making it even more important to stretch for long-term research objectives.

Economic growth and defense leadership, if based on evolutionary improvements to yesterday's research results, are not sustainable–the rest of the world is not standing still in seeking economic advantage from new information technologies. By the Committee's own calculation, basic research spending may be as low as five percent of the total Federal information technology R&D budget. As a result, promising long-term research is being passed over in order to meet the goals of near-term technology development.

It is time to swing the pendulum back in the other direction and to strike a proper balance. We need more basic research–the kind of groundbreaking, high-risk/high-return research that will provide the ideas and methods for new disciplinary paradigms a decade or more in the future. We must make wise investments that will bear fruit over the next forty years.

Finding: The Federal information technology R&D funding profile is incomplete.

The current funding portfolio is not properly balanced. It is deficient in the support of multiple-institution, long-duration projects. Funding for projects of longer duration and larger scope is critical to the Federal research program. Projects of larger scope allow for multiple-investigator, inter disciplinary collaboration, intramural research in academia and Federal research institutes, and joint industry-Government-academia experiments or proofs of concept. Projects of longer duration allow exploration of research problems with multiple-year horizons, which may lead to unexpected and significant discoveries.

It is important that Federal investments include a range of complementary funding modes, including classical single principal investigator (PI) research, multiple-PI experimental research, and multiple-institution/multiple-year efforts. Such diversity in funding approaches and tactics is important. It provides complementary modes for research, ensuring a broad perspective in addressing problems, thus increasing opportunities for discoveries.

2.2 Recommendations for Research

The current boom in information technology is built on basic research in computer science carried out more than a decade ago. There is an urgent need to replenish the knowledge base. Our legacy to future generations should be based on an intelligent, well-planned, highly disciplined investment in information technology that is commensurate with its role in the Information Age.

Recommendation: Create a strategic initiative in long-term information technology R&D.

The Advisory Committee recommends that the President create a strategic initiative to support long-term research in fundamental issues in computing, information, and communications. The initiative should endeavor to increase the total funding base to over two billion dollars per year after five years. The Federal funding agencies should use the resulting budget increases to encourage research that is visionary and high risk. The goal should be to recapture in the universities and research labs much of the excitement that existed at top-rated departments in the past.

The Administration's proposed Federal budget for FY2000 demonstrates a commitment to sustained growth in IT research through its Information Technology Initiative (IT^2). This commitment is an important first step in what must be a continuing effort on the part of the Federal government to increase research dollars and to create a new management system designed to foster innovative research. But the effort cannot stop here. Further increases and continued oversight are needed to remedy the shortfall in long-term research investments that have accrued. In order to maintain U.S. leadership, 21st Century budgets should be guided by the principle that Federal support for information technology must be increased to bring it in line with its prominence in the economy and its importance to solving critical societal problems.

Recommendation: Increase the investment for research in software, scalable information infrastructure, high-end computing and socioeconomic issues.

Four areas of the overall research agenda particularly need attention, and must be a major part of the strategic initiative:

- *Software.* The science and methods for efficiently creating and maintaining high-quality software of all kinds, for ensuring the reliability of the complex software systems that now provide the infrastructure for much of our economy, for improving the interaction between people and computer-based systems and devices, and for managing and using information.
- *Scalable Information Infrastructure.* Techniques for ensuring that the national information infrastructure–communications systems, the In-

ternet, large data repositories, and other emerging systems–is reliable and secure, and can grow gracefully to accommodate the massive numbers of new users and applications expected over the coming two decades.

- *High-end Computing.* Continued invention and innovation in the development of fast, powerful computing systems and the accompanying communication systems needed to implement high-end applications ranging from aircraft design to weather and climate modeling.
- *Socioeconomic issues.* Research on understanding the effects of information technology on our society, its economy, and the workforce should be funded. Furthermore, research should be focused on strategies for ameliorating information technology's potentially harmful effects and amplifying the benefits.

The Committee recommends increasing investments in research in these specific areas, to meet critical national economic and defense needs and maintain U.S. leadership, as detailed in chapter 3.

Recommendation: Fund projects of larger scope and duration.

We stand at the dawn of a new century, a century where leadership in information technology may well be economically and militarily decisive. To meet this 21st century challenge, we believe we must diversify the modes of research supported by the Federal government. Computer science and information technology are collaborative research fields. Hence, special emphasis should be placed on involving and supporting researchers at many institutions in large-scale research projects that can explore technologies farther into the future with teams of researchers that may be interdisciplinary and multi-institutional.

Chapter 5 of this report discusses ways to diversify research support to accomplish this goal:

- Expand support for research carried out by teams of two to five researchers, possibly at different institutions, to address a single research project.
- Fund large centers for "Expeditions to the 21st Century," which would involve large teams of researchers in explorations of future information technologies and their societal effects, and
- Establish "Enabling Technology Centers," which would conduct research on the application of information technology to particular problems of national importance.

These approaches would allow the Government to adapt its funding mechanisms appropriately to critical research goals.

3. Technical Research Priorities

3.1 Software Research

Software is the new physical infrastructure of the information age. It is fundamental to economic success, scientific and technical research, and national security. Software is increasingly important for commerce, communication, information access, and the Nation's physical infrastructure. The continuing emergence of cheaper and faster microprocessors has allowed more and more functions to be performed by software.

While we may think of the phone system, the Internet, or even a camera or a car as devices, these devices cannot function without software. We rely on software to work without fail, to be modifiable as requirements change, and consistently to provide more functions with better performance.

However, because the demand for software has grown at such an explosive rate, demand now far exceeds supply. Furthermore, the Nation depends on software that is often fragile, unreliable, and extremely difficult and labor-intensive to develop, test, and evolve. Our ability to construct the software to manage information and our ability to analyze and predict the performance of the enormously complex software systems that lie at the core of our economy are painfully inadequate. We are neither training enough professionals to supply the needed software[2] nor adequately improving the efficiency and quality of our construction methods. The recent Government emphasis on short-term applied R&D has hurt software research, reducing the generation of new ideas and approaches that would address the demand.

3.1.1 Findings

Finding: Demand for software far exceeds the Nation's ability to produce it.

The explosive growth of information technology has fueled an unprecedented demand for new software. Software is needed to support millions of legacy systems as well as new products, to provide fundamental services on the Internet, and to solve important national problems. The resources needed to develop this software have not kept pace with the demand, producing what might be called a "software gap."

An interaction of factors has caused this software gap: accelerated demand for software, increased complexity of systems, labor-intensity of development, variable quality in the labor pool, labor shortages, and lack of adequate science and technology to support robust development. Since hardware improvements create demand for more software, the demand will continue to accelerate. Today's systems and applications software are

2 Bureau of Labor Statistics. "Employment Outlook 1996-2006." *Monthly Labor Review*, November 1997.

substantially more complex than in the past. Software systems are now among the most complex human-engineered structures.

This situation threatens to inhibit the progress of the current boom in information technology. It may also threaten the health and welfare of the Nation by reducing the rate at which solutions to software-intensive problems, like aviation safety and crisis management, can be solved.

The Federal Government must take steps to address the situation by investing in software research. Note that "software research" means more than just research on new strategies to improve software development productivity. Software research also includes research into new tools to manage information and support commerce, development of new software to make computers easier to use, creation of software for embedded devices and new applications, and research to make software systems interoperate more effectively. It also means more than just investing in refining standard approaches to software development. It means funding well-defined but radical approaches that are high risk but could have a major impact. It will require conceptual breakthroughs to invent new, "outside-the-box" approaches. The Federal Government should also help increase the pool of information technology professionals capable of developing good software.

Finding: The Nation depends on fragile software.

The Nation needs robust systems, but the software our systems depend on is often fragile. Software fragility is its tendency not to work properly–or at all. Fragility is manifested as unreliability, lack of security, performance lapses, errors, and difficulty in upgrading.

Examples can be found everywhere, from our huge information systems for air-traffic control to the personal computers on our desks, from the Pentagon to the Internal Revenue Service (IRS). One aspect of our software's lack of security is evidenced by the fact that teenage hackers have easily penetrated our telephone and military systems; in reality the software-security problems we face go far beyond calculated external attacks. The Federal Aviation Administration (FAA) and IRS systems have proven difficult to upgrade. Large telecommunications networks have crashed, and banks have been robbed electronically. Even after large, expensive testing efforts, commercial software is shipped riddled with errors ("bugs"). Software producers rely on their users to discover the remaining errors in actual use, making it even more likely that our systems will fail.

Not only are our products inadequate–so are our processes. The Standish Group[3] reports that 73 percent of software projects are late, substantially

3 The Standish Group International, Inc., is a market research and advisory firm specializing in mission-critical software and electronic commerce. They are located in Dennis, Massachusetts.

over budget, canceled, or outright failures. The well publicized, potentially catastrophic "Year 2000 problem" is an example of known unreliability and illustrates the difficulty and cost of upgrading to meet new circumstances. Companies spend enormous amounts to back-up their on-line data in order to compensate for the risk that their own systems will corrupt it.

The Nation cannot afford to let the current situation continue. We must commit to develop the science, technologies, and methods needed to build robust software systems–ones that are reliable, fault-tolerant, secure, evolvable, scalable, maintainable, and cost-effective.

Finding: Technologies to build reliable and secure software are inadequate.

During the past 40 years, computing hardware has seen an increase in performance of at least eight orders of magnitude. Our ability to develop software has not kept pace with the opportunities those hardware advances provide. Our ability to construct the systems we require and our ability to anticipate the performance of the enormously complex software systems that lie at their core are inadequate.

Large software systems are beyond our capability to describe precisely. Consequently, there is little automation of their construction, little re-use of previously developed components, virtually no ability to perform accurate engineering analyses, and no way to know the extent to which a large software system has been tested.

Having meaningful and standardized behavioral specifications would make it feasible to determine the properties of a software system and enable more thorough and less costly testing. Unfortunately such specifications are rarely used. Even less frequently is there a correspondence between a specification and the software itself. Often software behavior and flaws are observable only when the program is run, and even then may be invisible except under certain unusual conditions. Programs written in such circumstances frustrate attempts to create robust systems and are inherently fragile.

Software development relies on individual genius and creativity, and, as with all design-based disciplines, will continue to do so. But it has become clear that the processes of developing, testing, and maintaining software must change. We need scientifically sound approaches to software development that will enable meaningful and practical testing for consistency of specifications and implementations. This requires long-term research in languages, theories, simulation, analysis, and testing that could lead to standardized multilevel mechanisms similar to those which have created the success in computer-aided design for digital hardware.

The construction and availability of libraries of certifiably robust, specified, modeled, and tested software components would greatly aid the development of new software. Libraries of software components for developing business applications are beginning to be constructed commercially, but only in limited circumstances.

Finding: The diversity and sophistication of software systems are growing rapidly.

The enormous advances in computing capability, the integration of computing and communications, and the striking reductions in the costs of those technologies have increased the opportunities for new and different software-based applications. The algorithms and the information structures needed for these new application domains require increased research in order for the applications to flourish. For example, real-time control is essential for transportation systems, for observational input to weather and climate systems, and for telemedicine. Scene recognition systems play an important role in national security. Ubiquitous computing and collaborative decision-making have very broad and important military and civilian uses. Whether the purpose of software is systems-oriented (e.g., an operating system, a compiler, or a protocol for wireless communication) or applications-oriented (e.g., a billing system, a speech interface, or a simulation), the software technology must continue to advance.

Finding: More and more often, common activities of ordinary people are based on software.

Twenty years ago, most computer use was job-related. Today, people use computing services all the time–for banking and financial matters, for entertainment, to arrange travel, to correspond with friends and family, to reach other people by phone, and so forth. Not only must the services they use be reliable and available, but they also must be sufficiently easy to use, both physically and intellectually. As software processes replace manual activities, important characteristics of the older processes must be preserved (three examples: historical records must be preserved, errors and omissions must be tolerated, and it must be possible to get assistance and to pause a partially completed activity and resume later).

Computers provide communication and information access that previously required specialized devices, such as telephones, music players, televisions and VCRs. Making these devices more intelligent promises to make user interfaces much richer and more pleasant to use. These new devices will allow us to use all our senses when working with computer applications much as we do in the real world. Since computers are much more complicated than consumer devices, we should not introduce complexity as we integrate these new modes of interaction into computer user interfaces. For example, playing music with a computer or new computer-based client device should be simpler and better than playing music with a CD player.

The needs of inexperienced users are not well understood. New modes of interaction must be developed, along with a better sense of what constitutes ease-of-use. More research is needed in human-computer interfaces and in human-centered information management. Using computers–particularly computers embedded in devices such as teller machines, "smart" phones, electronic kiosk systems, and the like-must be natural and instinctive. Additionally, the integration of separately developed services must appear seamless to the user. Research in middleware (the standard interfaces and standard services that enable interoperability, including those supporting distributed computing), in scripting of complex applications, and in overall systems integration is required.

Finding: The Nation is underinvesting in fundamental software research.

Over the past decade the U.S. has underinvested in research to create fundamentally new software technologies.[4] Most of today's software technology is based on research performed 15 or more years ago. If we fail to invest in far-reaching, high-risk research now, on what ideas will the commercial advances of the year 2015 be based?

In recent large Federal information technology initiatives, the Committee found systematic underinvestment in the software side of the research agenda, leaving a growing gap between the theoretical potential and actual performance of hardware. The Federal High Performance Computing Program (HPCC) plan, for example, stated, "advances in software will be critical to the success of high performance computers with massively parallel architectures."[5] HPCC researchers, however, were encouraged to focus on the immediate use of their results on Grand Challenge problems or in semi-production parallel computer systems. This successful, short-term strategy emphasized software to solve particular scientific or systems problems, but it failed to develop the fundamental software technology needed to solve wider classes of problems such as interoperability across different computing architectures.

Additionally, underinvestment in long-term software research has interrupted the pipeline that produces software research professionals. The Committee strongly believes that increases in research funding are needed to adequately develop the next generation of promising young researchers.

---

4 Council on Competitiveness. *Endless Frontier, Limitless Resources.* Washington, DC. April 1995: 112-113.

5 Executive Office of the President, Office of Science and Technology Policy. *The Federal High Performance Computing Program.* Washington, DC. September, 1989.

3.1.2 Recommendations

Software is increasingly important to the fabric of our society. By failing to improve the quality of the software we develop, and by failing to provide adequate tools to manage information, we put the Nation at risk. Long-term research is needed to strengthen our software enterprise. That research is not being adequately supported. This research should be closely coordinated with research undertaken in response to the President's Commission on Critical Infrastructure Protection designed to protect the Nation's infrastructure from intentional attacks.

The Committee recommends that a variety of additional investments be made to enable fundamental improvements in the Nation's software quality and its development processes. In particular, major improvements must be made to methods for software development, verification and validation, maintenance, user interfaces to computing systems and electronically represented information, software for high-end computing, and software to support emerging ubiquitous and collaborative computing.

Technology for building software is only part of the solution, since we also need to know what software to build. Current computer systems are too difficult to use. New theories and techniques to simplify the way people interact with information appliances are urgently needed. This will involve new techniques for seeing, hearing, and understanding the environment, techniques for representing and organizing this information within computer systems, and new techniques for analyzing, summarizing, and explaining information.

Chapter 1 described how information technology will transform science, medicine, commerce, government, and education. These transformations will require new algorithms, new tools, and new ways of using computers. Research ideas and prototypes will fuel the transformations. The huge demand for commercialization of current prototypes is draining the pipeline of talented university faculty and researchers–essentially eating the seed corn. The Committee recommends expanding and refocusing university research funding to address longer-term research issues.

Major Recommendation: Make fundamental software research an absolute priority.

The Committee recommends that the Government make fundamental research in software both for computer systems engineering and for applications one of the Nation's highest R&D priorities. We recommend that a focus on software research be a mandatory element of all the Expeditions proposed in Chapter 5. In addition, significant added research funding within current modes of funding should be focused on fundamental software research.

Recommendation: Fund more fundamental research in software development methods and component technologies.

The Committee recommends that research in software methods be aggressively pursued, especially in the area of automated support for software development and maintenance. Such research should:

- Explore and create component-based software design and production techniques, and the scientific and technological foundations needed for a software component industry.
- Explore and create theories, languages and tools that support automated analysis, simulation, and testing of components and their aggregation into systems.
- Create a national library of certified domain-specific software components that can be reused by others.
- Explore and create techniques for using measurably reliable and secure components and their aggregation into predictably reliable and secure systems.
- Explore and create protocols, languages, and data structures to promote interoperability of applications running concurrently across wide-area networks.

Recommendation: Support fundamental research in human-computer interfaces and interaction.

The purpose of many computer applications is to inform people. Moreover, most computer applications require information and guidance from us. Yet the interaction between people and computers is still primarily limited to very rudimentary actions and outputs, such as keyboard strokes, simple gestures using pointing devices such as a mouse or touch-sensitive screen, and textual and simple graphical output on displays. People are capable of far richer interaction. More to the point, humans are perceptual creatures, using all their varied sensory systems in concert to form a harmonious interpretation of the environment. The sensory-motor systems of the human brain are tightly integrated, yet in our computer programs, computer inputs (from human motor systems) and computer output (which go to human sensory systems) are treated as independent and somewhat unrelated activities.

People and computers both deserve better. Fundamental research in human sensory-motor systems, perception, attention, pattern recognition, and decision-making has the potential to make dramatic improvements in the interaction of people and machines.

Today, simulators for commercial aircraft, military systems and games provide the richest examples of what is possible. Full-immersion environments are beginning to explore these areas. However, without sustained, fundamental research into the possible interactions, we will not make much progress beyond today's limited systems. Moreover, these investigations will have rich implications: reducing errors and misinterpretations, enhancing complex control settings, whether in industrial process control, air traffic control, or military applications, as well as helping to overcome some of the access problems faced by people with sensory, motor, and perceptual handicaps.

There are many facets to human-computer interfaces and interactions. We must provide easy access for all people, regardless of economic circumstances, physical impairment, or intellectual limitations. To that end, user interfaces must become considerably richer, depending less on textual interfaces and manual dexterity. They should take more advantage of the decreased cots and increased miniaturization of audiovisual devices, increase the use of natural language, and provide more assistance to the user through techniques ranging from help systems to inference. Progress in ease-of-use is a good example of the coupling of software advances with those of the underlying hardware devices.

Recommendation: Fund more fundamental research in information management techniques to (1) capture, organize, process, analyze and explain information and (2) make information available for its myriad uses.

We already have on-line access to vast quantities of information. Now the challenge is finding the right information quickly and easily, and using it effectively. Ideally, the information will come to us in anticipation of our needs. The Committee recommends increased funding for basic research on capturing, organizing, processing, and using information. Information management is based on the classic computer science disciplines of new and better data structures and algorithms, but also includes theories and new approaches to digital libraries, databases, knowledge discovery, data visualization, and information-intensive applications. Software tools that augment our intelligence and increase our productivity will be key components of the Nation's prosperity in the future. The research in this area should:

- Extend both existing information technologies and those of the future so that non-textual information (pictures, audio streams, animations, etc.) is fully incorporated and integrated.
- Develop improved technologies for filtering information, for data mining, and for tracking the lineage and quality of information, while protecting privacy and intellectual property rights.
- Create digital libraries that can enhance the Nation's scientific and educational processes, and that include mechanisms for the long-term preservation of valuable information.
- Create new technologies for incorporating rich information management services into complex, distributed, demanding applications.

- Develop tools and technologies for modeling and understanding the semantics of information.
- Create tools that make it easier for people to manage, visualize, and exploit information and knowledge.

Recommendation: Make software research a substantive component of every major information technology research initiative.

In the past, major initiatives, such as HPCC, have suffered from underinvestment in software and information management research. When considering new initiatives like the ones proposed later in this Chapter, we must ensure that the software research necessary for success (not just the software development) is included.

For example, we cannot build a scalable information infrastructure without an adequate investment in the software that will provide both system-level and user-level services on that network. For this reason, the Committee recommends in Section 3.2 substantive new software investments to support scalability, distribution, and reliability.

Similarly, there is a continuing need to raise the level at which users of high-end computing create applications software and migrate their software solutions to new architectures. Thus an adequate investment in software to support those activities is called for in Section 3.3.

In short, Federal research support programs have a history of underestimating the software research and development investment needed for success. This tendency must be reversed if new information technology initiatives are to be as effective as the Nation needs.

Budget Recommendation: Increase current funding for software research as follows over fiscal years 2000-2004.

The recommended software research budget increases must account for two important factors. First, the need to increase the average funding per senior investigator by at least $100,000 per year, these increases to start immediately. Second, based on the technical objectives in the above recommendations, the need to increase the number of researchers by approximately 1000 and to support that larger community. The latter increase must be phased in over the five-year period.

The following numbers are in addition to current (FY 1999) funding levels.

Funding increases for Software Research ($ millions)					
	FY 2000	FY 2001	FY 2002	FY 2003	FY 2004
Software Engineering and Component Technologies, Human-Computer Interaction and Information Management	112	268	376	472	540

The Committee recommends that a significant portion of the new monies be used for thematic programs. Rather than establishing large centers with structured management, thematic programs should fund a diversity of groups that are small enough to be well-focused but are large enough to reach "critical mass." Groups might have several cooperating senior investigators, together with staff, graduate students, and undergraduates. Funds must be awarded for long enough periods to address the truly difficult but important software problems described above.

3.2 Scalable Information Infrastructure

Over the next several decades, the transformations described at the beginning of this report will dramatically change our country. The Internet is at the heart of this transformation.

3.2.1 Findings

Finding: Our Nation's dependence on the Internet as the basis for its information infrastructure continues to grow at a dramatic rate.

Within the next two decades, the Internet will have penetrated more deeply into our society than the telephone, radio, television, transportation, and electric power distribution networks have today. For many of us, the Internet has already become an integral part of our daily lives. Communications by electronic mail is commonplace, and many prefer it to the telephone or fax machine.

Many more people will soon depend on the Internet to communicate with friends, family, and businesses; for banking and financial transactions; to purchase goods and services; for entertainment; as well as for vital medical and governmental services. As we come to rely on using the Internet every day, and as billions of dollars of financial transactions are conducted electronically, this information infrastructure becomes even more critical to our Nation's well being. Our ability to manage and control it will become as important as our ability to manage our electricity distribution system. It is rapidly becoming an essential component of our economic and social infrastructure.

America leads the world in developing and applying information and communications technologies, and it is crucial that we continue to invest in them. At stake are the technologies that will determine our Nation's ability to sustain its economic well being, to compete successfully in the global marketplace, to maintain world leadership in basic and applied scientific research, and to provide for national security and defense.

Finding: The Internet is growing well beyond the intent of its original designers, to such an extent that we no longer understand it and cannot confidently continue to extend it.

First, the Internet is growing rapidly in size. A network that connected 2,000 computers in 1985 now connects 70 million devices and traffic is doubling in size every hundred days. More than 50 million Americans use the Internet today, and there will likely be one billion Internet users worldwide early in the next century. No one has ever built a network of such size that provides this level of access to information resources. The combination of common services, a common global name-space, and interactive access creates a new cultural environment on a global scale.

Second, the Internet is growing in capability. We are using it in more and more varied ways. Electronic mail servers are increasingly used for "anytime" communications. Satellite and wireless systems will soon provide many users with "anywhere" connectivity. World Wide Web directory and search services help users find a growing fraction of all the information known to humanity. Electronic payment services handle more of the Nation's commerce. Streaming audio and video is being shown and distributed. Building blocks for new applications are being developed, enhanced, and used: digital signatures, secure transactions, advanced modeling and simulation software, shared virtual environments for collaboration, intelligent agents for discovering and retrieving knowledge, speech recognition, and low-cost networked sensors.

Finally, extending the Internet poses challenges we may not find so easy to meet. We cannot confidently extend today's technologies to new networks and information systems that are orders of magnitude larger, more complex, and that can handle more types of information than today, while expecting to achieve the quality and reliability demonstrated by today's U.S. telephone systems. We understand how to build some, but not all, of the individual elements of the future infrastructure. We do not know how to make these elements work together in reliable, efficient, and robust ways. Even greater challenges will arise when we try to build scalable services, such as search engines that can index exabytes of data, payment services that can process billions of transactions per day, and scalable applications, onto the existing infrastructure.

Finding: Learning how to build and use large, complex, highly-reliable and secure systems requires research.

The Internet-based information infrastructure of the future must provide high performance-reliable, robust, and secure connectivity and functionality among an increasing number of networks and computing systems. The combinatorial character of such highly capable networks of computing systems filled with disparate software defies easy comprehension. We need to analyze this large, complex system in order to understand its behavior better.

We need a better scientific and engineering base to build network links and computing system nodes that can handle radically enlarged and diversified scales of usage and traffic in the expanding information infrastructure. Many technical obstacles–including ease of management, ease of use, integration, privacy, reliability, and security–stand between the current state of the art and the global-scale, heterogeneous, ubiquitous (anytime, anywhere) infrastructure implied by having one billion Internet users.

Networking research must allow us to keep the capability of the Internet ahead of its unprecedented growth. Moore's Law, which has long and successfully guided the semiconductor industry by projecting when performance and technology requirements are needed, can suggest future Internet scalability and performance needs. But the rate of Internet growth is even faster than Moore's Law would predict, and telecommunications researchers will need to develop revolutionary rather than just evolutionary new ideas and theories in order to keep ahead of the Internet's growth curve.

To support the growing demand and dependence on the information infrastructure, research advances in scaling are needed in many different dimensions:

- Scaling to provide robust, reliable, high-speed access. This will improve the quality and richness of the content that users can enjoy and that the information infrastructure can support.
- Scaling to provide assured quality of service. These advances will improve the quality of a user's interaction with the information infrastructure through multi-model communications including speech, sight, or touch, and incorporating isochronous streaming data such as voice and video.
- Scaling to provide ubiquitous access. These advances will enable more people to have continuous access to information and services regardless of their location.
- Scaling of infrastructure services to handle users and requests reliably. These services include authentication, resource directories, search engines, and banking. These advances will improve the quality of information.

- Scaling of the security of the infrastructure to safeguard intellectual property rights, to protect against failures and attacks, and to provide privacy of access when needed. These advances will increase the trustworthiness of both the information and the infrastructure.
- Scaling to support huge servers that can service billions of requests per day against exabyte data stores. These stores need to always be available, must never lose data, and must be able to search the entire data store quickly.

This research should not stop at network creation, but encompass the operation of the complete end-to-end system. Applications research is needed to better understand how to take best advantage of these networks and their large-scale information servers. One example would be scientific and engineering applications that are most efficient when they use different types of computing systems distributed over a national network. Another research area is network-aware applications that can interact with network management software to improve both application and network performance. A third area focuses on designs for large high-traffic servers. A fourth research area is efficient management of data flow among distributed data stores and the applications that access those data. Managing the flow of information from original data through processing to publication and usage will create an information environment that is the core of our scientific, commercial, and governmental systems. We need to learn how to build and use such a system if we are to take full advantage of information technology's potential.

3.2.2 Recommendations

The following recommendations are listed in their logical order and touch upon all aspects of the scalable information infrastructure, ranging from analysis and modeling, to implementation through middleware and scaling, to deployment on testbeds. The Committee considers these recommendations to be of equally high priority in a balanced research program.

Recommendation: Fund research in the behavior of the global-scale network and its associated information infrastructure. This should include collecting and analyzing performance data as well as modeling and simulating network behavior.

It took more than 30 years for the telephone industry to develop useful mathematical models that allowed them effectively to configure, manage, and plan the future growth of telephone networks. Unfortunately, those models do not apply to the Internet, and indeed have become less useful in their own right as telephony has expanded to include fax and Internet communications. The telephone network, though not ideally suited for this task, is the principal method of accessing the Internet. However, the shortcomings of the telephone network are increasingly limiting our future deployment capability. Fundamental differences between telephony and the Internet include:

- The packet-switched Internet Protocol (IP) as opposed to the reserved end-to-end band-width of telephony.
- The wide range of applications that run over the Internet (from e-mail to the Web to teleconferencing) as opposed to voice-only telephony.
- The wide range of networking technologies (wireline, wireless, satellite, cable, broadcast) and vendors as opposed to the mostly twisted-pair technology of the then largely monopolistic telephone industry.
- The wide range of performance requirements of both networks and end-node devices, again as opposed to the simpler requirements of telephony.
- The considerably broader statistical distribution of characteristics such as service times compared to those found in conventional voice telephony.

We are not yet able to model this complex system, either at the level of an individual node or the behavior of Federal, commercial, or corporate subnets of the Internet. We must develop a new set of models. This will require a long-term effort in developing new data-collection techniques, mathematical and statistical modeling and simulation methods and tools, and analytical techniques. It will also require encouraging data sharing among network service providers.

Recommendation: Support research on the physics of the network, including optical technologies, wireless technologies including satellites, wired technologies including cable, and the related bandwidth issues.

Improvement in transmission technologies drives many advances in networking. Improving how fast, how far, or how clearly we can transmit a data packet has a profound impact on what the information infrastructure can do. These advances primarily require research in physics and signal processing: waveforms, signal quality, and manufacturing techniques. Particularly pressing issues surround optical, wireless, and local-loop transmission. Successful research today will help meet the Internet's demands for quality, speed, and ubiquitous access tomorrow.

Recommendation: Support research to anticipate and plan for scaling the Internet.

The most pressing problem in the Internet today is scalability. In whatever way it is measured, it is becoming larger and more diverse:

- Number of nodes: Millions of nodes and links are added each year, and it is not unreasonable to expect the Internet to have more than 100 billion nodes and links by 2025. These will include networked systems embedded in objects ranging from cars to homes to ordinary appliances.

- Number of users: Every person will be able to communicate with every other person and every node on the network virtually instantaneously.
- Geographical dispersion: the Internet is global.
- Heterogeneity of the networking environment (including wired such as optical, cable, etc; wireless, including satellite transports; symmetric and asymmetric two-way communications; and broadcast).
- Heterogeneity of its end nodes (from small embedded to large computers and data stores and other devices from a variety of vendors).
- New types of traffic such as voice, audio, video, or other streamed content.
- Availability, bandwidth, efficiency, reliability, security, and survivability.
- Networks with low to high latency characteristics, as well as the ability to manage them.
- New features such as differentiation of service.
- New uses such as information management.
- New applications such as telemedicine.

We can summarize the research challenges simply. How do we design and build systems and software for deployment on an Internet that is many thousands of times larger and more diverse than today's network? How do we test a system that cannot be built today? How do we manage such a system once it is deployed? How do we add new capabilities without degrading or interrupting the operation of the underlying system on which so much depends? How do we create a system that can evolve capabilities that we cannot imagine today? Answering these questions requires new mathematical and statistical theory dealing with new classes of data distributions, modeling, simulation, experimentation, and testbeds.

Recommendation: Support research on middleware that enables large-scale systems.

Middleware for the scalable information infrastructure is the collection of shared software and rules that will improve the software development process, help make the infrastructure operate properly and efficiently, and make large new software systems possible. It will provide reusable software that performs common functions, making software development more efficient. This will let applications software developers concentrate on their applications and not on data transmission or computing issues. It will enable the development of applications that communicate with each other. It will also enable the development of middleware that adapts to a changing environment through services that coordinate and supervise the infrastructure itself. Middleware research topics fall into the two broad categories:

Information management

Scalable and widely application information management services require that we build systems to organize and manage vast amounts of information. The needed technologies include those for:

- Managing exabytes of data and billions of requests for information each day.
- Managing text, semi-structured information, relational data, multimedia data, and active data.
- Translating between information representations.
- Discovering and monitoring resources.
- Analyzing, summarizing, integrating, and fusing information from diverse heterogeneous sources.
- Presenting information in meaningful ways.
- Tracking the lineage or provenance of information, so users can know the sources of original data and how information was derived.
- Tracking and controlling access to information, protecting privacy, and charging for information where appropriate.

Information and services survivability

Our Nation's security, commerce, education, and well-being depend increasingly on our information infrastructure. It is thus critical to ensure the survivability of that infrastructure in the face of malicious attacks or viruses, equipment or software failures, and overload. Survivability means that services will be available when needed and information will be delivered in a timely fashion. Services must operate correctly and the information they deliver must be of high quality or be identified if otherwise. Survivability includes long term preservation of information; a document of historical importance should be preserved, even as underlying storage technologies and information representation evolve and even if it is not accessed for hundreds of years.

Survivability technologies include:

- Authentication and security mechanisms for a large, heterogeneous, and evolving infrastructure.
- Mechanisms for detecting system intrusion, and information and software corruption.

- Mechanisms for detecting, mitigating, responding to, and recovering from, or for preventing, human error in the creation and use of the infrastructure.
- Mechanisms for assuring information quality.
- Scalable information and service replication strategies.
- Mechanisms for monitoring services to ensure correct operation within given quality-of-service bounds.
- Repositories for guaranteed long-term preservation of information.

Recommendation: Support research on large-scale applications and the scalable services they require.

Very large applications have unique needs that today's technologies cannot meet. The process of developing these applications has been found to be difficult to manage, and product quality is difficult to engineer. Today's commercial-off-the-shelf technologies do not scale for use in these large applications. In addition, university-based researchers are not well equipped to work on these problems, which require substantial labor and infrastructure. Research focused on these scalability issues is needed.

Several of these applications come from Federal agencies with science and engineering missions, including the Department of Energy's Accelerated Strategic Computing Initiative (ASCI) for nuclear stockpile stewardship, NASA's Earth Observing System Data and Information System (EOSDIS–one of the most complex data handling systems ever to be built), and the FAA's air traffic control system. These applications combine communications-intensive, data-intensive, and computation-intensive requirements.

While these applications are small in number and, by today's standards, are large in scale, they will be quite common in 20 years. In the meantime, today's demanding applications will have grown in size and will continue to make scalability demands, so the research requires a long-term approach.

The technologies developed to address large-scale Federal science and engineering applications will be useful in addressing non-science Federal applications such as improving the capabilities and efficiency of large Federal entitlement programs, and in addressing non-governmental applications that are not being addressed today. We examine two such applications in some detail.

Application: A National Digital Library

A National Digital Library would integrate all electronic knowledge sources, including books, journals, music, films, and informal "documents." The Nation's scalable information infrastructure would manage this shared information base. It might begin with a digital library for the country's scientific literature, which while largely produced at taxpayer expense is not widely available even to scientists, engineers, and students. Such a library would have a strong positive impact on scientific and engineering research and development and on education by providing users with more up-to-date information, reducing effort and cost, and preserving this body of knowledge for future generations. The university-led Federally-funded Digital Libraries projects, begun in 1994, have developed a base upon which to develop this library. Some of the challenges a National Digital Library presents are:

- Mechanisms for protecting and managing intellectual property.
- Strategies for managing the data that describe the library's holdings and mechanisms for translating document formats.
- Technologies for multi-lingual interoperability, allowing users to access information in different languages.
- Interfaces that let users visualize and understand the wealth of available information.
- Digital journals with on-line collaborative authoring, submission, reviews, publishing, and reader annotation.
- Technologies for organizing large bodies of diverse knowledge (for example, clustering, summarization, and classification).
- Technologies for efficiently digitizing paper documents.

Application: The Next Generation World Wide Web

Today's Web and its browsers can be viewed as a primitive information infrastructure. It is has evolved rapidly from a simple design that did not anticipate either today's size or the wide class of applications for which it is now used. Designing and developing a Web for the year 2010 is a research challenge that requires addressing capability, flexibility, power, robustness, and scalability needs and applications far different from today. While a Next Generation Web shares many research issues with information management and digital libraries described above, it has its own unique needs. These include research and development in translating among the protocols and policies of different countries and providing a graceful transition path.

Recommendation: Fund a balanced set of testbeds and research infrastructure that serve the needs of networking research as well as research in enabling information technologies and advanced applications.

The Federal government has funded two testbeds as part of its Next Generation Internet (NGI) initiative. One connects more than 100 sites at end-to-end speeds 100 times faster than the Internet of 1997, and the other connects about 20 sites at end-to-end speeds 1,000 times faster than 1997's Internet. The former is used primarily for developing "revolutionary" applications and the latter for research in networking technologies. While interconnected, these testbeds have different characteristics.

The 100x testbed connects several Federal research networks, including NSF's vBNS, DoE's ESnet, NASA's NREN, and DoD's DREN. Each of these networks is a service provided by a U.S. company–MCI WorldCom, Inc., Sprint Communications Company, and AT&T, respectively. In addition to being used by researchers in almost 100 universities and major Federal research facilities to develop applications, the testbed provides a national-scale environment for experimenting with new capabilities such as multicast protocols and protocols enhanced to include differentiation of service, privacy and security. It also serves as a testbed for the interoperability, robustness, and scalability of networks provided by different vendors, and the management of such interconnected networks. The testbed uses networking technologies more advanced than those in the commercial Internet. These technologies have been made stable and robust largely through collaborative efforts by the Federal agencies and the providers. Users appropriately expect high availability and high performance from every NGI site nationwide, because their research depends on network availability and performance.

The 1,000x testbed is used to explore networking technologies for a future generation (beyond the next generation) Internet. It must stress the limits of the new technologies being developed. It is smaller than the 100x testbed, is more fragile, and may experience outages.

Information infrastructure facilities are increasingly serving a similar role for scientists and engineers that large-scale shared physical instruments have for decades. As is the case for national telescopes, microscopes, and particle accelerators, agencies like NSF and DoE provide peer-reviewed access to leading edge IT facilities to enable a wide range of disciplines to do research. It is clear that such facilities are only useful to maintaining our Nation's research leadership if they are continually improved to keep them at the state of the art. However, this need to continually improve specialized IT facilities, such as high-end computers, high performance networks, and large discipline databases, must be managed by the Federal agencies in a manner that does not directly conflict with funding the basic research that is enabled by the facilities.

We recommend a periodic review of testbed and research infrastructure activities. This will help ensure that funding and usage is consistent with the research missions for which they are designed. The participants in this review should include representatives from the Federal Large Scale Networking Working Group, the Federal agencies that provide these resources, user communities, and the appropriate private sector communities.

Recommended funding levels

Scalable Information Infrastructure is a relatively new and especially fast moving field, making it difficult to estimate its research and testbed needs.

The large number of research topics explicitly identified above, including many topics never before or at best minimally addressed in the past, and the growing demand for new Internet capabilities, suggests that research in scalable technologies can be highly effective. The Federal government's responsiveness to the rapid evolution of this field supports this position. The Government has modified its plans (for example, by increasing the size of its testbeds to meet demand), accelerated its research schedule, and consistently addressed issues that the private sector will not address (such as interoperability across providers).

The considerations in proposing the scalable information infrastructure budget included the growth rate of the Internet and acknowledged cost savings such as in-kind contributions from the private sector. The Committee recommends that funding for scalable information infrastructure research and testbeds grow from the current approximately $200 million per year to $500 million per year in a linear fashion between FY 2000 and FY 2004. The emphasis should be on networking research. The research should be a mix of small (now on average less than $100,000 per year, growing to $200,000 per year), medium (approximately $1 million per year growing to twice that amount), and large (more than $4 million per year) research projects, plus networking research as part of two expedition centers (each approximately $5 million per year). The relative ratio of the number of projects should be approximately 1 large: 4 medium: 20 small projects. Costs for applications development testbeds should be shared with the applications communities whenever possible.

The following numbers are in addition to current (FY 1999) funding levels.

Funding increases for Scalable Information Infrastructure Research ($ millions)					
	FY 2000	FY 2001	FY 2002	FY 2003	FY 2004
Scalable Information Structure	60	120	180	240	300

3.3 High-End Computing

Since its creation under the High Performance Computing Act of 1991, the Federal High Performance Computing and Communications (HPCC) Program has contributed greatly to U.S. technological leadership. The Program has created and disseminated technologies to speed the pace of innovation, enhance national security, promote education, and help us understand the global environment. Furthermore, advances in high-performance technology have become an important component of mid-range computing and communications systems. Through the dedicated effort of many scientists and engineers, HPCC has succeeded in laying a foundation for future growth.

Historically, high-performance computing was a specialized market segment within the computer industry, serving the focused needs of science, engineering, and national defense. High-performance supercomputers were based on unique architectures and technologies, optimized to the requirements of these compute-intensive applications and were expensive relative to mainstream computers. However, over the last few years, the use of high-performance computing has undergone a major transition from purely compute-intensive applications to include data-and communications-intensive ones. As the scope broadened, the description "high-performance computing" was broadened to "high-end computing," and now encompasses an increasing number of applications beyond science, engineering and defense.

With the advent of powerful microprocessors used in workstations and servers, a large portion of the high-end computing market shifted from supercomputers built with specialized processors to ones built with microprocessors. Due in part to Federally funded research efforts more than ten years ago, these systems, based on commodity parts, are now available at a fraction of the cost of earlier supercomputer systems. But for various technical and economic reasons, highly parallel systems (with hundreds or thousands of processors)–with the potential of ever-increasing performance into the teraops range (10^{12} operations per second) and beyond–have not yet been used widely for critical high-end applications.

Research in high-end computing has produced far-reaching benefits to society. Many of the present uses of high-end computing are well-documented in previous reports.[6] They include:

- Designing new cancer fighting and anti-viral drugs.
- Understanding the causes and sources of air, water and ground pollution, and devising solutions to these problems.
- Forecasting local weather and predicting long-range climate changes.
- Designing safer, fuel-efficient vehicles.
- Ensuring the safety and effectiveness of the Nation's nuclear stockpile.
- Designing new aircraft, such as the Boeing 777.

---

6 Brooks, Jr., Frederick P., and Sutherland, Ivan E., co-chairs. *Evolving the High Performance Computing and Communications Initiative to Support the Nation's Information Infrastructure.* Prepared by Committee to Study High Performance Computing and Communications: Status of a Major Initiative. Washington, DC: National Academy Press, 1995.
Lax, Peter D., [chairman], *Report of the Panel on Large Scale Computing in Science and Engineering*. Under the sponsorship of Department of Defense (DoD) and National Aeronautics and Space Administration (NASA). Washington, DC, 1982.
Department of Defense, "National Security High End Computing Integrated Process Team for National Security High Performance Computing" briefing, 1997.

- Making scientific discoveries in a broad range of disciplines from elementary particles to cosmology.
- Simulating the effects of attacks on weapons systems, such as the New Attack Submarine.

High-end computing has three distinct and highly inter-related markets, each with its own dynamics and issues:

- The broader data-intensive and communications-intensive market, with applications such as data mining, web serving, digital libraries, and transaction processing. This market is doing very well, and growing very rapidly.
- The scalable, microprocessor-based compute-intensive market. This market is doing quite well at the mid-range of the market, with symmetric multiprocessor systems and parallel systems of moderate-size. However, the high end of this market, which highly parallel systems are expected to fill, remains in the research and experimental stage primarily because we lack adequate software technology, application-development tools, and, ultimately, well-developed applications.
- The vector-architecture-based sector. This market has hundreds of installed systems but is shrinking, primarily because most new applications are being developed for the more modern and economical microprocessor-based systems. Thus, the majority of the market for vector systems consists of older applications that have not been converted and tuned to high-end parallel systems referred to at the end of the previous paragraph. Currently, it is not known whether all vector-architecture applications, some of which are critically important to selected Federal missions, can be efficiently executed on high-end commodity processor-based parallel systems.

3.3.1 Findings

Finding: High-end computing is essential to science and engineering research.

The successes of the HPCC program have led to widespread use of computational simulation and modeling as a means to understand natural phenomena and to explore and optimize engineering designs. Over the next few decades, computation will become even more essential to the Nation's fundamental and applied science and engineering endeavors, both civilian and military. Currently, the extreme high-end of computing is done under the auspices of focused mission agencies such as the Department of Energy (DoE), while widespread academic research is primarily sponsored by the National Science Foundation (NSF). Both for the sake of fundamental scientific research and to enable applications to benefit from the research, the research community needs access to systems at the leading edge of capability. The power of these systems should be comparable to that of systems-like the

DoE Accelerated Strategic Computing Initiative (ASCI) systems available to the mission agencies. Furthermore, if the scientific community is to continue to benefit from high-end computing, it is important to ensure that there are enough computer and computational scientists to collaborate in making leading-edge computation an effective research tool.

Finding: High-end computing is an enabling element of the United States national security program.

Government customers account for at least half the market for supercomputer systems, with national security programs historically driving the market. These programs are very important to the Nation. Such Government organizations face the challenge of converting their applications to new high-end architectures without disrupting their missions. This transition has proved to be technically difficult and is exacerbated by the lack of mature software tools. The DoD HPC Modernization Program, DoE's ASCI, and the National Security Agency are examples of national-security users of high-end computing.

Through the DoD HPC Modernization Program, the Office of the Secretary of Defense is investing more than a billion dollars over five years in high-end computing to provide the United States military with a technological advantage. This program will provide advanced hardware, computing tools, and training to DoD developers using the latest technology to support the warfighter. The incorporation of high-end computing will allow the United States to maintain its technological supremacy in weapons systems design into the foreseeable future, while decreasing the total life-cycle costs of fielding new defense systems.

The DoE ASCI program is critical to the Stockpile Stewardship Program. ASCI is designed to assist in ensuring the safety, reliability, and performance of the United States nuclear stockpile in the absence of underground testing. In addition to addressing National security needs, the computer technology and products developed by the ASCI program will be applied to a broad spectrum of national needs. The computational techniques developed by the ASCI program have great potential, if widely adopted by the civilian researchers, to bring economic and scientific benefits to the United States in environmental studies, biology, drug design, automobile and aircraft development, consumer product safety, and information management and access.

For decades, the National Security Agency (NSA) has influenced high-end computing and has been an early and sophisticated user of the most advanced commercial computer, storage, and networking systems. In its mission to protect U.S. information systems and produce foreign intelligence information, the NSA has stimulated both industry and academia with some of the most challenging problems in the Nation. NSA research is directed to the discovery and application of methods seeking order-of-magnitude improvement for deriving intelligence by use of mathematical and signal processing methods. Activities range from invention and prototyping of new concepts, to improving the ability to use leading-edge commercial products.

Finding: New applications of high-end computing are ripe for exploration.

In addition to advancing the long-term research agenda in traditional high-performance computing, the Committee believes that the Nation needs to develop new uses of high-end computing to promote a better understanding of our world and to improve services to all citizens. This expansion of high-end applications will attract computer, software, and application vendors to support a healthy high-end computing marketplace. Today the high-end computing market is a shrinking fraction of the information technology industry. Among the areas in which high-end computing could provide enormous benefits are:

- "Intelligent" systems: using sophisticated data mining algorithms against very large data bases to make expert decisions in a variety of application areas, from health care to market research.
- Design: using computer assisted design and engineering technologies in the many, more mundane, but far more broadly applied sectors such as fashion and architecture rather than just in extremely important but rarefied fields such as aircraft design and nuclear weapons stewardship.
- Transportation: managing route optimization and fuel efficiency for aircraft more effectively, at a savings of billions of dollars per year.
- Crisis management: simulating crisis scenarios, particularly weather and groundwater flow, critical to the management of emerging crises as well as to effective training and preparedness.
- Infrastructure support: managing the large-scale networks that are increasingly essential to the Nation's economy and government.

Finding: United States suppliers of high-end systems suffer from difficult market pressures.

Despite the economic and societal benefits of solving problems like those described above, the market for high-end computing systems is mixed. The low-and mid-range parts of the high-end market–workstations and serves–are doing well financially, but the highest end, sometimes referred to as "supercomputing" to reflect its origins, has not been growing. Its sales are less than $2 billion per year. This part of the market (for instance the fastest 100 computers installed worldwide–the Top100), historically dominated by U.S. vendors, has gone through a period of intense consolidation in the past several years, leaving only SGI (with its Cray subsidiary) and IBM as significant U.S. participants. No well-established U.S. company remains with its primary focus on technical, high performance computing. Recently established companies Tera and SRC have yet to show a profit.

At the same time, foreign competitors with substantial government subsidies have enlarged their efforts to capture the very-high-end market, which creates a potential reliance on foreign vendors for future specialized high-end systems. For instance, as of November 1998, whereas only 10 percent of the more broadly based Top 500 fastest computers installed worldwide are of non-U.S. manufacture, the foreign market share of the higher-end Top 100 rises to 25 percent. The Committee believes that this trend will continue unless the Nation invests in research and development programs in cooperation with industry and academia to increase U.S. leadership.

Finding: Innovations are required in high-end systems and application-development software, algorithms, programming methods, component technologies, and computer architecture.

These innovations need to be aimed at breakthroughs in sustained performance on real applications, programmability, ease-of-use, and scalability. Inadequate funds for software and algorithm research and development, as well as excessive focus on short-term research, have created an unduly limited software base usable by only the most intrepid users.

Current Federal programs that address innovative high-end architectures and technologies are too small, having shrunk dramatically in the last decade. Petaflops-scale computing is required for many of the applications listed above. To realize petaflops computing, we need advances in architecture, system software, algorithms, programming methods, and hardware technology.

Finding: The high-end computing capability available to the civilian science and engineering community is falling dangerously behind the state of the art.

Information infrastructure facilities increasingly serve a role for scientists and engineers similar to the role that large-scale shared physical instruments (for example, national telescopes, microscopes, and particle accelerators) have served for decades. Agencies like the NSF and DoE provide peer-reviewed access to leading-edge IT facilities to enable a wide range of disciplinary research.

For the past decade, science and engineering researchers working on civilian applications have typically had access to computing capability that lagged the national security community by a factor of two to five. This situation resulted from smaller budgets for the civilian-access machines and the timing of the national security requirements that required the first available systems. Today, the civilian science and engineering community finds its computing capability a factor of ten to twenty behind the largest installed systems, those used in the ASCI program. Furthermore, the decision to reduce the number of NSF supercomputer centers from four to two without increasing the funding for high-end systems in the remaining centers has substantially reduced the overall capacity available to academic researchers.

It is clear that these facilities are useful for maintaining our Nation's research leadership only if they are continually improved to keep them at the state of the art. However, Federal agencies need to manage this need for continual IT facility improvement (for example, of high-end computers, high-performance networks, and large disciplinary databases) in a manner that does not directly conflict with funding the basic research enabled by the facilities.

3.3.2 Recommendations

In high-end computing as in other areas of information technology, we need more fundamental research–the kind of groundbreaking, high-risk research that will provide the ideas and methods for new disciplinary paradigms a decade or more in the future. Our greatest needs are improving systems software and algorithm-level software support at the high end, exploring innovative architectures and devices, and making it possible for the academic research community and the Federal Government to conduct essential research and development on computers of the highest possible performance. Research on software, architecture, and hardware can be effectively driven by an initiative to reach sustained performance in the petaflops and petaops range by the year 2010.

To have the greatest effect, the NSTC Subcommittee on Computing, Information, and Communications should coordinate all research on and access to high-end computing systems across all the Federal agencies. The program-planning processes should be coordinated, and the programs that result from the following recommendations should be closely coupled and mutually supporting. For example, architecture and software efforts are mutually dependent, and the recommended petaops milestone can drive both the hardware and software efforts.

Recommendation: Fund research into innovative computing technologies and architectures.

There is evidence that current scalable parallel architectures may not be well suited for all applications, especially where the computations' memory address references are highly irregular or where huge quantities of data must be transferred from memory to support the calculation. To address these limitations, the Committee recommends expanding research and development funding for new computer architectures. In particular, we need substantive research on the design of memory hierarchies that reduce or hide access latencies while they deliver the memory bandwidths required by current and future applications. Progress in this area would improve the performance of almost all programs and would make high-end software development less difficult.

The Committee recommends funding research into promising new technologies such as optical, quantum, biological, and neuromorphic computing. Ultimately, silicon chip technology will run up against the laws of physics. We do not know exactly when this will happen, but as devices approach the size of molecules, scientists will encounter a very different set of problems fabricating faster computing components. Potential areas for investment are improved simulation and analysis tools for investigating innovative architectures and full-system experiments to explore and validate architectural design alternatives for real-world application workloads. The Government must begin to invest now in these new technologies to ensure that computing capability continues to improve well into the 21st century. Architectures for high-end computing must be developed in concert and iteratively with hardware and software technology investigations. The high-end computing research funding strategy should consider the coupling between these domains. For many of the larger projects, industry participation will be critical as only industry can provide important practical insight. As innovative research succeeds, industry will bring the new architectures and technologies into the open marketplace.

Recommendation: Fund R&D on software to improve the performance of high-end computing.

The Committee recommends substantial investment in software to improve the performance and efficiency of high-end computers. The software investments fall into three categories: system software, algorithm development, and software to manage integrated systems in a balanced fashion. Investments in system software-languages, compilers, runtime libraries, operating systems, file systems, I/O drivers, debuggers, programming interfaces, performance tuning tools, and so on will lead to improved efficiency and performance and will make high-end systems usable by a much larger community. Algorithm development is necessary to ensure the efficient use of leading-edge machines. Work on system software and algorithms will permit the effective exploitation of parallelism and will contribute to the efficient use of memory hierarchies.

For maximum performance, we require an integrated computer environment consisting of software to perform resource allocation, parallel input and output, ultra-high-speed and high-capacity intelligent storage, scientific data management and visualization. We need to do research on software that will improve overall system efficiency by eliminating potential bottlenecks between the high-end computer and its peripheral equipment. For example, petaflops computers will generate exabytes (10^{18} bytes) of data that the system must move rapidly to high-speed storage and then make readily available to users. Scalability of these tools is critical since they must keep pace with the rapid development of processors used in high-end computers. These efforts will not only increase usability, but also facilitate transition of applications, including those required for national security, to new computer architectures.

Cost effective and efficient use of emerging terascale systems (10^{12} operations per second) requires major advances in software research, development, and deployment. Without effective building blocks for software, algorithms, and libraries, high-end hardware is an expensive, underutilized technology. And adequate software will emerge only from an aggressive, stable, sustained, and long-term software research program. The software problem will become even more critical with the possible realization of a petaop capability (10^{15} operations per second) by 2010, perhaps earlier. The time to explore petascale software aggressively is now, not after we build hardware prototypes.

Even more alarmingly, limited funding for high-performance software research is causing an exceptional group of researchers to leave this research area in favor of other areas where funding is more readily available. If the Nation is to increase, or even retain, the pool of intellectual talent exploring terascale and petascale software, the Government must fund long-term software research.

Recommendation: Drive high-end computing research by trying to attain a sustained petaops/petaflops on real applications by 2010 through a balance of software and hardware strategies.

The Committee reviewed the CIC/HECC (High-End Computing and Computation) working group's proposed petaflop activity. The preliminary studies that group sponsored have determined that we will need substantial technological advances to achieve the desired performance levels by 2010. The goal of building systems capable of achieving sustained performance on important applications at the petaops or petaflops level can be an effective way to drive research on the hardware technologies, new architectures, software systems, and algorithms that will be needed to make such systems possible. We, therefore, recommend that funding for this activity should be increased to ensure that the necessary technological advances–in algorithms, systems software, application-level software components, and hardware–are achieved.

The goal of sustained petaops and petaflops performance must be understood as a technology driver, not as a goal unto itself. It is conceivable but unlikely that sustained, real-world performance in this range could be reached by faster hardware technology and increased hardware parallelism alone, but these sustained performance levels are almost certain to require a tightly coupled, well-balanced effort on both software and hardware. It would be counterproductive to produce a computer system capable of petaops performance without the software and algorithms needed to make it useful for the full range of high-end applications. An appropriate program will, for example, accommodate a technical comparison and evaluation of current architectures with promising new ones such as those based on processor-in-memory components. It should also address the software challenges of programming computers with tens or hundreds of thousands of processors and deep memory hierarchies.

Adequate funding would permit a petaflops program to accelerate the rate of progress towards petaflops computing by directly investing in and developing key hardware and software technologies that may prove crucial to long term computing objectives, even if they are not immediately suitable for short-term, medium-scale computing systems.

Recommendation: Fund the acquisition of the most powerful high-end computing systems to support science and engineering research.

High-end computing is becoming critical to an increasing breadth of science and engineering research. Recently the Nobel Prize in Chemistry was awarded to two pioneering computational researchers whose efforts led to the development of widely used methods in computational chemistry. Research like this helps us unlock the behavior of new materials, advance our understanding of biological structures, and understand and enhance the environment in which we live.

If the United States is to continue as the world leader in basic research, its scientists and engineers must have access to the most powerful computers. Therefore, the Committee recommends that the Federal government continue to provide these computing systems to the research community through major, shared-facility centers. To increase long-term, fundamental research across all scientific and engineering disciplines, the first priority should be to increase the computing capacity of the centers that can best serve the entire research community. These are the NSF Partnerships in Advanced Computational Infrastructure (PACI) centers and the specialized centers that support specific disciplines like NSF's National Center for Atmospheric Research.

The first priority for high-end acquisitions should be to bring the performance level of academic computing close to that of mission agencies; DoE's ASCI program is the current benchmark for computational performance. The second priority should be to increase the capacity of focused-mission centers. Mission agencies like DoD, DoE, and NASA should seek additional funds to acquire high-end systems. We encourage the NSF PACI centers and mission-specific agencies to cooperate in providing computing capability to scientists and engineers independent of sponsoring organization.

Efforts should be made to expand the use of high-end computing in information technology research. For example, the simulation of proposed petaflops and petaops architectures will be essential to developing software for those systems, yet such simulators cannot even run simple computational kernels without using computers of the highest possible performance. The Committee, therefore, recommends funding additional high-end capacity to support the Research Expeditions to the 21st Century described in Section 5.3. These high-end systems will enable truly innovative research and thereby assist in developing a cadre of researchers who can exploit high-end computing in the 21st century. These high-end systems will also create an environment for new applications and will expand the critical, domestic high-end computing market.

Recommendation: Expand the NSTC CIC High End Computing and Computation (HECC) Working Group's coordination process to include all major elements of the government's investment in high-end computing.

The Committee recommends that all of the Federal Government's major research, development, and acquisition investment budgets and plans in high-end computing be provided to the NSTC CIC Subcommittee's HECC Working Group. Currently, some large investments in high-end computing such as the DoE's ASCI program, the Department of Defense's High Performance Computing Modernization Program (HPCMP), and some of NSA's investments are excluded. Thus, these programs and their contributions to high-end computing are generally excluded from the Federal Government's attempt to coordinate high-end computing efforts across agencies. Exclusion of any large investment makes it very difficult to coordinate and plan research and development of high-end computing.

Budget Recommendation: Increase current funding for high-end computing research and acquisitions as follows over the fiscal years 2000-2004.

The budget recommended for the high-end effort was developed using a three-step process. First, four subpanels were formed–innovative technologies and architecture, high-end software, sustained petaops/petaflops, and high-end acquisition. Second, based on the findings in this chapter of the report, each subpanel independently determined the technical objectives underlying the recommendations. Third, each subpanel determined a budget required to accomplish these objectives. The budget below totals the research budgets of the first three subpanels, keeping the acquisition budget separate.

The following numbers are in addition to current (FY 1999) funding levels.

Funding increases for High-End Computing Research ($ millions)					
	FY 2000	FY 2001	FY 2002	FY 2003	FY 2004
High-End Software, Architectures, and Petaops Computing	180	205	240	270	300
High-End Acquisitions	90	100	110	120	130
Total	270	305	350	390	430

This table covers only those efforts described in the first four recommendations above. It excludes efforts currently excluded from the HECC crosscut and coordination process addressed in the fifth recommendation.

4. Socioeconomic Research and Policy Priorities

Our National well-being depends on understanding the potential social and economic benefits of on-going advances in information technology. We must also understand the problems that are arising from the increasing pace of information technology-based transformations. To realize the promise of the new technologies, we must invest in research to identify, understand, anticipate, and address these problems. We need more data, and we need to understand social, economic, and policy issues in much greater depth. The research that is required to develop this knowledge should be broad-based, long-term, and large-scale, in its scope and in its promise for continuing the beneficial transformation of our society and economy.

We must ensure that all Americans are well-equipped to live in this changing world. To that end, we must improve education beginning at the K-12 level and extending through post-secondary education. To improve the skills of our information technology workforce and to increase the information technology literacy of our overall workforce, we must also concentrate on education and retraining for all of the information technology skills that are important to our society. Furthermore, we must concentrate on programs that will enable the United States to use its diverse human resources to the fullest capacity.

The promise of the information age must include all citizens, regardless of ethnicity, gender, physical ability, income level, or geographic location. Access to information technology must be accompanied by the assurance of a high degree of personal privacy. Programs to provide broader access should be based on sound understanding of the social and technical conditions and emerging developments to ensure that we reach out to people wherever they are through the most efficient, effective means.

Information technology can significantly improve the flow of information to all people and institutions in the Nation, and can thus be a powerful tool for democratization, equity, and access. However, to achieve the potential, we need concrete objectives and comprehensive metrics. We must conduct careful research to measure the effects of the transformations against the objectives, and we must develop appropriate policies based on knowledge gained from the research.

4.1 Findings

Finding: The use of information technology–in particular the growing popularity of the Internet and the emergence of global commerce–has introduced a series of important and complex policy issues.

In addition to the many technical challenges discussed in the preceding chapters of this report, there are many barriers to realizing the benefits of information technology that are social, political, or legal in nature. For example, consider the host of problems associated with implementing our common vision of transforming the practice of health care, laid out in Chapter 1 of this report.

Several challenging technical problems remain to be solved through IT research, before telemedicine can become a common practice. However, the solution of problems arising from several socioeconomic and policy issues will, most likely, present even more formidable barriers to the deployment of telemedicine. A few of those issues include the privacy and security of patient records, use of records by insurance companies, current requirements for state-by-state licensing of doctors, and the development of mechanisms for creating trust between patient and doctor in an on-line environment. Similarly, widespread use of electronic commerce will require mechanisms for insuring even greater security and trust between parties in an on-line environment. These are some of the important questions that can be addressed only by further research on the economic and social effects of computing and communications.[7]

Further, the use of information technology, particularly in a widely accessible networked environment, has led to new policy issues. Among the areas in which concerns have arisen are personal privacy, intellectual property rights, digital signatures, and the control over the availability of particular kinds of content (for example, pornography or inflammatory rhetoric.)

An example of an important set of complex policy issues of particular importance are those related to privacy. The management and analysis of

---

7 National Research Council, Computer Science and Telecommunications Board. *Fostering Research on the Economic and Social Impacts of Information Technology. Report of a Workshop.* Washington, DC: National Academy Press, 1998.

information is critical to many applications. Information about users and their usage of applications will assist service providers in business, government and the medical profession to provide informed, cost-efficient, and highly-personalized service. However, this will be possible only if individuals are guaranteed that their information is truly protected. If there is no trust in the system, its benefits will be severely limited. Guaranteeing the privacy of information is a prerequisite for many of the exciting applications that are possible to build. Our citizens must work together with government and business to ensure that only the most appropriate regulations are put into place. There is a very fine line between helping citizens to feel safe and creating an undesirable and onerous bureaucracy.

On an international scale, the European Union Privacy Directive heightens the urgency of information privacy policies. Under European law, firms doing business in the E.U. are prohibited from sending personal information to countries, such as the U.S., that do not adhere to the privacy principles embodied in the European directive. This is likely to have a dramatic impact on U.S. electronic commerce.

Finding: Policy decisions and IT investments are being made on the basis of incomplete research and data concerning the effects of IT on our society.

Integration of IT into our personal and professional activities has not come about by accident: decisions have been made that have brought about the many transformations we are currently experiencing. As the pace of IT research and integration accelerates, more decisions will be made which have ever-expanding consequences and effects on our daily lives. Consequences will range enormously in their scope and scale, from determining the format through which our children are first introduced to IT in early primary school, to determining policies which have long-lasting effects on our national transportation, communications infrastructures and economic practices. In too many cases, decisions about IT implementation have not been properly informed by adequate research. In fact, it is often the case that the implementation of information technologies has a considerably different set of consequences than were originally intended or anticipated.

Much more social science research on the impact of IT on our society is needed to inform ongoing debates and policy decisions on IT-related issues. Such research can also help IT researchers develop technical solutions for some difficult policy problems (e.g., development of new metadata tagging standards and micropayment technologies for managing intellectual property). Moreover, insights derived from social science research may be able to contribute to the better designs of information systems. The design of groupware, for example, should be driven by research on how groups of people share information and make decisions.

Equity and Access

Finding: All of our citizens must have access to information technology.

In his remarks at the MIT commencement ceremony on June 5, 1998, the President reiterated his vision of limitless possibilities for all Americans brought about by the very technical advances that we are recommending in this document. One of the challenges before the Nation is to extend those possibilities to *all* Americans. In an increasingly competitive global economy, our Nation cannot afford to squander our human resources by providing opportunities only to those Americans who are favored by geographic or economic circumstance. Access to and use of information technology, particularly in educational settings (K-12 through university) is a prerequisite to building the skills that will allow our citizens to function productively in the 21st century information society. It is a critical stepping stone for instilling interest and developing the skills of the budding information technology researchers who are essential to sustaining our national research capabilities.

Information technology can help eliminate social and economic inequities and mitigate problems arising from physical disabilities. Information technology tools and applications can provide opportunities that transcend barriers of race, gender, disability, age, income, and location. The enabling quality of the technology, in addition to the cultural values cultivated through its best-known application, the Internet, carry a democratizing potential that already has transformed our social interactions and economic opportunities, both at home and abroad. Public libraries have always played a critical role in ensuring equity of access to information resources and the role of public libraries is even more vital today in guaranteeing free and equitable access for all to Internet resources and communications access. It is important that we study the impact, costs and potential funding sources to support this continuing essential role for ensuring equity of access by public libraries across the country.

Yet we see a gap in access to the Internet and the accompanying participation in the information technology revolution among racial groups. For instance, whites are more likely than African-Americans to have Internet access within or outside of their households, to have access to a PC at work, and to have used the Web.[8] We expect there are similar gaps with other minority groups, such as Hispanics and Native Americans. Recent research released by the National Telecommunications and Information Administration (NTIA) suggests that the racial gap in Internet use is increasing.

---

8 Hoffman, D. L., and Novak, T. P. "Bridging the Racial Divide on the Internet." *Science*, April 17, 1998: 390-391.

The consequences to our Nation of a persistent racial divide on the Internet will be severe. If a significant segment of our society fails to have equal access to the Internet, U.S. firms will have difficulty hiring personnel with the necessary technological skills, and competitiveness will suffer. Employment opportunities and income differences among whites and minority groups will be exacerbated, with further adverse consequences for the Nation. The Internet may provide equal opportunity and help level the playing field, but only for those with access.

Finding: Full participation in information technology research requires access to high-band-width connectivity.

Information technology research and education rely increasingly on high-speed network connectivity for access to information, collaboration between geographically separated colleagues, and access to advanced technology resources such as astronomical telescopes and high-power electron microscopes. Scientists, educators, and students at rural, small, and ethnic universities are often unable to participate in new research and educational opportunities because they are not connected to thevBNS[9] The experimental program to stimulate competitive research (EPSCoR) exists to promote the full participation of institutions in eighteen states and Puerto Rico that have been underrepresented in research partnerships. However, small or rural universities exist in other states as well. Full participation, particularly in research partnerships such as the proposed Expedition centers, is not possible without high-bandwidth connectivity.

Workforce

Finding: The supply of information technology workers does not meet the current demand.

By virtual unanimity, chief executive officers of a cross-section of America's leading corporations have identified the need to strengthen the technological workforce as the single greatest challenge to U.S. competitiveness over the next decade.[10] It is crucial that we produce a continuous supply of well-trained, high-quality professionals in engineering and computer and information science, not merely skilled users, but researchers, creators, and designers of advanced technology. In this fast-moving field, those people must continue to update their knowledge and skills.

Today we fall far short of meeting these needs, and projections for the future are not encouraging. While the information technology sector and demand for skilled personnel are growing rapidly, the pipeline for com-

---

9 vBNS is NSF's very high performance Backbone Network Services.

10 Council on Competitiveness. *Competing Through Innovation: A Report of the National Innovation Summit.* Washington, DC: March, 1998.

puter engineering and computer science graduates is not filling fast enough. Beginning with skills that require a BS-level of training, qualified information technology workers are in extremely scarce supply. The largest fraction of open positions exists in those requiring graduates with advanced degrees. In its Interim Report to the President, the Committee recommended increasing the number of H-1B visas as a short-term measure, but this solution is untenable for the long term.

There is evidence that large pools of potential information technology personnel in the U.S. workforce exist and employers are recruiting new workers from outside information technology fields. However, there is also the problem of what economists call "appropriability", a type of "market failure." Many information technology employers who are finding hiring more difficult are unable to provide the required education or training due to resource limitations. Others are unwilling to do so because of competition for their newly trained employees. Meanwhile, many potential information technology workers may be motivated to get the education and training needed for information technology careers but unable or unwilling to pay the (often substantial) costs involved– they are often otherwise employed and precluded from fulltime study. It should be recognized that a complete solution to the workforce problem would involve recruiting from underrepresented groups as well as nationwide excellence in K-12 education. These issues are discussed more fully in the following findings.

Finding: A diverse workforce literate in information technology is critical for ensuring that our Nation is prepared to meet the challenges and opportunities of the Information Age.

It is critical to tap all of this country's talent for the information technology workforce. There is, without any doubt, a vast untapped talent pool in the United States among women and minorities, currently under-represented in engineering and information sciences. African Americans, Latinos, and Native Americans constitute a fourth of the total U.S. workforce, 30 percent of the college-age population, and a third of the birth rate. Yet members of these minorities collectively comprise only 6.7 percent of the U.S. computer and information science labor force (all degree levels), 5.9 percent of the engineering workforce, 1.7 percent of the U.S. computer science faculty, and 4.9 percent of the engineering faculty.[11] Women participate in both computer science and engineering at low rates for all degrees and subsequently in academia and industry.

---

11 Campbell Jr., George, Ronni Denes, and Catherine Morrison, eds. *Access Denied: Race, Ethnicity and the Scientific Enterprise.* New York: Oxford University Press. Forthcoming, August 1999.

These facts, trends, and issues represent a significant opportunity to build the information technology workforce we need. However, we can achieve the goal of a stronger technological workforce, remain economically competitive, and prosper only by developing and employing the talents of the currently under-utilized majority of the population.

Finding: Both K-12 and post-secondary education are inadequate to meet the challenges of the information age.

If our citizens are to reap the benefits of the information technology revolution, personally and professionally, and if the Nation is to have the information technology-literate, well-educated, and highly skilled workforce it needs, every student must learn some aspects of information technology and many must become highly educated experts. That is not happening everywhere now. Many children do not have the opportunity to acquire the information technology skills they will need in the future. Many adults have no access to continuing education or the retraining that would enable them to hold good information technology jobs. Furthermore, the educational pipeline is not preparing enough people for the information technology workforce.

4.2 Recommendations

Our recommendations are organized into two categories: recommendations involving education, training, and workforce policies and those calling for additional research. To begin to solve some of the problems described above, new research and substantive policy changes will be needed. Some of the research we recommend could well be supported within the agencies that traditionally fund science and engineering, but other appropriate sponsorship for this research should also be considered.

Overarching Recommendation: Expand Federal initiatives and government-university-industry partnerships to increase information technology literacy, education, and access.

In implementing the research agenda recommended in this report, the Government must increase its efforts to provide broader opportunities for developing information technology literacy, access, and research capabilities. The Government must also expand opportunities for information technology literacy and access as it addresses issues related to workforce skills and development. Specific recommendations are provided below.

Recommendation: Expand Federal research into policy issues arising from information technology.

As the information revolution introduces changes in society, the United States must continue to protect its citizens' information privacy, the rights to intellectual property, the ability to provide verifiable signatures, and the ability to exercise control over appropriate access to information. New policies and procedures are certain to be needed. As an example, we consider information privacy–an issue that is increasingly visible.

It is appropriate that industry take the lead in establishing and implementing privacy procedures. Government regulation might be required if industry fails to act properly, or in very sensitive areas like health care and protection of minors.

Policies for the protection of individuals' information privacy on the Internet should be evaluated and established. Industry education programs should be developed that carry the message that protecting consumer privacy on-line will actually be good for business. The government and industry should jointly fund this effort.

The Committee recommends that research in privacy issues be aggressively funded to help address the following questions:

1. What privacy policies would both protect individual privacy in electronic environments and permit the conduct of business on the Internet? For example, common United States practice is to permit users to "opt-out" of information sharing, but our notification procedures are often lacking. The E.U.'s directive requires full disclosure (notice) and informed consent ("opt-in"). Other areas of study include the conditions under which individuals should be able to change information recorded in databases and how noncompliance with privacy policies would be redressed.

2. What mechanisms can best address the protection of individual privacy on the Internet? For example, should the U.S. try to develop regulatory mechanisms for enforcing privacy policies, or are self-regulatory efforts more effective? Should an industry trade group take the lead? Should a nonprofit organization be developed? How can the U.S. best foster dialogue on privacy?

3. How is the E.U.'s Privacy Directive likely to affect the development of electronic commerce, both locally and globally?

Recommendation: Fund information technology research on socioeconomic issues.

Since an understanding of the socioeconomic issues that arise because of technological change is intrinsically a multi-disciplinary investigation, there should be research on social, economic, and policy issues. The studies could be tied to particular aspects of information technology, such as the Internet, or they could be more broadly based. As appropriate, research on socioeconomic issues should be carried out in Expedition or Enabling Technology centers, including some of those with a primarily technical focus. These centers should define and legitimize research areas, provide the scale of research activity necessary for experimental research in socioeconomic topics, allow for research teams of sufficient size for multidisciplinary research, and allow sufficient time for the maturation of a research area.

Possible areas of research for such centers include (but are not limited to):

- Transformations of social institutions.
- Governance.
- Electronic commerce research.
- Social and economic simulation and modeling.
- Sustainable use of large information infrastructures.
- Electronic groups and communities research.
- Barriers to information technology diffusion.
- Human interaction and communication laboratories.
- Innovative scientific development of a national information infrastructure to support research on social and economic issues.
- Digital government.

Recommendation: Expand the participation of underrepresented minorities and women in computer and information technology careers.

Systematic research efforts are needed to identify specific barriers experienced by underrepresented groups in their pursuit of careers in information technology. The research program should include investigations of pre-college preparation issues; barriers to undergraduate enrollment, financial aid and admissions policies; retention in information technology degree programs; graduate education issues, career access; and career advancement obstacles. Experimental programs should be developed to test the effectiveness of different approaches toward reducing those barriers. In particular, it will be important to determine if (and how) the tools of IT can themselves be used to break down these barriers. A significant proportion of this research could be carried out at the Enabling Technology Centers, and at an Expedition Center focused on socioeconomic and workforce issues.

Based on the data obtained from this research, and from other effective programs already in place, the IT community should increase its efforts to recruit women and minorities into IT careers. We note that several private-sector programs have been successful in producing minority and women graduates, particularly in engineering. They have developed a high-quality research and support infrastructure that provides such services as alternative assessments to identify potential recruits among educationally disadvantaged high school students, essential academic enrichment programs for those students, scholarships, mentoring, internships and professional development. The National Science Foundation has also developed several programs aimed at improving education in under-served communities, and at increasing participation of minorities in science and engineering. Those successful programs should be extended to encompass information technology and related fields. Government support for exemplary programs operated through public and private foundations should be increased substantially.

Recommendation: Create programs to remove the barriers to high bandwidth connectivity posed by geographic location, size, and ethnic history of research, educational institutions, and communities.

Programs to facilitate high bandwidth connectivity, such as the NSF vBNS connectivity grant program, should be continued and expanded to address problems posed by geographical location, size or ethnic history. It is essential that such limiting factors are removed so that all institutions with efforts related to the network testbeds are included, both to provide the necessary research and to contribute needed students to the information technology workforce. Special supplements to connection funding, such as the present EPSCoR supplements in the NSF Connections program are one approach to this problem. Other approaches such as research to develop and apply new technology to facilitate access and mitigate policy/regulatory concerns should be addressed, as well. In addition, we should support community efforts to provide connectivity through local institutions, such as libraries, schools, places of worship, and community centers.

High-bandwidth connectivity in every state is needed. Similar to historic large-scale national infrastructure programs such as the Rural Electrification Act or the interstate highway program, the construction of an "interstate information superhighway" system must be set as a high-priority, long-term National goal. We do not propose, however, that the government subsidize the deployment of a national high-bandwidth infrastructure at this time. Independent of any related policy considerations, the expense would be prohibitive using current technologies. Rather, the first and essential step toward attaining that long-term goal is to mount the type of research effort proposed in this report. By doing the research necessary to decrease connectivity costs, nationwide high-bandwidth connectivity will become possible.

Recommendation: Accelerate and expand education in information technology at all levels– K-12, higher education, and lifelong learning.

To prepare our citizens for the Information Age, education and training are needed for computer literacy, for entry into jobs and careers that require information technology skills, and for continuing to update knowledge and skills in the face of rapid technological change.

As part of preparing students to live in an information technology world, they must learn the basic use of computers and the Web to acquire information, communicate with others, and perform other daily tasks. In addition, computer science and information science material must be incorporated into the K-12 curricula as a natural augmentation of education in math and science. In order to make education in information technology available to all K-12 and college students, effective mechanisms must be developed to train highly qualified computer and information science educators, and to encourage them to teach in under-served communities.

Lifelong-learning is an extremely important part of the information technology profession– information technology professionals must continually update their knowledge and skills to keep up with this fast-changing field. Advances in personal computers and Internet access can enable learners to acquire high-quality information technology courses and training anytime and anywhere. Information technology education and training should be made readily available to motivated learners.

Collaborations among schools, government, and industry will be an important factor in information technology education and training. An example of a program that facilitates collaboration between government and industry to enhance information technology education and training is the National Science Foundation Advanced Technological Education Program (ATE) begun in 1994.[12] Other programs and incentives for such collaborative efforts should be developed and supported.

Recommendation: Strengthen the use of information technology in education.

Information technology has great promise as a vehicle for improving both the quality and the availability of education at all levels. Although many ideas have been advanced and tried, considerably more research and experimentation are needed to realize that promise. At the same time, it is essential that the use of those aspects of information technology that are known to be effective for education become more widely available and used.

The President's Committee of Advisors on Science and Technology (PCAST) recently released the "Report to the President on the Use of Technology to Strengthen K-12 Education in the United States." They recommend that the Federal government play a leadership role in supporting an aggressive research agenda to evaluate the efficacy in both the short and long run of the educational approaches that are being fostered by the new technologies. This Committee strongly endorses that recommendation.

---

12 Advanced Technical Education (ATE) grants are specifically designed to improve science, mathematics, engineering, and technology education for community college and high school students preparing for careers as technicians in such fields as biotechnology, electronic, and communications.

Interest in "electronic learning" is growing rapidly among universities, community colleges, and private education providers. Yet information technology educators and trainers often don't use it. The use of information technology would be particularly beneficial for teaching people about information technology. In particular, we believe these "anytime, anywhere" approaches offer the potential for high-quality education and training to be accessible by large pools of would-be students (including high school students) who could not otherwise participate. Such approaches could be powerful responses to needs for both increased numbers undertaking initial education in information technology fields, and for information technology professionals to constantly update their skills while continuing their employment.

Recommended Funding Levels

The following numbers are in addition to current (FY 1999) funding levels.

Funding increases for Socioeconomic Research ($ millions)					
	FY 2000	FY 2001	FY 2002	FY 2003	FY 2004
Socioeconomic Research	30	40	70	90	100

The committee recommends ramping up to $100 million over five years. In the first year, we recommend spending an additional $30 million, as follows:

1) Add $10 million to fund more socioeconomic research. This includes workforce data collection and multidisciplinary socioeconomic research and one to two Enabling Technology Centers at $3-5 million each.

2) Add $10 million to increasing research on educational technologies.

3) Add $10 million to fund high achieving and pilot programs in engineering or science that target students from economically disadvantaged populations.

The large increase in the third year represents our recommendation to fund an Expedition Center.

5. Creating an Effective Management Structure for Federal Information Technology R&D

Given the importance of information technology to the Nation's future, it is critical that the Federal research investment in this area be well managed. An effective management structure for information technology research can ensure that the Federal research portfolio is appropriately balanced, can encourage innovative long-term research, can help the Administration articulate research priorities, and can help assure Congress that its goals are being met. The management structure should ensure a diversity of funding mechanisms, including single-investigator grants, multiple-investigator projects within and across disciplines, and integrated research centers. Such diversity in approach and funding modes will increase opportunities for unexpected discoveries, encourage a broader attack on problems, and ensure fewer missed opportunities.

Unlike other disciplines that are usually incremental and evolutionary, computer science and engineering is a transformational and revolutionary technology. Since tomorrow's applications will be ones that we cannot even envision today, it is critical that we as a Nation invest in core computer science and engineering to enable those revolutionary advances. The Committee feels strongly that information technology research must not be driven entirely according to the extrapolated demands of today's commercial, scientific, and/or national security applications.

An effective management structure should accomplish the following:

- Ensure important national goals for information technology are addressed.
- Provide a focused and informed policy setting.
- Allow for strategic management and planning to ensure adequate investments.
- Identify the entire Federal information technology R&D endeavor, so that an assessment of the portfolio can be made.
- Guarantee that Federal research funds are managed by using open, competitive selection processes, thereby ensuring that the best ideas and the best investigators receive support.
- Provide mechanisms to connect complementary efforts across research disciplines and funding agencies.
- Review and assess Federal efforts.
- Provide a stable, but flexible overall program that sustains a healthy base of computing, computation, and communications science technology.

5.1 Agency Roles

Currently, funding for information technology R&D comes from several different agencies, with no single agency having information technology as its primary responsibility. This system has worked surprisingly well in the past and the Committee views the diversity of research funding as a crucial strength.

However, the Federal funding portfolio has not kept pace with the importance of information technology to the Nation, especially with respect to support for fundamental research in computer science and engineering. The Committee recommends that the proposed budget increase be used primarily to fund fundamental information technology R&D, with a significant portion of the funds directed to the National Science Foundation (NSF). The Committee believes that NSF should use this increased funding to reinvent and significantly enhance its traditional role as the champion of core computer science and engineering research.

In parallel, other agencies conducting information technology research should renew their commitment to fundamental research and engineering, with specific emphasis on making software a substantive component of every major information technology initiative. For example, the Defense Advanced Research Projects Agency's (DARPA) flexible and creative management style and its technical expertise and history of successful investments in information technology R&D has and should continue to make significant contributions. Participation by a variety of agencies, which tend to have different styles and emphases for supporting research or procuring technology, is important to establish and maintain an effective U.S. information technology R&D program. This diversity allows many views to compete and fosters pursuit of intellectual questions via a range of complementary modes.

Such information technology research will be motivated by long-term mission goals, but should be evaluated and funded with fundamental information technology challenges in mind–developments that will generalize to other domains despite their genesis in a specific mission agency. The Committee believes, for example, that biomedically motivated basic research in information technology should be supported by the National Institutes of Health (NIH) and viewed both as important information technology research and as fundamental biomedical research. Similar arguments can be made for DARPA, DoE, NASA, and the other research agencies that will be continue to be involved with mission-motivated basic research in information technology.

Recommendation: Strongly encourage NSF to assume a leadership role in basic information technology research. Provide NSF the necessary resources to play this role.

In its leadership role, NSF should foster interagency collaborations to ensure adequate funding levels in basic research. While NSF can only control its own budget, it should work with other agencies with similar research needs to assess major requirements and needs being addressed, identify shortcomings of the fundamental research portfolio, and initiate efforts to address deficiencies.

This leadership role for basic information technology research may require innovations internal to the NSF, as may be necessary to ensure that NSF is able to define, support, and coordinate a broad range of modes of research support. Example modes of support are centers of diverse sizes and multiple-investigator projects with longer terms. NSF may also have to make changes to ensure that it fosters high-risk, high-payoff long-term research and invests sufficiently in the core of computer and information science and engineering. We also believe that there needs to be significantly more information technology representation on the National Science Board than exists now.

To successfully carry out the program of research proposed in this document, NSF will need to support a portfolio of modes of research that differ from the current mix of mostly single-investigator research grants together with a small number of centers. For example, multi-investigator projects of longer duration (5-7 years of funding) are needed in order to carry out high-impact experimental research agendas, whether or not such grants are available in other parts of the Foundation. Some of these projects will include long-term basic technology research that has an engineering orientation and involves explicit technology transfer. Some projects should be supported in response to topic-specific solicitations in order to support a DARPA-like concentration of effort on compelling research areas. Because of the nature of the field, the portfolio must successfully meld science investigations with technology exploration.

Roughly 40-50% of the proposed budget increases for information technology should go to NSF, to support basic research in information technology focused in the Computing and Information Science and Engineering (CISE) Directorate, with the rest allocated to other research agencies. The majority of the NSF increase should go to the new programs and modes of funding; the remainder should go to the traditional programs within the CISE Directorate, expanded as appropriate to projects of larger size and longer duration.

5.2 Policy and Coordination of Information Technology R&D

Given the importance of information technology to Nation's economy, security, and well being, the Committee believes there should be high-level management attention and focus by the Administration for information technology R&D.

Recommendation: Designate a Senior Policy Official for Information Technology R&D.

Information technology is of critical importance to the Nation; it is the basis for a $700B domestic industry that drives many of today's innovations and scientific discoveries and offers even greater potential for innovation in the next century. The Committee believes that information technology R&D requires a high-level policy voice, similar to that afforded to other priority research areas in the White House. Positioning an individual at this level would help ensure the Administration receives timely advice on information technology issues and foster a strategic approach in determining R&D investments. This individual would be responsible for leading the White House effort to establish Federal policies to support, encourage, and help coordinate long-range information technology development to maintain U.S. leadership in this vital part of our economy.

Recommendation: Establish a senior-level policy and coordination committee to provide strategic planning and management.

The principal goal of coordination is to ensure that the Federal information technology research portfolio is sufficiently aggressive and properly balanced to meet the pressing needs of the Nation. Since many Federal agencies fund information technology R&D, a high-level venue for cross-agency coordination is desirable. The coordinating committee should consist of agency directors, since it is they who have policy and budget making authority. This coordination committee should advise and report directly to the President's Assistant for Science and Technology to increase the overall effectiveness and productivity of Federal R&D efforts to develop and apply information and communications technology. The coordination committee should address significant national policy matters that cut across agency boundaries. The committee should establish objectives for research programs and review them to ensure that they are meeting those objectives. In addition, it should ensure that the overall Federal program is well-balanced and has good coverage of important topics.

This coordination should include the entire Federal information technology R&D endeavor, including the efforts resulting from the proposed budget increase. A problem with the present management structure is the difficulty of getting a clear and complete picture of the total Federal investment in information technology, since there are Federal computing, information, and communications R&D programs that are not part of the formally coordinated Computing, Information, and Communications (CIC) programs. The coordination in information technology should encompass all major investments in R&D, as well as investments in people, supplies, and the like. For instance, high-end computing acquisitions are a necessary part of a coordinated program.

Recommendation: Extend the HPCC program coordination model to major Federal information technology R&D activities.

To make this coordination successful, the entire enterprise should have strong staff support in the White House and in the agencies themselves. The HPCC program, with the National Coordination Office and working groups with agency representation, is an effective model of interagency collaboration. The Committee recommends extending this model to all major Federal information technology R&D activities, with the NCO facilitating interagency coordination and supporting the management structure recommended by the Committee.

5.3 Support and Implementation

To achieve the goal of reinvigorating long-term, high-risk research in information technology, Federal funding agencies will need to support many different kinds of projects, including single-investigator grants, multiple-investigator projects within and across disciplines, and integrated research centers. To encourage the visionary planning needed to produce true innovation, funding for these activities will need to be provided for longer periods–long enough for high-risk projects to be carried through to completion. Furthermore, because effective research in computer science involves collaborations of all sorts–with other computer scientists, with researchers in other disciplines and at other institutions, with industry, and with the community–the modes of support should encourage these collaborations, without turning them into strict requirements. Research projects should have the flexibility to organize themselves in the best ways to achieve the goal of technological innovation.

Recommendation: Diversify the modes of research support to include more projects of broader scope and longer duration, placing a renewed emphasis on research carried out in teams.

During the 1970s and 1980s, DARPA computer science researchers were encouraged to imagine dramatically different futures and to carry out projects that would explore those futures. Researchers were given enough resources and time so they could concentrate on problems rather than worry about their next proposals. The results were dramatic advances in speech recognition, robotics, chip design, high-performance computing, machine vision, artificial intelligence, and virtual reality. The Committee would like to see this spirit renewed and replicated across all Federal information technology R&D programs.

Currently, most NSF fundamental research support is for single-investigator grants. Although this kind of support will remain an essential component of the research portfolio, information technology can often make the most dramatic progress when carried out in teams of moderate size. DARPA has used this model effectively for several decades, and it is rapidly becoming part of the portfolios of other funding agencies including NSF. The committee believes that team research in information technology should be at least as important as single-investigator research, including the National Science Foundation. Furthermore, inter-institutional team projects help build a coherent national research community.

Recommendation: Fund collaborations with applications to drive information technology research, but take measures to ensure that research remains a primary goal.

A principal goal of information technology research is to produce results that will someday have commercial or social applicability. Thus, it is important to understand applications to which information technology may be applied and to use this knowledge to drive research into new technological strategies. Therefore, the Committee recommends that the Federal Government continue to fund collaborations between information technology researchers and applications developers. However, the funding programs must be carefully structured to ensure that the research remains a principal focus.

In the early 1990s, a principal focus of the Federal High Performance Computing and Communications (HPCC) Program was the use of large-scale applications efforts to drive research on parallel computing. In this vision, teams of applications developers and computer scientists would have worked together to rebuild major applications on parallel computing platforms developing many useful computational support technologies. Unfortunately, many of the projects came under pressure to meet progress and performance deadlines for the applications, leading them to jettison all but the most immediately applicable computer science research.

To reemphasize long-term fundamental research, we must acknowledge that the time to applicability will be too long in many cases to directly affect an application project that is driving the research. For example, if a computer architecture researcher works with a scientist on computational fluid dynamics to understand how to design better computers for that application, it is unlikely that machines of the resulting architecture will run that specific application in a two to three-year time frame. In a funding program designed to foster long-term research, this situation must be completely acceptable. Indeed, it is essential that the funding programs be designed to give equal value to the information technology research as to progress on the application.

Recommendation: Fund centers for Expeditions into the 21st Century.

Projects of moderate size cannot alone address the entire spectrum of long-term research needed to sustain our national leadership. To foster research with truly dramatic impacts, the Committee recommends the creation of two types of center-sized activities: "Expeditions into the 21st Century," which will endeavor to intersect with the long-term technological future, and "Enabling Technology Centers," which will focus on research with application to problems of national importance.

"Expeditions into the 21st Century" will be centers, perhaps virtual, that bring together scientists, engineers, and computer scientists from academia, government, and industry to "live in the technological future." The mission of these expeditions will be to report back to the Nation what could be accomplished by using technologies that are quantitatively more powerful than those available today. In essence, these centers will create "time machines" to enable the early exploration of technologies that would otherwise be beyond reach for many years. Just as the Lewis and Clark expedition opened up our Nation and led to unanticipated expansion and economic growth, the ideas pursued by information technology expeditions could lead to unexpected results and nourish the industry of the future, creating jobs and benefits for the entire Nation.

There are existing successful examples of the use of this "living in the future" approach. In the private sector one of the most famous examples is the Xerox Palo Alto Research Center (Xerox PARC) where researchers created an experimental network of computers for use by individuals. This effort pioneered many of the revolutionary technologies that led to today's personal computers, including graphical user interfaces, pointing devices, laser printing, distributed file systems, and WYSIWYG word processing. In the university community, the MIT Media Lab has been conducting similar explorations. Finally, there is the example of ARPAnet, which evolved into today's Internet.

The Committee recommends funding several Expeditions, each with a different focus. The focus may be on either a discipline-based theme, such as bioinformatics or multi-scale engineering, or on an infrastructure-based theme, such as distributed databases or tele-immersion. To establish a context, each Expedition should be based on assumptions not true today, for example, ubiquitous computing or a vast amount of simulation, as described in Gelernter's *Mirror Worlds.* Each Expedition need not be limited to a single such assumption, but an Expedition should invest sufficient resources to make exploration of its assumption areas, those parts of the map of the future, possible.

Each Expedition would be required to carry out several activities, including:

- *Technology testbeds.* Expeditions would establish testbeds in which the future technologies that are the focus of their activities could be effectively explored in a realistic setting. The computing environment established by Xerox PARC is a good example of such a testbed.
- *Economic and societal impact studies.* Expeditions should include research on the impact of future technology on society. Such studies should involve researchers from disciplines other than computer science.
- *Education.* Each center should have an extensive educational program to inspire students of all backgrounds to think about the long-term technological future and its implications.
- *Outreach.* Expeditions should reach out to the research community, to industry, and to the public. The purpose of these programs is to accelerate the realization of the center's vision in the form of knowledge and products. Outreach to the research community should include open meetings where the results and effects of center activities are publicly presented.

Criteria for selecting proposed expeditions for funding should include:

- Elucidation of a clear and exciting vision for the future.
- Novelty of the technological area of exploration.
- Identification of significant computer science and engineering research challenges.
- A coherent plan for carrying out and managing the proposed activities.

The Committee recommends creating up to five Expeditions, to be selected by a competition to establish both a core activity, and a standard process to allow additional researchers to participate. Competitions should be held at regular intervals e.g., every three years, to ensure a flow of new ideas. Each of these "virtual centers" should include investigators from many research institutions. Each center would receive an initial funding for five years, renewable after five years for a full term of 10 years. To encourage truly aggressive efforts, very high annual funding levels should be possible, perhaps up to $40 million per center.

Recommendation: Establish a program of Enabling Technology Centers.

The Committee also recommends establishment of centers of excellence in computer science and engineering research applied to particular applications of information and communications technology. These Enabling Technology Centers (ETCs), located at university and/or Federal research institutes, will provide integrated environments for academia, industry, and Government to focus on the application of next-generation information technology to important national problems. There are many application domains where information and communications technologies could make a difference. These include computational science and engineering; health care; delivery of Government services/Digital Government; crisis management; environmental monitoring; life-long learning; law enforcement and public safety; arts, culture, and the humanities; intelligent transportation systems; improving the quality of life for persons with disabilities; and distributed work (e.g. telecommuting, collaboration by geographically distributed teams).

Enabling Technology Centers should be focused on applied technology and development. Researchers at the centers would conduct R&D on information technology problems arising from the center's application domain, develop new curricula for students and mid-curricula for both students mid-career professionals, participate in testbeds, and identify barriers to more widespread adoption of information technology in the application domain of concern.

In addition to research and development, these centers should perform several functions, including:

- *Education and training.* Researchers will develop educational programs and curricula the intersection between computer science and engineers and a particular applications domain.
- *Testbeds.* Researchers will participate in the experimental deployment of information systems "in the field."
- *Research on factors inhibiting deployment of information technology in the application domain.* Researchers may discover that the use of information technology in particular application domains is being slowed by legal and regulatory barriers, lack of end-user training, the absence of compelling cost-benefit analysis, a lack of technical standards or other mechanisms for interoperability, etc.
- *Community building.* The centers could help build communities of researchers, companies, Government officials, users, and other stakeholders by convening conferences and workshops, developing research agendas, and supporting intellectual infrastructure such as electronic print archives, collaboratories, databases, and case studies of successful and unsuccessful uses of information technology.

Funding for Enabling Technology Centers should be shared between mission-oriented agencies with an interest in a particular applications domain and an agency with a broad mandate to support information technology research (e.g., the National Science Foundation).

Grants need to be of sufficient duration and size to support a critical mass of researchers interested in a particular applications domain. The Committee recommends each center be funded on the Science and Technology Center funding model. The full term of an ETC would be 10 years. Annual funding of up to $10M per center is recommended, with up to 15 centers simultaneously in operation. Competitions should be held every three years.

5.4 Annual Review

Recommendation: Establish an annual review of research objectives and funding modes.

Both the Coordination Committee and the President's Information Technology Advisory Committee (PITAC) should be instrumental in reviewing research objectives. PITAC's role in advising the President through NSTC serves to provide high-level private sector advice on information technology. The Coordination Committee will provide high-level advice from within the government.

The Coordination Committee, with advice from the Advisory Committee, should conduct an annual review of research programs to ensure that they are achieving the goals set out for them. In particular, the Coordination

Committee should ensure that the modes of support proposed for the programs–centers, multiple-investigator interdisciplinary projects, and testbeds, along with a renewed emphasis on software and long-term research–are not being compromised and that these modes of research support are meeting the goals set out for them.

This review is intended to serve as a high-level check to ensure the Federal R&D portfolio is properly balanced, comprehensive, and properly coordinated. To maximize the opportunities for full and frank exchanges among the principals, the degree of formality and process associated with the preparation for and conduct of this review should be strictly limited. Given that the purpose is to coordinate a scientific endeavor, the review might more closely resemble a scientific workshop than a traditional committee meeting.

5.5 Funding

Budget Recommendation: Increase current funding for IT R&D as follows over the fiscal years 2000-2004.

Table 5.1 summarizes the recommended funding increments for each of the four research priority areas identified in this report. Descriptions of the research topics and associated funding tables were provided in Chapters 3 and 4 of this report. The funding recommendations are in addition to those for current (FY 1999) Federal IT R&D programs and represent, rather, new or expanded activities.

Table 5.1
Recommending Funding increases for IT R&D ($ millions)

	FY 2000	FY 2001	FY 2002	FY 2003	FY 2004
Software	112	268	376	472	540
Scalable Information Infrastructure	60	120	180	240	300
High End Research	180	205	240	270	300
High End Acquisitions	90	100	110	120	130
Socioeconomic	30	40	70	90	100
Total	472	733	976	1192	1370

Can the Proposed Increases be Invested Effectively?

One question that arose during the public comment on the interim report was whether the IT research community could effectively invest an increase of the size we are proposing. To answer that question, we have developed a simple investment model described here. Note that this is not a proposal on how invest the funds, but rather an exercise to determine whether we can reasonably expect the community to be able to take on the challenge.

There are three principal ways that funding increases might be used to affect change in the research community:

- To add new researchers taken from the pool of graduating Ph.D.'s over the next five years.
- To upgrade the support base for existing faculty.
- To provide shared infrastructure for the community, typically supercomputers and high-bandwidth networking.

New researchers will be critical to increase the capacity for both research and teaching in IT. The current Ph.D. production in the field is approximately 1000 per year. Of these, less than half the choose academic careers. If research investments can increase that number by half or more, the Nation can add 2500 new researchers to the academic base over the next five years. The new funds can be used to add an additional 2500 researchers and to upgrade the support base for the current IT research community. We assume a range of annual support levels from $100K-$250K per entry-level researcher within that community. Assuming 1000 new researchers at $250K and 1500 new researchers at $100K, we arrive at a total of $400M per year at the end of five years.

Our analysis also establishes that there is unused research capacity in the existing research community. Currently the top-rated universities have an average support level of $400K to $500K per researcher per year, while universities not ranked in the top five have average support levels substantively lower, in the range of $150K per researcher per year. One goal might be to bring the next twenty-five departments up to the quality and support levels of the top five and another fifty departments up to about $250K per researcher per year. Assume that 25 organizations (with about 40 principal researchers per organization) will need to have their support base per principal researcher increased from about $150K on average to about $500K to $250K per principal researcher. These target levels are quite reasonable in terms of today's costs. For example, $250K will typically cover the cost of two graduate students, a programmer, some equipment, and summer salary for the researcher. The above calculation leads us to estimate an expenditure of $650M per year after five years.

These numbers reflect the vital importance of upgrading the current research base to improve the overall effectiveness of current funding and to insure that the universities remain competitive with the IT industry in providing computational and networking environments sufficient to be attractive to new graduate students. Without this latter investment, the needed increase in researchers may not materialize in the universities. These investments will insure that the universities can attract the needed researchers and that the existing research base can be made competitive with the state-of-the-art in industry.

Following the recommendations of the report, we have budgeted $330M for large shared infrastructure such as supercomputers, external networks and testbeds to link the researchers and associated maintenance. This leaves us with the following totals, which are consistent with the budgets proposed in the report, for the fifth year:

Funding entry-level researchers	$400M
Provided Supercomputers and external network infrastructure	$330M
Upgrdae 25 sites to $500K per researcher support base	$350M
Upgrade 50 sites to $250K per researcher support base	$300M
Total	$1380M

Conclusion

In both the public and private sectors today, U. S. investments in technology R&D have slowed to a relative trickle. American businesses, in an ever-shrinking and more highly competitive world, have devoted less and less of their precious resources to long-term R&D, directing their efforts instead to reducing costs and getting new products into the pipeline today at the expense of the future. The Federal government has mirrored this trend because of dramatically increasing pressures on the research and development budgets, with only modest increases in funding levels. Funding agency managers have responded by making the natural and correct decision to favor the short-term needs of their missions over the need for long-term research in information technology. The U.S. Government's lead in research in high end computing and computation–so crucial to keeping our military edge in the competitive era of the Cold War–seems to have come down along with the Berlin Wall.

As a result, the once robust technological edge the U.S. has enjoyed over the rest of the world is actually built on an increasingly fragile technological substructure. If the trend away from long-term research continues, the flow of bold new ideas that has fueled the information economy in this decade is likely to slow to a trickle by 2010. To keep its competitive edge, the United States must rededicate itself to cutting edge high-tech research and development–or risk being overcome by nations with a clearer plan and a stronger view of the future. This is a risk the Nation cannot afford to take.

The initiatives proposed in this report would represent a major step toward restoring long-term Federal research in information technology to levels that will ensure continued prosperity and new technological solutions to national problems in the next millennium. The time to act is now and the Federal government has a unique role to play in supporting research in this critical area. The Brooks/Sutherland report stated this well:

> Very few companies are able to invest for a payoff that is 10 years away. Moreover, many advances are broad in their applicability and complex enough to take several engineering iterations to get right, and so the key insights become 'public' and a single company cannot recoup the research investment. Public investment in research that creates a reservoir of new ideas and trained people is repaid many times over by jobs and taxes in the information industry, more innovation and productivity in other industries, and improvements in the daily lives of citizens. This investment is essential to maintain U.S. competitiveness.[13]

This committee strongly recommends that the Federal government embark upon the kind of leading-edge, visionary research necessary to continue the revolution that has transformed the lives of our citizens in ways not thought possible just thirty years ago. The recommendations of this committee also stress the needs to: 1) to upgrade the knowledge base and skills of our workforce, so that our citizens will be prepared to face a new century fully prepared for the technological challenges that are yet to come; and 2) to give all American the opportunity to participate in the information age, so that our citizens will be able to fulfill its promise. These steps, if taken now, will bring handsome returns to the Nation over the coming decades.

---

13 Brooks and Sutherland, 23.

CHAPTER 18

THE ELECTRONIC INTRUSION THREAT TO NATIONAL SECURITY AND EMERGENCY PREPAREDNESS (NS/EP) TELECOMMUNICATIONS

An Awareness Document

March 1999

Third Edition

Office of the Manager
National Communications System
701 South Courthouse Road
Arlington, VA 22204-2198

TABLE OF CONTENTS

EXECUTIVE SUMMARY

ES-1 KEY FINDINGS

1 INTRODUCTION

1.1 OVER VIEW

1.2 SCOPE

1.3 REPORT ORGANIZATION

2 BACKGROUND

2.1 INTRODUCTION

2.2 TELECOMMUNICATIONS AND INFORMATION SYSTEMS

2.2.1 The Public Network

2.2.2 The Internet

2.3 DEPENDENCE ON TELECOMMUNICATIONS AND INFORMATION SYSTEMS

2.3.1 National Dependence on Telecommunications and Information Systems

2.3.2 NS/EP Dependence on the Public Network

2.4 JOINT GOVERNMENT – INDUSTRY ACTIVITIES

2.5 ELECTRONIC INTRUSION

3 TELECOMMUNICATIONS AND INFORMATION SYSTEMS AS INTRUSION TARGETS

3.1 INTRODUCTION

3.2 ELECTRONIC INTRUSION VULNERABILITIES

3.2.1 Vulnerabilities Resulting From Increased Competition

3.2.2 Vulnerabilities Resulting From Software-Driven Technology

3.2.3 Vulnerabilities of Operations Support Systems

3.2.4 Firewalls

3.2.5 Human Resources

3.3 SOFTWARE-BASED TOOLS AND TECHNIQUES

3.3.1 Denial of Service Attacks

3.3.2 Use of Internal Backdoors

3.3.3 Sniffers

3.3.4 Rootkit

3.3.5 Spamming and Electronic Mail Bombing

3.3.6 Malicious Software

3.3.7 Mobile Code: Java and ActiveX

3.3.8 Embedded Code

3.4 TRENDS

3.5 POTENTIAL IMPACT

3.6 CONCLUSION

4 ECONOMIC COMPETITOR AND ADVERSARY USE OF ELECTRONIC INTRUSION

4.1 INTRODUCTION

4.2 FOREIGN STATES

4.2.1 Russia

4.2.2 China

4.2.3 South Korea

4.2.4 Cuba

4.2.5 Japan

4.2.6 France

4.2.7 Germany

4.2.8 Iraq

4.2.9 Israel

4.2.10 Bulgaria

4.3 TERRORIST USE OF ELECTRONIC INTRUSION

4.4 ORGANIZED CRIME USE OF ELECTRONIC INTRUSION

4.5 CONCLUSION

5 HACKER USE OF ELECTRONIC INTRUSION

5.1 INTRODUCTION

5.2 HACKERS

5.3 PROBING FOR VULNERABILITIES

5.4 AVAILABILITY OF TOOLS

5.5 CONCLUSION

6 THE INSIDER THREAT TO TELECOMMUNICATIONS AND INFORMATION SYSTEMS

6.1 INTRODUCTION

6.2 INSIDERS

6.3 PROFILE OF A MALICIOUS INSIDER

6.4 METHODS OF ATTACK

6.5 CONCLUSION

7 THREAT ANALYSIS

7.1 INTRODUCTION

7.2 THREAT SOURCES

7.2.1 Foreign Intelligence Services

7.2.2 Terrorist Groups

7.2.3 Organized Crime Groups

7.2.4 Hackers

7.2.5 Insiders

7.3 CAPABILITIES
7.4 IMPLICATIONS OF THE CHANGING THREAT
7.5 IMPACT ON THE NATION'S NS/EP POSTURE

8 COUNTERING THE THREAT
8.1 INTRODUCTION
8.2 COUNTERMEASURES
8.3 AWARENESS
8.4 INFRASTRUCTURE PROTECTION GUIDANCE
8.5 INFORMATION SHARING
8.6 COMPUTER INCIDENT RESPONSE TEAMS
8.7 TECHNOLOGY
8.8 LEGAL
8.9 CONCLUSION

APPENDIX A MALICIOUS SOFTWARE DESCRIPTIONS
A.1 OVERVIEW
A.2 TROJAN HORSE
A.3 WORM
A.4 LOGIC BOMB
A.5 COMPUTER VIRUS
A.6 BACTERIA

APPENDIX B LIST OF ACRONYMS B-1

APPENDIX C GLOSSARY C-1

APPENDIX D REFERENCES D-1

TABLE OF FIGURES

FIGURE 3-1: TAXONOMY OF MALICIOUS SOFTWARE

TABLE OF TABLES

TABLE 5-1: EXAMPLES OF ELECTRONIC INTRUSION SOFTWARE TOOLS

TABLE 8-1: CHARACTERISTICS OF ENHANCED INTRUSION DETECTION TECHNOLOGIES

EXECUTIVE SUMMARY

This report examines the electronic intrusion threat to national security and emergency preparedness (NS/EP) telecommunications and information systems.[1] Electronic intrusion is defined as gaining unauthorized access to automated information systems (AIS) including software, hardware, firmware, and the information these systems store and process. Electronic intrusion also includes exceeding or abusing authorized access to that system. The threat posed by electronic intrusion continues to grow due to increased global connectivity, the dramatic worldwide growth of computer literacy, the increased sophistication of intrusion tools and techniques, and the ready availability of detailed intrusion information and user-friendly intrusion tools on the Internet. The increasing complexity of information system software and the massive interconnection of telecommunications and information systems have resulted in a wide range of unintended, often unrecognized, vulnerabilities intruders can exploit.

Telecommunications and information systems are high-priority targets because of the United States' extensive dependence on information infrastructures for its economic and national security and because of the types of information they carry and their central role in supporting NS/EP requirements. Electronic intrusion attacks against NS/EP telecommunications and information systems can be carried out to obtain sensitive information or disrupt or disable vital systems. The dependence of NS/EP telecommunications and information systems on the public network (PN) increases their vulnerability to attack.[2] For example, adversaries can exploit the PN's pervasive interconnection with the national and global information infrastructures to gain access to NS/EP telecommunications and information systems. In addition, adversaries can attack the PN itself as a means to diminish NS/EP capabilities. The complexity of the PN and its interconnection with a multitude of information networks have made it very difficult to find and correct intrusion vulnerabilities. The likelihood

---

1 NS/EP telecommunications services are used to maintain a state of readiness to respond to and manage any event or crisis (local, national, or international). NS/EP telecommunications and information systems include the public network and all designated National Communications System (NCS) primary assets.

2 The PN is the backbone of the NII and supports virtually all NS/EP telecommunications and information systems requirements. The PN includes any switching system or voice, data, or video transmission system that provides communication services to the public (e.g., public switched networks, public data networks, private line services, wireless services, and signaling networks). The PN is a combination of several distinct entities interconnected over many years to provide the reliable long-haul communication networks that the United States has grown to rely on in the private and public sectors during both peacetime and wartime. Each PN interconnection involves software and hardware components that represent a technical trade-off between user convenience and restrictive security in the interest of greater efficiency.

that electronic intrusion will be detected and traced to those responsible is relatively low and, consequently, little threat of retaliation or apprehension exists to deter possible intruders.

ES-1 Key Findings

- The United States depends increasingly on the National Information Infrastructure (NII) to control its critical infrastructures, operate its economy, and coordinate critical Government functions. The backbone of the NII is the PN, which is controlled by complex computer-based operating, switching, and signaling systems. Each of these systems has been attacked successfully by electronic intruders, demonstrating these systems' vulnerabilities. Adversaries seeking to harm the United States could cause severe disruption through coordinated attacks on key systems within the PN. The enormous expansion of connectivity among systems within the PN and their increasing dependency on automation will continue to increase these vulnerabilities and broaden exposure to electronic intrusion. [3]
- The Internet is a primary vehicle for attacks on Government and proprietary systems. Other means for gaining unauthorized access include poorly protected dial-up modems, intrusion through private branch exchange (PBX) systems, and direct intrusion into telecommunications subsystems, such as operations, administration, maintenance, and provisioning (OAM&P) systems. New services and advanced technologies, such as advanced intelligent networks (AIN), synchronous optical networks (SONET), asynchronous transfer mode (ATM) switching, and local number portability (LNP), have introduced unforeseen vulnerabilities into the network.[4]
- The use of commercial off-the-shelf (COTS) operating systems and applications software to replace proprietary software has reduced the cost of operating large information networks, increased software standardization, and fostered the rapid expansion and interconnection of telecommunications and information systems. However, the use of COTS software has also increased vulnerabilities within these systems. When a vulnerability is discovered that allows an intruder to compromise a particular type of software, this technique can be used against all telecommunications and information systems using that software. Additionally, an increasing amount of code for these pro-

---

3 Office of the Manager, National Communications System (OMNCS), *Assessment of PSN Components Critical Roles and Interdependencies in Call Processing*, Arlington, VA: OMNCS, June 1997, p. 53.

4 National Security Telecommunications Advisory Committee (NSTAC), Subgroup on Widespread Outage, *Report on the Likelihood of a Widespread Telecommunications Outage*, Washington, DC: NSTAC, December 1997, p. 4.

grams is being written overseas because of cost considerations. Consequently the potential exists for insertion of malicious software or disguised backdoors by foreign competitors or intelligence services from the nation where the software is being produced. A similar problem exists concerning the overseas production of silicon chips, which may contain embedded code.

- The primary electronic intrusion threat continues to be trusted insiders who can exploit system privileges to abuse or exceed their authorized access to system applications, network operations systems, stored information, and interconnected information systems. Destructive and malicious activities by insiders are often undetected because of the insider's trusted position.
- Several countries are developing formal information warfare (IW) programs, and many of these countries pose a sophisticated electronic intrusion threat to NS/EP telecommunications and information systems. Russia, China, and France have acknowledged IW programs; and according to one estimate, at least 33 countries have developed sophisticated electronic intrusion programs for intelligence collection.[5]
- Economic competitors have increased their use of electronic intrusion to obtain and collect economic intelligence. The National Counterintelligence Center (NACIC) has concluded that at least 23 countries are collecting economic intelligence within the United States. Electronic intrusion is a primary technique foreign collectors use to obtain economic information for intelligence analysis. [6]
- Terrorist organizations are increasingly adept at using electronic information systems and advanced technologies. Terrorist organizations are aware of the U.S. dependency on complex infrastructures and have been known to recruit hackers or privileged insiders to attack telecommunications and information systems. [7]
- Criminal organizations now consider computer systems as lucrative targets for fraud related activities, the theft of proprietary information, and the theft of funds and securities transmitted through electronic commerce systems. Russian organized crime has proven

---

5 National Intelligence Council, *The Foreign Information Warfare Threat to U.S. Telecommunications and Information Systems,* Undated Briefing; and Wayne Madsen, "Intelligence Agency Threats to Computer Security," International Journal of Intelligence and Counterintelligence, 6:4, Winter 1993, pp. 446-487.

6 National Counterintelligence Center (NACIC), *Annual Report to Congress on Foreign Economic Collection and Industrial Espionage,* Washington, DC: NACIC, June, 1997, p. iii.

7 President's Commission on Critical Infrastructure Protection (PCCIP), *Critical Foundations: Protecting America's Infrastructure,* Washington, DC: USGPO, October 1997, p. 18.

particularly adept at using computers for bank fraud and has been implicated in electronic fraud cases in Russia, the United Kingdom, Germany, the Netherlands, Hong Kong, and the United States. Additionally, the Cali drug cartel, Russian organized crime, and other organized criminal activities have targeted law enforcement systems to gather intelligence concerning law enforcement activities.[8]

- Hackers remain a significant threat to the PN and interconnected NS/EP telecommunications and information systems. This threat is increasing because of the wide availability of malicious software and hacker tools with graphical user interfaces (GUI) through the Internet and hacker Web sites. Software tools such as the Security Administrator Tool for Analyzing Networks (SATAN), which automatically probes networks for security flaws and vulnerabilities, allow relative novices to conduct sophisticated attacks against targeted systems. Additionally, detailed data on telecommunications and information system vulnerabilities are often available through hacker Web sites and Internet relay chat (IRC) channels. By using available tools and information, even a relatively inexperienced hacker can conduct effective attacks against targeted telecommunications and information systems.
- The pervasive interconnection of information systems makes it impossible for network administrators and telecommunications carriers to completely understand the composition and vulnerability of their networks. Unbounded systems are characteristically distributed and interoperable and lack global visibility and common security practices. The Internet–a nonhierarchical network of systems, each under local administrative control only–is a primary example of an unbounded system, and is the prototype for many evolving information networks. By nature, unbounded systems are insecure and the opportunities for intrusion into them are unlimited. Information security practices designed for bounded systems, in which all parts are known and are controlled by a unified administrative and security program, are ineffective in addressing the security needs of unbounded systems. Security mechanisms must be adapted to meet the needs of evolving unbounded systems. [9] The proliferation of unbounded systems, along with their close interconnection with the PN, poses a significant security challenge to interconnected NS/EP telecommunications systems.

---

8 Center for Strategic and International Studies (CSIS), *Russian Organized Crime: Global Organized Crime Project*, Washington, DC: CSIS, 1997, pp. 36-37.

9 R.J. Ellison, D. A. Fisher, et al, *Survivable Network Systems: An Emerging Discipline*, Technical Report CMU/SEI-97-TR-013, Pittsburgh, PA: Carnegie Mellon University, November 1997, pp. 5-6.

- Information security practices have not kept pace with the electronic intrusion threat. In particular, little emphasis is placed on defending telecommunications and information systems against insider attacks. These types of attacks pose the greatest threat to the PN and interconnected telecommunications and information systems; however, in many cases they have received less attention than remote attacks.

ES-2. Conclusions

For the foreseeable future, electronic intrusion will remain a serious threat to the PN, NS/EP telecommunications and information systems, and interconnected infrastructure systems. The ability to protect our critical telecommunications and information systems is hampered by the lack of a clear understanding of the vulnerabilities resulting from the expanding interconnection of such systems, the use of COTS software, and the growing dependence of all critical infrastructures on the PN as the backbone of the NII. The United States increasingly relies on complex, networked information infrastructures for its national and economic security and the welfare of its citizens. Economic, political, and social dependence on these systems extends from national-level activities to individual communities and their residents. The United States is moving to an information-based economy; and the rapid growth in electronic transactions has greatly increased dependence on the PN, interconnected banking and finance systems, and electronic commerce services. The infrastructures that depend on integrated telecommunications and information systems to perform their functions include electric and natural gas utilities; transportation systems; and essential Government services. Increasingly, these telecommunications and information systems depend on the PN for connectivity, distributed system management, and data acquisition activities. Any protracted loss of critical information infrastructure capabilities could severely harm national security and the national welfare. To meet the potential threat, Government and industry must work together to improve information security practices, intrusion detection capabilities, and network restoral and reconstitution.

The electronic intrusion threat requires joint action by the Government and industry to be effectively addressed. Numerous ongoing efforts exist to address threats to the PN, including changing laws, enhancing coordination mechanisms, and disseminating information regarding stronger protection mechanisms. Although these measures are significant, and their goals sound, the nature of the electronic intrusion threat is such that countering it will continue to be an uphill battle. The implementation of PDD-63 provides a significant opportunity to coordinate, and maximize the benefits from, diverse efforts to address the electronic intrusion threat. The joint National Communications System (NCS) and National Security Telecommunications Advisory Committee (NSTAC) activities to address complex problems regarding communications for Federal NS/EP activities can both contribute to, and benefit from, the broader effort to protect the Nation's critical infrastructures.

1 INTRODUCTION

The magnitude of the threat from various forms of intrusion, tampering, and delivery of malicious code is extraordinary. We know with specificity of several nations that are working on developing an information warfare capability.... These countries recognize that cyber attacks-possibly launched from outside the US – against civilian computer systems in the US - represent the kind of asymmetric option they will need to "level the playing field" during an armed crisis against the United States.

Testimony by DCI George Tenet before the
Senate Committee on Government Affairs,
24 June 1998

1.1 Overview

This report examines the electronic intrusion threat to national security and emergency preparedness (NS/EP) telecommunications and interconnected information systems. It serves as an essential component for risk assessments and provides a baseline for countermeasure development. The threat analysis in this report was developed from multiple unclassified sources drawn from the intelligence community, law enforcement, industry, and information assurance activities. Additionally, this report describes the techniques involved in computer intrusion and telecommunications and information systems targeting, discusses the motives of those who pursue such activities, and identifies adversaries who could use electronic intrusion to attack the public network (PN) and interconnected telecommunications and information systems.

1.2 Scope

This report is intended to raise awareness of the dependence of NS/EP activities on a diverse range of supporting telecommunications and information systems, many of them outside the traditional NS/EP telecommunications assets, and to discuss the threats that may affect these systems. A threat is defined as the capabilities, intentions, and attack methods of adversaries bent on exploiting the vulnerabilities of an information system or an information-based network, or any circumstance or event with a potential to cause harm in the form of destruction, disruption, and/or denial of service.[10] This report examines electronic intrusion in the context of the threat it poses to the PN and the telecommunications and information systems linked to it.

As with previous versions, this report is based entirely on open source information to increase its availability throughout the Government as well as the private sector. No proprietary or classified information has been used in preparing this document, and judgments made in the report are based on publicly available data.

---

10 National Security Center, Glossary of Computer Security Terms, NCSC-T G-004, Version-1, Washington D.C.: U.S. GPO, October 21, 1988, p. 47.

In addition, this report reviews the opportunities that intruders may be afforded by global interconnectivity and the availability of inexpensive and powerful technological capabilities, and discusses the implications of these trends for vital NS/EP telecommunications and information systems, including the actions that could be taken to guard against electronic intrusion.

The objectives of the report are to:

- Describe the electronic intrusion capabilities of U.S. adversaries, foreign competitors, terrorist organizations, organized crime groups, and individual intruders.
- Explain how the threat posed by electronic intrusion and the growing dependence on automated information systems (AIS) have increased the risks to critical infrastructures.
- Briefly discuss telecommunications and information system vulnerabilities and potential risks to national security and national welfare to demonstrate the potential effect of threat capabilities.

1.3 Report Organization

Sections 2 through 8 of this report discuss the following:

2. Background—discusses the basic systems that compose telecommunications and information systems, the United States' dependence on those systems, and some of the Government and industry efforts underway to protect those systems.

3. Telecommunications and Information Systems as Intrusion Targets —discusses the vulnerabilities of telecommunications and information systems to electronic intrusion, and examines the use of electronic intrusion as a mechanism for attacking them. This section also briefly discusses intruders and their motivations and techniques

4. Economic Competitor and Adversary Use of Electronic Intrusion —discusses use of electronic intrusion by economic competitors and adversaries of the United States, including the use of electronic intrusion by terrorists and organized crime groups.

5. Hacker Use of Electronic Intrusion—discusses the motivations, trends, and techniques of the electronic intrusion activities of hackers.

6. The Insider Threat to Telecommunications and Information Systems—discusses the threats to NS/EP telecommunications and information systems from employees and other insiders.

7. Threat Analysis—analyzes of the electronic intrusion threats discussed in this report and their implications for NS/EP telecommunications and information systems.

8. Countering the Threat—examines current activities geared to improving telecommunications and information network security.

2 BACKGROUND

There are two reasons to be especially concerned about information warfare. First, there is the growing dependence on worldwide information infrastructure through telecommunications and computer networks. Second, both nations and terrorist organizations can with relative ease acquire the techniques to penetrate information systems.

Testimony of DCI John Deutch before the
Senate Subcommittee on Governmental Operations,
25 June 1996

2.1 Introduction

To conduct its NS/EP activities, the Federal Government depends on telecommunications and information systems that are part of the PN and other interconnected systems. Virtually all requirements for NS/EP telecommunications and information systems within the United States are supported by the PN, which has been the target of electronic intrusion attacks. Electronic intrusion is defined as gaining unauthorized access to AISs, including software, hardware, and firmware, and the information these systems store and process. Electronic intrusion also includes exceeding or abusing authorized access to such systems. Electronic intrusion can result in the following:

- Compromise or theft of data and information
- Adulteration of data and information
- Destruction of data
- Denial or disruption of services
- Economic loss or theft of services.

Electronic intrusion attacks resulting in compromise, adulteration, or destruction of data, or denial of service pose a significant threat to NS/EP telecommunications and information systems. The targeting of NS/EP telecommunications systems by an adversary could compromise critical national security information, diminish the Federal Government's ability to react to a crisis or emergency, or result in the loss, theft, or adulteration of confidential personal or financial information. To perform their functions, NS/EP telecommunications systems must provide reliable communications throughout the spectrum of possible emergencies.

2.2 Telecommunications and Information Systems

The PN includes any switching system or voice, data, or video transmission system that provides communication services to the public (e.g., public switched networks, public data networks, private line services, wireless services, and signaling networks). It is the backbone of the NII and supports virtually all NS/EP telecommunications and information systems requirements. It is a combination of several distinct entities interconnected during many years to provide the reliable communication networks that the United States has grown to rely on in the private and public sectors during both peacetime and war. Each PN interconnection involves software and hardware components that represent a technical trade-off between restrictive security and user convenience in the interest of greater efficiency.

Because the Internet is a critical component of the PN, it is important to highlight some of its characteristics. The Internet is a global network of computers joined by high-speed, digital telecommunications that use a common rule set known as the Transmission Control Protocol/Internet Protocol (TCP/IP). The Internet has been described as "the collection of loosely connected networks worldwide that are accessible by individual host computers through a variety of gateways, routers, dial-up connections, Internet access providers, and Internet service providers (ISP)."[11]

2.2.1 The Public Network

Two factors have made possible the vast array of PN services available today: 1) the increased competition brought about by the divestiture of AT&T in 1984 and, more recently, the Telecommunications Act of 1996; and 2) the migration of the PN from a manual, mechanical, hardware-driven technology to one that is increasingly driven by software. These factors have had a positive effect on our society, economy, and national security. However, they also introduce potential vulnerabilities that could have a negative effect on those very services on which we have come to rely so much.

Due to the diversity of the PN's components and a growing number of service providers, a unified security architecture has not been implemented. The rapid expansion of PN services and the growth of interconnected computer networks have increased the potential risk to the PN and all interconnected systems. The PN is increasingly complex and is more dependent on a growing number of networked computer subsystems for its operation. As a result, the number of targets susceptible to electronic intrusion and the range of intrusion options have increased significantly. The complex software used in these systems is rarely free of defects and is often difficult to configure and operate. Consequently, unrecognized security flaws can result in vulnerabilities that may be exploited by intruders. Additionally, the complexity of the software makes it difficult if not impossible to determine if malicious code or backdoors have been surreptitiously placed in the software. Finally, the interconnection of the PN with pre-

---

11 James Ellis, et al., *Report to the President's Commission on Critical Infrastructure Protection,* Pittsburgh, PA: CERT Coordination Center, www.cert.org/pres_com/cert.rpcci.body.html, January 1997.

viously isolated infrastructures enables the propagation of an attack through their systems. The combined effect of these factors is an increased potential for undetected electronic intrusion, and an increasing vulnerability to electronic intrusion attacks across the various infrastructure systems. Although software and security standards have been developed to protect individual components and subsystems of the PN, the ability to identify and counteract network intrusions varies throughout the telecommunications industry. The state of system security has been further weakened by the entry of new competitive local exchange carriers (CLEC) into the market with limited experience in telecommunications security.[12] Intruders are likely to enter through subsystems or carriers that have the weakest security and then attack other components of the PN as an apparently trusted insider.

Another factor that must be accounted for is the changing nature of information systems. Increasingly, information systems are interconnected with other information systems in a manner that is frequently unintended and often unrecognized. In part, this situation has resulted from the explosive expansion of the Internet and the move to distributed information systems by virtually all elements of society. The pervasive interconnection of information systems makes it impossible for network administrators and telecommunications carriers to completely understand the composition and vulnerability of their networks. Unbounded systems are characteristically widely distributed and interoperable, and lack global visibility and common security practices. The Internet–a non-hierarchical network of systems, each under local administrative control only–is a primary example of an unbounded system, and the prototype for many evolving information networks. By nature, unbounded systems are not secure and the opportunities for intrusion into them are unlimited. Information security practices designed for bounded systems, in which all parts are known and are controlled by a unified administrative and security program, are ineffective in addressing the security needs of unbounded systems. Therefore, security mechanisms must be adapted to meet the needs of evolving unbounded systems.[13] The proliferation of unbounded systems, along with their close interconnection with the PN, poses a significant security challenge to interconnected NS/EP telecommunications systems.

---

12 Network Reliability and Interoperability Council (NRIC), *Network Interoperability: The Key to Competition*, Washington, DC: Alliance for Telecommunications Industry Solutions (ATIS), July 15, 1997, pp. 109-110, p. 124.

13 R.J. Ellison, D. A. Fisher, et al, *Survivable Network Systems: An Emerging Discipline*, Technical Report CMU/SEI-97-TR-013, Pittsburgh, PA: Carnegie Mellon University, November 1997, pp. 5-6.

2.2.2 The Internet

Many daily operations depend on Internet connections, and new Internet connections are continuously being created. Although attempts are made to discern the size of the Internet, the dynamic nature of the network makes precise measurements difficult. Since 1987, the Internet Domain Survey has attempted to count the number of Internet hosts. The current method is to count the number of IP addresses which have been assigned a name.[14] According to this survey, approximately 5,846,000 computers were connected to the Internet in January 1995. By July 1998, this number had grown to 36,739,000. [15]Within the next decade, it is likely that Government, industry, academia, and individuals will be as dependent on the Internet as they currently are on telephones, facsimiles, and desktop computers. In light of this prediction, it is critical that Government and the ISPs implement Internet security and survivability measures to maintain stability and prosperity. [16] Due to the close interconnection between telecommunications and information systems and the Internet, security solutions should be jointly agreed to and implemented.

Formerly, most Internet access was achieved through dial-in ports, but the current trend is to increase use of interconnected networks for a wide range of activities, including Government and industry operations, and support for the delivery of human services such as health care and education. This growing interconnection could increase the likelihood that proprietary information such as financial data, intellectual property, and strategic plans could be compromised. While the Internet can enhance the ability of organizations to conduct daily activities in an efficient manner, this convenience comes with increased vulnerability. [17]

The Computer Emergency Response Team (CERT) Coordination Center at Carnegie Mellon University (CMU) has seen a steady increase in the use of sophisticated, distributed attacks against the Internet. CERT officials examining this problem believe that a relatively small number of attacks against key Internet nodes could result in cascading failures that would be difficult to stop. Because many of the Internet's vulnerabilities are intrinsic to its architecture and protocols (i.e., TCP/IP), they cannot be eliminated easily. An increasing number of Internet vulnerabilities are associated with the

---

14 This methodology estimates the size of the Internet based on computer connections rather than on number of users or volume of traffic.

15 *Internet Domain Survey:* July 1998, Network Wizards, http://www.nw.com/-zone/WWW/new-survey.html.

16 R.J. Ellison, D. A. Fisher, et al, *Survivable Network Systems: An Emerging Discipline,* Technical Report CMU/SEI-97-TR-013, Pittsburgh, PA: Carnegie Mellon University, November 1997, pp. 5-6.

17 Ibid.

software applications used on the Internet. In 1995 the CERT Coordination Center received an average of 35 reports regarding new vulnerabilities each quarter. By 1996 this average had doubled, and vulnerabilities identified in earlier versions of products continued to appear in newer versions of those products. This indicates that vendors and users are placing a low priority on security features. This situation is unlikely to change unless users insist on more secure products.[18]

2.3 Dependence on Telecommunications and Information Systems

Virtually every facet of American life is dependent upon information-driven systems. Telecommunications and information systems are the backbone of all critical U.S. infrastructures including transportation, banking and finance, and electric power. Uninterrupted communication of data is essential to conduct Government operations, emergency services, and daily commerce. The prolonged loss of telecommunications and information systems services and connectivity could significantly diminish the Nation's economic and national security. The Nation's information dependence, combined with increasing vulnerabilities, a decreasing number of infrastructure nodes and links, and the growing capability of U.S. adversaries to conduct electronic intrusion attacks, has made telecommunications and information systems attractive and lucrative targets. The security of the information carried by telecommunications and information systems is also extremely important. Information carried by information networks includes vital national security information, financial and securities transactions, law enforcement sensitive information, and proprietary information critical for economic growth. It is essential to assure the integrity, confidentiality, and reliability of this information.

2.3.1 National Dependence on Telecommunications and Information Systems

Information networks have rapidly expanded to meet the information needs of Government, industry, and individual citizens. Increasing interconnectivity and the rapid movement toward enterprise-wide telecommunications and information systems have made these systems more susceptible to intrusion. The introduction of advanced information technology has increased the capabilities of individual users, provided greater individual autonomy, and increased anonymity within vast information networks. At the organizational level, corporate intranets have consolidated previously isolated local or functional networks into organization-wide information networks, and employees have been given enhanced access to internal systems. Intranets frequently have few internal security barriers due to an

18 James Ellis, et al., *Report to the President's Commission on Critical Infrastructure Protection*, Pittsburgh, PA: CERT Coordination Center, www.cert.org/pres_com/cert.rpcci body.html, January 1997.

assumed level of trust within the organization and the desire to increase collaborative interaction within the organization. [19] Consequently, intranets often have significant intrusion vulnerabilities that can be exploited by insiders or by external intruders once they have gained entry. An example of an intranet vulnerability occurred when a virus affected the operations of National City Corporation, a $50 billion commercial bank holding company headquartered in Cleveland. The virus spread throughout the company's Novell NetWare environment to nearly all of the bank's 300 file servers and 10,000 client workstations across six cities in four states. The virus infected the bank's network from eight newly purchased notebook PCs. Although the systems administrators checked the new equipment, this virus was too new for the company's antivirus software to detect.[20]

Organizations often use the Internet to connect physically distant intranets to a corporate network. The Internet provides global interconnectivity to data networks and enhances the ability of organizations to conduct their activities in an efficient manner. The Internet is a central component of the National Information Infrastructure (NII) and will likely assume increased importance as the demand for access to networked telecommunications and information systems increases. However, the Internet was not designed with information security practices in mind. Well-known, easily exploited software and architectural vulnerabilities allow intruders to remotely access targeted telecommunications and information systems from anywhere. The pervasive interconnection of telecommunications and information systems through the Internet, corporate intranets, and other components of the PN, gives intruders an unprecedented capability to remotely attack these systems. Interconnection between the Internet and other elements of the PN creates the potential for remote attacks against telecommunication assets to originate through the Internet. An example of this includes the recent Solar Sunrise case in which hackers obtained access to Department of Defense (DoD) information systems by exploiting Internet connections.

The number of organizations that cited their Internet connection as a frequent point of attack rose from 37 percent in 1996 to 54 percent in 1998. The number of respondents citing their Internet connection as a frequent point of attack is now equal to the number of respondents citing internal systems as a frequent point of attack.[21]

---

19 Richard Power, *CSI Round Table: Intranet Security*, San Francisco, CA: Computer Security Institute, 1997.

20 Beth Davis, "Security Survey: Is It Safe?," *Information Week*/Ernst & Young Information Security Survey, World Wide Web, InformationWeek Online, www.informationweek.com/647/47iuss.htm, September 8, 1997.

21 Computer Security Institute, *Issues and Trends: 1998 CSI/FBI Computer Crime and Security Survey*, www.gocsi.com/prelea11.htm, March 1998, p.2.

As critical infrastructure[22] systems are becoming more dependent on the Internet, their vulnerability to remote attack is increasing.[23] Examples of the increased use of Internet technology by critical infrastructure systems include two major ongoing efforts: the Open-Access Same-Time Information System (OASIS) and the Utility Communications Architecture (UCA). OASIS is a key element of the Federal Energy Regulatory Commission's (FERC) effort to increase competition in the generation, distribution, and sale of electric power. Under the terms of recent FERC rulings, electric power transmission system owners must post their capacity, availability, and rates on Internet Web servers for open access by market participants. Under the UCA initiative, utilities are replacing the proprietary languages currently used in many supervisory control and data acquisition (SCADA) systems with a uniform set of software-based controls that will use TCP/IP-based packet switched networks. This architecture is being adopted to reduce costs and encourage the integration of control systems.

While both the OASIS and UCA will provide significant benefits to electric power suppliers and customers, they will also create new vulnerabilities that the industry has little experience in addressing. The information provided through the OASIS Web site will permit adversaries to more easily determine the criticality of individual facilities, and will provide links between the energy management systems of the individual electric power companies and the OASIS host. Intruders could potentially exploit these links. The migration from proprietary control systems to TCP/IP-based control systems under the UCA will mean that intruders who are already familiar with TCP/IP will now have the technical knowledge to attack SCADA systems. Finally, the increasing use of PN assets to provide connectivity for SCADA systems supporting electric power systems and other infrastructure industries could provide a gateway for attacks designed to cascade through interconnected infrastructure systems. The potential damage that can result from such attacks will grow as telecommunications and information systems supporting infrastructures become more interconnected and interdependent.[24]

Economic factors are driving electric utilities and other infrastructure systems to increase their use of information system driven automated processes. Increased competition has led electric utilities to introduce new layers of complexity into their control and management architectures. Computing power is being injected throughout the system, from intelligent metering

---

22 The PCCIP identified eight critical infrastructures: transportation; oil and gas production and storage; water supply; emergency services; government services; banking and finance; electrical power; and telecommunications.

23 PCCIP, *Critical Foundations: Protecting America's Infrastructure*, Washington, DC: USGPO, October 1997, pp. 3-4, 11, and A-5 to A-7.

24 Electric Power Research Institute (EPRI), *UCA and DAIS Information Security Analysis*, Palo Alto, CA: EPRI, August 1994.

and "local operating networks" that control electrical devices throughout a home at one end to using data warehousing and sophisticated customer models to track subtle changes in demand at the other. Significant cost savings are being realized by automating many monitoring and control functions, particularly at the substation level, but much of this automation is being implemented without effective or consistent levels of security.[25]

The dependence of critical infrastructures on telecommunications and information systems and networks has become a matter of national concern. The creation of the President's Commission on Critical Infrastructure Protection (PCCIP) by Executive Order (EO) 13010, on July 15, 1996[26], and the Senate's hearings on security in cyberspace (June-July 1996) highlighted the potential risks to businesses and Government agencies posed by electronic intrusion.[27] The PCCIP's final report concluded that the information and communications infrastructure was of such central importance that an attack resulting in a widespread outage would severely affect all of the critical infrastructure systems and could result in unanticipated cascading failures. Based on this conclusion, the protection of the PN is critical to preserving all of the identified critical infrastructures. [28]

2.3.2 NS/EP Dependence on the Public Network

NS/EP operations rely on information networks such as the PN for emergency communications. The policy of the United States is to have sufficient capabilities at all levels of Government to meet essential defense and civilian needs during any national security emergency. As defined in EO 12656, Assignment of Emergency Preparedness Responsibilities, a national security emergency is "any occurrence, including natural disaster, military attack, technological emergency, or other emergency, that seriously degrades or seriously threatens the national security of the United States."

Commercial carriers provide the great majority of U.S. Government telecommunications services. Consequently, emergency response organizations rely heavily on the PN to ensure public safety and welfare in times of crisis or disaster. The Federal Government must, under all circumstances, be capable of satisfying priority NS/EP telecommunications requirements and achieves that capability through use of commercial, Government, and privately owned telecommunications networks.

---

25 NSTAC, Information Assurance Task Force, Electric Power Risk Assessment, Arlington, VA: NCS, 1997, p.ES1.

26 Executive Order 13010, *Critical Infrastructure Protection,* Washington, DC: The White House, July 15, 1996.

27 United States Senate, Committee on Governmental Affairs, *Security in Cyberspace,* Hearings before the Permanent Select Committee on Investigations, Senate Hearing 104-701, Washington, DC: USGPO, 1996.

28 PCCIP, *Critical Foundations: Protecting America's Infrastructure,* Washington, DC: USGPO, October 1997, p.4

Efforts to ensure the availability of NS/EP communications should include measures to reduce network vulnerabilities that could lead to disruptions caused by electronic intrusion. New trends in technology, economics, and regulatory practice can increase network vulnerability. Technological advancements and increasing customer access to PN administrative services may increase the vulnerability of NS/EP telecommunications that depend on the PN. Even though service providers and other similar organizations are within the planning process for NS/EP telecommunications and are directly accountable to various Federal, State, and local authorities, customer premises equipment (CPE) remains outside the NS/EP process. Those involved in the NS/EP telecommunications planning process can take every measure within their power to protect the segment of the communications path for which they are responsible, but unless end-user organizations also take additional measures to protect their own CPE from electronic intrusion, they cannot be assured of the security of their telecommunications and information systems.

2.4 Joint Government - Industry Activities

The telecommunications industry, through its representatives in the President's National Security Telecommunications Advisory Committee (NSTAC), and the Office of the Manager, National Communications System (OMNCS), have responded to the threat to the NII by creating the NSTAC and Government Network Security Information Exchanges (NSIE). The NSIEs were established in response to an April 1990 request from the Chairman of the National Security Council's (NSC) Policy Coordinating Committee for National Security Telecommunications and Information Systems (PCC-NSTIS). The PCC-NSTIS requested that the Manager, National Communications System (NCS), identify what actions should be taken on the part of Government and industry to protect critical national security telecommunications from the threat from computer intruders. Working together, the Manager, NCS, and the NSTAC established a structure and a process for addressing network security issues. In addition, the NSTAC established its Network Group (NG) and Information Infrastructure Group (IIG), which also work closely with the Government through the OMNCS.

Central to this process are separate, but closely coordinated, Government and NSTAC NSIEs. Government member organizations include departments and agencies that are major telecommunications services users, represent law enforcement, or have information relating to the network security threat. Industry member organizations include telecommunications service providers, equipment vendors, systems integrators, and major users. NSIE representatives are individuals who are engaged full time in the prevention, detection, and/or investigation of telecommunications network software penetrations, or who have security and investigative responsibilities as a secondary or collateral function. Both Government and NSTAC NSIE representatives are subject matter experts in their fields. The NSIEs provide a forum to identify issues involving penetration or manipulation of software and databases affecting NS/EP telecommunications. The NSIEs' primary focus is to exchange information and views on threats and vulnerabilities affecting the PN's software. Periodically, the NSIEs also assess the risk to the PN from computer intruders. The last risk assessment was completed in 1995. The NSIEs are currently developing their next risk assessment, which will be published in 1999.

The NG is another NSTAC entity created in response to the initial request from the PCC-NSTIS in 1990 to address the threat to telecommunications from computer intruders. While the NSIEs focus on exchanging information that will help network security practitioners take explicit measures to protect the PN from intruders, the NG addresses higher-level issues affecting the overall security of the PN. Since it was established in 1992 as the successor of NSTAC's Network Security Task Force, the NG has addressed issues such as computer crime legislation, network security standards, and network security research and development.

The NSTAC began addressing information assurance and infrastructure protection issues in May 1995, when it established the Information Assurance Task Force (IATF). In April 1997, the Industry Executive Subcommittee restructured its organization and the IATF and its charge were incorporated into the Information Infrastructure Group (IIG). As the focal point for information assurance (IA) activities within the NSTAC, the IIG has conducted IA risk assessments, examined the implications of IA risk assessments for overall infrastructure protection, investigated a cooperative Government-industry approach to cyber security, and considered the NS/EP implications of electronic commerce (EC). The IIG has focused its IA risk assessments on three critical infrastructures - electric power distribution, financial services, and transportation - and has completed risk assessments on the first two. The Electric Power Information Assurance Risk Assessment Report (March 1997) examined the extent to which the electric power distribution infrastructure depends on information systems and described how associated vulnerabilities placed those groups at increased risk to denial-of-service attacks. The Financial Services Risk Assessment Report (December 1997) analyzed the risks to the financial services industry derived from its dependence on information technology. The report noted that while current security measures are adequate, the increasing dependence of the financial services industry on a deregulated telecommunications industry and the growth of Web-based financial services would likely increase the vulnerability of the industry to electronic intrusion attacks.

In September 1997, the IIG initiated its study of the transportation information infrastructure by sponsoring a workshop that addressed intermodal information dependencies, Government-industry information sharing, transportation information infrastructure vulnerabilities, and Government understanding of the transportation industry's infrastructure vulnerabilities. The workshop revealed that there was a significant need for increased awareness within the transportation industry concerning the dependence of transportation systems on the PN and interconnected information systems. The IIG recently initiated work to address two additional issues: the short-term, technical, and time sensitive issue relating to cyber security training and forensics; and the long-term, policy oriented, high-level issue of the NS/EP implications of electronic commerce (EC). In addressing the short-term issue, the IIG found that a stronger partnership between Government and industry is required to establish appropriate levels of trust and understanding and to spur cooperation in addressing cyber security issues. The IIG is currently in the process of addressing the long-term EC issues.

2.5 Electronic Intrusion

The threat of intrusion is difficult to assess because intrusion incidents are hard to detect and often are not recognized; failures or disruptions resulting from intrusion activities are frequently attributed to operator or user error; threat incidents are not always documented for further study or investigation; and even when attacks are identified as incidents, they often go unreported.[29]

Electronic intrusion encompasses the unauthorized access to AIS software, hardware, firmware, processes, activities, and information. Such activities include attacks by hackers, disgruntled employees, terrorists, organized crime groups, and foreign intelligence agents. Once intruders obtain access to an information system, depending on their access privilege level, they can compromise sensitive information, alter system attributes, adulterate database records, insert malicious software, send false commands to disable or disrupt the system, or conduct fraudulent activities. The multifaceted threat posed by electronic intrusion far exceeds the simple notion of mischievous hacking. Recognizing the threat that electronic intrusion poses, the United States has recently increased the penalties for intrusion into Government and Government-interest telecommunications and information systems under the NII Protection Act of 1996 and the Economic Espionage Act of 1996. While this legislation has been helpful, more work needs to be done to bring U.S. laws into the electronic age, particularly in terms of computer crime.

The next section of this report will discuss the reasons that telecommunications and information systems are targeted by adversaries and economic competitors, the vulnerabilities of these systems, and the tools and techniques used to exploit those vulnerabilities.

---

29 PCCIP, *Critical Foundations: Protecting America's Infrastructure*, Washington, DC: USGPO, October 1997, pp. 14-18.

3 TELECOMMUNICATIONS AND INFORMATION SYSTEMS AS INTRUSION TARGETS

Practical computer security is a series of actions and counteractions, of attacks and defenses. As with chess, success depends upon anticipating your opponent's moves and planning countermeasures ahead of time. [30]

Simson Garfinkel and Gene Spafford,
Practical UNIX and Internet Security

3.1 Introduction

While motives for attacking automated systems are diverse, there are some fundamental characteristics of electronic intrusion that appeal to a wide variety of potential adversaries:

- The resources (hardware and software) required to attack automated systems are relatively inexpensive and readily available.
- The risk of being caught is low. First, attacks often go undetected. Then, even if their actions are detected, intruders can attack from almost anywhere, and can disguise their locations so the chances of being located are minimized.
- Even if intruders are caught, the likelihood of prosecution and conviction is low and the penalties are generally light (although this has begun to change recently).

Telecommunications and information systems, in particular, are being targeted at an increasing rate. As will be discussed in Section 4, hackers, foreign intelligence agents, members of criminal organizations, and others are becoming more aware of the value of telecommunications and information systems as targets. Such systems have value as targets because:

- Large numbers of people depend on them to provide essential services, so attacks can have far-reaching effects.
- These systems offer access to a diversity of information attractive to a variety of intruders.
- These systems can be very expensive to repair. For example, an intrusion can cost an organization hundreds of man-hours. Even when an attack ultimately turns out to be simple to repair, the victim organization must first diagnose the problem. It could take only one hour to fix the problem, but 100 hours to find it, and maybe another 100 hours to ensure that all instances of the problem have been found.

---

30 O'Reilly Online Catalog, Simson Garfinkel and Gene Spafford, *Practical UNIX and Internet Security,* Second Edition, World Wide Web, www.ora.com/ora-com/sysad/puis-exc.html, 1996.

The potential benefits of attacking these systems are becoming more widely appreciated as a cost-effective and valuable means of countering the capabilities of those nations that depend on a highly developed technology infrastructure, such as the United States. The number of intrusions, network disruptions, and computer-based crimes will continue to increase because of the rise in computer literacy, increased media attention, and the opportunities for financial gain.

3.2 Electronic Intrusion Vulnerabilities

Telecommunications systems are increasingly controlled by automated information systems. These systems have become more complex, creating increased vulnerability within the telecommunications industry as a whole. The following business drivers affect the level of risk within telecommunications networks: increased economies of scale, decreased levels of administration, increasingly complex services, strong customer focus and customization, and open access. Many of the changes within the telecommunications industry have made its networks susceptible to numerous malicious acts.[31]

One important factor affecting PN security is the enactment of the Telecommunications Act of 1996. The Federal Communications Commission's (FCC) Network Reliability and Interoperability Council (NRIC) analyzed the security-related barriers to implementation of the Act and identified risk management security issues and concerns. Technical analysis of the security-related barriers by the Focus Group 1 Operation Task Group and input from the NSTAC Network Security Group, which examined the security implications of the Act on NS/EP telecommunications, identified the following security issues:

- Increased access points and collocation of critical assets will likely decrease core infrastructure diversity and increase single points of failure.
- Increased number of interconnected service providers with inferred trust relationships will degrade overall security and network integrity.
- Embedded operations channels of advanced signaling and transport protocols (e.g., SONET Data Communications Channel (DCC); Asynchronous Transfer Mode (ATM) operations, administration, and maintenance (OAM) cells; SS7 Network Management Messages) give virtually unlimited access to everything and everyone connected to them, given the current state of security standards and practices in such advanced technologies.

---

31 Hank Kluepfel, *Toward a More Secure Telecommunications Infrastructure: Mitigating the Risks*, Bellcore, Washington, DC: March 31, 1994, p. 7.

- Increased numbers of persons and processes with access privileges will present major risk challenges.
- Insecure Internet and intranet technology used for interconnection access to network operations and signaling systems will provide unintentional backdoors to PN mission critical systems, protocols, and information.
- Lack of regulatory, legal, or competitive motivation to invest in security safeguards will increase risks to the PN.
- Lack of personnel security requirements for individuals employed in key positions within the telecommunications carriers will increase the risk to the PN.[32]

The NRIC Security Subgroup also noted the lack of standards for interconnected data communications networks and gateways that support PN interconnection and partitioned access and recommended that the telecommunications industry develop a standard security baseline defining security requirements. The subgroup stated that existing standards do not adequately address the security concerns related to signaling and Operation Support System (OSS) access and detection of malicious code. It also found that SS7 gateway screening (firewall) capabilities were inadequate to address a sophisticated attack targeting the PN. The subgroup recommended that standards for gateway screening address mediated access and detection of malicious code. Finally, the subgroup recommended that a certifying body be created to develop appropriate security standards and test telecommunications network products for conformance with security standards.[33]

The OMNCS has also studied the security implications of the Telecommunications Act of 1996, focusing primarily on how PSN security might be affected by the technical and operations changes required to implement network unbundling and local number portability (LNP). The security concerns identified in OMNCS's report, Security Implications of the Telecommunications Act of 1996, are consistent with those identified by NRIC and NSTAC. Most of the vulnerabilities described in the report result from network unbundling. To the extent that LNP components become available on an unbundled basis, they will also be affected.[34]

---

32 Network Reliability and Interoperability Council (NRIC), *Network Interoperability: The Key to Competition*, Washington, DC, July 15, 1997, pp. 109-110.

33 Network Reliability and Interoperability Council (NRIC), *Network Interoperability: The Key to Competition*, Washington, DC, July 15, 1997, pp. 111.

34 OMNCS, *Security Implications of the Telecommunications Act of 1996*, Washington, DC; May 8, 1998, page 34.

3.2.1 Vulnerabilities Resulting From Increased Competition

The PN in the United States was built when computer intruders were not considered a threat. Local and long distance service was provided by a single carrier, which facilitated consistency in equipment, systems, processes, and procedures; also, systems and employees operated on the basis of mutual trust. In the apparent absence of a need for strong security measures, vendors built systems with minimal security features, and operators often did not adequately implement the security features that were available. The divestiture of local exchange service by AT&T in 1984 resulted in two significant changes to the telecommunications infrastructure: because each provider implemented its own equipment, systems, processes, and procedures, there was no longer uniformity throughout the infrastructure; and systems designed for access by a single provider had to be reconfigured to allow access by many different providers. With the implementation of the Telecommunications Act of 1996, the diversity among providers has further increased.

Security Capabilities of the New Service Providers. The primary intent of the Telecommunications Act of 1996 was to foster competition within the telecommunications industry. Because most new service providers, or CLECs, would have only limited distribution and switching capabilities, they would need to buy these services from the incumbent local exchange carriers (ILEC). The Telecommunications Act sought to create a "level playing field" by requiring the ILECs to provide unbundled services to the CLECs by granting them unhindered access to the supporting ILEC's network elements and OSSs. These systems are ubiquitous and affect every aspect of the operations, administration, maintenance, and provisioning (OAM&P) systems within the ILEC. New providers often do not have experience in providing telecommunications services, may not understand the security environment in which they are operating, and may not have the resources required to adequately protect their networks from hackers. Although the State Public Utility Commissions are responsible for governing the entrance of new providers into the telecommunications market, their criteria for certifying CLECs generally do not take security into account.

Consequently, hackers may be able to exploit a CLEC's vulnerabilities to gain access to an ILEC's OAM&P systems.

Competitive Pressures. The presence of the new service providers in the marketplace has created a more competitive environment, causing both the incumbent service providers and the new service providers to look for ways to reduce their operating expenses. Security is not generally perceived as a revenue generating opportunity, and this perception has the potential to affect the level of security offered by all providers. Reducing expenditures on security tools, training, and staff can result in human error that may affect security. Many service providers also reduce their expenses by contracting out functions previously handled in-house, exposing their systems to their contractors' vulnerabilities. Further, the client company does not have visibility and control over its contractors' hiring practices. For example, a telecommunications service provider may terminate an employee for exceeding or abusing access to the company's systems. That employee may be hired by one of the service provider's contractors and once again have access to their former employer's systems.

3.2.2 Vulnerabilities Resulting From Software-Driven Technology

The PN offers services, capabilities, and features that could only be imagined 20 years ago, with new ones emerging every day. Software-driven technology is a critical element in making this rapid progress possible. However, the increasing complexity of today's software makes it difficult, if not impossible, to identify potential vulnerabilities and discover malicious code. The desire for easy-to-use open systems has resulted in systems that are user friendly, but extremely difficult to administer and configure for secure use. Frequently, security features are viewed by software developers as a hindrance with little market value and are often ignored. Consequently, the software development community has generally failed to apply the lessons learned from intrusion attacks when developing new software, and as a result, new software is issued that has the same vulnerabilities as the previous version.

The PN's long history of reliability demonstrates that service providers rigorously test their software to ensure that it functions properly. Even so, the testing protocol generally does not address security issues. The far greater challenge, however, is negative testing (i.e., testing software to ensure that it does not do what it is not supposed to do). Some applications may have millions of lines of code, and it is not feasible to test the number of cases required to definitively conduct negative testing. Consequently, it is simply impossible to absolutely ensure that the software is free of all potential errors.[35] Because such large software programs are developed in modules by many programmers, it is possible for any of them to insert malicious software (or for an intruder to do so) that would not be detected in the final product.

Software applications are also simplifying operations, and although this is desirable, it can also have unintended consequences. For example, customers can now directly control certain aspects of their service. Customers with 800 service can now redirect incoming traffic from one destination to another; for example, they can redirect calls from their east coast service center to their west coast service center if a blizzard keeps their east coast

---

35 OMNCS, *Software Integrity, An NSIE White Paper,* Prepared by the Government and NSTAC NSIEs, Washington, DC: OMNCS, July 1997, p.2.

staff from getting to work. This saves the service provider time and money and gives their customers more immediate and customized control over their service. Yet that same ability to redirect calls could be exploited by a hacker; for example, a hacker could redirect calls from a high-volume 800 number to an E-911 destination, making that service unavailable to those who need it.

Finally, the new services, capabilities, and features made possible by software-driven technology are entering the marketplace at such a rapid pace that it is difficult, if not impossible, to identify, much less address, all the potential implications for PN security. This problem is exacerbated by the failure of most software developers to incorporate security features as an integral part of their software. Security features that may be included are often an added feature that may not operate seamlessly with other software functions. As a result, significant vulnerabilities are often created that can be exploited by adversaries.

Service providers have become increasingly aware of the computer intrusion threat and are taking measures to address that threat with varying degrees of success. However, the magnitude and volume of these changes in the marketplace and the technology, and the speed with which these changes occur, make achieving a high level of security a formidable challenge.

3.2.3 Vulnerabilities of Operations Support Systems

The PN's increasing dependence on internetworked computer systems for its operations has increased its susceptibility to electronic intrusion attacks. Public network components are becoming more interdependent, and critical call processing functions are becoming centralized into fewer network elements. OSSs assist in the automation of network administration and maintenance from a centralized location. These systems are growing in power and sophistication, and they can remotely manage many other functional PN components.[36]

Because threats to the PN can affect NS/EP assets, and the PN depends on its OSSs, the reliability and integrity of these OSSs are of significant concern. The network has been engineered to recover from individual outages in a predictable manner. This recovery is facilitated by vendor-specific OSSs that routinely monitor network performance, run diagnostic programs to identify developing problems, and initiate actions to correct those problems and restore service. Intruders who understand these processes could use OSS attributes to turn the PN on itself, manipulating network control systems to act in a manner that introduces instability into the subsystems. For example, an intruder could alter the software that runs the

---

36 Robert Kane, "Losing the Keys to the Kingdom of Domain," World Wide Web, Intrusion Detection Inc., www.intrusion.com/keys.html, 1996, p.2.

diagnostic programs, causing the OSS to initiate actions that disrupt, rather than restore, service. Another intrusion technique is to disrupt OSS functions at critical junctures, causing network subsystems to fail and outages to occur. Such attacks on OSSs pose a significant threat to the PN.

3.2.4 Firewalls

Firewalls are an essential component of a well-conceived information security program. However, merely having a firewall will not protect a system from attacks; 81 percent of the respondents reporting Internet intrusions in the 1998 Computer Security Institute Computer Crime and Security Survey use firewall technology. Many firewalls are not configured properly to prevent system intrusions and they may create a false sense of security. Furthermore, even the best firewall cannot fully protect a system all by itself–a firewall must be complemented by various other security measures.[37]

3.2.5 Human Resources

The growth of the Internet has placed the greatest strain on the least expandable resource of all–the support staff. A glance at employment advertisements in any major newspaper shows that systems administrators are in great demand. Most systems administrators today have less than 3 years' experience. Not surprisingly, "the average systems administrator has a far lower level of technical expertise than 5 years ago," as CERT's Barbara Fraser has noted. Current estimates are that more than 400,000 open positions require system administration and software engineering skills.[38]

If support personnel in general have been in scarce supply, the situation for security staff is even worse. In a 1996 Computer World magazine survey of information systems managers, more than half reported security as the number one skill set they needed.[39] The 1996 Ernst & Young Security Survey states, "there is a short supply of adequately trained information security personnel and tools/solutions to address the multi-tiered, client/server, distributed computing architecture found in almost all organizations."[40] In too many organizations, the systems administration and customer support staff have had to incorporate security as an additional duty. Customer demands for service will almost always take precedence

---

37 Computer Security Institute, *Issues and Trends: 1998 CSI/FBI Computer Crime and Security Survey,* San Francisco, CA: Computer Security Institute, March 4, 1998.

38 Paulina Borsook, "Hackers Bring the Net Down to Earth," *Network World,* January 1, 1996.

39 Melanie Menagh, "First Line of Defense," *Computerworld,* February 10, 1997.

40 Ernst & Young/*Information Week, 3rd Annual Information Security Survey: Trends, Concerns, and Practices,* Cleveland, OH: Ernst & Young, 1996, p. 11.

over concerns about security. Many of the mundane tasks critical to security, such as reviewing audit trails or monitoring access controls, are not performed. [41]

This shortage of trained systems administrators and system security personnel has increased the vulnerability of information systems to attack and has decreased the likelihood that an attack will be noticed and reported. Systems administrators not only implement security programs, but they are also the most likely to detect an anomaly that may indicate an attack. They perform a tripwire function by noting and reporting an attack, and are critical to effective investigations because they have in-depth knowledge of the system being attacked. The current shortage of systems administrators with a knowledge of information security practices is unlikely to be resolved soon. Colleges and universities offer few, if any, courses on information systems security. The limited number of personnel with strong technical skills are being hired by organizations that appreciate the scope of the electronic intrusion problem and are willing to pay high salaries for competent system security technicians.

3.3 Software-Based Tools and Techniques

Information technology applications provide the tools to conduct electronic intrusions and attacks on telecommunications and information systems and networks. Intrusion tools are becoming increasingly powerful and readily available. Automated tools using graphical user interfaces (GUI) enable even relatively inexperienced intruders to conduct sophisticated attacks. The following sections discuss various types of software-based tools and attacks.

3.3.1 Denial of Service Attacks

One of the most devastating attacks on NS/EP telecommunications systems is the denial-of-service attack. In these attacks, the intruder's primary goal is to deny the victim access to a resource. Attacks of this nature against critical networks can be life threatening for those who depend on systems supporting functions such as emergency services, flight safety, and war readiness.

A SYN (i.e., synchronized) attack is an attack against a computer that provides service to customers via the Internet. SYN refers to the type of message that is used between computers when a network connection is being made. In a SYN attack, the intruder executes a program from a remote site that congests or disables the service on the victim computer. The attack can prevent one system from being able to exchange data with other systems, or it could prevent the system from using the Internet at all.

---

41 Paulina Borsook, "Hackers Bring the Net Down to Earth," *Network World*, January 1, 1996.

A SYN attack against an ISP usually disrupts Internet service to all the provider's customers.[42]

The smart attack sends forged Internet control message protocol (ICMP) echo request packets (i.e., "ping" packets) to IP broadcast addresses. This attack can cause network congestion or outages because of the large number of ICMP echo reply packets being sent to the victim site. ICMP is usually used to convey status and error information regarding operating systems. An overload of this process congests the system, resulting in degraded network performance, or may render the system inoperable.[43]

Two new denial-of-service attack tools that exploit vulnerabilities in the TCP/IP protocol are Land and Teardrop. Land exploits some implementations of TCP/IP that are vulnerable to packets that are crafted in a particular way (i.e., spoofed). Land works by producing phony communications between a computer outside the victim's network and operating system, routers, network printers, or other equipment. Reports of repeated computer, router, and network failures at corporate, educational, and Government sites have been attributed to Land attacks. Attacks on large routers could disable e-mail, Internet browsing, and other networking capabilities. With Land, any remote user who can send spoofed packets to a host can crash that host. Teardrop exploits some implementations of the TCP/IP fragmentation re-assembly code that do not properly handle overlapping IP fragments. With Teardrop, any remote user can crash a vulnerable machine.[44]

3.3.2 Use of Internal Backdoors

A backdoor, also known as a trapdoor, is an undocumented way of gaining access to a computer system or particular software program. A backdoor may be a legitimate feature, installed by the vendor to allow remote maintenance of the system, or it may be put in by a system programmer who wants to break into that computer after he or she is no longer employed by the company. A backdoor may also lead to hidden areas of a system or network that are neither documented nor available to authorized users. Intruders can use remote network dial-up to access a backdoor and gain unauthorized access to a system. Because intruders who use backdoors are able to evade security features and gain access privileges, their actions

---

42 James Ellis, et al., CERT Coordination Center, *Report to the President's Commission on Critical Infrastructure Protection,* Pittsburgh, PA: CERT, World Wide Web, www.cert.org/pres_comm/cert.rpcci.body.html,January 1997.

43 CERT, SEI, CMU, CERT Advisory CA-98.01.smurf, Pittsburgh, PA: CERT, January 5, 1998.

44 CERT, SEI, CMU, *CERT Advisory CA-97.28,* Pittsburgh, PA: CERT, December 16, 1997; and Strategic Forecasting L.L.C., "Land Attack Emerges as New Threat to Network Security Reports," Computer Security Alert, World Wide Web, www.stratfor.com, December 11, 1997.

often escape the notice of security administrators until they have caused some type of damage or malfunction.

3.3.3 Sniffers

Sniffers are programs that monitor information packets as they are sent through networks and capture selected information for the intruder. A sniffer is an invaluable intrusion tool because it allows the attacker to look for user IDs and passwords as they traverse a network, often in unencrypted text. An intruder must first gain access to a host on the network on which to install the sniffer program. Once the sniffer is installed, it runs continuously or at selected intervals without a live connection from the intruder's own computer. This improves the efficiency of subsequent attacks while reducing the intruder's risk of being detected.

Sniffers such as tcpdump have become particularly effective with the increased use of the Internet. The volume of traffic and the number of remote sessions–logging into a host from a site across an Internet link where the user ID and password are passed in the clear–offer a large amount of sensitive data.

3.3.4 Rootkit

Rootkit, a package of software utilities and documentation designed to guide a novice through the process of gaining control over a target machine, began to appear on hacker Web sites in mid-1995. Rootkit includes a network sniffer, a backdoor login to disable auditing, Trojan horse system utilities, and an installation tool to match checksums to originals. The last feature allows intruders to cover their presence on the host by making key configuration files appear as if they have never been altered. Intruders who can install Rootkit successfully on a target server not only possess "root" privileges but can also hide their identity. [45] According to reports from UNIX users, Rootkit attacks are becoming more prevalent.[46] The newer versions of Rootkit are even harder to detect than the original version. [47]

3.3.5 Spamming and Electronic Mail Bombing

In a spamming attack, the attacker generates an enormous volume of e-mail messages that overload a server's processing, disk space, or network bandwidth. The term is probably derived from a Monty Python comedy skit in which the characters use the word "Spam" over and over until it drowns out all other dialog. Spamming attacks have been directed against

---

45 Jon William Toigo, "Six Steps Toward a More Secure Computing Environment," *Uniforum IT Solutions,* July 1996.

46 David O'Brien, "Recognizing and Recovering From Rootkit Attacks," *Sys Admin: The Journal for UNIX Systems Administrators,* Volume 5, Number 11, November 1996, pp. 8-20.

47 Ellen Messmer, "No Defense Against Latest Hacker Tool?" *Network World,* March 24, 1997.

such visible targets as the White House, the U.S. Senate, and the French Ministry of Education, and they exemplify the rise of denial-of-service attacks. Programs–with names like KaBoom, Avalanche, UpYours, and UnaBomber–that make these attacks even simpler are available from several Web sites.

Recent attacks show that individuals who perform spamming attacks have discovered that they can cover their tracks by exploiting a simple feature in e-mail clients and by taking advantage of the availability of many unprotected mail servers. The spammer simply locates a mail server that allows communication to simple mail transport protocol (SMTP) clients without a login and then types that server's name in his e-mail client program. The target server then sends copies of the spam message to hundreds or thousands of addressees without the spammer ever entering a user ID or password. [48]

3.3.6 Malicious Software

With the advent of publicly available software development tools for creating malicious software, the risk for this type of attack has increased. Most of this software can easily be downloaded from the Internet and is simple to use. Programs such as virus-authoring tools can create malicious software programs and require little or no computer programming knowledge by the author. The latest versions of these programs offer step-by-step information via a menu-driven process that easily constructs a ready-to-use virus. With global network connectivity, there are no geographic limitations on the creation of these programs and few on the insertion of malicious software.

Figure 3-1 illustrates the general principles for introducing and proliferating malicious software. For additional information on malicious software see Appendix A of this report.

---

48 Nick Wingfield, "Email Vendors Fight Spammers," (reprinted from *C/net* News), World Wide Web, DFN-CERT Webpage, www.news.com/News/Item/0,4,8247,00.html, February 25, 1997.

Figure 3.1
Taxonomy of Malicious Software

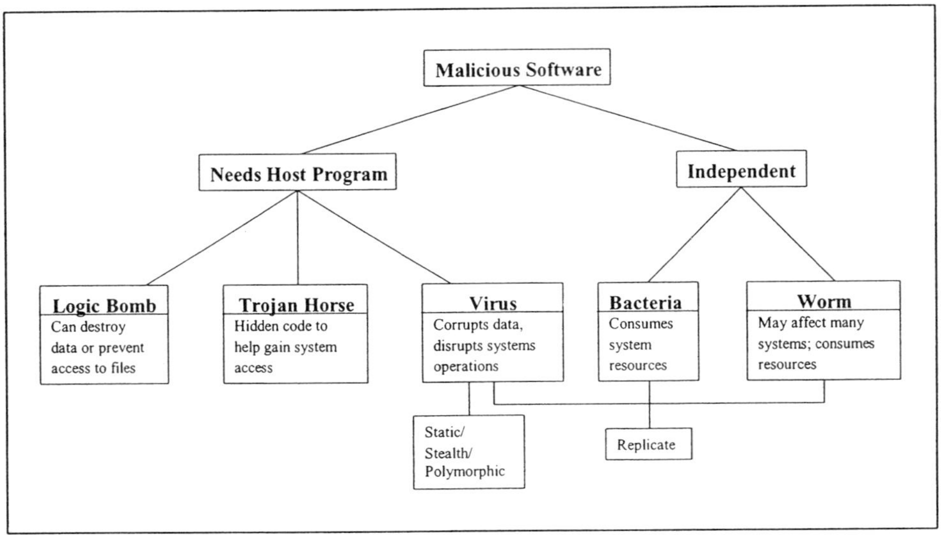

There is increasing concern about the transmission of malicious code through e-mail systems. Researchers at the Secure Programming Group at Finland's Oulu University recently discovered a security flaw in Microsoft's Outlook and Outlook Express mail programs, and in the Netscape Messenger mail program. The security deficiency allows an attacker to use an extremely long file name to cause a buffer overflow, which causes the system to crash. Malicious code can be contained in the long file name and is executed when the system is rebooted. Unlike previous malicious code attachments affecting e-mail, this attack is especially pernicious because an attachment does not have to be opened for the malicious code to be delivered to the computer. Simply downloading the message can execute the malicious code payload and cause the computer to crash.[49] Additionally, the ability of many e-mail applications to attach binary files to messages has provided a rich new medium for propagating viruses. Macro viruses such as the Microsoft Word "Concept" virus can infect a large population by being forwarded in a document attached to e-mail messages. Several years ago, floppy disks were the primary medium through which viruses were spread. But today, with the pervasiveness of e-mail and the reliance

49 Brian McWilliams, "Security Bug Affects Unopened E-mail Attachments," *PC World Today*, July 27, 1998, www.pcworld.com/pcwtoday/article/0,1510,7559,00.htm.

on network file servers, viruses can be disseminated far more broadly and rapidly, which increases the potential impact of such malicious code.

3.3.7 Mobile Code: Java and ActiveX

Mobile codes are programs that move from one processor in a network to another. Sun Microsystems' Java applets (i.e., applications) and Microsoft's ActiveX are two popular examples of mobile code that have received considerable attention in the press and technical literature.

Java applets are designed to operate in a closed environment, known as the Java Sandbox. The use of the closed environment should prevent an application from accessing unauthorized systems within a computer. The dilemma is that resourceful programmers have developed malicious software that enables the applets to escape the closed environment and access local disks or network connections. [50]

ActiveX allows programs to communicate with functions within standard applications, such as word processors and spreadsheets. ActiveX requires a trusted relationship to be identified before downloading to the operating system.[51] However, this security measure may be beyond the capability of unsophisticated end users.

The fundamental problem with mobile code is that it can be used to download and run potentially hostile programs on computers without the knowledge of the authorized users. These codes may be downloaded when a Web site is accessed, and in some cases these codes can even exploit security holes in e-mail. Furthermore, multiple machines can be attacked simultaneously by using mobile code. [52] For these reasons, the introduction of mobile code increases concern about the trustworthiness of networks.

3.3.8 Embedded Code

The increasing complexity of PN software systems makes it extremely difficult to detect malicious code that is placed in software at the time of its manufacture, or surreptitiously placed in semiconductors within a targeted piece of equipment. The movement of coding functions overseas for large software programs and the increasing use of foreign semiconductors gives adversaries the opportunity to embed malicious code in software and chips destined for PN components. The embedded code could be

---

50 National Institute of Standards and Technology (NIST), Information Technology Laboratory, Computer Security Division, *Internet Security Policy: A Technical Guide*, Gaithersburg, MD: NIST, July 21, 1997, p. 44.

51 Ibid.

52 Brent Mendel, "Mail Hack Affirms Mobile Code Fear," *Internet Week with LanTimes Online*, September 14, 1998, http://www.lantimes.com/98/98sep/809a001a.html.

used to create backdoors for exploitation by a foreign intelligence service, disable key network components, disrupt communications, or randomly adulterate data.

3.4 Trends

The threat of electronic intrusion is changing and growing. In a recent survey conducted by the Computer Security Institute (CSI) and the Federal Bureau of Investigation (FBI), 64 percent of the 520 organizations responding stated that their systems had been attacked. Fifty-four percent of those responding in 1996 cited Internet connections as the source of intrusions into their networks; 2 years earlier this number had been 37 percent. Attacks reported included brute force password guessing (13.9 percent), network scanning (21 percent), denial-of-service (22 percent), IP spoofing (12.9 percent), and data alteration (16.2 percent). The most pervasively attacked information systems were those at financial institutions, high-technology firms, medical facilities, Government institutions, and utility providers.[53]

According to the Defense Information Systems Agency (DISA), the number of officially reported attacks has increased significantly. Defense installations reported 53 attacks in 1992, 115 in 1993, 225 in 1994, and 559 in 1995. However, it is difficult to determine the exact number of attacks, as is illustrated by the experience of DISA's Vulnerability Analysis and Assessment Program. Under this program, personnel attempt to penetrate various DoD sites to test how well they are protected. Between 1992 and 1996, DISA successfully gained access 65 percent of the time, but only about 4 percent of the attacks were detected, and only about 27 percent of those detected were reported to DISA.

Attackers have installed malicious software, crashed systems, and stolen, modified, and corrupted data and software. The problems these attacks cause extend beyond the security breaches; they are also very costly, as shown by the attack on Rome Laboratory in 1994. In this instance, DoD spent more than $500,000 to assess the damage to systems, ensure the reliability of the information in the systems, patch the vulnerabilities, and attempt to identify the attackers.[54]

Electronic intruders learn how networks and systems work, and what their vulnerabilities are, by studying the open source materials developed by the telecommunications carriers to document their networks and systems. Experienced hackers use this detailed knowledge to develop social engineering techniques and software intrusion tools. These experienced hackers

---

53 Computer Security Institute, *Issues & Trends: 1998 CSI/FBI Computer Crime and Security Survey,* Volume II, No.2, San Francisco, CA: CSI, March 1996.

54 U.S. General Accounting Office (USGAO), *Information Security: Computer Attacks at Department of Defense Pose Increasing Risks,* GAO-AIMD-96-84, Washington, DC: USGPO, May 1996, p. 20-21.

freely share their knowledge, techniques, and tools with less experienced hackers through publications, hacker conferences, Web sites, and Internet Relay Chat (IRC) channels, enabling novice hackers to quickly and easily launch the same attacks as experienced hackers.

The average level of technical capability among the systems administrators who must address this threat has not kept pace with that of the hacker community. Because the demand for skilled systems administrators is growing more quickly than they can be trained, many positions are being filled by less experienced personnel, which gives the hacker community a further advantage. This overall situation, coupled with the United States' dependence on the PN, creates a potentially high-risk situation.

The financial resources required to launch an electronic attack are minimal; the equipment is affordable and readily available in retail stores and mail order catalogues. In fact, much of the equipment can be found today in the homes of millions of individuals, here and abroad, who have personal computers and modems and are growing increasingly adept at using them. Statistics compiled by the Federation of American Scientists - CyberStrategy Project reflect that by 1997, more than 50 million households in the United States had personal computers, more than 20 million of those had modems, and more than 5 million of those subscribed to on-line services. By 2000, those numbers are expected to rise to 70 million, 40 million, and 10 million, respectively,[55] October 15, 1998.

Many organizations and agencies have noticed an increasing number of attacks on computer systems, many of which affect multiple operating systems. A recent worldwide computer security survey of 4,226 information security officers (ISO) revealed that computer intrusions are increasing; intranets increase the risk of attack; viruses are a continuing threat; and computer-based industrial espionage is real. Most respondents indicated insider attacks still account for the majority of intrusions into computer systems.[56] This increase in attacks corresponds with three other trends: the availability of increasingly sophisticated automated intrusion tools, the virtually exponential increase in the number of attractive targets, and the proliferation of globally connected systems. These intrusion tools not only enable highly skilled intruders to target multiple systems simultaneously, rather than one at a time, but they also give novice intruders the ability to achieve this same level of efficiency. The increase in the number of computers and new technologies has created a vast array of targets in both Government and private industry. This combination of powerful tools, more targets, and

---

55 Federation of American Scientists CyberStrategy Project, http://www.fas.org/cp/index.html D

56 Beth Davis, "The Fifth Annual Information Week/Ernst & Young Information Security Survey," *Information Week*, Issue 647, September 8, 1997.p. 182; and Bob Violino, "The Security Facade," *Information Week*, October 21, 1996, p. 38.

virtually ubiquitous access has been a significant factor in the increase in the number of attacks against telecommunications and information systems.

3.5 Potential Impact

The potential impact of any particular electronic intrusion attack against the PN depends on the technical sophistication of the attacker, the objective of the attack, and the ability of the attacker to exploit vulnerabilities. The most serious threat and highest potential impact is associated with a coordinated IW attack, possibly aided by a knowledgeable insider. A coordinated electronic intrusion attack could be conducted by an individual or a small group. A small group with the ability to carefully target key software functions, databases, and system components could induce major disruptions, possibly resulting in cascading failures throughout interconnected information networks. Even an unsophisticated hacker could cause major problems if he altered or destroyed data, or modified network management functions. Criminals attacking the PN would most likely be interested in stealing information, obtaining services through fraudulent means, or altering financial transactions.

3.6 Conclusion

The ongoing deregulation of the telecommunications industry has increased PN vulnerabilities. Although system architects have developed software and security standards, these standards are not applied consistently across all systems. Further, the ability to identify and counteract network intrusions at key points throughout the telecommunications industry varies from organization to organization. A clear understanding of risks associated with computer-based telecommunications will help guide policy makers and systems operators as they develop ways to support the security of NS/EP telecommunications systems.

Use of open network architecture (ONA) systems has always been a risk to organizations that require the continued availability of information. Data can be altered, disrupted, captured, or destroyed at numerous points in the information chain. System architects and network operators must consider a balance between ease of use and system security. Threats to AISs are constantly changing, and the integrity of the network is only as secure as the weakest link in a shared environment. It is not only the integrity of the system that is at stake but also the information within the system itself. Information is at risk any time it is collected, transported, and stored. Understanding the threats to information and systems is a critical factor in creating a secure environment and managing risk.

The following sections of this report provide information on individuals and groups engaging in computer intrusion attacks against the PN and interconnected NS/EP telecommunications and information systems.

4. ECONOMIC COMPETITOR AND ADVERSARY USE OF ELECTRONIC INTRUSION

Information Based Warfare (IBW) is an approach to armed conflict focusing on the management and use of information in all its forms and at all levels to achieve a decisive military advantage in especially the joint and combined environment. IBW is both offensive and defensive in nature, ranging from measures that prohibit the enemy from exploring information to corresponding measures to assure the integrity, availability, and interoperability of friendly information assets. While ultimately military in nature, IBW is also waged in political, economic, and social arenas and is applicable over the entire national security continuum from peace to war and from "tooth to tail." Finally, IBW focuses on the command and control needs of the commander by employing state of the art information technology such as synthetic environments to dominate the battlefield.

Definition developed by the School of Information Warfare of the National Defense University.[57]

4.1 Introduction

This section discusses the use of electronic intrusion by economic competitors and adversaries. These include foreign states and organizations, terrorists, and organized crime groups. Foreign states involved in electronic intrusion desire data on political, economic, military, and commercial capabilities and intentions. Some foreign states have formal IW programs, while others have more limited intelligence collection activities that use electronic intrusion. Terrorist groups are becoming aware of the potential of the information infrastructure, not only as a target, but also as a means of communicating their message. Organized crime groups have also begun to notice the potential impact of telecommunications and information systems on their operations.

4.2 Foreign States

In today's high technology world, adversaries are no longer limited to using traditional methods of attack (e.g., bombs and other weapons of physical destruction); they have the option of using nontraditional methods of attacking their enemies. One such method is information warfare. Although the effect of an electronic attack on a telecommunications or information system is not as dramatic as the physical impact of a bomb, the results can be more destructive to a modern society. The National

---

57 Charles B. Everett, Moss Dewindt, and Shane McDade, *The Silicon Spear: An Assessment of Information Based Warfare (IBW) and U.S. National Security,* World Wide Web, Institute for National Strategic Studies, National Defense University, www.ndu.edu/ndu/inss/siws/ch2.html, November 1996.

Counterintelligence Center (NACIC) has concluded that the governments of at least 23 countries are targeting U.S. firms.[58] The American Society for Information Science's (ASIS) 1997 survey noted that high-tech companies, such as those in Silicon Valley, are the most popular targets of foreign countries. Other frequent targets include manufacturing and service industries. According to press reports, the most lucrative information obtained includes research and development strategies, manufacturing and marketing plans, and customer lists.[59]

4.2.1 Russia

Russia's interest in information systems and IW tactics has increased dramatically within the last several years. Although the Russian Ministry of Defense (MOD) does not have an "official" definition of IW, several unofficial definitions were provided by Russian officers at the General Staff Military Academy, one being the following:

> Information warfare is a way of resolving a conflict between opposing sides. The goal is for one side to gain and hold an information advantage over the other. This is achieved by exerting a specific information/psychological and information/technical influence on a nation's decision-making system, on the nation's populace and on its information resource structures, as well as by defeating the enemy's control system and his information resource structures with the help of additional means, such as nuclear assets, weapons and electronic assets.[60]

Russia is well aware of the characteristics and the scope of IW. Regarding this new development in modern military affairs, Colonel Aleksandr Pozdnyakov, doctor of philosophy and a deputy department head at the General Staff Military Academy, states, "effectiveness of modern weaponry is determined not only by firepower, but also by information controllability."[61] Thus, it appears that as the technological advancements in information systems increase in Russia, "the growing role of information-technology warfare is rapidly lowering the barrier between war and peace. The armed forces of likely adversaries are in a state of constant information warfare, and military

---

58 National Counterintelligence Center (NACIC), *Annual Report to Congress on Foreign Economic Collection and Industrial Espionage,* Washington, DC: USGPO, October 1997, p. 18

59 Jack Nelson, "U.S. Firms' '97 Losses to Spies Put at $300 Billion," *Los Angeles Times,* January 12, 1998.

60 Tim Thomas, "Russian Views on Information-Based Warfare," Fort Leavenworth, KS: Foreign Military Studies Office, World Wide Web, leav-www.army.mil/fmso/opart/pubs/airpower.htm, July 1996, p. 2.

61 Foreign Broadcast Information Service (FBIS), "Academy Department Head on Importance of Information Security," FBIS-UMA-95-239-S, December 13, 1995.

informatics works to accomplish tasks characteristic of war even in peacetime."[62]

The Russian military is studying software virus warfare as one of the more crucial components of information warfare. This type of warfare can be disguised easily and is therefore difficult to detect. The successful use of viruses as a weapon may diminish the willingness to use traditional means of warfare. As one Russian officer noted, "there is no need to declare a war against one's enemies and to actually unleash more or less large military operations using traditional means of armed struggle."[63]

Viruses can be seen as precise and forceful tools that can cause chaos in an initial period of war. As one Russian officer indicated, there are several viruses that can have catastrophic effects if used properly, including the forced quarantine virus, the overload virus, and the sensor virus.[64] Because viruses can hinder a computer's ability to function, Russia is researching ways to detect and destroy viruses that could destroy its own information systems. One example is the antivirus device used to detect stealth viruses. The Russian military has developed a complicated mathematical procedure that compares the files on a disk with the file structures and virtual free space to uncover a stealth virus.[65]

Russia has also shown an interest in using electronic intrusion for economic espionage. According to press reports, Russia's National Center for Data Exchanges uses electronic means to secretly monitor foreign data networks and databases to steal trade secrets and intellectual property.[66] The priority that Russia has placed on IW has forced the Russian military and government to re-evaluate their fundamental priorities in waging war. As one Russian military theorist stated, "it is necessary to place paramount importance on technological indicators of new weapons, which are capable of largely predetermining the end result of military operations."[67] Because

---

62 Ibid.

63 Tim Thomas, "Russian Views on Information-Based Warfare," Fort Leavenworth, KS: Foreign Military Studies Office, World Wide Web, leav-www.army.mil/fmso/opart /pubs/airpower.htm, July 1996, p. 6.

64 FBIS, "Academy Department Head on Importance of Information Security," FBIS-UMA-95-239-S, December 13, 1995.

65 Tim Thomas, "Russian Views on Information-Based Warfare," Fort Leavenworth, KS: Foreign Military Studies Office, World Wide Web, leav-www.army.mil/fmso/opart /pubs/airpower.htm, July 1996, p. 6.

66 Jack Nelson, "U.S. Firms' '97 Losses to Spies Put at $300 Billion," *Los Angeles Times*, January 12, 1998.

67 Tim Thomas, "Russian Views on Information-Based Warfare," Fort Leavenworth, KS: Foreign Military Studies Office, World Wide Web, leav-www.army.mil/fmso/opart /pubs/airpower.htm, July 1996, p. 6.

computer intrusions and IW are powerful weapons, "it's no accident" as Pozknyakov says, "that the following motto has become so popular–'he who controls information controls the world'."[68]

4.2.2 China

The Chinese military doctrine and strategy is in the process of transforming from mechanized warfare to information warfare. Qian Xuesen, a renowned military expert, expresses the view that, "After going through the stages of bare-handed fighting, cold steel, hot weapons, and mechanization, military operations are now entering the information age. We should pay attention to information warfare in the context of nuclear deterrence."[69] As Major General Wang Pufeng states, the Chinese believe that in the near future IW will control the "form and future of war."[70]

The revolution in China's military philosophy "proposes to focus on hi-tech limited wars and profound thinking on new operational methods."[71] In fact, a new training program has been accepted by the Headquarters of the General Staff of the People's Liberation Army (PLA) to dramatically change the current conventional war tactics to a warfare strategy involving high technology. This new military ideology has been established to cover the Army, Navy, Air Force, Artillery Corps, the Commission of Science, Technology and Industry for National Defense, and the Armed Police Force.[72] The Chinese military is also developing innovative strategies for network systems to be able to disguise true information with false information in modern high-tech warfare.[73]

The following IW strategies are derived from research done by various Chinese military experts. They relate national IW and national defense IW, strategic IW and tactical IW, and offensive IW and defensive IW. According to open source reporting, China's strategies emphasize:

- The primary goal of IW is to attack the enemy's command and control systems. The struggle to control information is the focus of weapons systems and the countermeasures taken against these systems.

---

68 FBIS, "Academy Department Head on Importance of Information Security," FBIS-UMA-95-239-S, December 13, 1995.

69 Ibid.

70 FBIS, "General Views Importance of Information Warfare," FBIS-CHI-95-129, July 6, 1995.

71 FBIS, "Article Discusses Information Warfare," FBIS-CHI-95-239, December 13, 1995.

72 FBIS, "New PLA Training to Stress High-Tech Warfare," FBIS-CHI-95-240, December 14, 1995.

73 FBIS, "China: Defense Military Computer Network Interconnects PLA Army, Navy, Air Force," FBIS-CHI-97-324, November 25, 1997.

- The extensive use of information to harass the enemy. One must fight with such accuracy and speed that the enemy will be unable to perceive where the actual battlefield is. This allows one to effectively wield superiority in the area of information.
- The primary method of operation to be employed in war is to attack the enemy's command authorities, staff headquarters, theater of operations general headquarters and unit headquarters. This attack aims to disable all of the enemy's information systems.
- The destruction of the enemy's "eyes and ears," while effectively protecting the unimpeded flow of information via one's own "eyes and ears."
- The use of multinode, multipath, multifrequency network systems that use information deception and information concealment procedures to ensure one's survival.
- The introduction of equipment with "digitized technology." Information technology will be used to conduct electronic warfare, command and control warfare, and warfare characterized by attacks with computer viruses. In this environment a l-ounce integrated circuit chip in a computer could be more useful than a ton of uranium.[74]

Chinese military experts studying IW believe that in a future war, the primary target will be a nation's information infrastructure, including financial, military, electronic communications systems, and computer networks.[75] In this new type of war, "one possible battle plan is to seize, utilize, and control an information edge by using information-based weapon systems."[76] One such weapon is the software virus. Computer viruses are one of the most popular IW weapons used by the Chinese to destroy data programs. It is believed that "conducting warfare with computer viruses is more effective than using nuclear weapons."[77]

Computer viruses will find "extensive application in the future hi-tech war because they can be transmitted to the enemy through wires or by wireless means, securely planted in the computer components ordered by the enemy, projected into the enemy's computer system by advanced means

---

74 Wang Xusheng, Su Jinhai, and Zhang Hong, "China: Information Revolution, Defense Security," *China Computerworld (No. 30, p. 21)*, World Wide Web, Infowar.com Webpage, wysiwyg://34/http://www.infowar.com/mil_c41/mil_c41_121897a.htm l-ssi, August 11, 1997.

75 FBIS, "PRC Army Daily on Weaknesses of Information Warfare," FBIS-CHI-96-014, January 22, 1996.

76 Ibid.

77 FBIS, "China: Defense Military Computer Network Interconnects PLA Army, Navy, Air Force," FBIS-CHI-97-324, November 25, 1997.

when a war is under way, and used to attack the enemy's command and control systems and battle platforms."[78] Virus-contaminated chips could be used to inject viruses into the enemy's computer networks, which could then be remotely activated to disrupt the enemy.

The Chinese military is well aware of how a nation's dependence on computers also measures the nation's modernization and the technological strength of its military. China's revolution in military and national security doctrine requires that telecommunications and information technology be used to undermine and destroy its enemies. Previous methods must be abandoned and more advanced approaches to warfare must be used to survive. Shen Weiguang, a Chinese official working at the State Council Special Economic Zones Office, sums up China's IW capability by concluding:

> The level of information science of our armed forces is not that developed; however, our country, our society, which is pursuing reform and opening up, has great potential for waging information warfare, including qualified people and technology. We have the experience of waging people's war; here lies our advantage.[79]

China has been one of the few countries that has been brought into court under the 1996 Economic Espionage Act, which makes theft of proprietary information a felony punishable by a $10-million fine and 15-year prison sentence. Retired Eastman Kodak manager, Harold C. Worden, pleaded guilty in November 1996 to stealing information and passing it to China and was sentenced to a year in prison.[80]

4.2.3 South Korea

In May 1997, Robert Kim, a U.S. Navy intelligence analyst accused of spying for South Korea, pleaded guilty to one count of espionage. As a civilian employed by the U.S. Office of Naval Intelligence, Kim had access to classified information in the joint Navy-Coast Guard computer system. Press reports stated that Kim supplied South Korea with a classified software program developed for maritime vessel tracking. At the time, the United States was negotiating the sale of the system to South Korea. Kim was also charged with supplying U.S. intelligence information regarding North Korea and China.[81]

---

78 FBIS, "PRC: 'Digitized Forces' Developed for Electronic Warfare," FBIS-CHI-96-097, May 17, 1996.

79 FBIS, "Article Discusses Information Warfare," FBIS-CHI-95-239, December 13, 1995.

80 Jack Nelson, "U.S. Firms' '97 Losses to Spies Put at $300 Billion," *Los Angeles Times*, January 12, 1998.

81 Robert Jackson, "Ex-Analyst Admits Spying for S. Korea in Plea Bargain," *Los Angeles Times*, May 8, 1997, p. 13.

Because of the decrease in its competitive products, difficulty in developing new technology through its own research and development capabilities, and the high cost of buying foreign technology, South Korea has engaged in systematic efforts to gain access via computer intrusion methods to the sensitive technological secrets of competitor countries. According to press reports, South Korea uses not only its intelligence agencies, but also its academic exchange programs to engage in industrial espionage. South Korean government institutions and companies target U.S. technologies, such as computers, communications, electronics, data communications and processing, semiconductor technology, digital signal processors, digital communications, ATM technology, and fiber optics. In fact, the South Korean government has facilitated the transfer of technology from other countries by giving South Korean companies access to foreign databases that contain industrial, scientific, and technological data from foreign networks.[82]

4.2.4 Cuba

Cuba's IW capabilities, specifically, its computer virus capability, can pose a threat to our Nation's critical infrastructures. In 1997, the press reported that according to a declassified report, the Cuban Military Intelligence Directive had a project to gather information to develop a computer virus that could infect U.S. civilian computers. The press reports also stated that the Cubans invested in the world market with the intent to buy unclassified data on computer networks, viruses, satellite communications (SATCOM), and other related areas of communications technology.[83] Although the specific information Cuba gained is not known, it can be speculated that its information-gathering efforts could be a serious threat to any country's national defense.

Lourdes, a Russian-built and manned high-technology communications center in Cuba, has been in operation since the Cold War. Walter Deeley, who was the communications security director at the National Security Agency (NSA) in the 1980s, told NBC News that the agency worried that the Soviets could use Lourdes to insert malicious software into the U.S. communications system.[84] Such a scenario would be one example of how

---

82 *Counterintelligence News & Developments,* Volume 2, World Wide Web, www.nacic.gov/cind/cindijun9.htm, June 1996.

83 Brock N. Meeks, "Information Warfare and the Real Threat," World Wide Web, MSNBC News, www.msnbc.com/news/123040.asp, 1997.

84 Brock N. Meeks, "Information Warfare and the Real Threat," World Wide Web, MSNBC News, www.msnbc.com/news/123040.asp, 1997.

Cuba could use electronic intrusion and IW tactics to conduct information warfare against the United States.

4.2.5 Japan

Japan can be considered one of the world's most prominent players in computer espionage and electronic intrusions. Internet gateways currently provide Japan's official and semi-official intelligence agencies access to databases around the world. Japanese universities are connected to the largest academic research and development computer networks in the United States, Canada, Western Europe, and Australia, which can be seen as a threat given the country's ability to penetrate networks.

With the end of the Cold War, there has been an intense international trade war in which information from computers and databases can be used as weapons if exploited by a foreign state. The intelligence effort within Japan is coordinated by the government's Ministry of International Trade and Industry (MITI). Japan's IW efforts have been recognized for a considerable time. As noted in 1994, much of MITI's intelligence gathering is conducted through the Japan External Trade Organization (JETRO).[85] Japan's ability to engage in business IW and corporate espionage can be viewed as a dangerous weapon by competitor nations. Computer intrusions carried out by the Japanese are the weapon of choice, used primarily to obtain secrets regarding economic information, bio-industries, and advanced technology.[86]

The Japanese ideology on computer espionage and IW is that the key to success is to have not only the ability to hack into computer systems and insert viruses but also the ability to destroy the computer and communications functions controlling a country's transmissions for financial, transportation, and energy sectors. This capability could ultimately debilitate the economic and civil life of any country.[87] For Japan, being able to achieve this ambitious goal against enemy countries would be an essential tool in undermining economic competitors.

---

85 William E. DeGenaro, *Steal This Country: How Foreign Spies Are Destroying American Jobs,* Presented at the Fifth National Operations Security Conference in McLean, Virginia, 1994, p. I-D-4.

86 FBIS, "Japan: Japan Seen Lagging in Global Economic Intelligence War," FBIS-EAS-97-279, October 8, 1997.

87 FBIS, "Japan: Journalist on Security, Intelligence Issues," FBIS-EAS-97-065, April 7, 1997.

4.2.6 France

France has excellent IW capabilities and is considered one of the more advanced countries in terms of technical competence. Press sources have stated that France at one time targeted more than 70 major U.S. corporations, including Boeing, IBM, Texas Instruments, and Corning Glass.[88] The French General Directorate of External Security (DGSE) has targeted U.S. economic and proprietary data since 1964. DGSE's Service 7, which is the primary agency responsible for intercepting network information from foreign countries, has conducted technical operations against telecommunications systems throughout the world. As a result of these successful operations, sensitive data was collected. The press also reported that DGSE targeted Loral Space Systems and Hughes Aircraft for information on telecommunications satellite technology, Lockheed Missile and Space Company for data on the Milstar communications satellite system, GTE Telecommunications Products for microwave technologies, and TRW for military telecommunications technologies.[89]

4.2.7 Germany

The Bundes Nacrichten Dienst (BND), the German Federal Intelligence Service, views the use of computer viruses as an effective means to target information systems. Such attacks on computers "would then damage highly interlaced industrial countries which have become so dependent on intact databases in such a way that large parts of the functional and social systems would collapse."[90] It appears that the German government has advanced computer espionage programs. According to press reports, a computer facility near Frankfurt gives German agents the capability to intrude into data networks and databases of companies and governments world wide.[91] This computer operation is called Project RAHAB and was formed by the Federal Intelligence Service. The mission was developed in large part because of the success of the Chaos Computer Hackers Club of Hamburg, which had illegally gained access to hundreds of computers by using the National Aeronautics and Space Administration's (NASA) SPAN network.[92] Since

---

88 Jack Nelson, "U.S. Firms' '97 Losses to Spies Put at $300 Billion," *Los Angeles Times*, January 12, 1998.

89 "Telecommunications Satellites Said To Be Targeted for Espionage by France," *Common Carrier Week*, May 17, 1993.

90 FBIS, "Germany: Study Examines Changes to Armed Forces in Information Age," FBIS-TAC-97-279, October 8, 1997.

91 Peter Schweizer, *Friendly Spies*, New York, NY: Atlantic Monthly Press, 1993, p. 158.

92 *Clifford Stoll, The Cuckoo's Egg*, New York, NY: Doubleday, 1989, p. 285.

the operation began, RAHAB computer technicians have accessed computers in Russia, Japan, France, Italy, Britain, and the United States.[93]

4.2.8 Iraq

In August 1990, shortly after Iraq's invasion of Kuwait, a large-scale effort was launched worldwide to infiltrate various sensitive U.S. Government and military computers. Although most of the intrusions originated in the Netherlands, an Iraqi intelligence operation against NATO that involved a German citizen was uncovered during the same time period.[94]

After the destruction of the headquarters of the Iraqi Intelligence Service (Mukhabarat) in downtown Baghdad during the Persian Gulf War, it was discovered that Iraq maintained a sophisticated computer operation and had connections to the Internet through a gateway host in Bahrain, operated by the Bahrain Telecommunications Company. This suggested that Iraq had an ability to capture sensitive data from computers in other countries.[95]

4.2.9 Israel

Israel is known to have excellent technical capabilities that could be used to compromise the security of networks worldwide. In 1991, Israeli intelligence may have known about the activities of a young Israeli hacker during the Persian Gulf War. At that time, Israel detained 18-year-old Deri Schreibman, who had penetrated U.S. defense computers and copied information on the Patriot missile defense system. However, Israeli police did not charge him for fear that a court case would reveal the ease of penetrating very sensitive U.S. defense computers from Israel and compromise Israel's more sophisticated computer espionage activities against U.S. computers.[96]

4.2.10 Bulgaria

During and after communist rule, Bulgaria became known as a "breeding ground" for computer viruses. One virus, "Dark Avenger," destroys data on PCs. The text of the virus reads, "This program was written in the city of Sofia." Other destructive virus programs designed in Bulgaria are the "Dark Lord," Darth Vader," and "Kamikaze"viruses. There is evidence that by 1991 Bulgaria had produced some 30 separate viruses and more than 100 clones and strains, and it was releasing these viruses into computer systems at a

---

93 Peter Schweizer, *Friendly Spies*, New York, NY: Atlantic Monthly Press, 1993, p. 160.

94 Ibid.

95 Wayne Madsen, "Intelligence Agency Threats to Computer Security," *International Journal of Intelligence and Counterintelligence*, 6:4, Winter 1993, p. 437.

96 Wayne Madsen, "Intelligence Agency Threats to Computer Security," *International Journal of Intelligence and Counterintelligence*, 6:4, Winter 1993, p. 434.

rate of one per week. This indicates that the viruses originating in Bulgaria might have been developed by a clandestine virus writing factory within the former communist Bulgarian intelligence service.[97] Also, it is presumed that the Bulgarian virus researchers may be selling their services to the highest bidder.[98]

4.3 Terrorist Use of Electronic Intrusion

In today's information age, the scope of terrorism is expanding. Not only have the potential impacts of a cyber terrorism attack increased, but these potential impacts have also become more well known. Publicity about these vulnerabilities, along with the potential for greater damage, may make these targets increasingly attractive for individuals determined to perpetrate terrorist acts. Terrorist groups could use IW to raise funds, promote awareness of the group's ideology, and to attack the NII. Electronic intrusion could be used not only for gathering intelligence, but also for the "disruption or destruction of the information infrastructure, including basic services such as power supply, police databases, social security transfers, medical networks, transportation signals, money transfers, and telephone switching systems."[99]

Additionally, cyber attacks allow terrorists to perpetrate the attack from almost anywhere in the world, while remaining in a safe haven. Not only does this make it more difficult to identify the responsible parties, but extraditing those who are identified may be problematic. Another factor that may affect the growth of cyberterrorism is the decreasing cost of information technologies. As this cost falls, more foreign governments and non-government organizations will be able to afford higher quality information systems. This technology may provide a wide variety of state and non-state actors the ability to engage in more sophisticated attacks. This presents a potential threat to nations that depend on increasingly complex information infrastructures, especially the United States.

International terrorist groups do not need to have advanced intrusion skills themselves attack information systems. They can hire hackers to do

---

97 David Ferrache, *A Pathology of Computer Viruses,* London, England: Springer-Verlag, 1992, p. 30.

98 Wayne Madsen, "Intelligence Agency Threats to Computer Security," *International Journal of Intelligence and Counterintelligence,* 6:4, Winter 1993, p. 426.

99 Andrew Rathmell, Richard Overill, Loranzo Valeri, and John Gearson, "The IW Threat From Sub-State Groups: An Interdisciplinary Approach," (Paper presented at the Third International Symposium on Command and Control Research and Technology), World Wide Web, Infowar.com Webpage, wysiwyg://53/http://www.infowar.com/mil_c4i/icsa/icsa3.html-ssi, June 17-20, 1997.

it for them.[100]An individual believed to be a member of the Pakistani terrorist group Harkat-UI-Ansar paid a hacker, known as Chameleon, $1,000 to provide information on U.S. Government computer networks. While Chameleon admitted to cashing the check, he denied sending the alleged terrorist any information.[101] Chameleon was perhaps most notable for his affiliation with a hacker organization known as the Masters of Downloading (MOD). This group gained national attention when it gained access to several U.S. military computers. According to some press reports, the data accessed through these computers included sensitive network management tools and network topology maps of the Secret Internet Protocol Router Network (SIPRNET).[102]

The use of the Internet and other information systems can give terrorist groups a global and near real-time command and control communications capability. In addition, cyberterrorists can spread their ideology or misinformation to millions of people. Terrorist groups, such as the Liberation Tigers of Tamil Eelam (LTTE) and various racial identity groups, have established Internet Web sites that include their ideologies, manifestos, and communiqués.[103] The implications of the presence of terrorist Web sites can be serious. Not only does the Internet connect millions of individuals globally, but it also provides an affordable medium for terrorists to enlist support for their causes.

4.4 Organized Crime Use of Electronic Intrusion

Organized crime groups are rapidly learning the benefits of using computers to commit their crimes. Russian organized crime groups are increasingly using computers to commit bank fraud. Computers are used to illegally transfer money to the West, obtain fraudulent lines of credit, and exchange black market currency. According to the Russian Organized Crime Task Force, more than 300 attempts were made to break into the Central Bank of Russia between 1993 and 1995.[104] The Russian criminals inserted fictitious

---

100 United States Senate, Senate Governmental Affairs Committee, Permanent Subcommittee on Investigations, "Testimony of John Deutch, Director of U.S. Central Intelligence," World Wide Web, www.fas.org/irp/congress/1996_hr/s960625d.htm, June 25, 1996.

101 Niall Mckay, "Cyber Terror Arsenal Grows", *Wired News,* World Wide Web, www.wired.com/news/news/politics/story/15643.html.

102 John Vranesevich, "AntiOnline's Coverage of Chameleon Raided By The FBI," *AntiOnline,* World Wide Web, www.antionline.com/SpecialReports/chameleon/index.html.

103 Frank J. Cilluffo and Curt H. Gergely, "Information Warfare and Strategic Terrorism," *Terrorism Terrorism and Political Violence,* Vol. 9, No. 1, London, England: Frank Cass & Company, Ltd., Spring 1997, p. 88.

104 CSIS, *Russian Organized Crime: Global Organized Crime Project,* Washington, DC: CSIS,

information on payments into the bank's networks amounting to millions of dollars each quarter. These criminals usually rely on insiders to supply passwords and codes for the bank's network.[105]

The potential use of computer systems for criminal activities first received significant public attention in 1995. Vladimir Levin, a Russian biochemistry graduate student in St. Petersburg, accessed New York Citicorp's computerized cash-management system by using a sophisticated computer program. He was able to transfer more than $12 million to various banks worldwide. Although most of the money was recovered, this type of attack demonstrates how any individual from anywhere in the world may compromise various international economic systems.

Organized crime groups are also using computers to hinder police investigations. By breaking into law enforcement computer systems, criminal groups collect intelligence on police activities, destroy or alter data on investigations, and monitor the activities of informants.[106] For example, an international computer hacker organization headquartered in Dallas, Texas, successfully penetrated the networks of several telecommunications providers and acquired unlisted telephone numbers, personal addresses, credit information, and National Crime Information Center data, causing losses in excess of $500,000. The hackers then installed a sniffer that compromised at least 15 telephone company systems, including records, maintenance, and operational control systems, and installed illegal wiretaps. The advanced level of expertise of the hackers was comparable to that of telephone company experts and suggests that they could have disrupted telecommunications services if they had wanted to.[107] Also, according to a recent report, Colombian drug cartels had the phone records of millions of Colombian residents stored on an IBM mainframe. These records were checked against calls made to the U.S. Embassy and the Colombian Ministry of Defense in order to identify people who were cooperating with the government.[108] In Belgium, a computer system used

---

1997, p. 36.

105 Ibid., p. 37.

106 U.S. Senate; Security in Cyberspace: Hearings Before the Permanent Subcommittee on Investigations, Testimony of Jamie Gorelick, Committee on Governmental Affairs, Washington, D.C. USGPO, 1996, pp. 154-155.

107 Michael Vatis, Deputy Assistant Director and Chief, NIPC, FBI; Statement for the Record Before the Congressional Joint Economic Committee, Washington, D.C., March 24, 1998, pp. 6-7.

108 Dorothy E. Denning and William E. Baugh, Jr., U.S. Working Group on Organized Crime, *Encryption and Evolving Technologies: Tools of Organized Crime and Terrorism*, Washington, DC: National Strategy Information Center, 1997, p. 3.

for handling sensitive information on criminals was compromised when, according to press reports, a criminal organization paid a government employee to steal thousands of papers. These papers contained information from the computer files. It is unclear how long the theft had been going on or exactly what type of information had been sold.[109]

4.5 Conclusion

Economic competitors have increased their use of electronic intrusion to secure advanced technology and increase their economic advantage. The NACIC has concluded that at least 23 countries are collecting economic intelligence within the United States. Computer age communications connectivity, commercial enterprise activities, and availability of corporate data on office workstations, along with the growing number of home PCs, have made it extremely easy to copy and transfer valuable financial, business, scientific, technical, economic, and engineering information. Electronic intrusion is a primary technique used by foreign collectors to obtain economic information for intelligence analysis.[110]

The proliferation of sophisticated hacking tools and the subsequent intrusions into U.S. information systems and networks can be expected to increase because of the increase of nation-state involvement in formalized IW programs. Foreign Intelligence Services (FIS) are using new computer and communication technologies to target the United States and other developed countries. These FISs focus primarily on capturing information from targeted systems rather than modifying and destroying data or denying service to users. However, in times of conflict, these agencies could decide to use their capabilities to move from gathering intelligence to disrupting or destroying targeted systems. Foreign threats to U.S. Government and proprietary networks will continue to grow, focusing on technological, military, economic, and political data. Information terrorism activities can be expected to grow in popularity as nation-states and nonstate actors seek opportunities to shape political viewpoints or capture information useful to their causes.[111]

---

109 Jeremy Lovell, "Belgium Investigates Major Data Security Leak," World Wide Web, Infoseek News Channel, www.infoseek.com/Content?ar . . . v=N5&lk=ib&col=NX&kt=A&ak=news1486, December 15, 1997.

110 National Counterintelligence Center (NACIC), *Annual Report to Congress on Foreign Economic Collection and Industrial Espionage*, Washington, DC: NACIC, June, 1997, p. iii.

111 *Ibid.*, p. 7.

5. HACKER USE OF ELECTRONIC INTRUSION

"Our modern electronic infrastructure – computer systems that control everything from our power systems to our stock exchanges – is a potential target for attack by computer hackers and organized criminal enterprises. Such attacks pose a direct threat to our national security."[120]

[120]International Crime Control Strategy of the United States, Section VII: Responding to Emerging International Crime Threats, http://www.usdoj.gov/criminal/press/VII-Iresp.html, May 12, 1998.

International Crime Control Strategy of the United States

5.1 Introduction

The term "hacker" is defined as an individual who attempts to gain unauthorized access to a computer system. Because many of the systems hackers target are interconnected with and rely on the same telecommunications links that support NS/EP telecommunications and information systems, intrusions not specifically targeted at NS/EP systems may affect them adversely.

5.2 Hackers

The initial image of hackers portrayed by the media was that of computer-savvy teenagers and overzealous programmers who appeared to be simply curious about computer systems and network operations, and unlikely to engage in criminal or malicious activities. Unfortunately, this image is no longer accurate. The new generation of hackers appears to be motivated more by greed or malice than by simple intellectual curiosity. Hackers have begun to realize the value of the information contained in computer systems, and the potential profit that can be derived by stealing telecommunications services and committing computer fraud. In April 1998, a group of 15 hackers calling themselves the "Masters of Downloading" (MOD) broke into computers at the Defense Information Systems Agency (DISA) and bragged that they had stolen the means to cripple the military's communications network. The stolen program, known as the Defense Information Systems Network Equipment Manager (DEM), is key to controlling the military's Global Positioning System (GPS) of satellites used to target missiles and coordinate troop movements. The group also threatened to sell the stolen program to terrorist groups or foreign governments.[112]

112 The Boston Globe Online, "Hackers' Warnings on Info-war Appear Inflated," World Wide Web, Globe Newspaper Company, www.boston.com/dailynews/wi...rs_warnings_on_info_wa_appea.htm, April 27, 1998.

The MOD stole only the DEM program, which is unclassified, and did not gain access to the classified data that accompanied it. The DEM program cannot be used to control the GPS without this classified component, which is stored on computers that are not connected to the Internet.[113] However, if the intruders had been able to make undetected modifications to the program, there is the potential that the GPS could have been adversely affected.

Today's hackers insert malicious code and launch denial-of-service attacks for a wide variety of reasons, including greed, political motivations, theft of information, or sometimes just for the fun of it. While the motives of many hackers have shifted, their ability to cause significant damage to the PN, NS/EP telecommunications and interconnected information systems has greatly increased. The ready availability of sophisticated intrusion tools, automated denial-of-service attack programs, malicious code, and knowledge of PN vulnerabilities has drastically increased the threat posed by even the most inexperienced hacker. In 1990 the CERT Coordination Center received 252 reports of computer intrusion incidents. By 1997 the number of intrusions had increased to 2,134.[114] From October 1996 through October 1997, the Federal Computer Incident Response Capability (FedCIRC) handled 244 electronic intrusion incidents that affected more than 53,000 Government hosts and sites.[115] The increasing number of incidents and the growing sophistication of the attacks employed is indicative of the potential threat to NS/EP telecommunications and information systems.[116]

Hackers have demonstrated the ability to effectively use automated attack tools and exploit security flaws. These skills are readily transferable to the PN and NS/EP telecommunications systems. Hackers may act alone or with other hackers. The event known as Solar Sunrise is one example of hackers who coordinated their efforts. In February 1998, Israeli citizen Ehud Tenebaum, along with two other Israelis and two California teenag-

---

113 Jonathan Gregg, "Masters of What? Netly News: Pentagon Hackers Got Peanuts," World Wide Web, Time Online, cgi.pathfinder.com/time/daily/article/0,1344,11042,00.html, April 29, 1998.

114 CERT, SEI, CMU, CERT Coordination Center Statistics 1988-1997, World Wide Web, www.cert.org/stats/cert_stats.html, January 1998.

115 FedCIRC, *Summary of Incidents Handled by FedCIRC–October 1996-October 1997*, World Wide Web, fedcirc.llnl.gov/about/incidents1197.htm, January 1998.

116 U.S. Senate, Security in Cyberspace: Hearings Before the Permanent Subcommittee on Investigations, Testimony of Richard Pethia, Manager, Trustworthy Systems Program and Computer Emergency Response Team Coordination Center, Software Engineering Institute, Carnegie Mellon University, Washington, D.C.: USGPO, 1996, pp. 64-65.

ers, illegally accessed approximately 800 computers belonging to the United States and Israeli governments, as well as hundreds of other commercial and educational systems in the United States and elsewhere.[117] Some of the sites affected included Lawrence Livermore National Laboratories, NASA, the University of California at Berkeley, the Massachusetts Institute of Technology, the Pentagon, the Air Force, the Navy, and commercial sites.[118] Tenebaum, who calls himself "Analyzer," claimed to know how to gain entry into 400 DoD computer systems. He said that he began hacking as a challenge and concentrated on U.S. Government sites because he disliked organizations. Tenebaum was supposedly tutoring the California teenagers on how to target U.S. military systems.[119] Even though no classified material was compromised, the U.S. Government considered these incidents serious.

It is likely that computer crime will become the crime of choice because of the potential for remote attack and anonymity. Computer crimes are occurring globally; more than half of the computer crimes prosecuted by the U.S. Department of Justice involved international activities.[120] In an annual survey conducted by the FBI and the CSI in 1997, 70 percent of the respondents said they suffered losses from intrusion. The survey disclosed that losses resulting from computer theft and related crimes were estimated at about $110 billion in one 12 month period.[121]

Sophisticated computer hacking software makes it difficult to distinguish the unskilled intruder from an organized and state-sponsored actor attempting to retrieve military, political, and economic intelligence.

Additionally, today's hacker need not be technically competent to conduct a sophisticated electronic intrusion. The case of Matthew Singer (also

---

117 Department of Justice, "Israeli Citizen Arrested in Israel for Hacking United States and Israeli Government Computers," Press Release, 98-125, Washington, DC: DOJ, March 18, 1998; and Matt Richtel, "California ISP Says it Tracked Teenagers in Pentagon Hacking," World Wide Web, *The New York Times On The Web*, search.nytimes.com/search/d...oc+site+site+35489+1+wAAA+Analyzer, March 10, 1998.

118 Dan Reed and David L. Wilson, "Suspected Pentagon Hacker Found," World Wide Web, The Seattle Times Company, www.seattletimes.com/news/technology/html98/whiz_031988.html, March 19, 1998.

119 CNN Interactive, " Master Hacker 'Analyzer' Held in Israel," World Wide Web, www.cnn.com/TECH/computing/9803/18/analyzer/, March 18, 1998.

120 Scott Charney, *Computer Crime*, Presentation to the Inaugural Economic Crime Summit, Providence, RI, May 20, 1997.

121 Dana Blankenhorn, "If You're Hiring a Hacker, Stick With the Pros," World Wide Web, Net Marketing Home Page, www.netb2b.com:80...monthly/97/09/01/article4, September 1997, p. 4.

known as "Phantomd") illustrates the damage that can be done by a dedicated but relatively unsophisticated intruder with access to sophisticated tools. Singer was able to access hundreds, possibly thousands of computer systems through his extreme persistence, with the aid of sophisticated intrusion tools he obtained through hacker Web sites, and the information he gained from other experienced hackers in IRC chat sessions. He accessed computer systems supporting military research and development facilities, national laboratories, Fortune 100 companies, and the Supervisory Control and Data Acquisition (SCADA) system supporting the Oroville Dam north of Sacramento, California. He also obtained detailed information on computer system vulnerabilities from the Computer Security Department at Sun Microsystems and advanced intrusion tools from another research facility, dramatically increasing the effectiveness of his attacks against other information systems.[122]

When members of the FBI's National Computer Crime Squad arrested Singer on December 23, 1992, they were amazed to find that the individual they considered to be one of the most proficient and dangerous hackers was a mentally handicapped young man who had been unable to complete high school or hold a steady job. Singer's only real interest was computers, and with sheer persistence and user-friendly intrusion tools he was able to accomplish an enormous number of attacks. This case illustrates the threat hackers pose: their persistence often results in successful attacks because large complex systems are difficult to protect. Many security measures become meaningless when an intruder finds a weakness that can be exploited to gain root access.[123]

Hackers have a strong desire to show off their exploits and brag about breaking into a poorly protected network. They trade stories and information through e-mail, IRC chat sessions, local meetings, and annual conventions (e.g., "DEFCON" and "CHAOS"). This exchange allows information on vulnerabilities to be publicized in a relatively short time and gives new hackers knowledge, ideas, and a support base for conducting other activities. Intrusion tools, malicious code, and pirated software are also traded through these communications activities.

The more proficient computer hackers use "phreaking" techniques to steal telephone service. Phone phreaking is the act of stealing telephone or data line service. The most basic form of phreaking is using stolen telephone credit card numbers to masquerade as a legitimate subscriber,

---

122 David H. Freedman and Charles C. Mann, "Cracker," *U.S. News and World Report*, June 2, 1997, pp. 57-65.

123 Ibid.

which reduces the phreaker's or hacker's chances of detection. Many hackers and phreakers also trade stolen access codes for updated technical information, hacker software tools, or other useful materials. Several of the more famous hacker investigations conducted by Federal law enforcement have involved the discovery of stolen telephone access codes. Additionally, hackers and phreakers use stolen or cloned cellular telephones. Because cellular phones can be reprogrammed to present a false ID, hackers and phreakers use the phones to reduce their risk of detection.

Several of the more proficient hackers are employed by the very computer security industry that seeks to limit their access. The theory is that practical experience in identifying vulnerabilities is worth more than network security certification credentials. These individuals will presumably use their knowledge and skill to protect their employer's systems rather than to attack them. However, employers frequently discover that these employees are more likely to use their positions to further exploit the very systems they were hired to protect.

5.3 Probing for Vulnerabilities

Hackers use various tools to find vulnerabilities on computer networks. The most famous of these tools is the Security Administrator Tool for Analyzing Networks (SATAN). SATAN is an easy-to-use tool with a graphical user interface that allows a user to remotely obtain data on system vulnerabilities, network topologies, network services, types of software being run, and system hardware. In July 1998, the CERT Coordination Center announced that an improved hacker tool for widespread scans had become widely available on hacker Web sites. This tool scans a wider variety of vulnerabilities than many of the older tools, including *statd* vulnerabilities, Internet Message Access Protocol/Post Office Protocol 3 (IMAP/POP3) vulnerabilities, Berkeley Internet Name Domain (BIND) vulnerabilities, cgi-bin vulnerabilities, and vulnerabilities associated with X11 servers and NFS filenames.[124] Intruders are also continuing their efforts to exploit dial-up port and modem vulnerabilities. The use of remote access control systems on modem pools has decreased the war-dialer threat to many organizations, but unauthorized modems on individual workstations and maintenance ports can often be found despite these measures. Once attackers enter a system, they can install a war-dialing program and use a dial-up modem attached to the compromised system to search for other dial-up ports.[125]

---

124 CERT Coordination Center, *CERT Incident Note IN-98.02: New Tools Used for Widespread Scans,* July 2, 1998, www.cert.org/incident_notes/IN-98.02.html.

125 Steve Branigan, "Hacker Trends: '96 Version," Presentation to the NSTAC Information Assurance Task Force, June 1996.

Hackers study Government and private sector network architectures, satellite access communications and data links, encryption standards, and security practices to increase their knowledge and skills. Evidence of this type of activity surfaced with the MOD attack on the Defense Information Systems Network (DISN) in October 1997.[126]

Hackers continually attempt unauthorized intrusions into Government systems to locate system vulnerabilities. They also know that with some additional research on operating systems, along with selected social engineering of targeted networks, they can sharpen their skills and crack into more secure systems. A vulnerability in one system can give a hacker access to all the trusted networks connected to that one poorly guarded system. Within DoD, some officials are managing this problem by using firewall systems to protect closed and encrypted networks and guard against the possibility of security lapses on other trusted networks and systems.

The Internet Network Information Center (InterNIC) registration service, which registers Internet domains and assigns IP network numbers, provides a convenient way to gain information on the type of computer and operating system used on particular Internet hosts. Intruders use this information to match potential targets with known vulnerabilities. Although this information is derived from the manual registration applications rather than through automatic updates, it illustrates how an intruder can use a legitimate network tool to achieve unauthorized access to information systems.[127]

In addition, several hackers have conducted reverse engineering studies on several Internet Web server and Web browser products. These studies are conducted not only for the purpose of hacking, but also to extort thousands of dollars from software companies in exchange for full disclosure on an identified vulnerability. In several cases, the hackers have had some limited success in collecting their requested fees.

5.4 Availability of Tools

New developments in computer technology have lowered the barrier to entry and resulted in an increase in attacks by novice intruders. "We've seen many cases of individuals with absolutely no idea what they're doing using very sophisticated methods to break into systems," said

---

126 James Glave, "Have Crackers Found Military's Achilles' Heel?" World Wide Web, *Wired News* Online, www.wired.com/news/news/technology/story/11811.html, April 21, 1998.

127 Mitch Wagner and Gary H. Anthes, "Underground Tools Aid Fledgling Hackers," *Computerworld*,November 13, 1995.

Professor Gene Spafford of the Computer Operations Audit Security Technology laboratory at Purdue University. Within a few years, computer intrusion tools will be available for automatically searching the Internet and other possible targets to scan for interesting information, to retrieve information, and to avoid detection during the intrusion process. In the future, a single intrusion tool may locate, download, and install information as easily as the software available on the Internet.[128]

In the past, hackers used trial-and-error methods to exploit vulnerabilities until they discovered further vulnerabilities. Although hackers still use this method, beginners no longer need to spend hours testing and probing a system, primarily because of the advent of advanced hacker software tools. Some tools are robust enough to gather and record potentially exploitable avenues of approach on a target system. Table 5-1 presents some sophisticated software tools and their uses.

Table 5-1: Examples of Electronic Intrusion Software Tools

ELECTRONIC INTRUSION SOFTWARE PROGRAM	FUNCTION	CAPABILITY
Patch programs (e.g., *login/patch; netstat patch; bin/ls patch; Is patch; df patch)*	Allows a hacker to create backdoor access, hide files, hide presence on system, gain access by spoofing originating address.	Assists in covering a hacker's tracks while in the system, getting back in, and removing any trace after the hacker is gone.
Root access identification program (e.g., *Pinga*)	Allows a hacker to gain superuser status after login to the network.	Enables a hacker to view, alter; and install files anywhere on the network host.
Security vulnerability assessment software (e.g., *SATAN, Rootkit, Cracker Jack)*	Tests and probes a system to identify potential vulnerabilities for gaining access to the system.	Enables a hacker to gain unauthorized superuser access to a system.
Sniffer programs (e.g., *sni256*)	Captures first strings of login information. These strings usually contain user ID and password.	Enables a hacker to identify authorized user accounts so future access will not be suspect.
Current user log program (e.g, *Zap*)	Deletes a hacker login name from the log and listing of current users.	Prevents systems administrators from discovering a hacker's presence on the system.

128 Mitch Wagner and Gary H. Anthes, "Underground Tools Aid Fledgling Hackers," *Computerworld*, November 13, 1995.

5.5 Conclusion

Governments, companies, and individuals are becoming acutely aware of the damage from and cost of hacker intrusion incidents. As Internet connectivity increases and the global information infrastructure grows, incidents of computer intrusion will also increase. The FBI and international security organizations view economic espionage and computer crime as a growing problem. Hackers will most likely continue to develop sophisticated tools to support their activities. Their methods and techniques to gain unauthorized access to networks and systems can be expected to keep pace with advancing technology.

Hackers can undoubtedly be expected to continue attacking poorly protected systems. However, even the most robust OSSs and security practices cannot stop all intrusion attacks. As adversaries and economic competitors develop sophisticated IW programs, deterrence and vigilance against attack are important. Much of the recent publicity on theft of economic data and industrial espionage may spur additional hacker interest in acquiring valuable information. Hackers are likely to continue to launch denial-of-service attacks, disrupt operations, and collect data for economic gain. Therefore, it is vital that organizations continually test the integrity of their firewalls and OSSs. The risks to systems may increase as hackers become increasingly familiar with encryption methods, telecommunications architectures, SCADA systems, and SATCOM.

6 THE INSIDER THREAT TO TELECOMMUNICATIONS AND INFORMATION SYSTEMS

"Despite the media attention given to big name hackers and crackers, insiders are more dangerous to corporate and government computer systems."[129]

RSA Data Security Conference

6.1 Introduction

Although it is recognized that telecommunications and information systems are exposed to both external and internal threats, the insider threat to these systems is largely misunderstood and underestimated. In today's highly integrated network environment, insiders have greater access to proprietary information and mission-critical systems. With their knowledge of corporate network security practices and their access to corporate facilities, insiders have increased opportunity and capability to do harm. Malicious insiders who exploit their access to a company's telecommunications and information systems can have a devastating impact on the organization's network operations or its bottom line.

6.2 Insiders

For the purposes of this report, the term insider refers to those who exceed or abuse access to corporate resources to exploit, attack, or otherwise adversely affect corporate telecommunications and information systems. The definition of an insider encompasses an organization's direct employees and the employees of many of its business affiliates. It includes an organization's full-time, part-time, and temporary employees, contractors, business partners, network-connected competitors, vendors, and customers. As organizations change the way they conduct business, distinctions between their facilities, networks and information systems and those of their contractors, vendors, business partners, customers, and competitors are increasingly blurred. The dramatic changes in the business environment have led organizations to extend access privileges to people outside of their organizations.

The factors contributing to an insider's decision to exploit or attack a corporation's telecommunications and information systems are varied and complex. Primary motivational factors include revenge, financial gain, and fear. Malicious insiders may act alone, or in collusion with outside individuals or organizations (e.g., free-lance hackers, competitors, criminals, terrorist organizations, and foreign government organizations). Because of their institutional knowledge and authorized access to critical systems,

---

129 "FBI: Insiders More Dangerous Than Crackers," *Net Insider, http://www.newdimensions.net/fbi.htm January 15, 1998.*

insiders are attractive targets to outside individuals or organizations seeking to obtain proprietary information or to compromise key systems. These outside groups will recruit insiders using the primary motivational factors. They will try to appeal to an insider's desire for revenge or financial gain, or they will use fear tactics to coerce the insider to comply with their demands.

6.3 Profile of a Malicious Insider

There are six basic categories of malicious insiders: disgruntled employees, paid informants, compromised or coerced employees, former employees, "pseudo" employees, and business associates. The categories described below are not necessarily mutually exclusive; a disgruntled employee could also be a paid informant. Similarly, malicious insiders are likely to be influenced by more than one motivational factor; paid informants may attack for both financial gain and fear of reprisal from their cohorts.

- **Disgruntled Employees.** Disgruntled employees believe they have been treated unfairly by their employer. They may believe that they are underpaid, are not respected by peers or superiors, or have been unjustly denied promotion.
- **Paid Informants.** Paid informants sell information to information brokers, industrial spies, criminal organizations, and intelligence services.
- **Compromised or Coerced Employees.** Employees may be compromised by their experiences or by personal connections. They can be coerced through threats of harm to themselves or their family or friends.
- **Former Employees.** Former employees may retain the ability to access computer systems in their former organizations, and they are knowledgeable of security countermeasures and system vulnerabilities. Former employees may know user and password combinations, retain access to corporate buildings, and be able to defeat security measures such as dial-back modems. Additionally, former employees often maintain relationships with their former co-workers, which gives them the opportunity to discover changes in security procedures, personnel, and organizational structures.
- **"Pseudo" Employees.** Pseudo employees are a creation of the new corporate workplace, which relies on a temporary workforce, outsourcing, and partnerships with other companies. These arrangements often require organizations to open their facilities and telecommunications and information systems to individuals who may perform work for the company, but who are not employed directly by the company. In this new environment corporations do not control hiring, supervision, or general security policies; this increases

the risk associated with the insider threat. Pseudo employees may have the same knowledge of, and access to, systems and information as a company's actual employees, without being subject to the same scrutiny.

- **Business Associates.** Today's business environment has created other insiders: a company's vendors, customers, and its competitors. These groups may be given limited access to corporate telecommunications and information systems to facilitate efficient network operations. Consequently, a malicious insider may have further opportunities to exploit their access to those systems.

6.4 Methods of Attack

Insiders understand their organization's culture and its security policies, which allows them to identify the organization's weaknesses and leverage their position to obtain or compromise sensitive information. They are likely to have specific goals and objectives in exploiting or attacking telecommunications and information systems. Using their knowledge of the target system, organizational security practices, and plausible access requirements, insiders can exceed or abuse their access privileges with limited risk of detection.

Insiders use various of methods to attack telecommunications and information systems, ranging from social engineering to hacking. Their attacks differ in nature and scope, and can affect all systems. Insiders usually carefully plan and meticulously execute their attacks over a period of time. They use their familiarity with the institution and personal relationships with their co-workers to identify valuable targets and analyze methods to access systems. The insider may impersonate another employee with appropriate authorization, or use account information obtained from others, to gain surreptitious access to systems. A more technically sophisticated insider will use hacking techniques to overcome or avoid access controls. In addition to mounting their own attacks, insiders may use one of the many automated hacker tools available via the Internet. These tools allow less sophisticated attackers to use highly scripted and pre-fabricated programs to abuse or exceed their access privileges.

Insider attacks are most likely to be focused on systems or proprietary information that the insiders are most familiar with or have worked with previously. Because they understand these systems, they can readily identify pertinent information and easily manipulate the system to gain access to it. In addition, insiders often resent the systems they have worked with, especially if they were replaced or feel less important as a result of increased efficiency attributed to that system. Insiders compromise corporate systems in many different ways, including, but not limited to: stealing proprietary information, adversely affecting the system operations, or installing malicious programs that can be activated at a later time to affect system operations.

6.5 Conclusion

The growth of the insider threat is influenced by three factors: opportunity, capability, and motivation. As these three factors increase, insiders are more likely to exploit their access to corporate systems or information for personal gain or to seek revenge on the corporation.

Today's corporate environment continues to move towards increased outsourcing, international operations, and competitive business relationships. Consequently, the level of corporate interconnectivity has grown dramatically. Insiders have more direct access to critical corporate systems and resources, increasing their opportunity to perform malicious acts. Furthermore, the insider's capability to do harm is increasing, in terms of both skills and tools. The technical skills possessed by employees in general are becoming more advanced. In addition, there has been a significant increase in the power and sophistication of the hardware and software comprising corporate information systems, communications, and network analysis tools. These tools provide a powerful capability for the insider to do substantial damage to corporate telecommunications and information systems.

The final, and perhaps the most significant, factor affecting the growth of the insider threat is motivation. Employees no longer feel the same level of affiliation with their companies. Previously, mutual respect and loyalty was established through job security. Employees could reasonably expect to work at the same company for their entire career. In today's society, however, employees frequently move between jobs and companies, more work is being outsourced, and corporate downsizing is making employees uneasy about their futures. In this environment, employees are more likely to become disgruntled or motivated to commit malicious acts.

The technological, economic, and social conditions that have led to today's business environment are likely to persist, increasing the insider threat and posing new challenges for security professionals. As the insider threat grows, corporations will have to devote increased resources to manage it and to curb the potential corresponding increase in corporate espionage.

7 THREAT ANALYSIS

Information infrastructures are vulnerable to attack. While this in itself poses a national security threat, the linkage between information systems and traditional critical infrastructures has increased the scope and potential of the information warfare threat. For economic reasons, increasing deregulation and competition create an increased reliance on information systems to operate, maintain, and monitor critical infrastructures. This in turn creates a tunnel of vulnerability previously unrealized in the history of conflict.[130]

Defense Science Board

7.1 Introduction

This section analyzes the electronic intrusion threats discussed in the report and the implications these threats have for NS/EP telecommunications and information systems. The analysis is based on two dimensions of the threat: the source of the threat and the capabilities demonstrated by these sources. Each threat source has its own intentions, targets, resources, and approaches. For example, a threat source such as an independent hacker, with the objective of achieving notoriety, might intrude on a telecommunications company's systems and cause an extensive outage that would warrant a report on the evening news. A threat source such as an FIS may also target that same telecommunications company, but with the objective of obtaining proprietary information. In this case, the FIS would be unlikely to launch a denial-of-service attack because it would need to maintain a low profile to prevent the target from realizing that its protection mechanisms had been breached. Under different circumstances, however, the intent and approaches of these two threat sources could be reversed. For example, an independent hacker may be hired by one company to conduct industrial espionage against another; a denial-of-service attack would not accomplish this objective. During wartime, an FIS may need to destroy its enemy's communications capabilities by launching a denial-of-service attack on its telecommunications networks; in this case, maintaining a low profile would not be a consideration. These examples demonstrate that although various threat sources may have similar capabilities, and perhaps even similar targets, they often exercise those capabilities differently, depending on the circumstances, with significantly different results. The implications electronic intrusion threats have for NS/EP telecommunications and information systems can be equally diverse; consequently, all such threats must be taken into consideration, whether the source is an FIS, a terrorist group, an organized crime group, an economic competitor, or an independent hacker.

---

130 *Report of the Defense Science Board Task Force on Information Warfare-Defense (IW-D)*, Washington, DC: Office of the Under Secretary of Defense for Acquisition and Technology, November 1996, p. ES-1.

7.2 Threat Sources

This section summarizes the basic sources of threat - FISs, terrorist groups, organized crime groups, economic competitors, and independent hackers - and what is known about their intentions, targets, and approaches. As noted previously, there is a significant overlap among these sources of threat with respect to their intentions, targets, and approaches.

7.2.1 Foreign Intelligence Services

Foreign states involved in electronic intrusion seek data from the United States and other nations on political, economic, military, and commercial capabilities and intentions. The aims, methods, and level of institutionalization of such electronic intrusion programs vary widely. Some states use electronic intrusion as part of broader, more destructive IW efforts; others use it primarily to accomplish more limited objectives, such as intelligence collection. The formal IW programs of some countries include developing the ability to attack computer and communications transmissions for the financial, transportation, and energy sectors. Countries with a high level of interest in IW include not only adversarial nations (e.g., Russia and China), but also friendly nations (e.g., Japan). Interest in stealing trade secrets and intellectual property is not limited to countries with more formal IW programs: while Russia, China, and Japan share this objective, so does France. Bulgaria plays a role in electronic intrusion by serving as a breeding ground for viruses, which electronic intruders can use to damage critical systems. Several countries (e.g., Russia, China, and Germany) have shown interest in using computer viruses to attack critical systems.

Evidence indicates that the threat from FISs is growing. Governments of at least 23 countries are targeting U.S. firms. The 1997 survey by the American Society for Information Science (ASIS) noted that high-tech companies are the most popular targets of foreign countries.[131] Other frequent targets are the manufacturing and service industries. The most lucrative information obtained includes research and development strategies, manufacturing and market plans, and customer lists.[132] It should not be assumed that FISs are the only source of economic espionage. Such proprietary information would have the same value to domestic companies in competition with each other as it does to FISs and may lead them to engage in similar activities.

---

131 National Counterintelligence Center (NACIC), *Annual Report to Congress on Foreign Economic Collection and Industrial Espionage,* Washington, DC: USGPO, October 1997, p. 18.

132 Jack Nelson, "U.S. Firms' '97 Losses to Spies Put at $300 Billion," *Los Angeles Times, January 12, 1998.*

7.2.2 Terrorist Groups

Terrorist groups are increasingly adept at using electronic information systems and advanced technologies. Some groups have created Web sites to publicize their perspectives. They are aware of U.S. dependency on complex infrastructures and have been known to recruit hackers or privileged insiders to attack information systems.[133] Terrorists who engage in IW may attack U.S. interests directly and indirectly. The use of the Internet and other information systems can give such groups a global and near-real-time command and control communications capability. They may use information technology and IW techniques to conduct propaganda campaigns and raise funds to support their other activities, or they may seek the "disruption or destruction of the information infrastructure, including basic services such as power supply, police databases, social security transfers, medical networks, transportation signals, money transfers, and telephone switching systems."[134] Because terrorist groups have limited resources, and electronic intrusion can help them achieve their objectives at minimal cost, it is expected that their use of this method to further their goals will increase.

7.2.3 Organized Crime Groups

A growing number of criminal organizations are targeting computer systems to commit fraud, acquire and exploit proprietary information, and steal funds and securities transmitted through electronic commerce systems. Russian organized crime has proved particularly skilled in using computers for bank fraud and has been implicated in electronic fraud cases in Russia, the United Kingdom, Germany, the Netherlands, Hong Kong, and the United States. As electronic commerce grows, criminal organizations can be expected to target such systems for the same reason bank robbers target banks -because that is where the money is. Organized crime groups also use electronic intrusion to hinder police investigations. They collect intelligence on police activities, destroy or alter data on investigations, and monitor the activities of informants.

7.2.4 Hackers

In the past, hackers were motivated primarily by intellectual curiosity about computer systems and network operations. While they might steal

---

133 President's Commission on Critical Infrastructure Protection (PCCIP), *Critical Foundations: Protecting America's Infrastructure,* Washington, DC: USGPO, October 1997, p. 18.

134 Andrew Rathmell, Richard Overill, Loranzo Valeri, and John Gearson, "The IW Threat From Sub-State Groups: An Interdisciplinary Approach," (Paper presented at the Third International Symposium on Command and Control Research and Technology), World Wide Web, Infowar.com Webpage, wysiwyg://53/http://www.infowar.com/mil_c4i/icsa/icsa3.html-ssi, June 17-20, 1997.

telephone service for their own use, they were unlikely to engage in more serious criminal activities. Although their actions to steal telephone service or explore telecommunications systems may have inadvertently damaged a network element, they were generally not intent on disrupting the telecommunications networks that gave them access to the computer systems they were so interested in exploring. In contrast, today's hackers appear to be motivated by a broader spectrum of factors beyond mere curiosity (e.g., greed, revenge, politics) and their actions have become more malicious, to include attacks against not only the information systems connected to the PN, but also the PN itself.

Even hackers whose intent is not malicious pose a threat. For example, today's automated attack tools will allow novices to achieve objectives that only expert hackers could have achieved 15 years ago. Although these novices are far less knowledgeable than the hackers who developed the tools, they can use these tools to gain access to a system about which they know little, or nothing, and may damage it inadvertently. In addition to the possibility that these "casual" hackers may cause unintentional damage, their activities generate "noise," which can mask malicious activity. Organizations are forced to respond to these nuisance attacks when their resources could be used more effectively to improve protection measures and respond to malicious attacks. Finally, hackers often freely share vulnerability information and intrusion tools, which increases the problem of casual hacking and helps malicious intruders achieve their objectives.

7.2.5 Insiders

Insiders pose a significant threat to telecommunications and information systems. Because they have legitimate access to proprietary information and mission-critical systems and are familiar with corporate security practices, they have a greater opportunity and ability to do harm than outsiders, and are less likely to be detected. However, this aspect of the threat is largely misunderstood and underestimated, in part because dramatic changes in the workplace have created a new definition of "insider." In the past, this term meant "employee." Today, contractors, vendors, business partners, customers, and even interconnected competitors may have access to an organization's sensitive information and critical systems. While these new insiders have access to critical resources, their activities are less visible to the organization than those of its employees, and far more difficult to monitor and control. In addition, most media reports about electronic attacks on telecommunications and information systems describe intrusions from sources such as foreign governments, terrorist groups, organized crime, and hackers, further contributing to the impression that the threat comes primarily from outsiders. Intrusions by insiders are less likely to be reported in the media for two reasons: (1) organizations can respond more directly to an internal threat–they can terminate employees and contractors–and may not need to seek outside help to deal with the problem; and (2) organizations are extremely reluctant to reveal insider malfeasance for fear that their customers will lose confidence in them.

Malicious insiders can be motivated by many factors, such as revenge or financial gain. They may act alone or in collusion with outsiders, who may try to appeal to the insider's desire for revenge or financial gain. In some cases, outside groups may coerce an insider to assist them by threatening the insider or his family. In the past, employees had a degree of loyalty to their employers, with some expectation that their employers would reward this loyalty with job security. In today's business environment, with its increased competition, corporate restructuring, and mergers, employees have seen their benefits reduced and their opportunities for advancement diminished. Even competent, experienced employees with years of service to their company know they can lose their jobs, often without warning. This gives insiders greater motivation to seek revenge, as well as greater motivation to take advantage of opportunities for financial gain, and less reason to be loyal to their employers. This environment provides increased opportunities for external threat sources (e.g., FISs or organized crime groups), which have always recognized the advantages of recruiting insiders. As external threats become more interested in electronic intrusion, they are likely to increase their efforts to recruit insiders who can help them gain access to critical information and systems; diminished employee loyalty may make their recruitment efforts more successful. These factors suggest that the threat from insiders may increase.

7.3 Capabilities

It is reasonable to suggest that there is a difference among the threat sources with respect to the financial resources that may be available to them, e.g., FISs generally have the greatest resources, individual hackers the least, with organized crime and terrorist groups falling somewhere in between. In the more traditional threat arena, those with the greatest financial resources generally possess the most powerful weapons and capabilities: nation states may have nuclear warheads in their arsenals, while disgruntled individuals may have to settle for dynamite. It is also generally true that the more powerful the weapon, the more complicated it is to create and deploy; e.g., before an adversary can launch an intercontinental ballistic missile, it must create the infrastructure to design, maintain, manage, and launch it. This is not the case for weapons used to launch electronic intrusion attacks on telecommunications and information systems.

The financial resources required to launch an electronic attack are minimal; the equipment is affordable and readily available in retail stores and through mail order catalogues: In fact, much of the equipment can be found today in the homes of the millions of individuals, here and abroad, who have personal computers and modems and are growing increasingly adept at using them. Just as the equipment to launch electronic attacks has become easier to acquire, rapid advances in technology have made computer applications easier to use. For example, 5 years ago, someone who wanted to use the Internet to download a document first had to know the document existed, had to know its exact name and location, and then needed to execute numerous arcane commands to download it. Today, Internet search engines have become widely available. In under 5 easy steps, users can do a simple query, which will identify not just a single document, but several documents on the subject of interest. Similar advances have occurred in the development of user-friendly electronic intrusion tools. Electronic intrusion is no longer the exclusive domain of individuals with specialized knowledge and skills who are willing to sit at their computers for hours typing hundreds of cryptic commands. The cryptic commands have now been turned into automated programs with graphical user interfaces (GUI), so that attacks can be executed by almost anyone who can "point and click." Whereas 20 years ago, a lone hacker would spend many tedious hours trying to guess a password to get into a single system, today automated tools enable the least experienced hacker to identify multiple vulnerabilities in multiple systems in a matter of minutes. Ten years ago, hackers would share information on vulnerabilities and attack methods on a one-to-one basis, or more broadly, at hacker conferences or by posting it on restricted-access bulletin board systems. Today, they use the Internet to share such information much more broadly and quickly. This rapid dissemination of information often results in multiple similar attacks after a new vulnerability is discovered.

As a consequence of the ready availability of equipment and automated tools, powerful weapons for electronic intrusion into telecommunications and information systems are as accessible to a disgruntled individual as they are to a well-funded nation-state. This "equal access to powerful weapons" increases the potential danger from the threat in two ways: (1) there are more well-armed adversaries and (2) those adversaries are increasingly unpredictable. Foreign adversaries, terrorist groups, and organized crime groups are exploring the possibilities new technologies offer, and their objectives and approaches are undergoing dramatic changes, making it increasingly difficult to deduce their targets and intentions. Individuals with personal motives are even less predictable. This inability to anticipate targets and intentions diminishes the capability to protect against specific threats and respond to attacks, which may increase the impact of electronic intrusion attacks against telecommunications and information systems.

Another factor that increases the impact of any type of attack, whether electronic or physical, is information about the target. With knowledge of the target's mission critical facilities and an understanding of its vulnerabilities, an adversary can determine which elements will yield the greatest result and how best to attack those elements. For example, in traditional battle, destroying the enemy's command and control facilities will have a far greater impact than an attack on almost any other single element. An understanding of how those facilities are configured and protected would allow an adversary to identify vulnerabilities and develop the most effective attack plan. These same principles apply to an electronic intrusion attack against telecommunications and information systems. Much of the information on network architecture and system configurations is publicly available. Although proprietary information may not be *legitimately* available, intruders have demonstrated that this is not an insurmountable obstacle. As noted above, general information on the vulnerabilities of these systems is increasingly common knowledge, and scanning tools can readily identify vulnerabilities of specific targeted systems. Such critical information would be useful to those whose objectives might be to launch a carefully planned, well-targeted, multi-faceted attack.

7.4 Implications of the Changing Threat

The growing number of households in the United States with personal computers, modems, and Internet access reflects a change in the way our society depends on telecommunications and information systems. This dependence has grown far more rapidly than our understanding of its implications. While the advantages of this technology are readily apparent and fully embraced, its dangers are far less obvious. These dangers manifest themselves in ways that are inconsistent with our understanding of, and experience with, traditional threats.

One aspect of the electronic intrusion threat that differentiates it from the traditional threat is that frequently there is no clear point at which the target can definitively determine that it has suffered an attack. The intrusion may not be detected at all. If the intent was just to gather information, rather than cause harm, there may be no symptoms of the intrusion. Or, an undetected intruder could insert malicious code that would be executed at a specified date and time, or under certain conditions. If those "certain conditions" never occurred, the victim would never know about the intrusion. Those "certain conditions" may occur only in critical situations, when users depend on the system the most. For example, an intruder could insert malicious code in an E-911 system so that it would only malfunction if the call volume reached a certain level. Even if an anomalous event or situation is detected, it is often difficult to determine whether it was caused by an inadvertent software error, an unintentional human error, or an intentional malicious attack.

As noted earlier, the "equal access" to powerful, user-friendly attack tools is another factor that differentiates the electronic intrusion threat from the more traditional threat. This factor makes it difficult to prioritize threats and concentrate resources on protecting against those that present the greatest danger.

Even when an intrusion is detected, and is clearly an intentional malicious attack, it is often difficult to identify the source and the ultimate objective. Is the intruder working alone, or are several individuals working together? Is the attacker merely trying to block calls to a radio station so he can be the "ninth caller" and win the prize, or is it a member of an organized crime group trying to determine whether the FBI has a wiretap on his line? Is it just a couple of teenagers on a "joy ride" or is it the FIS of a nation with which our country is involved in an international crisis? Did the attack come from abroad or did it originate domestically? Without knowing the source, it is difficult to know how to (or whether to) respond or retaliate. Without knowing the intent, it is difficult to determine where to concentrate protective measures.

In the past, the most logical source of threat from foreign sources was assumed be a nation's declared enemies, and the most logical target of that threat was a nation's military resources. These assumptions are not valid with respect to the electronic intrusion threat. While it is no surprise that countries such as Russia, China, Cuba, and Iraq have targeted our country's communications and information infrastructure, friendly nations such as France and Israel have done so as well. Furthermore, a significant target of both adversary and friendly FISs has been proprietary, rather than classified, information.

The same factors that make the electronic intrusion threat difficult to understand make it difficult to assess. The penultimate question is, "Is it getting worse?" There are a number of indicators to consider in attempting to answer that question. The target's value is growing in proportion to our dependence on this technology, which increases the motivation to attack these systems. There is substantial evidence regarding the powerful intrusion and attack tools that have been developed and instances in which they have been used, reflecting increased capabilities. As more advanced intrusion detection tools have been developed and deployed, these tools have demonstrated their effectiveness, resulting in an apparent increase in the number of intrusion incidents. However, without a valid baseline to establish quantitative measures of the increase in the number of intrusion incidents, it is difficult to gauge how much of this reflects a genuine increase in intrusion activities and how much results from increased awareness and more effective intrusion detection tools. Despite these uncertainties, however, there is a general sense that the threat is growing.

All of these aspects of the electronic intrusion threat to telecommunications and information systems pose significant challenges to protecting our infrastructure. These changes demand that both the public and private sectors reconsider their approaches to determining what is at risk, how to protect it, how extensively to protect it, and from whom to protect it.

7.5 Impact on the Nation's NS/EP Posture

NS/EP telecommunications and information systems are used to support Government operations to maintain a state of readiness to respond to and manage any event or crisis (local, national, or international). This includes military operations as well as civilian operations, such as preparing for and responding to natural disasters. The continuity of Government aspects of NS/EP operations include ensuring that the president always has essential communications capabilities, whether he is in the White House, on the campaign trail, or traveling abroad. NS/EP telecommunications services also support operations that involve protecting the health and welfare of the populace, which include such functions as E-911 service. NS/EP operations rely heavily on critical telecommunications and information systems, and indeed could not fulfill their missions without these systems. Clearly, threats with the potential to adversely affect the availability and reliability of our Nation's telecommunications and information systems can have a significant impact on its NS/EP posture and measures must be taken to address them.

8 COUNTERING THE THREAT

Many of the nation's critical infrastructures have historically been physically and logically separate systems that had little interdependence. As a result of advances in information technology and the necessity of improved efficiency, however, these infrastructures have become increasingly automated and interlinked. These same advances have created new vulnerabilities to equipment failures, human error, weather and other natural causes, and physical and cyber attacks.[135]

White Paper on Presidential Decision Directive 63

8.1 Introduction

As technology advances, there is a growing need for additional research, detection, prevention, and awareness programs that support efforts to protect against the threat of electronic intrusion. Several agencies and programs, including the NCS, are seeking greater cooperation within Government organizations and between Government and the private sector to create a better understanding and greater awareness of intrusion threats. Protecting the information systems that support telecommunications and other critical infrastructures will require issues to be addressed on a number of different levels. Extensive work on critical infrastructure protection has been performed in a relatively short time by Government policy-makers, law enforcement officials, defense and intelligence communities, academia, and the private sector. These efforts include expanded research and development activities for improving computer and information network security.[136]

As noted earlier in this report, the NRIC Security Subgroup concluded that the essential first step in strengthening the security of the PN is to develop a standard security baseline for interconnected data communications networks and gateways supporting the PN. The Security Subgroup recommended that standards be developed for access to signaling and operations support systems; that the carriers improve their capabilities to defend against, detect, and react to intrusions and fraudulent activities; that improved standards for SS7 gateway screening be developed; and that a certifying authority be established to develop security standards and effectively test for conformance to security standards.[137] The NCS can play a vital role in the development of security standards, information assurance

---

135 The White House, *Protecting America's Critical Infrastructures:* PDD-63, Washington, DC: The White House, May 22, 1998.

136 NSTAC, *INFORMATION ASSURANCE: A Joint Report of the IA Policy Subgroup of the Information Infrastructure Group and the NCM Subgroup of the Operations Support Group,* Washington, DC: NSTAC, December 1997.

137 NRIC, *Network Interoperability: The Key to Competition,* Washington, DC: ATIS, July 15, 1997, pp. 110–112.

recommendations, and joint Government-industry solutions to impede the threat posed by computer crime and computer intrusion attacks.

Although it is a considerable challenge to stay ahead of intruders in an environment characterized by tremendous growth in complexity, vulnerabilities, and potential threats, significant progress has been made in a number of areas to help organizations manage the risks to their information systems and networks. Comprehensive information system security programs can be used to deter, detect, mitigate, prevent, and respond to electronic intrusion attacks. However, to justify the expenditure of resources for such a program, awareness and information sharing are required to foster understanding and stimulate sufficient interest throughout the public and private sectors. Once an organization determines that an information system security program is warranted, it faces the dilemma of balancing the benefits of global interconnectivity against inherent threats and vulnerabilities. To achieve this balance, many organizations employ a risk management process to find the most cost-effective solution. When security objectives are determined and the information security program is established to meet those objectives, it is vital that policies and practices are consistently followed, reviewed, and updated.

8.2 Countermeasures

Several initial steps can be taken to increase the security of telecommunications and information systems that support NS/EP systems and the PN. *The Report of the Network Security Task Force* identified several areas where network security could be enhanced for the PN. Although the report was written in 1990, the concepts are still valid.

One problem area that the task force identified was implementation of security measures. They noted that while security holes had been identified, many had not been fixed.[138] This continues to be a problem today. Infrequent updates of antivirus software and sporadic application of security patches contribute to the vulnerability of NS/EP telecommunications and information systems. Additionally, while the task force recognized that the features to enhance the performance of telecommunications systems make security more difficult, they also recognized that there are steps that can be taken to help mitigate these vulnerabilities. These steps include a mix of technical controls and monitors, personnel practices, operational constraints and management commitment.[139]

---

138 The Network Security Task Force, *Report of the Network Security Task Force*, NSTAC, October 1990, p. 5.

139 Ibid, p. 6.

The task force identified the following actions to improve security:

- Conduct intensive security evaluations and audits
- Ensure dial-access control
- Use existing security features
- Deploy new security technologies
- Control proprietary information
- Improve security staff skills
- Establish security awareness programs
- Develop and implement a security network architecture
- Demand, build, and purchase secure systems
- Establish an effective incident response strategy.[140]

Although these measures have been accepted as necessary elements of network security for the PN, they are not always uniformly implemented. Unfortunately, one poorly secured asset may easily affect the security of many of the other assets in the PN. Therefore, it is important to ensure that there are also measures in place to address the security of the PN once an intrusion has taken place.

One recent incident involving America Online (AOL) highlighted the need for proper security precautions. A fake e-mail was sent to InterNIC, the organization that maintains the domain name registry for the Internet. In this e-mail, someone impersonating an AOL official requested that the electronic address for AOL's domain be changed. Because AOL had chosen the lowest of three security levels for this type of transaction, the change was made automatically, with no review. E-mail meant for addresses at AOL was automatically diverted to the new address, a smaller ISP. The new company's computers were quickly overwhelmed. An official from AOL was unable to explain why the company had chosen the lowest security level. Most companies choose one of the higher levels, which require either a password or encrypted messages to make changes to the address.[141]

8.3 Awareness

The first step in putting together any effective response to the growing range of threats and vulnerabilities is to establish awareness of the magnitude of the problem. An information security program is far more likely to

---

140 Ibid., pp. 8–15.

141 Leslie Walker, "Fake Message Sends AOL E-Mail Astray," *The Washington Post Online*, World Wide Web, http://search.washingtonpost.com/wp-srv/Wplate/1998-10/17/0551-101798-idx.html.

succeed if there is consensus among decision makers that the risks to the organization's bottom line make security a top priority. "Most businesses just don't want to spend money on a threat they don't understand," observes Richard Heffernan, a security consultant.[142] A number of efforts since 1995 have highlighted the information security problem and raised overall awareness of the critical issues:

- The hearings by the Senate's Permanent Subcommittee on Investigations on "Security in Cyberspace" in mid-1996 examined the vulnerabilities of the Nation's information infrastructure to the full range of threats–from the British teen who attacked systems at critical DoD research centers to the prospect of full-scale coordinated IW.
- In July 1996, President Clinton established the PCCIP to develop a strategy for protecting and ensuring the continued operation of the Nation's critical infrastructures, including telecommunications, electrical power systems, gas and oil transportation, banking and finance, transportation, water supply systems, emergency services, and continuity of Government. In October 1997, the PCCIP published its recommendations in a report entitled *Critical Foundations: Protecting America's Infrastructure.*
- In July 1996, the Director of the FBI created the Computer Investigations and Infrastructure Threat Assessment Center (CITAC) to coordinate the criminal, counterterrorism, and counterintelligence authorities of the FBI relating to computer crimes and threats directed at the NII. CITAC's primary responsibilities included managing and coordinating computer intrusion investigations conducted by the FBI; identifying threats affecting the NII; increasing security awareness within the public and private sectors; and researching technology, legal issues, and policy affecting the FBI's ability to neutralize computer attacks. The DoD and other law enforcement organizations have initiated similar efforts.[143] The CITAC has been superceded by the National Infrastructure Protection Center (NIPC) under the terms of PDD-63.
- In its September 1996 report evaluating information security at 23 agencies, the U.S. General Accounting Office's (GAO) principal recommendation focused on the need for "increased awareness of the importance of information security, especially among senior agency executives."[144] In a subsequent report on information security in Sep-

---

142 Rochelle Garner, "The Growing Professional Menace,"*Open Computing Magazine*, July 1995.

143 Federal Bureau of Investigation (FBI), Office of Computer Investigations and Infrastructure Protection (OCIIP), *Computer Investigations and Infrastructure Threat Assessment Center (CITAC)*, Washington, DC: FBI, October 1997, pp. 1–2.

144 USGAO, Information Security: Opportunities for Improved OMB Oversight of

tember 1998, GAO studied 24 Federal agencies and identified significant information security weaknesses that placed a broad range of critical operations and assets at great risk of fraud, misuse, and disruption.[145]

- In November 1996, the Defense Science Board's Task Force on Information Warfare-Defense found that the threat posed by IW is not limited to the realm of national defense, and the effort to control the problem must encompass broader national security interests, including Congress, the civil agencies, regulatory bodies, law enforcement, the intelligence community, and the private sector. Among the task force's recommendations was the need for DoD to designate an accountable focal point for IW, to increase awareness, and to "raise the bar" to potential attackers by adopting some low-cost, high-payoff measures such as better access controls and escrowed encryption of critical data assets.[146]
- In early 1997, during an NSTAC executive session, U.S. Attorney General Janet Reno and NSTAC principals discussed the need to develop a more effective approach for addressing cyber crime. The NSTAC principals agreed with the Attorney General's statement that partnership between Government and industry was essential for combating cyber crime and welcomed the opportunity to investigate a joint industry-law enforcement approach to this issue.
- In September 1997, the NSTAC hosted a transportation information infrastructure workshop to assess that industry's reliance on telecommunications and information systems and subsequently presented an interim report to NSTAC in December 1997. Identifying the need for further input from industry associations and a better understanding of intermodal transportation trends, the NSTAC planned to complete the transportation risk assessment for the NSTAC executive session in June 1999.
- In late 1997, at the next NSTAC executive session, the Attorney General and NSTAC principals addressed several issues that might need to be addressed in the context of a Government-industry partnership on cyber crime and information infrastructure protection: Freedom of Information Act issues; antitrust issues; the reluctance to share proprietary information; and the need to respond quickly to electronic

---

Agency Practices, GAO-AIMD96-110, Washington, DC: USGPO, September 24, 1996, p. 37.

145 USGAO, *Information Security: Serious Weaknesses Place Critical Federal Operations and Assets at Risk*,GAO-AIMD-98-92, Washington, DC: USGPO, September 1998, p. 5.

146 Office of the Under Secretary of Defense for Acquisition and Technology,*Report of the Defense Science Board Task Force on Information Warfare-Defense*,Washington, DC: USGPO, November 1996, p. 3-1.

intrusions. The Attorney General invited NSTAC members to meet with her at any time to explore how the Department of Justice could work more productively with industry to address cyber crime and other critical issues.

- In June 1998, the NSTAC and Government NSIEs sponsored a workshop on the insider threat to information systems. The workshop addressed the current state of the insider threat, in terms of capabilities and intent, the factors that exacerbate the insider threat (e.g., technology, corporate downsizing, legal restrictions), and the policies and best practices to protect against the insider threat. Attendees included representatives from Government as well as the telecommunications, power, financial services, and transportation industries.
- In October 1998, the NSTAC sponsored its third R&D Exchange in concert with the White House Office of Science and Technology Policy (OSTP) and the Purdue University Center for Education and Research in Information Assurance and Security (CERIAS). The purpose was to stimulate discussion among security technology practitioners from Government, industry, and academia on the need for security technology R&D collaboration. Discussions concentrated on four broad areas: national R&D priorities; the appropriate roles of Government, industry, and academia; obstacles; and alternative approaches to collaboration.
- In December 1998, the Office of the Manager, NCS, published Public Switched Network Best Practices: Security Primer. This is a high-level primer that identifies a set of guidelines and recommendations covering significant security-related topics, and provides a list of publicly available security reports that address subjects relevant to PSN protection. These documents contain policies, generic requirements, recommendations, and guidelines that help encourage and enforce sound security practices that can help service providers determine what to secure, how to secure it, what needs to be considered up front, what needs to be done on an ongoing basis, and a number of other vital factors.[147]

In addition to Government efforts, colleges and universities are stepping up efforts to educate students regarding the ethics of computer use. The University of Delaware administers a test on the university's computer-use policies before students receive a password to the network. Administrators say that educational programs have helped them curb the growth

---

147 Office of the Manager, National Communications System, Public Switched Network Best Practices: Security Primer, OMNCS, Washington, D.C. December 1998.

of "nuisance problems," such as stolen passwords and inflammatory e-mailings, enabling them to focus on more serious computer crimes.[148]

Whatever its negative implications, the growth of the Internet has also moved security issues to the forefront. Recent surveys of information systems managers have shown that most managers rate security as their number one concern.[149] The FBI estimates that 80 to 90 percent of the computer intrusions they investigate originate via Internet connection. The American Institute of Certified Public Accountants (AICPA), for example, listed Internet security as the number one technology issue for businesses in 1997, and Ernst & Young found that security was the leading concern among information systems managers deploying intranets for their organizations.[150]

8.4 Infrastructure Protection Guidance

The most complete guidance is provided by PDDs 62 and 63, dated May 1998. These directives address the threat posed by adversaries of the United States targeting key U.S. infrastructure systems. Increasingly, adversaries have come to realize that our national dependence on complex, interdependent infrastructure systems provides an opportunity to cripple the ability of the United States to project military force and ensure the security and economic welfare of its citizens. Because of the United States' unrivaled military superiority, these adversaries are most likely to use asymmetric means of attack.[151] Exploitation of infrastructure vulnerabilities is one of the most significant asymmetric threats. In response to the potential threat posed by asymmetric attacks by adversary nations, terrorists, and criminals, the President has determined that specific actions must be taken to deter attacks, disrupt the activities of potential attackers, protect critical infrastructures, and respond to asymmetric attacks.[152]

---

148 EDUCOM, "Beefing Up Computer Security Efforts on Campus," World Wide Web, Edupage Mailing List, www.educom.edu/web/edupage.html, September 25, 1997.

149 "Security Concerns Tops the List of Challenges in Developing Web Applications" Press Release, Milpitas, CA: Strategic Focus, March 1997; and Steve Gold, "Internet Security Major Concern for IT Managers," World Wide Web, Newsbytes Network, www.newsbytes.com, December 19, 1996.

150 American Institute of Certified Public Accountants (AICPA), "CPAs Name Top Ten Technologies for 1997: Cyberspace Security Tops List," World Wide Web, Great Plains Software, Inc., www.gps.com/waynes_web/Perspectives/topten.htm , January 16, 1997; and Network World Media Lounge, "Network World Industry Surveys: Study Sponsored by Network World and Ernst & Young," World Wide Web, www.nwfusion.com/medialounge/press/surveys.html#anchor1101060, March 31, 1997.

151 The National Defense Panel defined asymmetric threats to the United States as the ability of our adversaries to exploit their strengths in attacking our national weaknesses. National Defense Panel,*Transforming Defense: National Security in the 21st* Century,Washington, DC: US Department of Defense, December 1997, p. 11.

152 The White House,*Combating Terrorism: Presidential Decision Directive-62*,Washington,

PDD-62, *Protection Against Unconventional Threats to the Homeland and Americans Overseas,*emphasizes the growing threat of unconventional attacks against the United States. PDD-62 details an integrated approach to increase national effectiveness in countering asymmetric threats and managing the consequences of these attacks to limit the damage that they can inflict. The PDD establishes the National Coordinator for Security, Infrastructure Protection, and Counterterrorism, who will oversee a broad variety of policies and programs including the following: counterterrorism, protection of critical infrastructures, preparedness, and consequence management for weapons of mass destruction (WMD).[153]

PDD-63,*Critical Infrastructure Protection,*calls for a national effort to ensure the security of the increasingly vulnerable and interconnected infrastructures of the United States. These critical infrastructures include telecommunications, electrical power systems, gas and oil transportation and storage, banking and finance, transportation, water supply systems, emergency services, and continuity of Government services. PDD-63 requires immediate Federal Government action, which includes risk assessment and planning to reduce exposure to attack. This PDD emphasizes the importance of the partnership between the Government and private sectors.[154]

- ***National Coordinator for Security, Infrastructure Protection, and Counterterrorism.***The National Coordinator reports to the President through the Assistant to the President for National Security Affairs. The National Coordinator will provide budget advice and ensure interagency coordination for policy development, implementation, and crisis management.
- ***Lead Agencies for Sector Liaison.***For each of the critical infrastructures a single U.S. Government agency will serve as the lead agency in coordinating infrastructure protection activities with the private sector. Each lead agency will designate a Sector Liaison Official who will work closely with designated private sector coordinators to develop measures needed to eliminate critical infrastructure vulnerabilities and develop required protective measures.

DC: The White House May 22, 1998.

153 Ibid.

154 White Paper,*The Clinton Administration's Policy on Critical Infrastructure Protection: Presidential Decision Directive* 63May 22, 1998, www.CIAO.gov Additionally, PDD-63 created a national structure to coordinate the critical infrastructure protection activities of the Federal Government. The activities for which this national structure is responsible include:

- *Lead Agencies for Special Functions.*These agencies perform functions related to critical infrastructure protection that must be performed chiefly by the Federal Government (national defense, foreign affairs, intelligence, and law enforcement). For each of the special functions, a lead agency will be designated to be responsible for coordinating Government activities. Each lead agency will appoint a senior officer to serve as the Functional Coordinator for that function.
- *Critical Infrastructure Coordination Group (CICG).*The Sector Liaison Officials and the Functional Coordinators of the Lead Agencies, as well as representatives from other relevant departments and agencies, including the National Economic Council, will meet as the CICG to coordinate the implementation of PDD-63. The CICG will coordinate with existing policy structures with related functions.
- *The National Infrastructure Protection Center (NIPC).*The NIPC will provide a focal point for gathering information on threats to infrastructures, and will provide the principal means of facilitating and coordinating the Federal Government's response to an incident, mitigating attacks, investigating threats, and monitoring reconstitution efforts. The NIPC's mission will include providing timely warning of attacks, comprehensive analyses, and law enforcement investigation and response.
- *Information Sharing and Analysis Center (ISAC).*The National Coordinator, working in concert with the Sector Coordinators, the Sector Liaison Officials, and the National Economic Council, will coordinate with the owners and operators of the critical infrastructures to encourage the creation of a private sector information sharing and analysis center. The ISAC will serve as a mechanism for gathering, analyzing, sanitizing, and disseminating private sector information and information received from the NIPC regarding critical infrastructure protection, including information about threats, vulnerabilities, intrusions, and anomalies.
- *National Infrastructure Assurance Council (NIAC).*Based on advice from the National Coordinator, the Lead Agencies, and the National Economic Council, the President will appoint a panel composed of major infrastructure providers and state and local officials to enhance the partnership between the public and private sectors in protecting critical infrastructures.
- *Critical Infrastructure Assurance Office (CIAO).*PDD-63 calls for a national plan coordination office, which the Administration has designated the Critical Infrastructure Assurance Office (CIAO). The CIAO will integrate the various sector plans into a National Infrastructure Assurance Plan, coordinate analyses of the U.S. Govern-

ment's dependencies on critical infrastructures, and assist in coordinating a national education and awareness program.[155]

While not created by PDD-63, OSTP plays a vital role in coordinating infrastructure protection. Created in 1976, OSTP is responsible for coordinating research and development activities for infrastructure protection for the Government through the National Science and Technology Council. Federally sponsored research and development will be coordinated subject to multiyear planning, and will take into account research being done in the private sector.

8.5 Information Sharing

Keeping pace with the complex technical, political, and business issues involved in securing networks is a considerable challenge, even when security is given a top priority. Organizations often find they need to draw on the experience and expertise of others with regard to security issues. A number of formal and informal information sharing mechanisms have been established to address electronic intrusions and defenses.

Perhaps the oldest such mechanism, the International Information Integrity Institute (I-4), hosted by SRI International, was created in 1986 to allow senior information security professionals to share information and experiences about controls and practices in a confidential environment. Through surveys, interviews with attackers and their victims, and other investigations, I-4 has developed a database of more than 3,000 computer security cases that may be the most extensive resource of this type in the world.

As mentioned previously, the OMNCS and the President's NSTAC established and sponsor closely coordinated Government and NSTAC NSIEs that meet jointly to exchange information on intrusions and technical and legal developments. The NSIEs supplement their meetings with workshops and reports available to wider audiences.

The National Computer Security Association (NCSA) has sponsored a series of consortia aimed at gathering and disseminating knowledge and expertise within focused communities of interest, including developers of security products and users of financial and medical information systems. Groups such as the American Bankers Association and the Electric Power Research Institute (EPRI) are developing forums for discussion of security issues within specific industries. Other forums, such as a monthly lunch meeting of the security officers of the member banks of the New York Clearing House, have been organized but operate more informally.

---

155 The White House, *Protecting America's Critical Infrastructures PDD-63*, Washington, DC: The White House, May 22, 1998.

One criticism often leveled at these groups is that they share little or no information outside their limited membership. This limited sharing is a result of the delicate balance between confidentiality and disclosure that must be maintained for effective sharing of information in this sensitive area. Organizations are willing to discuss details of incidents and protection measures within a limited community defined by common interests and trust. Although others outside the process do not benefit from the information, larger audiences would tend to inhibit disclosure to the point that the real value-the details, the "war stories," the open discussions-would be lost.

Crime statistics and security surveys reflect an increase in the number of computer crimes reported to law enforcement. This increase, coupled with the increasing dependence of the Nation's infrastructure on computer technology, demonstrates a need to enhance law enforcement capabilities. From a Federal standpoint, several agencies have stepped up the war on computer intrusion. Examples of this include the establishment of the FBI National Computer Crime Squad and the Justice Department's Computer and Telecommunications Coordinator Program, which trains U.S. attorneys in computer and telecommunications issues.

Private industry can lend its expertise to law enforcement to assist in detecting potential infrastructure attacks. However, many researchers, vendors, and end users lack understanding about what constitutes an intrusion and the risks associated with intrusions. In many organizations, intrusion detection and reporting depends on an employee with some level of awareness or understanding of computer security. Intrusion detection technologies can play a central role for a strategic indications, assessment, and warning (IAW) capability that examines national threats.

8.6 Computer Incident Response Teams

Incident response is a critical function performed by security personnel within organizations and requires a plan for handling emergencies. The benefits of incident response include limiting economic losses from service disruptions and protecting sensitive or classified information.

Effective organization and security staffing are just as important as an adequate response plan. The GAO estimated in its report of attacks on DoD computers that less than 1 percent of intrusions are reported within the Government.[156] One reason many security incidents go unreported is that users are not familiar with the agencies they are supposed to report to. As the experiences of CERT at CMU and the Department of Energy's Computer Incident Advisory Capability (CIAC) show, having a reputable focal

---

156 USGAO,*Information Security.- Computer Attacks at Department of Defense Pose Increasing Risks*,GAO-AIMD-96-84, Washington, DC: USGPO, May 1996, pp. 20-21.

point for reporting and responding to incidents encourages users to come forward when they see problems. Unfortunately, as recently as March 1997, CSI found that more than 60 percent of the organizations it surveyed did not have a computer emergency response team in place.[157]

As Information Week reported in May 1996, the number of companies forming internal incident response teams to deal with viruses, hackers, and information thefts is growing. Not all of these teams require a full-time commitment of staff-Chevron, for example, relies on a team of expert volunteers who have management support to "drop whatever they're doing and fly to wherever they're needed."[158] IBM, SAIC, and other companies are now offering commercial incident response services for customers who prefer to outsource that function, and the Federal Computer Incident Response Capability (FedCIRC) program has been established to provide a similar fee-based service for Federal Government agencies.

To help these teams share their expertise and cooperate in reacting to and more important, preventing-security incidents, the Forum of Incident Response and Security Teams (FIRST) has put in place a number of services, including a secure notification system. Since its creation in 1993, FIRST's membership has continued to grow, with commercial teams accounting for the majority of the most recent members.[159]

In addition to FIRST's efforts, the Internet Engineering Task Force (IETF) has established a security incident response working group to provide guidelines and recommendations to facilitate the consistent handling of security incidents throughout the Internet. The guidelines will address the roles of vendors and response teams in assisting organizations in resolving security incidents.[160]

8.7 Technology

Enhanced versions of intrusion detection technologies are promising because they offer indicators that can be analyzed to assess the overall threat, they can mitigate or aid in countering attacks, and they can support the

---

157 CSI, "Computer Crime Continues to Increase, Reported Losses Total Over $100 Million," World Wide Web, CSI Homepage, www.gocsi.com/preleas2.htm, March 6, 1997.

158 Bob Violino, "Crime Fighters: Corporate SWAT Teams Battle Mounting Security Threats, "*Information Week*, May 13, 1996.

159 Forum of Incident Response and Security Teams (FIRST), "What is FIRST?" World Wide Web, DFN-CERT Webpage, www.first.org/about/,September 10, 1997.

160 Internet Engineering Task Force, Working Group on Security Incident Response, "Guidelines and Recommendations for Incident Processing," World Wide Web, www.cert.dfn.de/eng/resource/ietf/grip/home.html, 1997.

development of prevention measures. Table 8-1 shows the attributes and capabilities of enhanced intrusion detection.

Table 8-1: Characteristics of Enhanced Intrusion Detection Technologies

CAPABILITIES	
Detects a wide variety of intrusion types for many technologies	Provides real-time detection of intrusion
Provides a very high certainty for recognition	Provides a network-wide view rather than simply local views
Uses analysis techniques that work reliably with incomplete data	Detects unanticipated attack methods
Scales to very large heterogeneous systems	Knows what data to collect for maximum effectiveness; network instrumentation
Provides automated response	Discovers or narrows down the source of an attack
Provides an integrated picture of network management and fault diagnosis	Provides cooperative problem solving for inferring intent and forming the big picture

Source: Teresa Lunt, Intrusion Detection Briefing: A Tutorial, Washington, DC: Defense Advanced Research Projects Agency.

Although these technologies enhance the ability to monitor the system, it is often difficult for the system administrator to effectively manage the data produced by intrusion detection systems. With limited amounts of time and resources, systems administrators often find it difficult to analyze the large amounts of audit and alert data generated.[161] Future intrusion detection systems will need to incorporate technology capable of reducing and analyzing this vast quantity of information.[162]

Security assistance and public awareness of network intrusion is also provided through CERT, which was formed by the Defense Advanced Research Projects Agency (DARPA) in November 1988 in response to growing threats to distributed information systems. The CERT charter is to work with the Internet community to facilitate its response to computer security events involving Internet hosts, raise the community's awareness of computer security issues, and conduct research targeted at improving the security of existing systems. CERT products and services include 24-hour technical assistance for responding to computer security incidents, product vulnerability assistance, technical documents, and seminars. In addition, the team maintains a number of mailing lists (including one for CERT advisories) and provides an anonymous file transfer protocol (FTP) server

---

161 Intrusion Detection Subgroup, *Report on the NS/EP Implications of Intrusion Detection Technology Research and Development,*The President's National Security Telecomunications Advisory Committee: Network Group, December 1997, p 22.

162 Ibid.

info.cert.org, where security-related documents, past CERT advisories, and tools are archived.[163]

8.8 Legal

Although the virtual nature of cyberspace continues to complicate the application of laws and legal concepts developed in the physical world, law enforcement has realized some gains over the past few years. Feedback from the NSIEs and other groups led the Department of Justice (DOJ) Computer Crime Unit to propose a series of changes to U.S. Code 1030, which sets forth jurisdiction and penalties for unauthorized use of computers. These changes were incorporated into the NII Protection Act of 1996, which was signed into law in October 1996. The Act accomplished the following:[164]*"U.S. Code., Title 18, Chapter 47, Section 1030: Fraud and Related Activity in Connection With Computers," (USGPO CD-Rom prepared by the Office of the Law Revision Counsel of the House of Representatives), World Wide Web, Cornell University Law Page,* www.law.cornell.edu/uscode/18/1030.shtml, January 16, 1996.

- Expanded jurisdiction of Federal computer crime law to include any computer "involved in interstate or foreign communication"
- Lowered threshold for resulting damages (now $1,000)
- Increased penalties for intentional misuse
- Made it a misdemeanor for an authorized user to cause reckless damage, regardless of intent.

Cooperation between commercial firms and law enforcement in reporting, investigating, and prosecuting attackers has been improving-slowly.[165] Commercial firms hesitate to report incidents to law enforcement because they fear negative publicity and they lack adequate internal controls for detection and auditing. Unfamiliarity with the basic rules of evidence is also a contributing factor. However, the growth of commercial uses of the Internet is beginning to provide a strong financial motivation for cooperation. Major ISPs now routinely refer cases to local law enforcement and the FBI. In other cases, ISPs have successfully sued "spammers" for damages resulting from lost service or use of their systems.[166] While progress has been made to improve the effectiveness of computer crime laws, and victim companies are growing less reluctant to report to law enforcement, more needs to be done to improve these laws and prosecute those who break them.

---

163 CERT, SEI, CMU, *Intenet-Related Organizations,*World Wide Web, http://freebie.cfx/tin/P/9605.html.

164

165 Chris Nerney, "Getting Civil With Hackers," *Network World,*August 12, 1996.

166 Donald Schutt, "How Do you Handle Spammers,"*Web Week,* Volume 3, Issue 10, April 14, 1997.

8.9 Conclusion

Although the interest in countering the threat to electronic intrusion is nothing new, until recently this interest was generally limited to Government departments and agencies responsible for national security and their vendors and service providers. In the past few years, this has changed dramatically and rapidly. Today, this interest and concern about the electronic intrusion threat is shared by civil agencies as well as the defense and intelligence communities, and by the private sector as well as the Government. Furthermore, classified systems and information are no longer the only resources considered worth protecting; both the public and private sectors are increasingly concerned about protecting those systems supporting administrative and commercial endeavors and containing personal and proprietary information.

One of the earliest examples of this shifting focus occurred in early 1990, with the joint Government industry network security activities initiated by the OMNCS and NSTAC to address the electronic intrusion threats and vulnerabilities affecting the Public Switched Network. By 1996, evidence of an intensified level of interest in electronic intrusion was pervasive: the Senate held hearings on "Security in Cyberspace"; President Clinton established the PCCIP to develop a strategy for protecting the critical infrastructures; the Director of the FBI created the CITAC (subsequently superceded by the NIPC) to coordinate the FBI's response to computer crimes and threats; GAO published a report evaluating information security at 23 agencies; and the Defense Science Board published the report of its Task Force on Information Warfare.

These activities (among several others), and the level of awareness they have generated, provided the foundation for PDD-63,*Critical Infrastructure Protection*.Published in May 1998, PDD-63 emphasized the importance of the partnership between the Government and private sectors and created a national infrastructure to coordinate the Federal Government's critical infrastructure protection activities.

All sectors of society - Government, industry, academia, and private citizens -have become increasingly aware of the Nation's dependence on our critical infrastructures, and on the interdependencies among these infrastructures. This growing awareness has led to a number of measures to counter the electronic intrusion threat to telecommunications and information systems:

- There is an increasing interest in finding ways to share information about electronic intrusion incidents, and ways to prevent and respond to them.
- Legislation is evolving to better define computer crime and establish penalties that are commensurate with the harm such crimes can cause.

- The law enforcement community is becoming more knowledgeable about the technology involved, which is improving the ability to effectively investigate and prosecute these crimes.
- Better legislation and improved law enforcement capabilities, along with the growing concern about electronic intrusion, are making victims of electronic intrusion more willing to report incidents to law enforcement for prosecution.
- Awareness and concern is increasing the demand for technology to prevent, mitigate, and counter attacks, and Government and industry are focusing research and development efforts to respond to this demand.

While these measures are significant, and are headed in the right direction, countering the electronic intrusion threat will continue to be an uphill battle. The implementation of PDD-63 provides a significant opportunity to coordinate, and maximize the benefits from, diverse efforts to address the electronic intrusion threat. The joint OMNCS and NSTAC activities to address complex problems regarding communications for Federal NS/EP activities can both contribute to, and benefit from, the broader effort to protect the Nation's critical infrastructures.

APPENDIX A MALICIOUS SOFTWARE DESCRIPTIONS

A.1 Overview

Malicious software inserted into computers and information networks may have catastrophic effects. This type of software may cause a loss of productivity, lockup of systems, corruption of files, interference, alteration or loss of data, and even system crashes. Developing these programs requires minimal equipment, cost, or expertise. Malicious software is readily available through hacker Web sites, and new variants can be created using tool applications such as the Virus Tool Kit.

A.2 Trojan Horse

A Trojan horse contains hidden code that executes potentially malicious acts when triggered by an external event and is frequently used in network attacks. A Trojan horse can provide backdoor access to intruders who wish to gain unauthorized access. To insert a Trojan horse, an intruder enters the system to replace system utilities. The intruder then installs the Trojan horse program, which may contain instructions for recording passwords entered by legitimate users, installing a virus, collecting system connectivity information, or performing other malicious acts. Intruders have become adept at surreptitiously getting authorized users to download Trojan horses either in the form of a hostile Java applet, executable attachments to e-mail, or other network files.

A.3 Worm

A worm is a self-replicating program that moves from one system to another along a network. A worm does not destroy software or compromise data. Worms were originally developed to make use of unused network resources to run large applications programs. The worm scans the network for unused resources and uses them to execute programs in small segments. A worm can severely harm a network by using all available computing resources and saturating communications links, similar to a denial-of-service attack. When a worm attacks, the network must be shut down before it can recover, which is a costly and time-consuming process. The vulnerability to a networked environment was demonstrated by the notorious Morris Internet worm of 1988. This attack resulted in the disruption of service to thousands of computers and their users across the Internet.[167]

---

167 ZDNET, *The Internet News Channel: Online Users Need to Beware of Password Poachers*, World Wide Web, Ziff-Davis Publishing Co., www5.zdnet.com/zdnn/content/0620/zdnn0006.html, 1997.

A.4 Logic Bomb

A logic bomb is a program that lies dormant until a trigger condition causes it to activate and destroy the host computer's files. A logic bomb, which may be hidden within a Trojan horse or carried by a virus, can be programmed to target specific users or files. When activated, the program prevents the victim from responding in time to prevent the disruption. Insiders have frequently used logic bombs as a means to obtain revenge or a personal advantage.

A.5 Computer Virus

Computer viruses are self-propagating malicious programs or pieces of code that are installed on a computer without the user's knowledge.[168] Viruses attach themselves to legitimate programs or files. The virus becomes active when users access the infected program or file. Once active, the virus has two basic functions: replication and execution. During replication, the virus identifies and gains access to other programs or files it is capable of infecting. During execution, the virus may execute additional malicious code. The damage caused during execution can vary substantially. Relatively minor damage can be limited to a reduction in hard drive disk space as a result of the propagation of the virus. Severe damage can include altered boot sector files or the loss of all the information on a hard drive.

Viruses are language dependent—they can only infect those operating systems (OS) or programs that understand and can execute the viruses' code. Viruses are generally found on DOS, Windows (3.x, 95), and NT operating systems. However, there are also some UNIX and LINUX viruses.[169] Based on their target environment, viruses can be categorized into four groups: boot sector, file, macro, and network viruses.[170] Many viruses incorporate components of one or more types to increase their efficiency and to decrease the possibility of detection. For example, a virus that ultimately attacks the boot sector of a disk may incorporate functionality that allows it to propagate as an attachment to a file or to propagate over the network. Similarly, a virus that corrupts files may be propagated as a macro program embedded in the data of a macro-enabled file (e.g., Word, Excel, or postscript files).

---

168 PC Webopaedia,"*Virus*,"World Wide Web, http://webopedia.internet.com-/TERM/v/virus.html.

169 NIST, Information Technology Laboratory, Computer Security Division, *Internet Security Policy: A Technical Guide*, Gaithersburg, MD: NIST, July 21, 1997, p. 39.

170 Antiviral Toolkit Pro Virus Encyclopedia, "The Classification of a Computer Virus," World Wide Web, http://www.metro.ch/avpve/classes/classes.stm.

Sophisticated viruses incorporate stealth and polymorphic capabilities that enable the viruses to cover their traces from the OS.[171] A stealth virus intercepts OS read/write calls to the infected object and temporarily replaces the infected portion of the file with uninfected information. A polymorphic virus uses various encryption techniques to mask the identity and signature of the virus.

It is difficult to quantify the number of virus attacks, because of the lack of reported incidents. However, recent trends indicate that the number is increasing. A recent 1997 survey conducted by the National Computer Security Association (NCSA) concluded that 33 out of every 1,000 computers contain viruses in any given month. This was a substantial increase over a 1996 survey, which concluded that only 10 out of every 1,000 computers were infected.[172]

A.6 Bacteria

Many virus researchers consider a bacteria program a type of computer virus. A bacteria program does not need a host program to run. Bacteria acquire as much central processing unit (CPU) time as possible, which significantly slows the host system. Bacteria can also fill up disk space with copies of itself, thereby disrupting or disabling the operation of the network or system.

---

171 Ibid.

172 Tech Report, "Trojan Horses, Hostile Java Applets Target Home PC," *USA Today*, World Wide Web, USA National News, www.usatoday.com:80/life/cyber/tech/ctb177.htm, September 3, 1997.

APPENDIX B LIST OF ACRONYMS

AICPA	American Institute of Certified Public Accountants
AIN	Advanced Intelligent Network
AIS	Automated Information System
ASIS	American Society for Information Science
ATIS	Alliance for Telecommunications Industry Solutions
ATM	Asynchronous Transfer Mode
BIND	Berkeley Internet Name Domain
BND	Bundes Nacrichten Dienst
CERT	Computer Emergency Response Team
CIAC	Computer Incident Advisory Capability
CICG	Critical Infrastructure Coordination Group
CITAC	Computer Investigation and Infrastructure Threat Assessment Center
CLEC	Competitive Local Exchange Carrier
CMU	Carnegie Mellon University
COTS	Commercial Off-the-Shelf
CPE	Customer Premises Equipment
CPU	Central Processing Unit
CSI	Computer Security Institute
CSIS	Center for Strategic and International Studies
CSTB	Computer Science and Telecommunications Board
DARPA	Defense Advanced Research Projects Agency
DCC	Data Communications Channel
DDN	Defense Data Network
DEM	DISN Equipment Manager
DGSE	French General Directorate of External Security
DIA	Defense Intelligence Agency
DISA	Defense Information Systems Agency
DISN	Defense Information Systems Network
DNS	Domain Name Service
DoD	Department of Defense

DOJ	Department of Justice
EC	Electronic Commerce
EDI	Electronic Data Interchange
e-mail	Electronic Mail
EO	Executive Order
EPRI	Electric Power Research Institute
FBI	Federal Bureau of Investigation
FBIS	Foreign Broadcast Information Service
FCC	Federal Communications Commission
Fed	CIRC Federal Computer Incident Response Capability
FERC	Federal Energy Regulatory Commission
FIRST	Forum for Incident Response and Security Teams
FIS	Foreign Intelligence Service
FTP	File Transfer Protocol
GII	Global Information Infrastructure
GUI	Graphical User Interface
HTTP	Hypertext Transfer Protocol
I-4	International Information Integrity Institute
IATF	Infrastructure Assurance Task Force
IAW	Indications, Assessment, and Warning
IBW	Information Based Warfare
ICMP	Internet Control Message Protocol
ID	Identification
IETF	Internet Engineering Task Force
IIG	Information Infrastructure Group
ILEC	Incumbent Local Exchange Carrier
IMAP	Internet Message Access Protocol
InterNIC	Internet Network Information Center
IP	Internet Protocol
IRC	Internet Relay Chat
ISAC	Information Sharing and Analysis Center
ISO	Information Security Officer

ISP	Internet Service Provider
IW	Information Warfare
IW-D	Information Warfare Defense
JETRO	Japan External Trade Organization
LAN	Local Area Network
LNP	Local Number Portability
LTTE	Liberation Tigers of Tamil Eelam
MITI	Ministry of International Trade and Industry
MOD	Masters of Downloading; also, Ministry of Defense
NACIC	National Counterintelligence Center
NASA	National Aeronautics and Space Administration
NCS	National Communications System
NCSA	National Computer Security Association
NAG	Network Group
NIAC	National Infrastructure Assurance Council
NII	National Information Infrastructure
NIPC	National Infrastructure Protection Center
NIST	National Institute of Standards and Technology
ERIC	Network Reliability and Interoperability Council
NS/EP	National Security and Emergency Preparedness
NSA	National Security Agency
NSC	National Security Council
NSIE	Network Security Information Exchange
NSTAC	National Security Telecommunications Advisory Committee
NSTIS	National Security Telecommunications and Information Systems
OAM	Operations, Administration, and Provisioning
OAM&P	Operations, Administration, Maintenance, and Provisioning
OASIS	Open Access Same-Time Information Systems
OCIIP	Office of Computer Investigations and Infrastructure Protection

OMNCS	Office of the Manager, National Communications System
ONA	Open Network Architecture
OSS	Operation Support Systems
OSTP	Office of Science and Technology Policy
OTA	Office of technology Assessment
PBX	Private Branch Exchange
PC	Personal Computer
PCC	Policy Coordinating Committee
PCCIP	President's Commission on Critical Infrastructure Protection
PDD	Presidential Decision Directive
PLA	People's Liberation Army
PN	Public Network
POP3	Post Office Protocol 3
PSN	Public Switched Network
RFC	Request for Comment
ROK	Republic of Korea
SATAN	Security Administrator Tool for Analyzing Networks
SATCOM	Satellite Communication
SCADA	Supervisory Control and Data Acquisition
SEI	Software Engineering Institute
SIPRNET	Secret Internet Protocol Router Network
SMTP	Simple Mail Transport Protocol
SONET	Synchronous Optical Network
SS7	Signaling System 7
SYN	Synchronization
TCP	Transmission Control Protocol
UCA	Utility Communications Architecture
U.S.	United States
USGAO	United States General Accounting Office
USGPO	United States Government Printing Office

VPN	Virtual Private Network
WAN	Wide Area Network
WMD	Weapons of Mass Destruction
WWW	World Wide Web

APPENDIX C GLOSSARY

ActiveX: A technology and set of programming tools from Microsoft for building interactivity into Web pages and application programs.

Assurance: A measure of confidence that the security features and architecture of an information system or network correctly mediate and enforce the appropriate security policies.

*Assessment:*The analysis of indications to determine the likelihood, nature, and potential of a threat.

*AsyncAronous transfer mode (A TM):*A cell-switched technology for digital communications based on a fixed-length 53 byte cell. ATM supports all types of traffic (voice, data, image or video) by combining circuit-switching and packet-switching technologies.

*Attack:*A set of actions that results in denial or degradation of service or a compromise of information, integrity, authentication, nonrepudiation, or other security feature.

Audit trail: A chronological record of computer system activities that is saved to a file on the system. The file can later be reviewed by the system administrator to identify users' actions on the system or processes that occurred on the system.

*Availability:*Ensuring that data transmissions or computing processing systems are not denied to authorized users.

*Backdoor:*A hidden software or hardware mechanism that can be triggered to circumvent system protection mechanisms. A backdoor is activated in an innocent-appearing manner, e.g., a special "random" key sequence at a terminal. Software developers often introduce backdoors in their code to enable them to reenter the system and perform certain functions. (Synonymous with "trapdoor.").

*Bacteria:*A program that reproduces itself so quickly that the host computer or network is overwhelmed.

*Classified information:*Information or material that is (1) owned by, produced for or by, or under the control of the U.S. Government; and (2) determined under Executive Order 12356, or prior orders, to require protection against unauthorized disclosure; and (3) so designated.

Confidentiality: Privacy of data during transmission, processing, or storage, usually through encryption or data separation.

*Countermeasure:*An action, device, procedure, technique, or other measure that reduces the vulnerability of an automated information system.

Critical infrastructure: Those infrastructures that are so vital that their incapacity or destruction would have a debilitating effect at a regional or national level. The President's Commission on Critical Infrastructure Protection (PCCIP) identified eight critical infrastructure systems: telecommunications, electrical power systems, gas and oil transportation and storage, banking and finance, transportation, water supply systems, emergency services, and continuity of Government services.

*Cyberspace:*Coined by William Gibson in his 1984 novel, Neuromancer. Usually applied to the universe of computer networks, including the Internet, on-line information services such as CompuServe, and isolated private systems.

Daemon:(pronounced "demon ") A program that maintains or performs specific computer tasks or functions such as printing files, monitoring incoming traffic, or providing outbound communication services.

*Data:*A representation of facts, concepts, information, or instructions suitable for communication, interpretation, or processing.

*Delivery or access mechanism:*A method by which malicious software places a payload into a target computer or computer network. Two principal means of delivery exist: dynamic or static.

*Denial-of-service attack:*An electronic intrusion or attack that renders the targeted computer server inoperable and/or the targeted service provider unable to continue operational service.

*Detection:*Comparing normal patterns of behavior and identifying abnormalities that could be intrusions; the process of identifying that an intrusion has been attempted, is occurring, or has occurred.

*Disinformation:*Providing deliberately incorrect or misleading information to counteract or discredit authentic information.

*Dumpster diving:*Sifting through refuse from an office or technical installation to extract confidential data, especially security-compromising information. The term was coined by early hackers, who acquired extensive information about how to defraud the long distance telephone network by retrieving internal AT&T manuals from trash dumpsters.

*Electronic data interchange (EDI):*A standard format for exchanging business data. An EDI message contains a string of data elements, such as a price or product model number, separated by delimiters. An EDI transaction often consists of what would usually be contained in a typical business document or form. The parties who exchange EDI transmissions are referred to as trading partners.

*Electronic intrusion:*Unauthorized access to networks and information systems or any other type of information system attack. Electronic intrusion includes activities to steal or corrupt sensitive information; to steal, modify, or destroy software; to circumvent system security countermeasures; to disrupt or disable an information system; to steal services or defraud providers; and other types of information system attacks such as interception, spoofing, desinformation, and denial-of-service.

*Encryption:*The conversion of plain text into unintelligible forms by means of cryptographic systems. Cryptographic systems use encryption algorithms to convert plain text into enciphered text.

*Exploitation:*Using a weakness or vulnerability in an automated information system to access or cause damage to or loss of an asset.

*Extranet:*A collaborative network that uses Internet technology to link organizations with their suppliers, customers, or other businesses that share common goals. Security and privacy could be achieved either by ensuring that the transmission lines were privately owned or leased, by tunneling through the Internet, or by using the Internet with password authorization.

*Firewall:*A firewall is either the program that protects the resources of one network from users from other networks or the computer on which it runs, usually an Internet gateway server.

*Global Information Infrastructure (GII):*The National Information Infrastructure (NII) concept applied globally.

*Hacker:*Traditionally, a person who enjoys learning details of a programming language or operating system through doing rather than simply theorizing. In common usage, though, "hacker" is synonymous with "cracker" (i.e., someone who breaks into someone else's computer system, often on a network). A cracker may do this for profit, malice, or because the challenge is there.

*Hypertext transfer protocol (HTTP):*The rules for exchanging files (text, images, sound, video, and other multimedia files) on the World Wide Web. HTTP is an application protocol that relies on the underlying Transmission Control Protocol / Internet Protocol (TCP/IP) suite. HTTP enables files to contain references to other files, whose selection will elicit additional transfer requests. A Web server contains, in addition to the files it can serve, an HTTP daemon, a program that is designed to respond to HTTP requests from Web browsers.

*Imbeds/implants:*Software or hardware covertly placed in a program or computer. These may accomplish a variety of tasks, from collecting covert technical intelligence to damaging an infected system.

*Information operations:*The continuous military operations within the military information environment that enable, enhance, and protect the friendly force's ability to achieve an advantage across the full range of military operations; information operations include interacting with the global information environment and exploiting or denying an adversary's information and decision capabilities.

*Information system:*The computers, networks, and software involved in the collection, storage, processing, transmission, and dissemination of information. This includes the individuals who create, analyze, and act on the information it transmits and the organizational processes it enables.

Information system security: protection of information systems against unauthorized access, loss, or corruption of information in storage, processing, or transmission. This includes those measures necessary to deter, detect, document, respond to, and counter any threat.

*Information warfare:*Actions taken to achieve information superiority by affecting adversary information, information-based processes, information systems, and computer-based networks while defending one's own information, information-based processes, information systems, and computer-based networks.

*Information warfare defense (I W-D):*The integration and coordination of policies and procedures, operations, intelligence, law enforcement, and technology to protect information and defend information systems. The objective of IW-D is to ensure access to timely, accurate, and relevant information when and where it is needed and to deny adversaries the opportunity to exploit friendly information and systems for their own purposes.

*Infrastructure:*The basic facilities, equipment, and operating instructions needed for a system to operate.

*Integrity:*Verification that data has not been modified in transmission or during computer processing.

*Internet:*A near-global network of computers joined by high-speed, digital telecommunications that use a common rule set known as TCP/IP.

*Internet Protocol (IP):*Part of the TCP/IP communications protocol. IP specifies the format and the addressing scheme of packets and provides the routing mechanism for information. Most networks combine IP with a higher-level protocol called Transport Control Protocol (TCP), which establishes a virtual connection between a destination and a source.

*Internet relay chat (IRC):*A system for chatting that involves special client and server software and informal conventions for participation. Chatting is the exchange of typed-in messages among a group of users who can participate from anywhere on the Internet. In some cases, a private chat can be arranged between two parties who meet initially in a group chat. Chats can be ongoing but are usually scheduled for a particular time and duration. IRC requires one site act as the repository (or "chat site") for the messages.

InterNIC:The organization responsible for registering and maintaining the .com, .edu, .gov, .net, and .org domain names on the Internet. (Today, this function has been contracted to Network Solutions, Inc.)

Intranet:A network that is contained within an enterprise, usually consisting of many interlinked local area networks. The network may also use leased lines over a WAN and connections through gateways to the Internet.

Intrusion:Unauthorized access to, and/or activity in, an information system.

Intrusion detection:The process of identifying that an intrusion has been attempted, is occurring, or has occurred.

Java:A programming language designed by Sun Microsystems for use in distributed environments. Java can be used to create complete applications that may run on a single computer or be distributed among servers and clients in a network. It can also be used to build small application modules (applets) for use as part of a Web page.

Local area network (LAN):A network of interconnected workstations sharing the resources of a single processor or server within a relatively small geographic area. Typically, this might be within an office or element of an organization.

Logic bomb:A form of malicious software that executes a specific task under specified conditions or at a specified time, either automatically or as the result of a remote command.

Looping:A technique in which hackers try to conceal their point of origin. Using this technique, hackers "leap frog" or loop through several computer systems before finally entering the system they intend to attack. The technique masks a hacker's actual origin from the system that is being attacked and from those pursuing him or her. Hackers will often ensure that the routing used to loop through the system crosses international and state borders. Crossing a border electronically has the same consequence as crossing it physically and will involve another country's or state's law enforcement agencies, which further complicates and slows efforts to pursue the hackers.

Malicious software/hardware:A complete technical package that carries out a mission preprogrammed by the attacker. Packages typically include components called a delivery mechanism, a trigger, and a payload. Various execution strategies exist for each component.

Modem:Communications device that converts digital signals to analog and vice versa. Modems work in pairs.

*National Information Infrastructure (NII):*In the words of Vice President Al Gore, "a seamless web of communications networks, computers, databases, and consumer electronics that will put vast amounts of information at users' fingertips."

*National security and emergency preparedness (NS/EP):*Capabilities required to maintain a state of readiness or to respond to and manage any event or crisis that causes or could cause injury or harm to the population, damage to or loss of property, or degrade or threaten the NS/EP posture of the United States.

*Network:*A network is composed of communications media and all components attached to them. These components may include computers, routers, multiplexers, switches, transmission systems, and management and support services.

*Password:*A protected word or string of characters that identifies or authenticates a user for access to a computer system, or a specific resource such as data set, file, or record.

*Payload:*The specific part of a virus that performs the action desired by the attacker. Conventional payloads erase data, display messages, or crash or freeze systems. A more sophisticated payload delivered via a Trojan horse could allow an attacker to bypass normal security measures and access the target information system.

*Phreaker:*An individual who hacks into a telephone system, usually to obtain free long distance calling and other services such as conference calling.

*Proprietary information:*Material and information relating to or associated with a company's products, business, or activities that have been clearly identified and properly marked as proprietary information, trade secrets, or confidential information. These items include financial information, trade secrets, product research and development, existing and future product designs, performance specifications, marketing plans or techniques, schematics, client lists, and computer programs.

*Public switched network (PSN):*A network operated by common carriers or telecommunications administrators for the provision of circuit-switched, packet-switched, and leased-line circuits to the public.

*Public network (PN):*The PN is the backbone of the NII and supports virtually all NS/EP telecommunications and information systems requirements. The PN includes any switching system or voice, data, or video transmission system used to provide communications services to the public (e.g., public switched networks, public data networks, private line services, wireless services, and signaling networks).

*Root access:*The superuser account, the top level of a hierarchical directory structure, or in programming, the top node of a tree. Root access to a system will allow access to all files and directories and full privileges to change and delete information.

*Reliability:*Assurance that systems will perform consistently and at an acceptable level of quality.

*Risk management:*The process of identifying, measuring, and minimizing events affecting an information system. Risk management is a process that involves continual reevaluation and adaptation to changes in the organizational, technological, and business environment.

*Security:*Freedom from danger, harm, or risk of loss. The tools for providing security focus on availability, confidentiality, and integrity.

*Security incident:*An attempt to violate a system's security. It may involve fraud, waste, or abuse; compromise of information; loss or damage of property or information; or denial-of-service. Security incidents include penetration of computer systems, exploitation of technical and administrative vulnerabilities, and introduction of computer viruses or other forms of malicious software. A security incident may also involve a violation of law.

*Sensitive but unclassified information:*Any information the loss, misuse, or unauthorized access to, or modification of which might adversely affect U.S. national interests, the conduct of Government programs, or individual privacy.

*Signaling System 7 (SS7):*An international standard protocol for communication and service provisioning over a common channel between telecommunications switches. SS7 is used to set up and control telephone calls and other switched services within and between common carrier networks.

*Sniffer:*A payload that is programmed to search for specific items in a computer program. Sniffers may seek out only passwords or other specified data sought by an attacker.

*Social engineering:*Hacker jargon for obtaining needed information (for example, a password) from an individual rather than obtaining it by breaking into a system. Social engineering can be used over an extended period of time to maintain a continuing stream of information and help from unsuspecting users.

*Spamming:*Bulk unsolicited e-mail, generated through mailing to existing Internet mailing lists or through names culled automatically from Usenet newsgroup postings. Spamming is usually conducted as a marketing ploy aimed at recipients or as a denial-of-service attack against a particular organization or newsgroup.

*Spoofing:*An attempt to gain access to a system by posing as an authorized user. Spoofing is synonymous with impersonating, masquerading, or mimicking.

*Stealth virus:*A dynamically delivered form of malicious software that is specifically designed to avoid detection. Stealth viruses typically try to avoid detection by being minute-only 1 or 2 kilobytes of data, which do not occupy a prominent place in the software program. Another means of avoiding detection is for a stealth virus to graft itself onto existing code, thereby not noticeably enlarging an existing file.

*Supervisory Control and Data Acquisition (SCADA):*Supervisory control and data acquisition systems are usually computer-controlled, network-based systems that allow companies to automate and remotely monitor operations and remotely conduct tests and maintenance. They are increasingly being used to control electronic power distribution, rail and other transportation systems, oil refinery operation, and natural gas distribution.

SYN-flood attack: Also known as "synchronization packet flooding." Moving or sending a large volume of repetitive e-mail packets to a designated computer server to render the server unusable. This type of electronic attack can be accomplished by sending hundreds or thousands of the same e-mail messages, containing huge unintelligible message files, e-mails that contain false or no return addresses, routed through random Internet service providers so that they cannot be traced or blocked. This is a typical denial-of-service attack.

*TCP wrapper:*Access control mechanism that allows/disallows and records access to TCP daemon. The wrapper sits between the inbound connection and daemon on the system, which controls access to the system. The wrapper reads the incoming traffic and originating site and compares the IP address to an access list that the sysop configures. The access list contains sites that are authorized or not authorized to connect the system. The wrapper records the time, date, and originating IP address of the inbound connection before it allows access to the system.

*Telecommunications:*The transmission, emission, or reception of signals, signs, writing, images, sounds, or intelligence of any nature, by wire, cable, satellite, fiber optics, laser, visual or other electronic means.

*Threat:*Capabilities, intentions, and attack methods of adversaries to exploit vulnerabilities of an information system, or an information-based network or any circumstance or event with a potential to cause harm in the form of destruction, disruption, and/or denial of service.

*Threat analysis:*The examination of all actions and events that might adversely affect a system or operation.

*Threat assessment:*Process of formally evaluating the degree of threat to an information system and describing the nature of the threat.

Transmission Control Protocol (TCP):Part of the TCP/IP communications protocol. TCP enables two hosts to establish a connection and exchange streams of data. TCP guarantees delivery of data and also guarantees that packets will be delivered in the same order in which they were sent.

Trigger:Portion of the virus that activates the payload. The trigger contains software code that tells it that it is actually in the targeted system. In the case of a virus, a trigger may control reproduction, focusing the virus toward a specific goal.

Trojan horse:A computer program with an apparently or actually useful function that contains additional (hidden) functions that surreptitiously exploit the legitimate authorizations of the invoking process to the detriment of security or integrity.

Unclassified information:Any information not designated as classified.

Uniform resource locator (URL):The unique address of a single page or file on the Web or an intranet. The address includes a domain name (an Internet server address) and a hierarchical description of a file location on the server.

Users:People or processes accessing an automated information system (AIS) either by direct connections (i.e., via terminals) or indirect connections.

User ID:A unique symbol or character string used by a system to identify a specific user.

Utilities:class of programs and programming aids used to facilitate tasks that are frequently performed, such as copying data and listing directories.

Virtual private network (VPN):A network that is constructed among a select set of organizations or users over a public transport, usually the Internet. VPNs use dedicated lines, encryption, or other security measures to ensure that only authorized users can access the network and that the data cannot be intercepted.

Virus: A computer program that embeds itself in other code and can replicate itself. Once active, it can take unwanted and unexpected actions that can result in either destructive or nondestructive outcomes in the host computer programs.

Vulnerability:A weakness in system security procedures, system design, implementation, hardware design, or internal controls that could be exploited to violate system security policy.

Vulnerability analysis: The systematic examination of systems to determine the adequacy of security measures, identify security deficiencies, and provide data from which to predict the effectiveness of proposed security measures.

Web browser: An application that provides a technique to look at, read, and hear all the information on the World Wide Web. The Web browser is a client program that uses the HTTP to make requests of Web servers throughout the Internet

*Web site:*A collection of Web files on a particular subject that includes an introductory file called a home page. Most organizations or individuals that have Web sites provide only their home page address. From this page, one can get to all the other pages on their site. A Web site is not necessarily synonymous with a Web server because a Web site may include files hosted on more than one server supporting Web, or HTTP, services.

*Wide area network (WAN):*A network connecting LANs in individual facilities over private, leased, or switched transmission systems.

*World Wide Web (WWW):*All the resources and users on the Internet that are using the HTTP. Tim Berners-Lee, who invented HTTP, offers a broader definition: "The World Wide Web is the universe of network-accessible information."

*Warnings:*An advisory of the results of the vulnerability and threat assessments, likely target(s), and recommended actions.

*Worm:*A program that propagates from computer to computer via a common network. As shown in Robert Morris' 1988 disruption of the Internet, a worm does not have to contain destructive software to cause problems. A worm may be designed to perform a specific task and may not necessarily affect other programs on the system.

APPENDIX D REFERENCES

American Institute of Certified Public Accountants, "CPAs Name Top Ten Technologies for 1997: Cyberspace Security Tops List," World Wide Web, Great Plains Software, Inc., www.gps.com/waynes_Web/Perspectives/topten.htm, January 16, 1997.

Antiviral Toolkit Pro Virus Encyclopedia,*The Classification of a Computer Virus*, World Wide Web, http//www.metro.ch/avpve/classes.stm.

Automated Systems Security Incident Support Team, Defense Information Systems Agency, Defense Communications System, DDN Security Coordination Center, "Subject: Ongoing Network Monitoring Attacks," *Defense Data Network Security Bulletin 94-02*, World Wide Web, csrc.ncsl.nist.gov/secalert/ddn/1994/sec-9403.txt, February 4, 1994.

Behar, Richard, "Who's Reading Your E-Mail?" *Fortune*, February 3, 1997.

Blankenhorn, Dana, "If You're Hiring a Hacker, Stick With the Pros," World Wide Web, Net Marketing Home Page, www.netb2b.com:80...monthly/97/09/01/article.4, September 1997.

Borsook, Paulina, "Hackers Bring the Net Down to Earth," *Network World*, January 1, 1996.

Branigan, Steve, "Hacker Trends: '96 Version," Presentation to the NSTAC Information Assurance Task Force, June 1996.

Center for Strategic and International Studies, *Russian Organized Crime: Global Organized Crime Project*. Washington, DC: CSIS, 1997.

Charney, Scott, *Computer Crime*, Presentation to the Inaugural Economic Crime Summit, Providence, RI, May 20, 1997.

Cilluffo, Frank J., and Curt H. Gergely, "Information Warfare and Strategic Terrorism," *Terrorism and Political Violence*, Vol. 9, No. 1, London, England: Frank Cass & Company, Ltd., Spring 1997.

CNN Interactive, "Master Hacker 'Analyzer' Held in Israel," World Wide Web, www.cnn.com/TECH/computing/9803/18/analyzer, March 18, 1998.

Cohen, Frederick B., *Protection and Security on the Information Superhighway*, New York, NY: John Wiley & Sons, Inc., 1995.

Cohen, William S., Secretary of Defense, *Report of the Quadrennial Defense Review*, Washington, DC: U.S. Department of Defense, May 1997.

Cole, Richard, The Associated Press, "FBI Hunts 'Master Hacker': Israeli Sought for Breaking into Military Computers," ABCNEWS.com, World Wide Web, www.abcnews.com/sections/tech/DailyNews/hackers0308.html, 1998.

Collin, Barry, "The Future of Cyberterrorism: Where the Physical and Virtual Worlds Converge," Institute for Security and Intelligence, World Wide Web, www.acsp.uic.edu/OICJ/CONFS/terror02.htm, 1997.

Computer Emergency Response Team, Software Engineering Institute, Carnegie Mellon University, *CERT Advisory CA-97.28,* Pittsburgh, PA: CERT, December 16, 1997.

Computer Emergency Response Team, Software Engineering Institute, Carnegie Mellon University,*CERT Advisory CA-98.01.smurf,* **Pittsburgh,** PA: CERT, January 5, 1998.

Computer Emergency Response Team, Software Engineering Institute, Carnegie Mellon University, *CERT Incident Note IN-98.02: New Tools Used for Widespread Scans,* Pittsburgh, PA: CERT, www.cert.org/incident_notes/IN-98-02.html, July 2, 1998.

Computer Emergency Response Team, Software Engineering Institute, Carnegie Mellon University, *CERT Coordination Center Statistics 1988-1997,* World Wide Web, www.cert.org/pub/cert-stats/ cert_stats.html, January 1998.

Computer Emergency Response Team, Software Engineering Institute, Carnegie Mellon University, *Internet-Related Organizations,* World Wide Web, www.cfcl.com/tin/p/9605.htm, 1997.

Computer Security Institute, "Computer Crime Continues to Increase, Reported Losses Total Over $100 Million," World Wide Web, CSI Homepage, www.gocsi.com/preleas2.htm, March 6, 1997.

Computer Security Institute, "Issues and Trends: 1998 CSI/FBI Computer Crime and Security Survey" World Wide Web, CSI Homepage, www.gocsi.com/prelea 11.htm, March 4, 1998.

Computer Science and Telecommunications Board, National Research Council,*Information Systems Trustworthiness: An Interim Report,* Washington, DC: National Academy Press, June 1997.

*Counterintelligence News & Developments,*Volume 2, World Wide Web, www.nacic.gov/cind/ cindijen9.htm, June 1996.

Critical Infrastructure Assurance Office, *The Clinton Administration's Policy on Critical Infrastructure Protection: PDD-63,* May 22, 1998, www.ciao.gov.

Davis, Beth, "The Fifth Annual Information Week/Ernst & Young Information Security Survey," *Information Week,* Issue 647, September 8, 1997, p. 182.

Defense Intelligence Agency, *CyberTerrorism: The Convergence of Our Worlds, Counter-Terrorism Perspectives for Senior Managers, Joint Military Intelligence Training Center,* Washington, DC: Institute for Security and Intelligence, July 16, 1997.

DeGenaro, William E., *Steal This Country: How Foreign Spies Are Destroying American Jobs,* Presented at the Fifth National Operations Security Conference in McLean, Virginia, 1994.

Denning, Dorothy E., and William E. Baugh, Jr., U.S. Working Group on Organized Crime, *Encryption and Evolving Technologies: Tools of Organized Crime and Terrorism,* Washington, DC: National Strategy Information Center, 1997.

Department of Justice, "Israeli Citizen Arrested in Israel for Hacking United States and Israeli Government Computers," Press Release, 98-125, Washington, DC: DOJ, March 18, 1998.

EDUCOM, "Beefing Up Computer Security Efforts on Campus," World Wide Web, Edupage Mailing List, www.educom.edu/web/edupage.html, September 25, 1997.

Electric Power Research Institute, *UCA and DAIS Information Security Analysis,* Palo Alto, CA: EPRI, August 1994.

Ellis, James, et al., CERT Coordination Center, *Report to the President's Commission on Critical Infrastructure Protection,* Pittsburgh, PA: CERT, World Wide Web, www.cert.org/pu...n/cert.rpcci.body.html, January 1997.

Ellison, R. J., D. A. Fisher, et al, *Survivable Network Systems: An Emerging Discipline,* Technical Report CMU/SEI-97-TR-013, Pittsburgh, PA: Carnegie Mellon University, November 1997.

Ernst & Young/*Information Week, 3rd* Annual Information Security Survey: Trends, Concerns, and Practices, Cleveland, OH: Ernst & Young, 1996.

Everett, Charles B., Moss Dewindt, and Shane McDade, *The Silicon Spear: An Assessment of Information Based Warfare (IBW) and U.S. National Security,* World Wide Web, Institute for National Strategic Studies, National Defense University, www.ndu.edu/ndu/inss/siws/ch2.html, November 1996.

Executive Order 13010, *Critical Infrastructure Protection,* Washington, DC: The White House, July 15, 1996.

Federal Bureau of Investigation, Office of Computer Investigations and Infrastructure Protection, *Computer Investigations and Infrastructures Threat Assessment Center (CITAC),* Washington, DC: FBI, October 1997.

Federal Bureau of Investigation, Chief, National Infrastructure Protection Center; Michael A. Vatis, Statement for the Record Before the Congressional Joint Economic Committee, www.fbi.gov/congress/vatis.htm, March 24, 1998.

Federal Computer Incident Response Capability, *Summary of Incidents Handled by FedCIRC–October 1996-October 1997,* World Wide Web, fedcirc.llnl.gov/about/incidents1197.htm, January 1998.

Federation of American Scientists CyberStrategy Project, www.-fas.org/cp/index.html, October 15, 1998.

Ferrache, David, *A Pathology of Computer Viruses,* London, England: Springer-Verlag, 1992.

Foreign Broadcast Information Service, "Academy Department Head on Importance of Information Security," FBIS-UMA-95-239-S, December 13, 1995.

Foreign Broadcast Information Service, "Article Discusses Information Warfare," FBIS-CHI-95-239, December 13, 1995.

Foreign Broadcast Information Service, "China: Defense Military Computer Network Interconnects PLA Army, Navy, Air Force," FBIS-CHI-97-324, November 25, 1997.

Foreign Broadcast Information Service, "General Views Importance of Information Warfare," FBIS-CHI-95-129, July 6, 1995.

Foreign Broadcast Information Service, "Germany: Study Examines Changes to Armed Forces in Information Age," FBIS-TAC-97-279, October 8, 1997.

Foreign Broadcast Information Service, "Japan: Japan Seen Lagging in Global Economic Intelligence War," FBIS-EAS-97-279, October 8, 1997.

Foreign Broadcast Information Service, "Japan: Journalist on Security, Intelligence Issues," FBIS-EAS-97-065, April 7, 1997.

Foreign Broadcast Information Service, "New PLA Training to Stress High-Tech Warfare," FBIS-CHI-95-240, December 14, 1995.

Foreign Broadcast Information Service, "PRC Army Daily on Weaknesses of Information Warfare," FBIS-CHI-96-014, January 22, 1996.

Foreign Broadcast Information Service, "PRC: 'Digitized Forces' Developed for Electronic Warfare," FBIS-CHI-96-097, May 17, 1996.

Foreign Broadcast Information Service, "PRC: Military Studies of Information Warfare Theory Reviewed," FBIS-CHI-96-035, February 21, 1996.

Forum of Incident Response and Security Teams, "What is FIRST?" World Wide Web, DFN-CERT Webpage, www.first.org/about/, September 10, 1997.

Freedman, David H., and Charles C. Mann, "Cracker," *U.S. News and World Report,* June 2, 1997.

Garfinkel, Simson, "FBI Uses Hackers' Tools to Sniff Out Hacker's Lair," World Wide Web, Mercury Center, simson.vineyard.ne...lips/96.SJMN.Wiretap.html, April 8, 1996.

Garner, Rochelle, "The Growing Professional Menace," *Open Computing Magazine,* July 1995.

Glave, James, "Have Crackers Found Military's Achilles' Heel?" World Wide Web, Wired News Online, www.wired.com/news/news/technology/story/11811.html, April 21, 1998.

Gold, Steve, "Internet Security Major Concern for IT Managers," World Wide Web, Newsbytes Network, www.newsbytes.com, December 19, 1996.

Gregg, Jonathan, "Masters of What? Netly News: Pentagon Hackers Got Peanuts," World Wide Web, Time Online, cgi.pathfinder.com/time/daily, April 29, 1998.

"Hackers Warnings on Info-war appear Inflated," *The Boston Globe Online*, World Wide Web, Globe Newspaper Company, www.boston.com/daily, April 27, 1998.

Hundley, Richard O., and Robert H. Anderson, "Emerging Challenge: Security and Safety in Cyberspace," *IEEE Technology and Society*, Winter 1995/1996.

International Crime Control Strategy of the United States, Section VII: Responding to Emerging International Crime Threats, http://www.usdoj.gov/criminal/press/VIIIresp.html, May 12, 1998.

Internet Engineering Task Force, Working Group on Security Incident Response, "Guidelines and Recommendations for Incident Processing," World Wide Web, www.cert.dfn.de/eng/resource/ietf/grip/home.html, 1997.

Jackson, Robert, "Ex-Analyst Admits Spying for S. Korea in Plea Bargain," *Los Angeles Times*, May 8 1997.

Joint Security Commission, *Redefining Security: A Report to the Secretary of Defense and the Director of Central Intelligence*, Washington, DC: USGPO, February 28, 1994.

Kane, Robert, "Losing the Keys to the Kingdom of Domain," World Wide Web, Intrusion Detection, Inc., www.intrusion.com/keys.ht ml, 1996.

Kluepfel, Hank, *Toward a More Secure Telecommunications Infrastructure: Mitigating the Risks*, Washington, DC: Bellcore, March 31, 1994.

Lovell, Jeremy, "Belgium Investigates Major Data Security Leak," World Wide Web, Infoseek News Channel, www.infoseek.com/Content?ar...v=N5&1k=1b&col=NX&kt=A&ak=ne ws1486, December 15, 1997.

Lunt, Teresa, *Intrusion Detection Briefing: A Tutorial*, Washington, DC: Defense Advanced Research Projects Agency.

Madsen, Wayne, "Intelligence Agency Threats to Computer Security," *International Journal of Intelligence and Counterintelligence*, 6:4, Winter 1993.

McKay, Niall, "Cyber Terror Arsenal Grows," *Wired News*, World Wide Web, www.wired.com/news/news/politics/story/15643.html, October 16, 1998.

McWilliams, Brian, "Details on Nationwide Computer Attacks", PC World News Radio, World Wide Web, www.pcworld.com/news/daily/data/0398/-980304180515.html, 1998.

McWilliams, Brian, "Security Bug Affects Unopened E-mail Attachments," PC *World Today,* July 27, 1998, www.pcworld.com/pcwtoday/article/0,1510,7559,00.html.

Meeks, Brock N., "Information Warfare and the Real Threat," World Wide Web, MSNBC News, www.msnbc.com/news/123040.asp, 1997.

Menagh, Melanie, " First Line of Defense," Computerworld, February 10, 1997.

Mendel, Brent " Mail Hack Affirms Mobile Code Fear," *Internet Week with LanTimes Online,* http://www.lantimes.com/98/98sep/809a001a.html D, September 14, 1998.

Messmer, Ellen, "No Defense Against Latest Hacker Tool?" *Network World,* March 24, 1997.

Mohamed, Airf, " Symantec Tops Virus Hunting Record," World Wide Web, Ziff-Davis UK Limited, United Kingdom, zdnet.com/uk/news/ns-2425.html, August 1, 1997.

National Computer Security Center, *Glossary of Computer Security Terms,* NCSC-TG-004, Version-1, Washington, DC: USGPO, October 21, 1988.

National Counterintelligence Center, *Annual Report to Congress on Foreign Economic Collection and Industrial Espionage,* Washington, DC: NACIC, June 1997.

National Defense Panel, *Transforming Defense: National Security in the 21st* Century, Washington, DC: U.S. Department of Defense, December 1997.

National Institute of Standards and Technology, Information Technology Laboratory, Computer Security Division,*Internet Security Policy: A Technical Guide,* Gaithersburg, MD: NIST, July 21, 1997.

National Intelligence Council, *The Foreign Information Warfare Threat to U.S. Telecommunications and Information Systems,* Undated Briefing.

National Research Council, *Growing Vulnerability of the Public Switched Networks: Implications for National Security and Emergency Preparedness,* Washington, DC: National Academy Press, 1989.

National Security Telecommunications Advisory Committee, *INFORMATION ASSURANCE: A Joint Report of the IA Policy Subgroup of the Information Infrastructure Group and the NCM Subgroup of the Operations Support Group,* Washington, DC: NSTAC, December 1997.

National Security Telecommunications Advisory Committee, Subgroup on Information Assurance,*Report on Electric Power Risk Assessment,* Office of the Manager, National Communications System, Washington, DC: NSTAC, May 1997.

National Security Telecommunications Advisory Committee, Subgroup on Widespread Outage,*Report on the Likelihood of a Widespread Telecommunications Outage*, Washington, DC: NSTAC, December 1997.

National Security Telecommunications Advisory Committee, *Report of the Network Security Task Force*, Washington, DC: NSTAC, October 1990.

Nelson, Jack, "U.S. Firms' '97 Losses to Spies Put at $300 Billion," *Los Angeles Times*,January 12, 1998.

Nerney, Chris, " Getting Civil With Hackers," *Network World*, August 12, 1996.

Net Insider, "FBI: Insiders More Dangerous Than Crackers," www.newdimensions.net/fbi.htm, January 15, 1998.

Network Reliability and Interoperability Council, *Network Interoperability: The Key to Competition*, Washington, DC: Alliance for Telecommunications Industry Solutions, July 15, 1997.

Network Wizards, "Internet Domain Survey: July 1998," http://www.nw.con/zone/WWW/new-survey.html.

Network World Media Lounge, "Network World Industry Surveys: Study Sponsored by Network World and Ernst & Young," World Wide Web, www.nwfusion.com/medialounge/press/surveys.html# anchor110106O, March 31, 1997.

O'Brien, David, " Recognizing and Recovering From Rootkit Attacks*Sys Admin: The Journal for UNIX Systems Administrators*, Volume 5, Number 11, November 1996.

Office of the Manager, National Communications System, *An Assessment of the Risk to the Security of Public Networks*, Prepared by the U.S. Government and National Security Telecommunications Advisory Committee Network Security Information Exchange, Washington, DC: OMNCS, December 12, 1995.

Office of the Manager, National Communications System, *Assessment of PSN Components Critical Roles and Interdependencies in Call Processing*, Arlington, VA: OMNCS, June 1997.

Office of the Manager, National Communications System, *Security Implications of the Telecommunications Act of 1996*, Washington, DC: OMNCS, May 8, 1998.

Office of the Manager, National Communications System, *Software Integrity*,An NSIE White Paper, Prepared by the U.S. Government and National Security Telecommunications Advisory Committee Network Security Exchange, Washington, DC: OMNCS, July 1997.

Office of the Manager, National Communications System, Network Security Information Exchanges (NSIE) White Paper, *The Insider Threat to Information Systems: A Framework for Understanding and Managing the Insider Threat in Today's business,* April 1998.

Office of the Manager, National Communications System, *Public Switched Network Best Practices: Security Primer,* OMNCS, Washington, D.C. December 1998.

Office of the Under Secretary of Defense for Acquisition and Technology, *Report of the Defense Science Board Task Forte on Information Warfare-Defense,* Washington, DC: USGPO, November 1996.

O'Reilly Online Catalog, Simson Garfinkel and Gene Spafford, *Practical UNIX and Internet Security,* 2nd Edition, World Wide Web, www.-ora.com/oracom/sysad/puis-exc.html, 1996.

PC Webopaedia, "Virus," World Wide Web, htt:///webopedia.internet.com/TERM/v/virus.html.

Power, Richard, *CSI Round Table: Intranet Security,* San Francisco, CA: CSI, 1997.

Power, Richard, Editor, *Computer Security Issues & Trends: 1996 CSI/FBI Computer Crime and Security Survey,* Volume II, No.2, San Francisco, CA: CSI, Spring 1996.

President's Commission on Critical Infrastructure Protection, *Critical Foundations: Protecting America's Infrastructure,* Washington, DC: USGPO, October 1997.

Quinn, Andrew, Reuters, "Teens Suspected as Possible Pentagon Hackers," TIME.com, World Wide Web, www.pathfinder.com/news/latest/rb/1998feb28/70.html, 1998.

Rathmell, Andrew, Richard Overill, Loranzo Valeri, and John Gearson, "The IW Threat From Sub-State Groups: An Interdisciplinary Approach," (Paper presented at the Third International Symposium on Command and Control Research and Technology), World Wide Web, Infowar.com Webpage,wysiwyg://53/http://www.infowar.com/ mil_c4i/icsa/icsa3.html-ssi, June 17-20, 1997.

Reed, Dan, and David L. Wilson, "Suspected Pentagon Hacker Found," World Wide Web, The Seattle Times Company, www.seattletime.com/news/techno logy/, March 19, 1998.

Richtel, Matt, "California ISP Says It Tracked Teenagers in Pentagon Hacking," World Wide Web, The *New York Times* on the Web, search.nytimes.com/~/Analyser, March 10, 1998.

roKK Industries, "Zap-c," World Wide Web, cips02.physik.uni-...king+security/files/zap.c, 1997. Schutt, Donald, "How do You Handle Spammers," *Web Week,* Volume 3, Issue 10, April 14, 1997.

Schweizer, Peter, *Friendly Spies,* New York, NY: Atlantic Monthly Press, 1993.

" Security Concerns Tops the List of Challenges in Developing Web Applications," Press Release, Milpitas, CA: Strategic Focus, March 1997.

Stoll, Clifford, *The Cuckoo's Egg,* New York, NY: Doubleday, 1989.

Strategic Forecasting L.L.C., "Land Attack Emerges as New Threat to Network Security Reports," *Computer Security Alert,* World Wide Web, www.stratfor.com, December 11, 1997.

Tech Report, "Trojan Horses, Hostile Java Applets Target Home PC," *USA Today, World Wide Web, USA National News,* www.usatoday.com.80/life/cyber/tech/ctbl77.htm, September 3, 1997.

"Telecommunications, Satellites Said To Be Targeted for Espionage by France," *Common Carrier Week,* May 17, 1993.

Thomas, Tim, "Deterring Information Warfare: A New Strategic Challenge," Fort Leavenworth, KS: Foreign Military Studies Office, World Wide Web, leav-www.army.mil/fmso/geo/pubs/infowar4.htm, May 1997.

Thomas, Tim, "Russian Views on Information-Based Warfare," Fort Leavenworth, KS: Foreign Military Studies Office, World Wide Web, leav-www.army.mil/fmso/opart/pubs/airpower.htm, July 1996.

Toigo, Jon William, "Six Steps Toward a More Secure Computing Environment," *UniForum IT Solutions,* July 1996.

United States Congress, Office of Technology Assessment, *Information Security and Privacy in Network Environments,* OTA-TCT-606, Washington, DC: USGPO, September 1994.

United States General Accounting Office, *Information Security: Computer Attacks at Department of Defense Post Increasing Risks,* GAO-AIMD-96-84, Washington, DC: USGPO, May 1996.

United States General Accounting Office, *Information Security: Opportunities for Improved OMB Oversight of Agency Practices,* GAO-AIMD-96-110, Washington, DC: USGPO, September 24, 1996.

United States General Accounting Office, *Information Security: Serious Weaknesses Place Critical Federal Operations and Assets at Risk,* GAO-AIMD-98-92, Washington, DC: USGPO, September 1998.

United States Senate, Committee on Governmental Affairs, *Security in Cyberspace,* Hearings before the Permanent Select Committee on Investigations, Senate Hearing 104-701, Washington, DC: USGPO, 1996.

United States Senate, Senate Governmental Affairs Committee, Permanent Subcommittee on Investigations, "Testimony of John Deutch, Director of U.S. Central Intelligence," World Wide Web, www.fas.org/irp/congress/1996_hr/s960625d.htm, June 25, 1996.

"U.S. Code, Title 18, Chapter 47, Section 1030: Fraud and Related Activity in Connection With Computers," (USGPO CD-Rom prepared by the Office of the Law Revision Counsel of the House of Representatives), World Wide Web, Cornell University Law Page, www.law.cornell.edu/uscode/18/1030.s html, January 16, 1996.

Vatis, Michael, Deputy Assistant Director and Chief, NIPC, FBI, Statement for the Record Before Congressional Joint Economic Committee, Washington, D.C., March 24, 1998.

Violino, Bob, "Crime Fighters: Corporate SWAT Teams Battle Mounting Security Threats," Information *Week,* May 13, 1996.

Violino, Bob, "The Security Facade," *Information Week,* October 21, 1996, p. 38.

Vranesevich, John, "AntiOnline's Coverage of Chameleon Raided By the FBI," *AntiOnline,* World Wide Web, www.antionline.com/SpecialReports/chameloen/index.htmld

Wagner, Mitch, and Gary H. Anthes, "Underground Tools Aid Fledgling Hackers," *Computerworld,* November 13, 1995.

Walker, Leslie, "Fake Message Sends AOL E-Mail Astray," *The Washington Post Online,* World Wide Web, http://search.washingtonpost.com/wp-srv/Wplate/1998-10/17/0551-101798-idx.html, October 17, 1998.

White House, *CombatingTerrorism: PDD-92,* Washington, DC: The Whit Hose, May 22, 1998.

White House, *Protecting America's Critical Infrastructures: PDD-63,* Washington, DC: The White House, May 22, 1998.

Wingfield, Nick, " Email Vendors Fight Spammers," (Reprinted from C/net News), World Wide Web, DFN-CERT Web page, www.news.com-/News/Item/0,4,8247,00.html, February 25, 1997.

Xusheng, Wang, Su Jinhai, and Zhang Hong, "China: Information Revolution, Defense Security," *China Computerworld (No. 30, p. 21),* World Wide Web, Infowar.com Web page, wysiwyg: //34/http://www.infowar.com/mil_c4i/mil_c4I_121897S.html-ssi, August 11, 1997.

ZDNET, *The Internet News Channel: Online Users Need to Beware of Password Poachers,* World Wide Web, Ziff-Davis Publishing Co., www5.zdnet.com/zdn...t/0620/ adnn0006.html, 1997.

Zuckerman, M.J. "U.S. Networks Most Vulnerable of Any Nation," *USA Today,* World Wide Web, lcs.usatoday.com/life/cyber/tech/ct033.htm, September 5, 1996.

CHAPTER 19

CYBERNATION:
THE AMERICAN INFRASTRUCTURE IN THE INFORMATION AGE

A Technical Primer on Risks and Reliability

CONTENTS

EXECUTIVE SUMMARY

INTRODUCTION

INFORMATION NETWORKS AND THE DOMESTIC INFRASTRUCTURE

THE TECHNICAL DIMENSION

FINDING SOLUTIONS

CONCLUSION

At a glance . . .

The automation—or *cybernation*—of the domestic infrastructure of the United States, in the transportation, finance, energy, and telecommunications sectors, which has been building for decades, has accelerated dramatically in recent years as advances in computers and information networks open new possibilities for improved service, lower cost, and greater efficiency. As a result, the United States has become a wired nation, a condition with implications that are not fully understood. [p.470]

The importance to the nation of infrastructure services makes attention to the reliability of their underlying information networks a necessity. The question is whether the marketplace will adequately anticipate and mitigate reliability deficiencies, or whether the nation will have to endure a major infrastructure problem in order to mobilize and act. [p.470]

The infrastructure of the United States has historically been very reliable. For most Americans, infrastructure disruptions have been more a nuisance than a nightmare. However, nothing guarantees that future disruptions will be similarly limited in national impact as past disruptions. [p. 477]

Three current trends raise concerns about the reliability of the automated infrastructure:

- Infrastructure services are becoming increasingly dependent on complex information networks which are potentially vulnerable to failure or disruption.
- The business environment is changing with deregulation, downsizing, increasing competition, and the entry of new companies into the market for providing infrastructure services.
- Infrastructure information networks are potentially becoming more accessible even as computer intrusions, already quite common, become increasingly sophisticated. [p. 474]

Network failures can be classified in terms of their *causes* and the *mechanisms* by which they are manifested.

- *Causes* range from natural phenomena such as weather, natural disasters, and other acts of God to deliberate destructive acts by persons intent on doing damage.
- *Mechanisms* range from chain reactions, in which small faults propagate and result in widespread disruptions, to the direct, independent failure of key components that in themselves represent major disruptions. [p. 483]

From a technical standpoint, it is not practical to focus exclusively on any one reliability threat. Like the interactions of prescription drugs, the remedy for one problem can interfere with the remedy for another. A holistic methodology for making the unavoidable tradeoffs is called for. [p. 488]

A technical agenda for addressing the reliability of infrastructure information networks consists of three steps:

1. Develop an analytical understanding of the specific reliability, vulnerability, and threat environment.
2. Establish a system engineering process which treats reliability as a primary parameter.
3. Maintain constant vigilance and continual learning to enhance reliability. [p. 485]

Neither the private sector nor the government can completely address infrastructure reliability alone. Developing consensus on the problem, as well as finding effective long term solutions, will require the sustained engagement of industry, utilities, the public, and government at all levels. [p. 472]

Areas for increased public policy attention include: [p. 493]

- Achieving consensus on what the minimum levels of reliability should be, what the threats re, what risks are acceptable, what protective measures should be taken, and how the costs should be met.

- Enhancing government/industry cooperation for identifying and characterizing reliability challenges, from whether and natural disaster prediction to intelligence collection on the threat of hostile attack.
- Focusing government and industry on the joint development of technical standards and methods to measures and certify reliability,
- Enhancing Federal/State government interaction to ensure consistent and appropriate attention is placed on infrastructure reliability.
- Defining the government research and development investment portfolio for network reliability.
- Working with other countries to develop compatible international legal regimes in cyberspace.
- Clarifying missions, responsibilities, and authorities of Federal Departments and Agencies in cyberspace.

The domestic infrastructure which underpins the economic life of our society increasingly depends on electronic networks for the flow of essential information. In sectors such as transportation, finance, energy, and telecommunications, computer networks have become indispensable in providing essential services that we take for granted. Air traffic data for the safe conduct of thousands of flights per day, financial transactions worth many millions of dollars daily, and control signals for operation of power distribution grids, railroads, pipelines, and the telephone system itself, all travel over electronic networks. Electronic networks have truly become the "nerves" of our infrastructure in yet another manifestation of the proliferation of information technology that characterizes the world of today.

How reliable are these networks? How can we ensure they are reliable enough? These pressing questions are not easily answered. Our critical infrastructure information networks face many reliability challenges, from natural disasters to human error, and from equipment failure to terrorists and computer hackers. From a technical standpoint, these are not different problems; they are different parts of the same problem. A systematic sector-by-sector analysis of threats and vulnerabilities, and a sustained system engineering process that emphasizes reliability are the technical ingredients of a successful approach to managing these risks.

As powerful a tool as technology can be, it is not the whole answer. Technology, and especially information technology, is best understood in its societal context. *People* represent both the strongest and the weakest links in the reliability chain. We should therefore not lose sight of the human element as we focus on the technical challenges of assuring infrastructure reliability. Instilling a culture of vigilance in the community responsible for the infrastructure is the most fundamental step in preventing reliability problems.

Deciding how much reliability we need for our infrastructure, against what threats, and at what cost are questions of public policy that will require the sustained consideration of stakeholders throughout society. This report seeks to promote a common understanding of the network reliability challenge in the technical and policy communities in private industry, public utilities, and government at all levels. The efforts of these diverse players, through the broad dialogue of democracy, will be necessary to effectively respond to this long-term challenge.

John H. Gibbons
Assistant to the President for
Science and Technology

EXECUTIVE SUMMARY

The infrastructure on which American society depends, in sectors such as transportation, finance, energy, and telecommunications, is becoming increasingly automated as advances in information technology open up new possibilities for improved service, lower cost, and greater efficiency. The automation—or *cybernation*—of the infrastructure has come about largely because it offers unmistakable economic and performance benefits. As a result, however, the United States has become a wired nation, with implications that are not fully understood.

The widespread application of information technology presents new challenges in what has historically been a highly reliable infrastructure. A changing public utilities business environment characterized by deregulation, corporate downsizing, increased competition, and new entrants to the market potentially places stress on the reliability of the national infrastructure. In addition, the nearly unconstrained application of computer technology in infrastructure control systems raises questions about the reliability of complex systems and their vulnerability to hostile intruders. Whether the forces of the marketplace will continue to provide infrastructure services with acceptable reliability in this environment remains to be seen. The importance to the nation of infrastructure services makes attention to the reliability of their underlying information networks a necessity.

Reliability challenges stem from both natural and manmade sources. To date, most of the national experience with major service interruptions caused by problems with infrastructure-related information networks comes from natural causes, accidents, or human shortcomings in design or operation. Infrastructure information networks will always be subject to these kinds of failures. However, as computer hackers increase in number and grow in sophistication, the threat of purposeful attacks by hostile actors looms increasingly large. Addressing this dual challenge in a measured way will be a long-term public policy priority.

This report describes the technical problem and sets forth a technical agenda for addressing network reliability. It uses the term *reliability* in its simplest sense -flawless, dependable operation, from the consumer's perspective, despite any reasonable challenge. It first defines a conceptual framework for characterizing network failures in terms of their causes and the mechanisms by which failures are manifested. It then outlines a technical agenda within this framework consisting of three steps: (1) developing an analytical understanding of the existing reliability, vulnerability, and threat environment; (2) establishing a system engineering process that treats reliability as a primary parameter; and (3) fostering a commitment to vigilance and a process of continual learning to enhance reliability.

> ***Although the nation is truly fortunate that major, sustained infrastructure outages have not occurred, this good fortune makes foreseeing and forestalling presumptive threats to infrastructure networks more difficult.***

Network failures can be classified in terms of their causes and the mechanisms by which they are manifested. Broadly stated, *causes* range from purely natural phenomena such as weather, natural disasters, and other acts of God to deliberate destructive acts by persons intent on doing damage. Between these two extremes lies a wide range of accidental or unintended occurrences with varying degrees of human involvement and varying human motivations.

Similarly, the *mechanisms* by which failures come about vary between two extremes as well. On one hand, a localized failure can become widespread through a chain reaction, in which a subsystem or component failure induces other failures, ultimately propagating through the network until overall performance is significantly degraded. The power outage that affected a large region in the western United States in summer 1996, for example, was attributed to a downed power line in Oregon which caused control system reactions that took generators in several States off-line in succession. On the other hand, major disruptions can occur that are not the result of chain reactions, but rather the independent disablement of critical subsystems. The destructive power of the Northridge, California earthquake in 1994, which interrupted electrical power and telephone service for millions of people, for example, caused outright failure of networks and did not depend on a chain reaction. The first failure mode can be likened to the domino effect; the second is more akin to upsetting the game table and knocking the dominoes to the floor. Of course, a great many events that have some qualities of each fall within these bounds.

Many of the recognized threats to the information networks supporting the domestic infrastructure have not actually been experienced. Although the nation is truly fortunate that major, sustained infrastructure outages have not occurred, this good fortune makes foreseeing and forestalling presumptive threats to infrastructure networks more difficult.

The first step in improving network reliability is to understand the existing reliability, vulnerability, and threat environment. This requires a detailed examination of the network architecture, physical layout, hardware and software, communications links, human factors, and operations. The finding of such an examination can guide a reliability engineering process.

The second step is to establish reliability as a primary tradeoff parameter in a system engineering process. Network configuration changes should not be implemented without a careful assessment of the tradeoffs involved. A strategy designed to counter one threat may increase vulnerability to another. Strategies narrowly focused on one aspect of reliability may introduce new vulnerabilities. The likelihood and severity of a particular problem may not justify the cost of a proposed solution. Additionally, the cost and performance implications of every strategy need to be understood. A structured methodology for making the unavoidable design tradeoffs between such primary factors as performance, cost, and reliability is essential.

The third essential step of the technical agenda is to foster a commitment to vigilance and a culture of continual learning to enhance reliability. As critical infrastructure information networks grow in size and complexity, there is an urgent need for institutionalized methods for capturing and applying the lessons of experience. Tools and procedures for detecting, reporting, and reacting to network problems all need to be developed and strengthened. An equitable, institutional means, within clear statutory limits, for the timely two-way flow of relevant intelligence information and incident data between government and the public utilities, which protects business-sensitive data as well as sources and methods, would do much to clarify the threat environment and allow for an effective response.

Although industry has a vested interest in assuring the reliability of the infrastructure, the federal government has an indispensable role as well. Neither the private sector nor the government can completely address infrastructure reliability alone. The national interest can only be served with the sustained engagement of industry, utilities, the public, and government at all levels.

INTRODUCTION

One of the great engineering marvels of the ancient world was the infrastructure of roads and aqueducts developed by the Romans. The road network, begun in about 300 BC, ultimately consisted of some 50,000 miles of hard-surfaced highway, some of which still survives today. Motivated by military needs, it also facilitated trade, agriculture, mail delivery, and made possible the establishment and administration of Roman rule in the far reaches of the empire. The aqueducts, ambitious projects even by today's standards, brought water to the city for public and private consumption, supplied baths and fountains, and provided for irrigation and sanitation. The well-developed Roman infrastructure contributed immeasurably to the prosperity and economic vitality that are among the hallmarks of ancient Roman civilization. But for all its advantages, this infrastructure also created new and serious vulnerabilities, providing attacking hordes easier access to Roman cities and becoming a target of direct attack itself. As a consequence, the Romans invested heavily in the construction of walls and other fortifications along their highways and around their cities, and they enacted laws and decrees aimed at protecting the structures associated with the water supply.

Today, the infrastructure of the United States is itself an engineering marvel. On a daily basis, the domestic telephone system carries hundreds of millions of calls, domestic and global financial networks conduct trillions of dollars worth of transactions, and our electrical grid serves hundreds of millions of consumers with a total generating capacity measured in the hundreds of millions of kilowatts of power. In sectors such as transportation, finance, energy, and telecommunications, our infrastructure is the machinery behind the American way of life.

However, the national infrastructure today is more than just a larger, more modern and complex version of the Roman road and aqueduct system. Today's infrastructure has a fundamental, indeed momentous, distinguishing characteristic: *it is automated*. From the routing of telephone calls to the distribution of electrical power, from the separation of aircraft to the electronic transfer of funds, the domestic infrastructure operates through automatic information networks. In all sectors, computer networks are an integral part of infrastructure operations—controlling processes, conducting transactions, dynamically adjusting capacity in response to usage, mediating communications among distributed components, and conveying information to human operators. This trend towards cybernation has been building for decades, but it has accelerated dramatically in recent years.

> ***Today's infrastructure has a fundamental distinguishing characteristic: it is automated.***

In today's dynamic business environment, increased reliance on automation makes good business sense. Automation with information technology enables new and better service offerings, more efficient operations and use of resources, and the potential for competitive advantages through greater responsiveness to customer demand. Information networks can provide managers with remote access for overseeing and managing their systems and can make it possible for them to tailor infrastructure services to specific customers.

It is worth considering whether the rapid and widespread adoption of information technology, for all its benefits, might also introduce vulnerabilities that could reduce the dependability that the public expects of the infrastructure. Everyday experience shows that when complicated computer systems fail, the failure is often both sudden and complete. "The computer is down" is a familiar lament for all who have endured the temporary inconvenience of computer problems in the workplace, the supermarket, the ticket counter, or the rental car agency. Such aggravations, however, are dwarfed by the potential problems that lie in the wake of comparable malfunctions in the computer networks supporting the nation's infrastructure.

The complex computer networks on which infrastructure operations increasingly depend are subject to failures just as any other manmade system. Human failings in design, construction, or operation, along with the effects of nature, aging, and natural disaster all have clear reliability implications. Any of these unavoidable problems could result in major disruptions of infrastructure services. Even more sobering is the possibility that remote access capabilities, so beneficial for customer service, could also allow computer terrorists—latter day barbarians at the gate—to deliberately disrupt infrastructure services by interfering with the information networks of the underlying control systems. In the extreme, disruptions or failures could threaten the well-being of society and undermine national security.

This report focuses on the reliability of the information networks that support the domestic infrastructure. It uses the term reliability in its simplest sense—flawless, dependable operation, from the consumer's perspective, despite any reasonable challenge. This report seeks to foster a common understanding among the technical and policy communities on the nature of the challenges to network reliability and the means to confront them. In doing so, it considers two questions:

- *How reliable are the critical infrastructure information networks?* An understanding of the technical problems and a sense of proportion about threats and vulnerabilities are essential to ensuring that the right priority is placed on addressing them.
- *How can society be certain that critical infrastructure information networks are reliable enough?* Achieving consensus on the appropriate levels of reliability, and on approaches for meeting them with acceptable costs, is an enduring challenge as the automated infrastructure evolves.

A framework for seeking answers to these questions is laid out in the pages that follow. Although this report concentrates on technical issues, technology alone is not a sufficient response to the reliability challenge. A complete approach must include operating procedures, training and awareness, personnel practices, and organizational factors in combination with technology. Indeed, technical solutions already available are not always effectively implemented. Furthermore, a framework for addressing infrastructure reliability as a public policy challenge is a prerequisite for delineating the issues and reaching consensus on the problems and the range of appropriate solutions.

INFORMATION NETWORKS AND THE DOMESTIC INFRASTRUCTURE

Electronic control systems are a common, though generally uncelebrated, feature of modern life. Familiar examples abound, from the simple thermostat to automobile cruise controls to automatic cameras. Each purposefully regulates a physical system to achieve a performance objective: if too hot, turn off the heater; too slow, add gas; too much light, reduce the aperture and increase the shutter speed. The technical discipline of automatic control is called *cybernetics*. It employs *feedback*—the use of measurements of present *output* to influence the next *input*—to converge on a desired operating condition. Electronic signals representing these measurements and control inputs are the coin of the cybernetic realm.

THE CYBERNETIC INFRASTRUCTURE

Broadly speaking, the control of any physical process with electronic signals is an application of cybernetics. In this sense, computer networks are the cybernetic control system of the domestic infrastructure. The "cyber" components of the infrastructure, including the computer hardware, software, communication links, and the abstract information embodied in them, make up the nervous system of the infrastructure on which the American public depends.

The sectors of the domestic infrastructure, as well as the cybernetic systems that support them, all have distinctive features, including:

- The energy sector provides power to meet the needs of the public in all aspects of modern life. It delivers energy in the form of electricity, oil, and natural gas, and has significant physical plant consisting of production facilities, distribution networks, substations, and rights-of-way. In this sector, the underlying cybernetic networks are made up of supervisory control and data acquisition equipment and associated communications links which control switches, relays, pumps, and valves throughout the distribution system. Its communications links often use the same distribution lines and rights of way as the infrastructure itself.
- Electronic transactions within the financial services infrastructure underpin the entire national economy, as well as the operations of the other infrastructure sectors. This sector depends on communications links and geographically distributed computer data bases, and uses electronic networks for the transfer of funds for consumer and business-to-business transactions, inter-bank transfers, stock, bond, and commodities markets, and government-to-government financial transactions.
- The transportation infrastructure, in addition to providing the mobility of personal travel for people, also delivers the manufactured goods and

agricultural products that are the lifeblood of commerce. In this sector, information networks are used for traffic control, navigation, and separation in the air, on the sea, on coastal and inland waterways, and on the ground.

- The telecommunications infrastructure is unique in that it not only is designed to deliver a service—the ability to communicate—but it also often comprises the signal channels on which the other sectors rely for the flow of their own cybernetic information. For the public telephone network itself, the cybernetic network is the signaling and call routing system, and the switches and signals that control individual connections. More broadly, it includes the internal corpo-rate information networks and computer data bases that telecommunications companies use to support the operations, administration, maintenance, and provisioning of the wide range of services they offer, from "plain old telephone service" to digital, wireless, broadband, and customized subscriber services.

Besides these individual characteristics, the different sectors of the domestic infrastructure have much in common, including:

- The sectors serve a wide variety of customers throughout society. Major interruptions in the services of any sector could have serious and widespread health, safety, and national security implications.
- There are numerous interconnections and mutual dependencies among the infrastructure sectors and among the information networks that support them. The public telephone network, for example, relies in part on the power grid, the power grid on transportation, and all of the sectors on telecommunications and the financial infrastructure. Most sectors employ the public telephone network for at least some of their cybernetic channels. Most control networks also have some connection to public networks, many to the Internet. Additionally, there are shared rights-of-way in many locations throughout the country.
- The infrastructure is inherently regional, national, and even global in scope. All sectors have components distributed over wide geographic areas.
- The infrastructure sectors are owned and operated predominantly by private industry, with various sector-specific interfaces with Federal, State, and local governments.
- Varying degrees of coordination exist among providers within a sector, but there is no complete central authority within or among sectors. Approaches to reliability vary by sector, ranging from voluntary self-regulation to various forms of partnership between the private sector and the government. Some degree of government regulation is the norm within all sectors, despite a general trend towards deregulation.

NETWORK RELIABILITY AND PUBLIC POLICY

The domestic infrastructure, although largely in private hands, is of such importance to the well-being of the nation that the government has an abiding responsibility in seeing that it best serves the national interest. For decades, government and industry have worked together to establish minimum levels of service, fair prices, and equitable access to infrastructure services for the American people. As the vulnerabilities of the automated infrastructure become understood, reliability takes its place as an explicit public policy objective as well. Simply put, the reliability objective is to provide, at reasonable societal cost, flawless, dependable infrastructure services that can withstand foreseeable challenges without interruption. Pursuing this goal will be a long-term focus of public policy.

> ***The concern for today is whether the marketplace will adequately anticipate and mitigate reliability deficiencies, or whether the nation will have to endure a major infrastructure problem in order to mobilize and act.***

The private sector has historically taken the lead in setting and meeting reliability goals in most infrastructure sectors. This approach has been highly successful. Because reliability deficiencies affect the corporate bottom line, either by disrupting revenue-producing services or eroding customer confidence and loyalty, industry can be expected to continue responding to credible reliability threats in the future. However, the industry is undergoing rapid changes in all sectors. Corporate reengineering, downsizing, the entry of new service providers due to deregulation, and the almost unconstrained application of computer network technology are all putting new pressures on formerly staid public utilities. The concern for today is whether the marketplace will adequately *anticipate* and *mitigate* reliability deficiencies in this highly dynamic environment, or whether the nation will have to endure a major infrastructure problem in order to mobilize and act.

The traditional policy tools available to the government for working with the marketplace to achieve national objectives—including legislation, regulation, licensing, tax and rate-setting regimes, and other inducements—all offer important options in the reliability arena. However, none of them can be effective unless and until there is consensus on what the minimum levels of reliability should be, what the threats are, what risks are acceptable, what protective measures should be taken, and how the costs should be met. At the present time, these questions are far from settled. Even the terms in which reliability thresholds should be expressed in the various sectors are open to debate. Additional policy emphasis can help address these difficult questions, but ultimately societal consensus will be a product of the broad democratic process. Part of the government's responsibility is to stimulate the public discourse and provide avenues for it to reach fruition in public policy. Government-sponsored forums, such as the Federal Communications Commission's Network Reliability and Interoperability Council in the telecommunications sector, can be instrumental in this regard by bringing together service providers, equipment manufacturers, standards-setting organizations, and consumer organizations in a public forum to develop recommendations for enhancing network reliability.

The Federal government, to create a climate that encourages infrastructure reliability through private sector initiative, can play a vital role by fostering innovation and commercialization, working with industry to develop technical standards, encouraging the development of methods to measure and certify reliability, and making it easy for industry to make reliability improvements. The government can also contribute by applying Federal capabilities to identify and characterize reliability challenges, from weather and natural disaster prediction to intelligence collection on the threat of hostile attack. Identifying appropriate activities in these areas, and putting them into practice, calls for continuous public policy attention.

Federal government interaction with State governments, and with State and regional regulatory commissions, is also essential to framing the question of how much reliability is needed in the infrastructure and to ensure that consistent and appropriate attention is placed on the reliability of the associated information networks. The Federal government can be helpful in providing uniform guidance to the States where appropriate, and in coordinating and focusing the resources of the Federal departments and agencies.

Public policy extends to government investment in research and development as well. Federal investment in science and technology—conducted at federal laboratories, universities, and in industry—has been instrumental in the unfolding information revolution. The advanced development of the integrated circuit, creation of the Internet, development of a global communications infrastructure, and other products of government research have done much to enable the automation of the infrastructure. Wise federal investments today in areas such as intrusion detection and prevention, robust network architectures, configuration management, and secure communications will pay similar dividends in the future.

Reliability is also served by laws that delineate the boundaries of permissible behavior. As an international arena that disregards national boundaries, cyberspace presents unique legal difficulties. Distance and geography do not impede the trespasser, bandit, or terrorist in this realm, and questions about jurisdictions, sovereignty, and the applicability of laws frequently arise. The Federal government has an obligation to work with other countries to develop compatible cyberspace legal structures and to foster worldwide cooperation among law enforcement agencies.

Similarly, cyberspace threats to the infrastructure are challenging existing boundaries between the national defense, intelligence, law enforcement, and regulatory roles of the U.S. government. Clarification of missions, responsibilities, and authorities in this new context are needed, and will necessarily involve all three branches of government, Executive, Legislative, and Judicial.

Throughout the policy realm, the sustained engagement of an informed public will ensure that the choices made will best serve the national interest.

THE TECHNICAL DIMENSION

The information networks which support the domestic infrastructure are complicated, evolving systems. Unlike Minerva, the mythological goddess who sprang fully grown from the head of Jupiter, they are not generally the product of a single top-down design. Rather, they have developed over time and continue to grow and change with the addition of new technology, new features, and new capabilities. This section identifies some of the main technical challenges inherent in networks of this type, and presents a conceptual framework for considering possible failure scenarios.

Most of today's cybernetic networks are actually combinations of networks, interconnected and interdependent. Interactions among these subsystems are critical to overall network performance, indeed they are the essence of network performance. Because the system also interacts with the real-world environment, the interactions among subsystems are not necessarily predictable and sequential, like the steps of an assembly line process, but can be essentially random, unsychronized, and even unanticipated. Many transactions are generated automatically by computers pursuing the logic with which they are programmed. The term "intelligent" is sometimes used to describe these networks, and it can be a fitting characterization, given the relatively autonomous nature of control systems.

Computer controlled subsystems often have little tolerance for variations—in sequence, content, or timing, for example—in the transactions they undertake with other subsystems. Small margins, like highway tailgaters, are vulnerable to dangerous chain reactions. Subsystems which have small margins—often called "tightly coupled" subsystems—are a reality within complex computer networks.

By their nature, infrastructure cybernetic systems must operate continuously, controlling physical components in real-time and ultimately serving human customers. They must deal with essentially random simultaneous inputs from a great many sources -countless telephone calls and personal financial transactions, the energy usage of legions of subscribers, the flight paths of thousands of aircraft, as well as the control inputs of managers throughout the networks. Control systems, optimized for statistically-important usage profiles, must also be able to handle unusual and even unlikely inputs. Telephone companies, for example, are well familiar with (and prepared for) the "Mother's Day" phenomenon, in which predictable usage spikes are prompted by national events. Nevertheless, the full range and diversity of input conditions is difficult to define beforehand. Experience teaches that the problem of unanticipated input conditions is formidable when designing for high reliability.

Invariably, human beings are key elements of infrastructure cybernetic systems. Nothing can replace human judgment, and no network is designed to be completely "hands off". However, the possibility for human error in the operation of complex systems is ever-present. The complexity and speed of interaction among network elements can easily exceed the ability of operators to assess and respond to problems. The alerts, indications, and displays available, along with the input actions allowed and the time available to make them, will frame any human intervention. Human beings also often provide the linkage path between two otherwise independent subsystems, creating unexpected feedback paths and opening up new possibilities for unanticipated subsystem interactions. Finding the right balance between human and machine control is a technical problem that goes to the heart of the network reliability challenge.

The physical integrity of hardware components and communications links is a basic requirement for a reliable network. Exposure to the elements, continuous duty, and simple aging impose unavoidable stresses on infrastructure hardware. The cybernetic system must ultimately interface with electromechanical devices—the switches, relays, valves, and motors that make the infrastructure function—and must be robust enough to accommodate performance variations in these components and continue to function despite degradation.

One of the motivations for employing information networks in infrastructure control systems in the first place is to make the control systems more accessible and responsive to management inputs. However, accessibility once afforded is not easy to constrain. For example, many networks are migrating to common technologies—especially Internet technology (never designed to be highly secure)—for reasons of cost and efficiency. The more common a networking technology becomes, the more widely known and exploitable are its weaknesses. Additionally, the move to fewer, more standardized configurations simplifies a potential attacker's problem. The use of public network as bridges between internal corporate networks is also becoming increasingly common. Lacking security discipline, employees sometimes place unauthorized and unprotected dial-up modems on internal networks, creating potentially serious and unrecognized vulnerabilities. Deregulation, particularly in telecommunications and energy, allows new entrants to legitimately gain access to control networks that were previously proprietary or carefully protected. Whether by design or not, infrastructure control systems are increasingly accessible to outsiders.

Designers must therefore be concerned about the full range of possible intrusions into infrastructure cybernetic systems. The potential perpetrators of such intrusions include the cyberspace equivalent of the graffiti artist as well as the hardened cyber-terrorist. Hackers are growing in number and sophistication and have increasingly advanced tools. They are also organized, freely exchanging tools, techniques, and information on vulnerabilities worldwide through the Internet. Deliberate measures to protect against this threat are a necessity, but its specific measure-countermeasure nature makes development of broadly applicable defenses extremely difficult. It is usually necessary to find specific defenses against specific attacks. These defenses, in turn, become targets for future attack. Presently there is much about this threat that is not known.

> ***An approach that seeks to manage risks by weighing cost, performance, and reliability tradeoffs along with the likelihood and severity of specific threats is the most sensible.***

Finally, infrastructure information networks are inherently dependent on software. Ensuring the reliability of software-based systems is among the most difficult of engineering challenges. The cost of exhaustive testing to validate software intended for complex real-time environments often proves to be prohibitive, when it is even technically possible. Complicating this challenge is the fact that companies today are increasingly contracting with others, often in other countries, for the development of software. Such "outsourcing" can leave system integrators with little insight into the development and validation of critical control software. Commercial off-the-shelf technology is also being adopted more frequently, even though software design details are usually not available and reliability may be uncertain. Long term maintenance of software is made difficult by changing preferences in programming languages and lack of support tools for obsolete or orphaned systems. Generational differences in aging equipment can give rise to insidious software incompatibilities. Competition can pressure developers to rush software to market without sufficient testing. It is even possible for malicious code, deliberately and surreptitiously included in critical software during production, to go undetected in installation. Additionally, every software performance enhancement carries the possibility of introducing logical errors, undoing previous algorithm corrections, changing software timing performance, and even introducing coding errors, all of which can increase system vulnerabilities. A misplaced bit in a data structure or an almost-correct algorithm may work most of the time, but when these errors manifest themselves, the result can be a dramatic transition from normalcy to catastrophe.

The information systems that have brought automation to the infrastructure are inherently complex. Complexity by itself, however, is not a good indicator of reliability. Human beings are capable of producing very complicated creations that are nevertheless quite reliable, such as jet engines, microchips, skyscrapers, and pharmaceuticals, to cite just a few examples. For information networks, complexity might make a network fragile and prone to failure, or it might be a source of robustness. In any case, these terms are relative and not precise enough for making engineering or policy judgments. The susceptibility of a network to major disruptions can only be gauged by carefully assessing a great many technical and operational factors in the context of the total threat environment. In the end, an approach that seeks to manage risks by weighing cost, performance, and reliability tradeoffs along with the likelihood and severity of specific threats is the most sensible.

HISTORICAL EXPERIENCE

Over the years, there have been many incidents in which network-related faults caused infrastructure problems, including these prominent representative cases:

- A widespread electrical power blackout affected 15 States in the western United States, as well as some regions of Canada and Mexico, in July 1996. The outages affected some two million people, and caused airport delays and subway breakdowns from Denver to San Francisco. The cause of the outage was traced to an overheated 500,000-volt transmission line in northern Oregon which sagged into tall trees, short circuited, and shut down. Two other 500,000-volt lines subsequently became overloaded and shut down, followed shortly thereafter by shutdown of the Pacific Intertie, the main power artery between the Northwest and California. The disturbance rippled as safety systems automatically shed load to try to keep the system in balance. Increased power demand caused generators in California to shut down; generators in other regions shut down because they suddenly had too much power and nowhere to send it. A similar blackout affecting the same region was experienced the following month.
- A series of breakdowns that disrupted local telephone service for some 16 million customers in Los Angeles, Baltimore, San Francisco, and Pittsburgh in June and July 1991 was attributed to a defect in a few lines of computer code in critical algorithms of the signaling system. The manufacturer traced the problem to a recent upgrade in its software which had not been put through its customary thorough testing because the change entailed only a few lines of new code.
- In September 1991, an internal power failure at a Manhattan telephone switching center cut off approximately half of the long distance traffic of

the nation's largest long distance carrier into and out of New York City. This incident had a particularly serious impact on air traffic because the affected switching center also carried some 90% of the communications of the New York air traffic control center. Although no aircraft accidents were attributed to the outage, about 400 flights were canceled at the three major New York airports and tens of thousands of passengers were inconvenienced over an eight-hour period. The outage was blamed on "a combination of equipment failure and human failure." Under an agreement with the local power company, the telephone company had adopted the practice of turning off city power and relying on its own generators in periods of high electrical demand. In this event, however, workers failed to follow established procedures and confirm proper operation of the generators. Unluckily, failed rectifiers prevented the generators from delivering power, leaving the switching system to draw power entirely from its backup batteries. Alarm bells and warning lights went unheeded for six hours and the batteries became depleted.

- In July 1994, a software upgrade to the computers of the over-the-counter Nasdaq marketplace caused that system to go down for over two hours, cutting volume for the day by about one third, and affecting stock exchanges, trading desks, and stock-index mutual funds throughout the country. Nasdaq, a stock exchange with no trading floor, relies on a nationwide computer network for a trading volume of hundreds of millions of shares daily. The software problems were manifested directly on the mainframe computers located in Connecticut. The backup system in Rockville, Maryland, which was being upgraded at the same time to maintain compatibility, also failed.
- The Northridge, California earthquake of 1994 caused long distance telephone service outages for about two million people for approximately eight hours as two major switching facilities at Sherman Oaks failed. Thirty-five cellular sites were also out of service. However, while the earthquake damaged many telephone exchange buildings, most continued in operation. Customers unable to access long distance service still had dial tone and could call numbers within their local dialing area including local emergency response organizations.
- In January 1990, a piece of interface equipment in one of the telephone toll switching systems of the nation's largest long-distance carrier in New York City developed a minor hardware problem. The control software, which had recently been upgraded network-wide, entered its fault recovery routine, suspending new call processing briefly. However, a flaw in the software effectively prevented the switch from coming back into service and disabled backups in the process. The problem cascaded through the network with switches throughout the system also going out of service. The result was blockage of some fifty percent

of all switched traffic for that carrier nationwide for a period of seven hours. Of approximately 148 million call attempts made, 83 million were completed. Ironically, the software that caused the problem was intended to speed restoration of call processing after suspension.

- In 1965, and again in 1977, the Northeastern United States experienced massive and costly electrical power outages. In both cases, the problem came about due to a cascading series of events in which operators and automated components followed the logic with which they were trained or programmed, shutting down or disconnecting generators as a protective measure in response to anomalous conditions.

In some cases, small problems have snowballed into major disruptions. In others, the abject failure or destruction of key components brought about major service interruptions directly.

This small sampling of the historical evidence shows that there have been many "major" interruptions of infrastructure services. Nevertheless, on almost any meaningful scale, the automated infrastructure of the United States has been highly reliable. In recent decades, few infrastructure disruptions have had large-scale effects on the population at the national level. Disruptions and outages that have occurred have generally been selective, affecting subscribers by region, for example, or by choice of provider, or by some other discriminator particular to the situation. When outages have threatened the safety of large numbers of people, such as when electrical power has been interrupted during major winter storms, the conditions created were addressable by local emergency services, government disaster assistance programs, and the dedicated emergency crews of the utility companies. For most Americans, infrastructure disruptions have been more a nuisance than a nightmare.

Besides illustrating the range of network-related infrastructure problems, the examples cited here also underscore the critical importance of the mutual dependencies among infrastructure sectors. Telecommunications, energy, transportation, and finance are all bound together—literally as well as figuratively given that they often depend on the same fiber optic bundle. It is also important to realize that each of these incidents, and many others like them, prompted remedial actions—engineering, procedural, and policy changes -to help avoid their recurrence. This provides a significant institutional legacy on which to build for the future protection of the infrastructure.

Nothing guarantees that future disruptions will be similarly limited in national impact as past disruptions. However, past experience certainly does provide insight into *how* networks are likely to fail in the future. These examples show that in some cases, small problems have snowballed into major disruptions. In others, the abject failure or destruction of key components brought about major service interruptions directly.

The propagating "chain reaction" failure mechanism is characteristic of complex systems with tightly coupled subsystems. This mechanism depends on the dynamics of the system itself, in that seemingly inconsequential events trigger multiple failures through an unanticipated domino effect. Further, the problems can be exacerbated by the very features and procedures intended to protect against failures. For this failure mechanism, the real problem is not the triggering event itself, but the interaction of anomalous operating modes among subsystems that it sets in motion. Like Mrs. O'Leary's cow and the Great Chicago Fire, the event which triggers a cascading system-wide catastrophe may have no intended nor apparent connection with the ultimate outcome.

Not all failures are the result of chain reactions, however. The direct, independent failure of certain components, or combinations of components, can result in the same level of disruption. Natural disasters are the most familiar cause of this type of network problem, but systemic design flaws can have the same apparent result.

For actual events, the failure mechanism depends both on properties of the system in question and on the particular circumstances of the incident. The two broad mechanism described here represent the conceptual extremes. In real life, failures usually have characteristics of each and fall somewhere between the two.

MALICIOUS ATTACKS

Albert Einstein once observed. "The Lord God is subtle, but malicious he is not." The same cannot be said of man. To fully understand network reliability challenges, it is therefore necessary to also consider the possibility of deliberate attacks.

Although the illicit penetration of computer systems is an increasingly familiar occurrence, deliberate acts of computer intrusion or sabotage are most often associated with the Internet rather than with infrastructure cybernetic systems. This distinction, blurred even now, will be lees clear in the future. Further, as personal on-line computer access is further woven into the fabric of society, it too will one day be considered an essential infrastructure service.

> ***On almost any meaningful scale, the automated infrastructure of the United States has been highly reliable. For most Americans, infrastructure disruptions have been more a nuisance than a nightmare.***

To date, although there is little historical experience in which cyber attacks of any kind caused serious disruptions of infrastructure services, experience that has accrued with computer hackers is helpful in understanding how this threat might be manifested.

- In 1988, the Internet "worm", a computer program designed to consume the memory and resources of computers, was deliberately released on the Internet. Thousands of computers were affected before the worm was brought under control. This incident is representative of a large class of attacks which involve the introduction of self-replicating malicious code. By their nature, such intrusions are not selective in their destructive effect.

- In 1994, more than 150 intrusions were made to the Air Force's Rome Laboratory by two hackers using specialized software that allowed their intrusions to masquerade as legitimate transactions. The attackers were able to seize control of Rome's support systems for several days, establish links to foreign Internet sites, copy and download critical data, and successfully attack systems at other government facilities, defense contractors, and private sector organizations. The Air Force, which did not even recognize the attack for at least three days, estimated the cost to the government at over $500,000, not including the value of the information that was stolen. This class of hacker intrusion is characterized by the use of sophisticated tools and techniques to seize control of a network and take actions normally reserved for trusted system managers. A spate of intrusions to government World Wide Web sites in 1996 serve as another example of this class of attack.

- In 1996, several Internet service providers were victims of deliberate attacks from unidentified sources using sophisticated software to overload servers with hundreds of messages per second. The messages, called synchronization requests, contained false return addresses, which confused and rapidly tied up the servers and rendered them unable to handle legitimate transactions.

- The World Trade Center bombing in New York City in 1993, and the Oklahoma City bombing in 1995, although not attacks on infrastructure *per se,* reveal the potential for infrastructure disruption that domestic terrorism holds. Destructive physical attacks on network components could cripple an infrastructure sector's cybernetic system and cause major service interruptions that would be difficult to alleviate.

The wide deployment of information technology potentially puts destructive capability, once the province of nations, into the hands of individuals.

These examples, by no means exhaustive, give an idea of the broad classes of malicious attacks that could be mounted in cyberspace. At least three distinct types of cyber attack can be postulated which could result in serious infrastructure disruptions:

Computer hackers. Small-scale, even unsophisticated, intrusions into information networks by "cyberspace joyriders" could induce major network problems if they altered or destroyed data, overloaded input circuits, or caused locally degraded operations. An intruder in this category is motivated by curiosity, technical challenge, mischief-making, or the aim of stealing services, but could nevertheless trigger a cascading failure with widespread effect.

Anarchist attacks. An attack on infrastructure cybernetic elements could be mounted to help achieve a broader criminal purpose, or by anarchists who may not have a clear objective or a cogent strategy other than to disrupt and destroy. The perpetrator in this category is a purposeful actor intent on doing damage, but who has probably not attempted a careful assessment of the precise effects the attack would have. Attacks on critical network components such as communications links, control nodes, or switching stations could render equipment inoperative either through physical damage, or through corruption of software or data. In this case, the affected component could itself be important enough that disabling it could cause a major disruption, or it could initiate a cascading failure that results in a major disruption.

Coordinated cyber attacks. Attacks in this category are focused, organized, and carefully calculated to yield a specific outcome. The perpetrator of this type of attack is motivated by strategic political goals, and may employ the same types of tools used by the anarchist described above, although in a more sophisticated manner. Destructive Trojan horses or "logic bombs" could conceivably be placed in operating systems software to induce a systemic failure. An insider could use specialized knowledge to maliciously attack a network and induce major disruptions through normal management controls. A coordinated attack on multiple vital components or communications links could directly disable the cybernetic system without the dynamic effect of cascading subsystem failures. Since this type of attack is carefully planned, it is more likely to employ tools with direct, decisive outcomes than to rely on a relatively unpredictable chain reaction.

Attacks in each of these three categories could cause the same level of disruption, the differences being in the motives and methods of the perpetrators and the mechanisms by which the failure is manifested. The wide deployment of information technology potentially puts destructive capability, once the province of nations, into the hands of individuals. At the same time, the growth and increasing interconnectedness of computer networks offers additional possibilities for outsiders to break into proliferation of sophisticated tools are what had previously been closed, internal systems. Increasingly advanced hackers and the forewarning that any of these scenarios could be dangerously real.

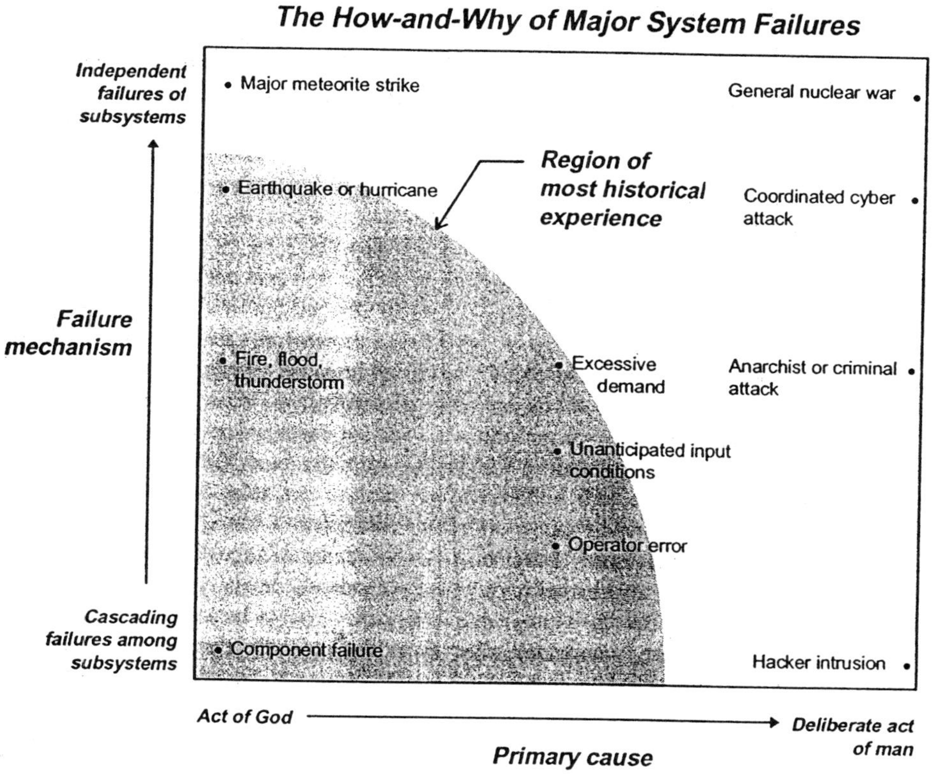

Figure 1. The failure scenarios chart.

This chart notionally depicts some possible scenarios in which major infrastructure disruptions are brought about by events affecting the supporting cybernetic system. It illustrates the ***how*** (failure mechanism) and the ***why*** (primary cause) of major system failures, and can be a useful tool for placing hypothetical scenarios in context. The specific details of a scenario, as well as properties of the particular network determine the precise placement of incidents on this chart.

A UNIFIED FRAMEWORK

The engineering challenge of building highly reliable networks is extremely broad. *Causes* of network failures cover the complete range from Mother Nature to human nature—from natural occurrences to the deliberate actions of hostile persons. Failure *mechanisms*also vary. Tightly coupled systems often fail through a cascading process in which a subsystem failure propagates in a chain reaction. In other cases, the independent failure of critical components is the dominant mechanism.

Figure 1 puts these two variables together in one framework that captures the *how* and the *why* of major network failures. This framework—conceptual rather than quantitative—is useful for comparing actual events with hypothetical scenarios, postulating and assessing scenarios to uncover potential vulnerabilities, and analyzing the risks and benefits of strategies for dealing with reliability deficiencies. Figure 1 makes the essential point that while challenges to reliability exist throughout the total space, historical experience to date is concentrated in one corner of it. Much about the network reliability challenge has not been experienced.

All of the events depicted by the failure scenarios chart represent network failures of potentially *equal severity.* An important discriminator among scenarios is their *likelihood.* Two probabilities need to be considered: the probability that the event will occur, and the probability, assuming the event did occur, that it will result in a major failure.

The overall probability, which is the product of the two, is an indicator of how seriously the threat should be taken.

For example, component failures are a certainty, but the probability that the loss of a single transformer, switching device, or sensor, for example, would trigger a chain reaction that disrupts a major portion of the network (although it does happen) is usually quite small. A meteorite strike, on the other hand, is a very unlikely event, but if it were to occur, there would be a high probability that it would destroy and disable infrastructure networks. This type of analysis can aid in understanding the relative priority of reliability threats.

There are many unanswered questions about how to characterize the cyber threat. Hacker intrusions are themselves an accepted reality, but the probability that a hacker could initiate a chain reaction that causes significant disruption of a network, even inadvertently, needs careful, detailed analysis. The probability of a coordinated cyber attack, and the probability that it would be effective if launched, also need examination. Producing a specific desired outcome by intruding into a complex network might be as sure as setting the hands of a precision timepiece, or it could be like expecting a Rube Goldberg invention to work as designed. A careful analysis of specific cases can shed light on vulnerabilities and can help identify the actors with the greatest chance of success. An effective response to cyber attack depends on understanding the specific attack methods, probabilities, and network failure modes.

FINDING SOLUTIONS

Addressing all the reliability challenges captured by the failure scenarios chart calls for a comprehensive technical methodology. The basic steps of such a technical approach include: first, developing an analytical understanding of the existing reliability, vulnerability, and threat environment; second, establishing a system engineering process which treats reliability as a primary parameter; and third, fostering a commitment to vigilance and a process of continual learning to enhance reliability. These steps, listed in Table 1, are discussed separately in the sections that follow.

Table 1. A Technical Agenda for Network Reliability

1. *Develop an analytical understanding of the specific reliability, vulnerability, and threat environment.*

2. *Establish a reliability engineering process.*

3. *Maintain constant vigilance and continual learning to enhance reliability.*

These three steps form the basis of a technical agenda for network reliability.

STEP 1: UNDERSTAND THE RELIABILITY ENVIRONMENT

Two metaphors give insight into the network reliability problem. On one hand, a network can be compared to a house of cards for which mutual dependencies are so great that removing one component can cause the whole precarious structure to collapse. On the other hand, a network can be compared to a chain-link fence which, although also made up of interdependent parts, is flexible and robust over a wide range of inputs. A breach in one section of the mesh does not cause complete failure of the fence. And once breached, a fence can be repaired—it does not need to be reconstructed in its entirety.

Neither of these metaphors is perfect, but both are illuminating. Clearly the chain-link fence metaphor is the more desirable of the two. But whether or not it applies in a given infrastructure sector is largely a function of the engineering and operational decisions made in the course of the network's life. Fundamentally, reliability is about the extent to which the operation of individual components affects the overall performance of the system. Robust systems are robust precisely because overall system operation is tolerant of component and subsystem faults. Since information networks already exist in every infrastructure sector, the first step is to characterize these networks and develop a full appreciation of their strengths and weaknesses. This can be aided by a thorough assessment in the following areas:

> ***Since information networks already exist in every infrastructure sector, the first step is to characterize these networks and develop a full appreciation of their strengths and weaknesses.***

Operational concept of the infrastructure sector. A detailed description of what the information network is used for, including a cataloging of the options for external entry, ability to execute commands remotely, and the nature and range of computer control available, is essential. How the system responds to failures, disruptions, or data corruption is also critically important information.

Network architecture and information flows. A detailed technical description of the network is needed, including the physical and logical layout, the flow of information, and the major nodes and interconnections. Particular attention should be given to how subsystems at all levels interact with each other under normal and degraded conditions, and how tolerant they are of faults, delays, malfunctions, or errors in other subsystems. This analysis would reveal how tight the coupling is among network subsystems. Formal mathematical methods, computer modeling and simulation, and transaction analysis are useful tools for describing and analyzing architectures and information flows.

Network components. Operating limitations or design flaws in supervisory control and data acquisition components, gateways, firewalls, routers, servers, or other critical nodes—especially those controlled by software—could constitute weaknesses that undermine overall network reliability. All components in the system should be examined from a reliability perspective for the applications in which they are used. Migration to common components and off-the-shelf equipment also potentially increases the likelihood of exploitable security weaknesses.

Signal protocols and transmission methods. Signals of interest can range from simple analog signals on dedicated circuits to complex digital protocols multiplexed onto high capacity data channels including fiber optic cable and satellite links. Signals may or may not be encrypted for transmission, and may be susceptible in varying degrees to monitoring, interception, interference, spoofing, or jamming. The design choices made have direct bearing on the network's vulnerability to physical and electronic disruption.

Human factors. People are at the heart of virtually all information networks, as system managers, operators, engineers, and technicians. The human interface -through controls, displays, alert and warning indications, and equipment layout—brings human judgment into the control loop. Operator carelessness, inattention, or procedural error are ever-present hazards. Well-intentioned workarounds of established procedures and system configurations can undermine reliability. The personal reliability of key people is critical to reliability of the network. A complete understanding of the human factors environment is vital to an overall assessment of system vulnerabilities.

Existing security environment. Reliability is heavily influenced by the existing network security environment. An understanding of the priorities and tradeoffs that led to the current configuration is necessary. There may be known vulnerabilities or methods of intrusion for which security measures have been designed. The security of password files, access to supervisory features, the integrity of access logs, and the ability of a system administrator to detect intrusions are all relevant factors. Security cannot be considered separately from analysis of the threat environment. Tools and techniques available to hostile intruders must be understood if they are to be effectively countered. Security provisions in particular are very dependent on implementation factors. For this reason, security discipline in all aspects of network operation should be assessed.

When compiled and combined with operational experience, the findings in these areas make up a comprehensive network reliability data base that can serve as the foundation of a sound system engineering process. Such compilations, however, should be carefully safeguarded to prevent access by those who would misuse such information to exploit system weaknesses.

STEP 2: ESTABLISH A RELIABILITY ENGINEERING PROCESS

In an ideal world, an information network could be designed "right" from the start and it would then function "correctly". However, in the real world, networks evolve and change with time and technology, often growing more by accretion than by design. The demands of the marketplace also evolve. In this dynamic environment, a fixed design is not likely to be sustained in the long run. The "perfect" network unfortunately cannot exist -there is a constant need to engineer solutions for specific problems.

Design principles

From the standpoint of reliability, three design principles should guide this ongoing engineering process:

Reduce the possibility of disruptions occurring. This "prevent defense" principle focuses on the static and external features of the design. It suggests strategies aimed at minimizing the possibility of component and subsystem failures, such as controlling the system's environment, thoroughly validating components, physical and cyber security, and personnel practices. It also fosters the recognition and avoidance of systemic weaknesses and single point failures.

Minimize the effect of disruptions which do occur. This principle emphasizes the dynamics of failure within a network. It focuses on the details of cascading failures, the interaction of subsystems and their failure modes, the detection of anomalies, and system tolerance to degraded components. In real-world occurrences, cascading network failures ultimately are contained at some level. An appreciation for "what stops the snowball," that is, what constrains cascading failures, gleaned from operational experience, may give valuable design insights, and should also be encouraged.

Design for efficient recovery. This principle focuses on operations and highlights the importance of recovery procedures, equipment re-initialization sequences, and the human interface in the aftermath of a disruption in which backups fail, prove inadequate, or become irrelevant. Providing the ability to restore normal operations from any of a huge theoretical number of failure states requires concerted attention at all stages of the engineering process. Recovery operations also present the greatest need to improvise; anticipating this need may suggest design features that do not affect design integrity but that could make a significant difference if reconstituting ever became necessary. For example, in the aftermath of the 1965 northeast power blackout, a US Navy destroyer, the USS Bristol, sailed from the Brooklyn Naval

Shipyard and provided ship's power to one of the electric company's substations. It is doubtful that this particular need could have been foreseen in its details, but it is not unrealistic to ask designers to consider the general need for expedient or provisional repairs.

Like the interactions of prescription drugs, the remedy for one problem can interfere with the remedy for another.

Design tradeoff studies

In implementing these design principles, strategies relevant to each region of the failure scenarios chart of Figure 1 are pursued. For example, to address cascading failures, increasing the margins among interacting subsystems so as to reduce the possibilities for unsuccessful transactions may be appropriate. To reduce the possibility that independent failures would result in major disruptions, measures such as distributing key components geographically, adding redundancy, and diversifying the operating software may all be worthwhile. Improving physical security or strengthening barriers against network intrusions could help prevent failures caused by deliberate attacks.

However, many strategies, useful singly, are at odds with each other and with network performance objectives. For example, diversifying the software by employing subsystems with different operating systems may avoid certain single-point failures, but it may also impose serious cost and performance penalties. Furthermore, protection features however well-intentioned may introduce additional vulnerabilities and new failure modes.

To illustrate, Figure 2 depicts a single example strategy (increasing component redundancy) overlaid on the failure scenarios chart. This strategy, pursued to reduce the effect of independent failures of critical components, may actually be harmful in other scenarios because it increases subsystem interactions and provides additional paths by which a failure can propagate. Deploying the redundant components may also create more intrusion options, thus exacerbating physical security concerns.

The crucial point is that any strategy, if pursued in isolation, may actually reduce overall reliability rather than enhance it. A focus on one type of threat or one region of the failure scenarios chart can increase the vulnerability to, and the effects of, another. Like the interactions of prescription drugs, the remedy for one problem can interfere with the remedy for another. Clearly, a holistic methodology for making the unavoidable tradeoffs is called for. The cost, performance, and reliability effects of technical options should all be weighed in the context of the likelihood and severity of the threat. Table 2 lists some of the questions which design tradeoff studies should address, including whether a proposed solution will work, whether it can be implemented, what the marketplace effects are, and how the national interest is met.

Design tradeoff studies are the centerpiece of the system engineering process. It is critically important that reliability be included along with such principal tradeoff parameters as cost and performance when any engineering change is considered.

It is not erroneous to talk of a system engineering process for networks that have no single owner who possesses complete configuration authority. The prerequisite is that there be a firm foundation of technical standards and practices so that decisions made at what is inevitably the subsystem level are not far from optimum for the total system. Industry coordinating committees, national and international standards organizations, and technical dialogue are all essential for getting designers, equipment manufacturers, service providers, and installers on the same page of the textbook.

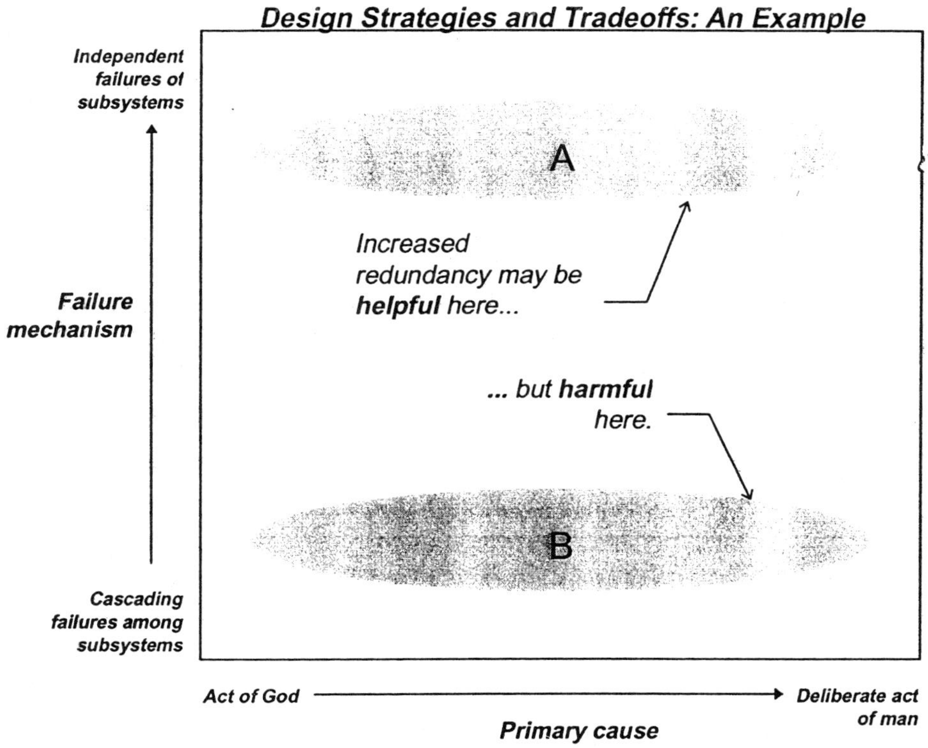

Figure 2. An example of the need for tradeoffs.

This chart illustrates the challenge of reliability engineering. In this example, a strategy for enhancing reliability that is helpful for scenarios which fall into region A might be harmful for scenarios in region B. Additionally, it might conflict with another strategy, impose cost or performance penalties, or introduce additional vulnerabilities. A methodology for making informed analysis and tradeoff is essential.

STEP 3: MAINTAIN CONSTANT VIGILANCE AND CONTINUAL LEARNING

The third essential step of the technical agenda is to foster a commitment to vigilance and a culture of continual learning to enhance reliability. Although this is easy to agree with in principle, it is often neglected in practice.

Specific needs include:

Early warning of hostile activities. Early warning of potential and actual hostile activity directed at the public infrastructure would allow both the service providers and the law enforcement community to take measures to prevent or minimize the disruption. An equitable, institutional means, within clear statutory limits, for the timely two-way flow of relevant intelligence information and incident data between government and the public utilities, which protects business-sensitive data as well as sources and methods, would do much to clarify the threat environment and allow for an effective response. Some threats, particularly cyber attacks, may be diffuse and only identifiable when data from a number of sources is aggregated. Numerous small and widely distributed anomalies that escape notice individually may, when compiled and correlated, be indicators of a systemic problem. In these cases, infrastructure network control centers could be important sources of strategic warning for the same reasons that global television news networks have become *de facto* sources of intelligence—on-scene presence and direct observation of real events. What is lacking, however, are the technical and institutional means to synthesize data from many sources and draw meaningful and timely conclusions. This is an area in which government and industry can work together for the benefit of all.

> ***Infrastructure network control centers could be important sources of strategic warning for the same reasons that global television news networks have become sources of intelligence—on-scene presence and direct observation of real events.***

Tools to detect and characterize network anomalies as they occur. At the operational level, network intrusions are difficult to detect because they can masquerade as legitimate transactions or go unnoticed in a busy network. In many networks today, successful intrusions are more likely to be detected by their effects rather than by any discernible telltale signature. Trustworthy tools for detecting anomalies early and judging their seriousness would help enable system administrators to take corrective action before major problems develop.

Management controls for reacting to disruptions. A range of system management controls including, for example, such measures as reconfiguration of the network, increased security in response to intrusion attempts, or decreased functionality to limit range of control can contain disruptions and prevent them from worsening. A means for selecting an appropriate response from among such control measures in an actual event is also needed.

Effective internal reporting procedures. Although self-monitoring, self-healing information networks may be on the horizon, ultimately the responsibility for recognizing problems and doing something about them rests with people. Effective internal reporting procedures need to be developed, implemented, and followed within the companies that operate infrastructure systems. Some industry-wide uniformity in reliability reporting, similar in concept to the standards of financial accounting and reporting established by the Financial Accounting Standards Board, can help put discipline into the process of capturing and applying lessons learned.

Mechanism for sharing reliability information among competitors. Private sector companies involved in providing infrastructure services should be able learn from each other. Mechanisms for sharing information—within legal limits—on vulnerabilities, incident data, technical solutions, and best practices among network managers which protect the confidentiality of proprietary or business-sensitive information are urgently needed. The Network Security Information Exchanges in the telecommunications sector are prominent examples of forums for intercompany sharing within carefully delimited boundaries. Such sharing is also important among sectors, especially where common networking technology is being employed.

Reliability certification. Legitimized, generally accepted, and broadly applicable methods for certifying levels of reliability in components and networks would directly enhance reliability by enabling companies to make more informed tradeoffs. They would also help foster a culture of reliability throughout the technical community. Scientifically-based consensus codes and standards which address network design and installation—as the National Fire Protection Agency's *National Electrical Code* addresses electrical practices—would lessen the danger of unreliable systems. Institutionalized product reliability testing and certification, comparable with the broader safety testing and certification that Underwriters Laboratories performs, would be of immense value in establishing a common yardstick by which products could be compared.

Reliability is ultimately a human enterprise and in most practical situations, human-machine interactions are part-and-parcel of the system.

Measures to strengthen human factors. Reliability is ultimately a human enterprise and in most practical situations, human-machine interactions are part-and-parcel of the system. Measures to ensure operators continually maintain situation awareness and other personal reliability standards need to be stressed at every point.

There are no simple solutions to reliability challenges. There are no permanent solutions. However, the three steps outlined in this section, if pursued with sustained commitment, represent a sensible approach to assuring that the dependability of the domestic infrastructure matches the importance placed upon it.

Table 2. Assessing Technical Options for Improving Network Reliability

Effectiveness

How well does the proposed solution address the region of interest in the failure scenarios chart?

What effects does the proposed solution have in regions of the failure scenarios chart other than the region for which it is designed?

Is the proposed solution practical? available? usable?

What effect does the proposed solution have on network performance? on reconstitution after failures?

What new vulnerabilities does the proposed solution introduce?

Will the proposed solution work in the long run?

Implementation approach

What legal, regulatory, or other regime is necessary to implement this solution?

What is the schedule for implementation?

How will the proposed solution figure in the long term management of the system?

Marketplace effects, costs, and benefits

Do the likelihood and severity of the problem justify the cost of solution?

Who bears the cost of adopting the proposed solution?

To what extent does the effectiveness of the proposed solution depend on its bringing about changes in the marketplace?

Will changes in the marketplace negate the effectiveness of the proposed solution?

Government and the national interest

Will its political implementation negate the effectiveness of the proposed solution?

How, if at all, does the government interact with the private sector to implement, operate, and administer the proposed solution?

Does the proposed solution affect government national security or emergency preparedness services?

This table lists some of the key questions relevant to assessing proposed solutions to network reliability challenges. These questions should be addressed in system design and tradeoff studies.

CONCLUSION

The automation—or cybernation—of the domestic infrastructure has been motivated largely by the cost, efficiency, and performance benefits offered by information technology. Because reliability is linked to profitability, the private sector will continue responding to credible reliability threats in the future. Today's concern is whether the marketplace will adequately *anticipate* and *mitigate* reliability deficiencies in a highly dynamic business and threat environment, or whether the nation will have to endure a major infrastructure problem in order to mobilize and act. The importance of infrastructure services to the well being of the nation make infrastructure reliability a public policy concern.

Threats to reliability come from both natural and manmade sources, from Mother Nature to human nature. From a technical standpoint, it is not practical to focus exclusively on any one reliability threat. Strategies to counter one threat can exacerbate vulnerabilities to another. Like the interactions of prescription drugs, the remedy for one problem can interfere with the remedy for another. A holistic methodology for making the unavoidable tradeoffs is called for.

Reliable networks, like scientific inquiry, must be based on real data and actual facts -facts about the nature of vulnerabilities, the evolving reliability challenges, and the real-world, real-time environment in which infrastructure information networks operate. There is no substitute for a reasoned, methodical approach to understanding this problem and seeking solutions to it. Cost, performance, and reliability objectives must all be balanced through an engineering process of analysis and informed tradeoffs.

The risks associated with infrastructure threats can never be eliminated entirely. A sound technical approach is one which recognizes the need to manage risks and keep them at societally acceptable levels. A technical agenda for comprehensively addressing the reliability problem consists of three steps: (1) developing an analytical understanding of the existing reliability, vulnerability, and threat environment; (2) establishing a system engineering process which treats reliability as a primary parameter; and (3) fostering a commitment to vigilance and a process of continual learning to enhance reliability.

Areas for increased public policy attention include:

- *Achieving consensus on the problem and approaches to solutions.* The traditional policy tools available to the government for working with the marketplace to achieve national objectives—including legislation, regulation, licensing, tax and rate-setting regimes, and other inducements—all offer important options in the reliability arena. However, none of them can be effective unless and until there is consensus on what the minimum levels of reliability should be, what the threats are, what risks are acceptable, what protective measures should be taken, and how the costs should be met. At the present time, these questions are far from settled.

- *Enhancing government/industry cooperation for identifying and characterizing reliability challenges.* Government and industry each have unique capabilities to apply in identifying and characterizing reliability challenges. Federal activities, from weather and natural disaster prediction to intelligence collection on the threat of hostile attack, can contribute greatly to industry's understanding of reliability challenges. Additionally, industry network control centers could be important sources of strategic warning if the technical and institutional means to synthesize data from many sources were in place. Identifying appropriate activities in these areas, and putting them into practice, calls for long-term public policy attention.
- *Focusing government and industry on the joint development of technical standards and methods to measure and certify reliability.* Legitimized, generally accepted, and broadly applicable methods for certifying levels of reliability in components and networks would directly enhance reliability by enabling companies to make more informed tradeoffs. They would also help foster a culture of reliability throughout the technical community. Scientifically-based consensus codes and standards which address network design and installation would lessen the danger of unreliable systems. Institutionalized product reliability testing and certification would be of immense value in establishing a common yardstick by which products could be compared. The most progress in this area can be made through government and industry working together.
- *Enhancing Federal/State government interaction to ensure consistent and appropriate attention is place on infrastructure reliability.* Federal government interaction with State governments, and with State and regional regulatory commissions, is also essential to framing the question of how much reliability is needed in the infrastructure and to ensure that consistent and appropriate attention is placed on the reliability of the associated information networks. The Federal government can be helpful in providing uniform guidance to the States where appropriate, and in coordinating and focusing the resources of the Federal departments and agencies.
- *Defining the government research and development investment portfolio.* Federal investment in science and technology has been instrumental in the unfolding information revolution. Wise investments today in areas such as intrusion detection and prevention, robust network architectures, configuration management, secure communications, and other critical areas will pay reliability dividends in the future.

- *Working with other countries to develop compatible international legal regimes in cyberspace.* Reliability is also served by laws that delineate the boundaries of permissible behavior. As an international arena that disregards national boundaries, cyberspace presents unique legal difficulties. Distance and geography do not impede the trespasser, bandit, or terrorist in this realm, and questions about jurisdictions, sovereignty, and the applicability of laws frequently arise. The Federal government has an obligation to work with other countries to develop compatible cyberspace legal structures and to foster worldwide cooperation among law enforcement agencies.
- *Clarifying missions, responsibilities, and authorities of Federal Departments and Agencies in cyberspace.* Threats to the infrastructure are challenging existing boundaries between the national defense, intelligence, law enforcement, and regulatory roles of the U.S. government. Clarification of missions, responsibilities, and authorities in this new context are needed, and will necessarily involve all three branches of government, Executive, Legislative, and Judicial.

Both government and the private sector have responsibilities for delivering dependable infrastructure services that can, at reasonable cost, withstand foreseeable challenges without interruption. Developing consensus on the problem as well as finding effective long term solutions will require the sustained engagement of industry, utilities, the public, and the government at all levels. Together all these stakeholders can assure the reliability of the most capable infrastructure in all of history.

CHAPTER 20

IMPACT OF ENCRYPTION ON LAW ENFORCEMENT AND PUBLIC SAFETY

Statement of
Louis J. Freeh
Director, Federal Bureau of Investigation

July 25, 1996

Before the
Committee on Commerce, Science, and Transportation
United States Senate

Thank you Mr. Chairman and members of the Committee for providing me with this opportunity to discuss with you an issue of extreme importance and of great concern to all of law enforcement, both domestically and abroad—the serious threat to public safety posed by the proliferation and use of robust encryption products that do not allow for timely law enforcement access and decryption.

First and foremost, the law enforcement community fully supports a balanced encryption policy that satisfies both the commercial needs of industry and law abiding individuals for robust encryption products while at the same time satisfying law enforcement's public safety needs. On the one hand, encryption is extremely beneficial when used legitimately to protect commercially sensitive information and communications. On the other, the potential use of such robust encryption products by a vast array of criminals and terrorists to conceal their criminal communications and information poses an extremely serious and, in my view, unacceptable threat to public safety. Recently, the President of the International Association of Chiefs of Police sent a letter to President Clinton expressing support for a balanced encryption policy that addresses the public safety concerns of law enforcement. Additionally, the National Sheriff's Association enacted a resolution last month also expressing their support for a balanced encryption policy and opposing any legislative efforts that would undercut the adoption of such a balanced policy.

Since 1992, when AT&T announced its plan to sell a small, portable telephone device that would provide users with low-cost but robust voice encryption, public policy issues concerning encryption have increasingly has been debated in the United States. Since then, people concerned about privacy, commerce, computer security, law enforcement, national security, and public safety have participated in the dialogue regarding cryptography. On the international front, this past December, the multi-national Organization for Economic Cooperation and Development (OECD) meeting in Paris, France, convened an Experts Group to draft global cryptography principles, thus reflecting an increased global interest in and concern about the use and availability of encryption that can be used to endanger a nation's public safety and national security.

In addition, several Members of Congress have also joined this public discussion by introducing legislation which essentially would remove existing export controls on encryption and which would promote the widespread availability and use of any type of encryption product regardless of the impact on public safety and national security. However, the impact of these bills, should they be enacted, has not been lost on other Members of Congress as reflected in the letters to the sponsors of both Senate encryption bills by the Chairman and Vice-Chairman of the Senate Select Committee on Intelligence. Senators Specter and Kerrey indicated in their letters that they had concerns regarding these bills and expressed the opinion, which I fully endorse, that there is a ". . . need to balance U.S. economic competitiveness with the need to safeguard national security interests." To that balance, I would also add public safety and effective law enforcement.

Without question, the use of strong cryptography is important if the Global Information Infrastructure (GII) is to fulfill its promise. Data must be protected—both in transit and in storage—if the GII is to be used for personal communications, financial transactions, medical care, the development of new intellectual property, and a virtually limitless number of other applications. Our support for robust encryption stems from a commitment to protecting privacy and commerce.

But we are also mindful of our principal mission responsibilities: protecting America's public safety and national security in the myriad of criminal, terrorist, and espionage cases that confront us every day. Notwithstanding the accepted benefits of encryption, we have long argued that the proliferation of unbreakable encryption —because of its ability to completely prevent our Nationts law enforcement agencies from understanding seized computer files and intercepted criminal communications which have been encrypted and then being able to promptly act to combat dangerous criminal, terrorist, and espionage activities as well as successfully prosecute them—would seriously and fundamentally threaten these critical and central public safety interests. The only acceptable answer that serves all of our societal interests is to foster the use of "socially-responsible" encryption products, products that provide robust encryption, but which also permit timely law enforcement and national security access and decryption pursuant to court order or as otherwise authorized by law.

Law enforcement is already beginning to encounter the harmful effects of conventional encryption in some of our most important investigations:

- In the Aldrich Ames spy case, where Ames was told by his Soviet handlers to encrypt computer file information to be passed to them.
- In a child pornography case, where one of the subjects used encryption in transmitting obscene and pornographic images of children over the Internet.
- In a major drug-trafficking case, where one of the subjects of one of the court-ordered wiretaps used a telephone encryption device which frustrated the surveillance.
- Some of the anti-Government Militia groups are now advocating the use of encryption as a means of preventing law enforcement from properly investigating them.

It is important to understand, as one can see from the cases I have cited, that conventional encryption not only can prevent electronic surveillance efforts, which in terms of numbers are conducted sparingly, but it also can prevent police officers on a daily basis from conducting basic searches and seizures of computers and files. Without an ability to promptly decrypt encrypted criminal or terrorist communications and computer files, we in the law enforcement community will not be able to effectively investigate or prosecute societyts most dangerous felons or, importantly, save lives in kidnappings and in numerous other life and death cases. We simply will not be able to effectively fulfill our mission of protecting the American public.

In a very fundamental way, conventional encryption has the effect of upsetting the delicate legal balance of the Fourth Amendment, since when a judge issues a search warrant it will be of no practical value when this type of encryption is encountered. Constitutionally-effective search and seizure law assumes, and the American public fully expects, that with warrant in hand law enforcement officers will be able to quickly act upon seized materials to solve and prevent crimes, and that prosecutors will be able to put understandable evidence before a jury. Conventional encryption virtually destroys this centuries old legal principle.

There is now an emerging opinion throughout much of the world that there is only one solution to this national and international public safety threat posed by conventional encryption—that is, key escrow encryption. Key escrow encryption is not just the only solution; it is, in fact, a very good solution because it effectively balances fundamental societal concerns involving privacy, information security, electronic commerce, public safety, and national security. On the one hand, it permits very strong, unbreakable encryption algorithms to be used, which is essential for the growth of commerce over the GII and for privacy and information security domestically and internationally. On the other hand, it permits law enforcement and national security agencies to protect the American public from the tyranny of crime and terrorism. We believe, as do many others throughout the world, that technology should serve society, not rule it; and that technology should be designed to promote public safety, not defeat it. Key escrow encryption is that beneficial and balanced technological solution.

American manufacturers that employ encryption in their hardware and software products are undoubtedly the technology leaders in the world. American industry has the capability of meeting all of societyts basic needs, including public safety and national security, and we, as responsible government leaders, should be sending a clear signal to industry encouraging them to do so. Key escrow encryption is "win-win" technology for societies worldwide. I know you agree that it would be irresponsible for the United States, as the world's technology leader, to move towards the adoption of a national policy that would knowingly and consciously unleash on a widespread basis unbreakable, non-key escrow encryption products that put citizens in the U.S. and worldwide at risk.

Unfortunately, in recent months, the nearly exclusive focus of the public discussion concerning the encryption issue has been on its commercial aspects, particularly with regard to removing export controls. This narrow focus ignores the very real threat that conventional, non-key escrow encryption poses both domestically and internationally to public safety. We continue actively to seek industryts cooperation, assistance, and great expertise in producing key escrow encryption products as a critical part of an overall, balanced, and comprehensive encryption policy that would logically include an appropriate relaxation of export controls for key escrow products.

As for export controls, we have had ongoing discussions with industry, and industry has articulated the view that export controls needlessly hurt U.S. competitiveness overseas. But once again we need to carefully consider the facts and balance a number of competing interests. Although some strong encryption products can be found overseas, they are simply not ubiquitous, and, as of yet, they have not become embedded in the basic operating systems and applications found overseas.

Importantly, when the U.S. recently let it be known that it was considering allowing the export of encryption stronger than that now permitted, several of our close allies expressed strong concerns that we would be flooding the global market with unbreakable cryptography, increasing the likelihood of its use by criminal organizations and terrorists throughout Europe and the world, and thereby imperiling the public safety in their countries. Ironically, the relaxation of export controls in the U.S. may well lead to the imposition of import controls overseas. The international implications and likely reactions of foreign governments to the U.S. unilaterally lifting such export controls must be fully considered.

Given the fact that the use and availability of robust encryption is an issue of concern internationally, it is important to understand what steps other countries are taking to address these concerns. Recently, France, Russia and Israel have established domestic restrictions on the import, manufacturer, sale and use of encryption products, as not to endanger their public safety and national security. The European Union is moving towards the adoption of a key recovery-based key management infrastructure similar to that proposed for use within the United States. This plan, based upon the concept of using a "Trusted Third Party," allows for encryption keys to be escrowed with an independent but non-governmental party, thus allowing for lawful government access to such escrowed key pursuant to proper legal authority.

Lastly, we have heard the oft-repeated argument that the tgenie is out of the bottle,t and that attempts to influence the future use of cryptography are futile. This is simply not true; and we strongly disagree. If strong, key escrow encryption products proliferates both overseas and domestically which will not interoperate (at least in the long-term) with non-key escrow products, then escrowed encryption products will become the worldwide standard and will be used by almost everyone, including the criminal elements, in countries participating in the GII. It is worth noting that we have never contended that a key escrow regime, whether voluntarily or mandatorily implemented, would prevent all criminals from obtaining non-key escrowed encryption products. But even criminals need to communicate with others nationally and internationally, including not just their criminal confederates but also legitimate organizations such as banks. Accessible, key escrow encryption products clearly will be used by most if widely available, inexpensive, easy to use, and interoperable worldwide.

In closing, if one considers the broad range of public safety responsibilities that fall upon the law enforcement community, there is only one responsible course of action that we as government leaders must embark upon—to promote socially-responsible encryption products, products that contain robust cryptography but which also provide for timely law enforcement access and decryption—that is, key escrow encryption. The entire law enforcement community believes not only that the removal of export controls for encryption products that are non- law enforcement accessible is unwise, but that such an action would jeopardize our national security and the interests and safety of law-abiding citizens worldwide.

We look forward to working with you and your staff on this difficult issue and would be pleased to answer any questions you might have.

CHAPTER 21

ELECTRONIC SURVEILLANCE—REPORT ON THE FBI'S PUBLICATION OF THE SECOND NOTICE OF CAPACITY

STATEMENT OF FBI ASSISTANT DIRECTOR JAMES K. KALLSTROM

Washington DC
FBI National Press Office
January 14, 1997

Introduction

Acting on a mandate from Congress, the Federal Bureau of Investigation today is announcing the amount of telephone system capacity the nation's law enforcement agencies may need for court-approved electronic surveillance to protect the public from terrorism, violence, drugs, and other grave offenses and to deal with unexpected crime emergencies through 1998 and beyond.

The Communications Assistance for Law Enforcement Act of 1994 (CALEA) requires the telephone companies to ensure that their systems and networks have both the capability and capacity to accommodate all federal, state, and local law enforcement agencies' court-approved intercepts in the face of new or changing telephone technology that could otherwise prevent electronic surveillance.

CALEA by itself will not increase the amount of electronic surveillance but only makes certain that law enforcement agencies will be able to carry out court-approved wiretaps, pen registers, and trap and traces no matter how sophisticated wireline or wireless telephone systems may become. As required by CALEA, the "Second Notice of Capacity" being announced today was developed after consultation with other law enforcement agencies, the telephone industry, and the academic community.

"CALEA does not give law enforcement any new electronic surveillance authority. Safeguards that have held law enforcement to strictly-enforced standards and conduct for a quarter of a century remain in effect, including the requirement of court approval of all telephone intercepts," said James K. Kallstrom, Assistant Director of the FBI.

"Under CALEA and as part of this notice process, telephone companies are to be given 'actual' and 'maximum' capacity estimates so that they will know the precise capacity requirements that their systems will need to have in the future to meet their statutory obligation to maintain some level of capacity," said Kallstrom.

"The FBI has conducted unprecedented research to unearth the facts needed to develop the very best estimates possible," Kallstrom said." The FBI estimates are prudent and conservative, and will enable law enforcement to adequately protect the public without placing any unfair burdens on the telecommunications industry."

"There should be no civil liberties concerns over these estimates because, as we want to stress again and again, the Act and the steps flowing from it do *not* give law enforcement any new powers to conduct electronic surveillance," Kallstrom said.

"The purpose of CALEA, and the FBI's purpose in developing the estimates, is to help protect the American people from the worst crimes—terrorism, drug trafficking, violence of all types," Kallstrom said. "Court-approved electronic surveillance is one of law enforcement's most important investigative techniques."

"Telephone systems are routinely used by the worst and most violent criminals to commit the worst and most violent crimes. Long-standing federal law allows for the tightly controlled use of court-authorized electronic surveillance in the investigation of the most serious violent offenses only when other investigative techniques will not work, when other investigative techniques are too dangerous, and when there is probable cause to believe that telephone communications are being used to carry out serious crimes," Kallstrom said.

In many cases, there is no substitute of electronic surveillance in gathering evidence, preventing crimes, solving crimes, and bringing the violent to justice. In the past 14 years, court-approved electronic surveillance has been directly responsible for obtaining the conviction of more that 26,000 dangerous felons," Kallstrom said." This does not include the hundreds of thousands of other investigations that have succeeded in part because of pen registers and trap and traces."

"We do not expect there to be any increase in the amount of electronic surveillance nationally because of CALEA itself, because CALEA does not give any new authority to law enforcement. Any increases would be related to growth of crime and unexpected crime emergencies that would have to be met with appropriate enforcement measures," Kallstrom said. "But as CALEA mandates, we must be prepared to meet grave crime problems that could erupt in any part of the nation—urban, suburban, rural."

"This second notice is the product of industry and government working together to find an understandable method of determining what level of electronic surveillance capacity will be preserved as telephone technology changes," said Kallstrom. "It is a fundamentally different approach than the Initial Notice that relied upon a percentage of 'engineered capacity.' That approach readily led to misinterpretation and the erroneous conclusion that the Initial Notice signaled a government intent to greatly expand the use of wiretaps. That conclusion was wrong and this new approach was developed, in part, to correct that misperception."

In addition to the federal government, 41 states, Puerto Rico, the Virgin Islands, and the District of Columbia have statutes allowing for the use of court-authorized wiretaps by law enforcement in the investigation of the most serious criminal acts. All states provide for law enforcement access to dialed telephone numbers using the less intrusive pen registers and trap and trace devices, this latter category constituting the vast majority of all interceptions (approximately 90 percent) covered by the capacity requirements set forth in this notice. Wiretaps account for only 10 percent.

According to Kallstrom, "Because of the nature of older telephone technology and of law enforcement's traditional interception methods, previous generations of telephone equipment and services had virtually a 100 percent capacity to accommodate all court-authorized electronic surveillance. However, many of the newer generations of telephone systems were designed in a way that inadvertently limited or provided no capacity for conducting any kind of electronic surveillance. Solutions must now be designed and developed that will have the effect of preserving the technical capacity to accommodate court-approved pen register, trap and trace devices, and wiretaps. CALEA seeks only to ensure that some small portion of the capacity of telephone equipment, facilities, and services can accommodate electronic surveillance. This notice is part of that process."

Congress authorized $500 million to be appropriated to aid telephone companies to make needed changes in existing telephone systems. To date, $100 million has been made available by the Congress and law enforcement for this purpose.

Capacity Needs and Notices

Section 104 of CALEA requires the Attorney General to publish in the Federal Register and give notice to telecommunications carriers of 1) the actual number of simultaneous communication interceptions, pen registers, and trap-and-trace devices that the Attorney General estimates that all authorized government agencies may need by October, 1998 ("Actual capacity") and (2) the maximum capacity required to accommodate all simultaneous communication interceptions, pen registers, and trap-and-trace devices that the Attorney General estimates that all authorized government agencies may need after October, 1998 ("maximum capacity").

A pen register records the numbers dialed in an outgoing telephone call. It does not include the content of the call, i.e., the actual telephone conversation. Trap and trace is a technique that identifies the telephone number and location of an incoming telephone call to a particular telephone. Like the pen register, the content or conversation is not intercepted. A wiretap captures the actual conversation or data going to or from a particular telephone. There are strict and precise legal requirements concerning precisely what can be intercepted during a wiretap. Wiretaps are subject to court supervision once the initial authorizing court order is obtained.

The FBI published on October 16, 1995, its Initial Notice of Capacity in the Federal Register seeking comments from all interested parties. In the Initial Notice, electronic surveillance capacity requirements were expressed as a percentage of "engineered capacity" for all telephone companies, and no distinction was made between wireline and wireless services. "Wireline" is the regular telephone service such as people have in their homes. "Wireless" is cellular and other telephone services that are mobile in nature. Unfortunately, because of the FBI's use of the engineered capacity approach, the Initial Notice led to gross misinterpretation and the erroneous belief that this notice signaled a government intent to use CALEA as a means to greatly expand the number of wiretaps conducted every year. This interpretation was not accurate and, as a result, a new approach was followed by the FBI in formulating this second notice.

The "Second Notice of Capacity" being offered today for comment is responsive to a number of telecommunications industry recommendations that were submitted in response to the Initial Notice. Specifically, the Initial Notice has been replaced with precise capacity numbers on a country-by-county or market-by-market basis throughout the country and a clear distinction has been made between wireline and wireless telecommunications carriers.

The Second Notice estimates the number of simultaneous communications interceptions (e.g., wiretaps) and call identifying information interceptions (e.g., pen register and trap and traces) that any or all of the authorized federal, state, or local law enforcement agencies together may require, if and only if, the need should arise and courts approve the applications, from telephone companies in a specific geographic area at any one time.

It is important to note that the terms "actual capacity" and "maximum capacity" do not mean the actual number of wiretaps, pen registers and trap and trace devices that will be conducted. Rather, the estimates of actual and maximum capacity set forth in this notice are only intended to establish sufficient interception-related capacity to accommodate those interceptions authorized by the various federal and state courts at any location throughout the United States should the need arise in the future. As there is no way to predict for most locations if and when the need will arise, the only safe approach from a public safety standpoint is to at least preserve this ability in case it is ever needed.

The concept of estimating future electronic surveillance capacity requirements is analogous to another public safety concept in use today - ensuring that a sufficient number of fire hydrants are positioned in the correct locations throughout the country to guarantee a hydrant's availability when needed by the fire department. Fire hydrants are placed in communities not because they necessarily represent the number of fires that are expected to occur, but are deployed so that, in the event a fire should occur in a particular location, there is a hydrant available for use by the fire department.

Likewise, the "Second Notice of Capacity" does not mean nor should it be misinterpreted as somehow representing the total estimated number of electronic surveillances that law enforcement plans to conduct. Just as it would be incorrect to add up all of the fire hydrants in existence today and assert that the total represents the number of fires that the fire departments anticipate they will encounter, so would it be incorrect to add up the actual or maximum capacity numbers set forth in this notice and assert the number reflects the total number of electronic surveillances that law enforcement agencies want to conduct.

The "Second Notice of Capacity" was released today on behalf of all federal, state, and local law enforcement agencies. It asks for comments from all interested parties during the next 30 days.

FBI Research and Analysis of Recent Interception Activity

In fulfilling the statutory obligation to estimate the future capacity requirements for new telephone equipment, the FBI believed it was essential to first establish a historical baseline of court-authorized interception activity. This was accomplished by contacting federal, state, and local law enforcement officials in each state; unsealing all types of federal and state electronic surveillance-related court orders; and surveying approximately 1,500 telecommunications carriers operating within the United States. The establishment of a historical baseline entailed an unprecedented, detailed, review, and analysis of the available information on all types of federal, state, and local law enforcement interceptions (wiretaps, pen registers, and trap and traces) throughout the United States during the period 1/1/93 to 3/1/95.

The Wiretap Report published annually by the Administrative Office of the United States Courts was a valuable source of historical information on criminal wiretap court orders. However, the Report provided insufficient information to establish an accurate baseline for specific locations or for specific telephone companies because it does not identify the actual number of telephone lines (or cellular transmission paths) covered in each court order and, by design, it does not address the vastly greater number of lines associated with pen register and trap and trace interceptions (approximately 90 percent of the total number of lines). It is the number of lines—not the number of court orders—that is a reliable predictor of simultaneous capacity needs for individual telephone companies.

Consequently, for the first time ever, comprehensive line- related information for all types of federal, state and local law enforcement court-authorized electronic surveillance interception activity was collected and analyzed to determine the day that peak simultaneous interception activity occurred during the period 1/1/93 - 3/1/95 for every area of the country.

CALEA requires that capacity requirements be expressed in terms of "simultaneous" interceptions. Law enforcement chose to consider interceptions occurring on the same day, rather than at exactly the same moment, as being simultaneous. This time frame was used because courts authorize electronic surveillance for a certain number of days as opposed to some other unit of time and because the time frame of one day was compatible with the historical data that was collected.

The interception activity for each county was examined on a day-to-day basis for the time period studied. An analysis was then used to identify the one day with the highest number of simultaneous interceptions to derive the historical baseline for that county. This approach provides a reasonable representation of past interception needs for the geographic area during the period reviewed. A majority of counties in the United States (57 percent) had no intercept activity during the period under review.

For telecommunications carriers offering local exchange services (wireline), county boundaries are used to define the geographic locations. For telecommunications carriers offering cellular and personal communications services (wireless), market service areas are used to define the geographic locations. The market service areas are currently used by the wireless telecommunications industry and include Metropolitan Statistical Areas (MSAs), Rural Service Areas (RSAs), Major Trading Areas (MTAs), and Basic Trading Areas (BTAs).

The Second Notice of Capacity lists in detail the extensive methodology used to arrive at actual and maximum capacity requirements as well as the historical simultaneous interception activity in each county and market service area throughout the entire United States. The majority of counties had no intercept activity, so a minimal estimate is included solely to preserve this ability should a crime occur that requires law enforcement to use court-approved electronic surveillance. Although unprecedented, these numbers are being provided to ensure that the requirements of CALEA are completely satisfied.

There is no direct correlation between the historical simultaneous intercept numbers included in this notice and the number of court orders for wiretaps published annually by the Administrative Offices of the United States Courts. First, unlike the annual report, these numbers include pen registers, trap and traces, and wiretaps, not just wiretaps. Pen registers and trap and traces constitute approximately 90 percent of the historical simultaneous intercept lines set forth in this notice. The annual report includes only the number of wiretap court orders and does not include the actual number of lines associated with each wiretap order nor does it include the number of pen register and trap and traces orders or their associated line-related information. Secondly, the annual report by design encompasses only the number of wiretap orders during a single specific year. The historical simultaneous intercept numbers represent the single busiest day in each county and market service area during the 26- month review period. For any given county or market service area, the historical experience could be any combination of pen register, trap and trace and wiretaps line—or the possibility of no wiretaps. Accordingly, while adding all the numbers in the historical intercept column results in a number—24,617—it is a number that does not correlate to the actual number of wiretaps conducted by law enforcement and is not relevant for comparison purposes.

The results of the historical analysis also indicated that there were significant areas in the United States where law enforcement did not conduct electronic surveillance during the 26 -month review period. For example, during this period, of the 3,146 counties in the United States, 1,788 reported no intercept activity. Likewise, of the 734 cellular market service areas, 467 reported no intercept activity. This is not to say that law enforcement did not conduct electronic surveillance either before or after the examined period, or would never need to conduct electronic surveillance in these locations in the future. In order to preserve law enforcement's ability to conduct electronic surveillance, a minimum baseline value of one interception was applied to cover instances where the **studied time period** did not provide historical interception information in a county or market service area.

The historical baseline data and the electronic surveillance capacity requirements set forth in this notice should not be misinterpreted or misconstrued as representing the crime rates in each county.

Estimates of Needed Intercepts

One of the purposes of this Second Notice of Capacity is to publish for comment estimates for possible future court-approved pen register, trap and trace, and wiretap activity. The county and market service area capacity estimates were determined by applying growth factors to the historical simultaneous interception activity, including those areas where there was no activity, to project a safe level of capacity as of October 1998 and thereafter.

To remain consistent with the approach to establish separate capacity requirements for wireline and wireless carriers, separate growth analyses were conducted for wireline and wireless services. The growth factors were developed from three historical records: 1) criminal wiretap orders set forth in the Wiretap Report (1980-1995); 2) Department of Justice (DOJ) survey of

Pen Register/Trap and Trace court orders (1987-1995); and 3) historical interception information covering a 26-month period (1/1/93-3/1/95).

Growth in criminal wiretap and DOJ pen register/trap-and-trace court orders was used as a reasonable predictor of growth in interception activity since no long-standing collection of historical line-related interception activity exists. A commonly-used statistical approach referred to as best-fit-line analysis was used to track the number of court orders over time and then project the number into the future.

The resulting growth factors for wireline carriers (applied to the historical baseline for the county) correspond to an annual growth rates of 5.92 percent for the years 1994-1998 and 4.55 percent for the years 1998-2004. The resulting growth factors for wireless carriers (applied to the historical baseline for the market service areas) correspond to an annual growth of 14.30 percent for the years 1994-1998 and 8.38 percent for the years 1998-2004. These growth rates are consistent with annual subscriber growth rates predicted for the wireline and wireless industries over the same period, which are projected at 3.5 percent and 12.0 percent respectively.

The growth factors derived are not necessarily indicative of the number of court-approved interceptions law enforcement will, in fact, perform in the future. Rather, they are part of a statistical method used for calculating future actual and maximum capacity requirements for the affected telephone companies as required by CALEA. To ensure law enforcement's future ability to conduct court-authorized electronic surveillance in all geographic areas of the country, capacity requirements were also established for those areas of the country in which no interception activity occurred during the period. This was accomplished by assigning each of those counties or market service areas with an historical interception factor of one and then increasing that number by the appropriate wireline or wireless growth rate. While this may have the appearance of artificially increasing the total number of all areas combined, it was necessary to ensure that the ability of law enforcement to conduct lawfully authorized pen registers, trap and traces, and wiretaps was not lost in a substantial number of counties and market service areas (2,255).

For wireline carriers operating in a county with no interception activity during the studied period, an actual capacity to accommodate two simultaneous court-authorized interceptions and a maximum capacity to accommodate three simultaneous court-authorized interceptions were estimated. For wireless carriers operating in a market service area in which no interception activity occurred during the studied period, an actual capacity to accommodate two simultaneous court-authorized interceptions and a maximum capacity to accommodate four simultaneous court-authorized electronic surveillance were established.

Using the chosen statistical method, actual and maximum capacity numbers were calculated for every county and market service area. These numbers represent the total number of simultaneous intercepts that individual telephone companies in every county and market service area must be able to accommodate by the dates specified in the law, should the need ever arise and law enforcement obtains the appropriate court order. This would include any combination of simultaneous pen registers, trap and traces, and wiretaps.

Again, it does not mean that there will be this much electronic surveillance. It is only establishing, through analysis of historical intercept data, some level of capacity across the country to be available should a situation or crisis arise in an individual investigation and all of the legal requirements for electronic surveillance can be satisfied.

Both the actual and maximum capacity columns set forth in this notice can be totaled but those numbers are not relevant for comparison purposes. For example, there are currently over 160 million telephone lines in the United States. Had new technology not have been developed, all of them would have continued to be fully capable of supporting a wiretap or other intercepts—that is, support 100 percent capacity to do 160 million plus simultaneous intercepts. If the actual and maximum capacity columns are totaled, the exceptionally modest sum of each respectively would be 39,767 and 57,749.

In essence, the entire national capacity is going from 160 million lines to less than 60,000 lines. Both numbers represent capacity to be available should a crime occur that warrants court authorized intercepts or should some other crisis such as a terrorist bombing or kidnapping occur. Neither number represents or has relevance to the number of intercepts that will occur in the future. That number will be determined by the happenstance of serious or grave crimes and severely but appropriately limited by what the law permits and resources allow.

Additional Comments and Final Notice

As previously noted, federal law and laws in 41 states allow for the use of court-authorized wiretaps in the investigation of the most serious criminal felony offenses only when other investigative techniques will not work or are too dangerous to employ and only when there is probable cause to believe that communications are being used in the furtherance of these serious criminal acts. However, law enforcement in all states utilizes pen registers and trap and trace devices to acquire telephone numbers that are dialed during significant investigations.

It is important to remember that in 1968 when Congress statutorily authorized the use of court-ordered electronic surveillance, there were no technological limitations on the number of simultaneous interceptions that a telecommunications carrier could implement for law enforcement. In the past as well as for the future, the limitations on the number of electronic surveillance interceptions conducted by law enforcement have been and will continue to be determined by two factors, 1) that amount of criminal activity being advanced through the use of our Nation's communications network and 2) adherence to the stringent legal requirements set forth in the federal and state statutes that govern the use of electronic surveillance.

Since 1970, telecommunications carriers have been statutorily required to provide law enforcement with the technical assistance needed to accomplish court-ordered electronic surveillance. CALEA's purpose was to further clarify this existing statutory obligation of telecommunications carriers to provide law enforcement with such technical assistance. CALEA was needed because some telecommunications carriers were technologically unable to provide law enforcement with access to all of the communications content or access to all of the call-identifying information that law enforcement was authorized by court order to intercept. This is so because of technological impediments within their systems, both in terms of network capabilities and in terms of the number of simultaneous interceptions that were needed to be performed (capacity). These technological impediments were brought on by recent advances in telecommunications technology.

CALEA applies to all telecommunications carriers as defined in section 102(8) of the Act. However, the "Second Notice of Capacity" now being offered for comment and its associated future "Final Notice of Capacity" only apply those telecommunications carriers that are of the most immediate concern to law enforcement, that is, those telecommunications carriers offering local exchange, cellular, and personal communication services (PCS). Law enforcement welcomes written comments from all telecommunications carriers, to include those telecommunications carriers offering services other than those set forth in this notice, as well as other interested parties.

After the 30-day comment period for the "Second Notice of Capacity" closes and a review of all relevant comments received is completed, a "Final Notice of Capacity" will be issued by the FBI which will fulfill the obligations of the Attorney General under section 104(a)(1) of CALEA. Telecommunications carriers will then have 180 days after publication of the "Final Notice of Capacity" in the Federal Register to submit a "Carrier Statement" as mandated by section 104(d) to the Government identifying any of its systems or services that do not have the capacity to accommodate these public safety requirements.

Even though CALEA requires law enforcement to estimate its future interception capacity requirements, which has the effect of specifying technological limitations on the number of simultaneous interceptions that can be conducted in a specific geographic area by telecommunications carriers, law enforcement believes that the capacity requirements set forth in this notice are reasonable, responsible and sufficient to maintain law enforcement's current ability to protect the public from the most serious crimes. If anything, a conservative approach was taken based on the historical data collected. The public must recognize, however, that it is possible in a given crisis situation that the needs of law enforcement may not be met by the capacities identified in this notice. It is possible that, as crime trends change or as new crime problems develop in specific areas of the country, these estimates may prove to be too conservative for particular counties or market service areas and the public safety will dictate revisiting capacity requirements in those areas. CALEA envisions that possibility and provides a process through which it can be addressed.

CALEA preserved law enforcement's long-standing ability to access with court-approved intercepted communications content and call-identifying information in an ever-changing telecommunications environment. The safeguards that have held law enforcement to strictly enforced standards and conduct for conducting electronic surveillance for a quarter of a century remain in effect, including the requirement of prior court approval for all electronic surveillance interceptions. It is important to remember that in 1968 when Congress statutorily authorized the use of court-ordered electronic surveillance, there were no technological limitations on the number of simultaneous interceptions that a telecommunications carrier could implement for law enforcement.

Written comments must be received on or before February 13, 1997, and should be submitted in triplicate to the Telecommunications Industry Liaison Unit (TILU), Federal Bureau of Investigation, P.O. Box 220450, Chantilly, VA 20153-0450.

CHAPTER 22

SPEECH BY
LOUIS J. FREEH, DIRECTOR OF THE FBI

1997 International Computer Crime Conference
New York, New York
March 4, 1997

Thank you Jim. Good morning, ladies and gentleman. It's really a pleasure to be up here and to welcome you to the conference, and also to tell you what a historic and great opportunity this is for all of the varying interests represented here.

It's appropriate that this conference be held in New York City, which is a global city in a global economy. It's appropriate, and we thank you, Mr. Moran, on behalf of Chase, for involving private industry, both nationally and internationally, in the principles and objectives of the conference, as well as our law enforcement friends and colleagues from around the United States and around the world.

I just returned from a very brief trip to the Mideast where I visited three countries and spent time with the leaders of all the countries involved in the current peace process. We met with Yasser Arafat, the Prime Minister of Israel, King Hussein, President Mubarak, and, over the course of several days, all of my counterparts, both in law enforcement services and security services. And part of our agenda, although not the priority part, did in fact deal with some of the issues that this conference is going to address—issues like technology crimes; law enforcement in the information age; threats to infrastructure; threats to national security; the new ways that criminals and terrorists have found to achieve their objectives; taking advantage of all of the technological changes; the transparency of borders; the ability to travel and send information instantaneously.

These are matters that are relatively mid-way on law enforcement security agendas—and appropriately so given the enormous security problems and the current crises with which those countries and authorities deal.

It's an obvious point, but one which I think we need to make: in the United States also, these critical issues will continue to occupy industry and law enforcement, but they are not at this time on the front burners for law enforcement or for national security people.

This should not be surprising. We are not imminently threatened with the collapse of infrastructures. We are not seeing intrusions at a frequent enough index that people are alarmed about them. Our experience in 1993, when we addressed the digital telephony issue of how to maintain and continue court-authorized wiretaps in a new technological environment, is a good illustration.

At that time, we could not go to the administration or go to Congress and say we're in a crisis—we can't do any more wiretaps. We could only tell them with a lot of confidence and good information that if we did not solve that problem in 1993, we were certain that in 2003 we would not be able to perform court-authorized wiretaps—neither in the criminal enforcement area, nor in the national security area.

Today, this kind of debate continues in Washington and around the world on encryption. Again, at this point we can't point to a proliferation of examples where encryption, unbreakable encryption, has caused the loss of lives or shut down major investigations. But we know, with great certainly, that if that problem is not dealt with very quickly, the time will come that, as robust encryption proliferates without any recovery systems, law enforcement and national security will clearly be at risk.

In a sense, this process really describes the history and the saga of law enforcement. In 1933, unarmed FBI agents transporting a prisoner were gunned down in a crossfire that became known as the Kansas City Massacre. Only then was Congress spurred to enact, within a week after that attack, the authority for FBI agents to carry firearms and make arrests.

It took the chance discovery of the Apalachin meeting up in New York and subsequent investigations in the mid 60's to demonstrate the existence of La Cosa Nostra in the United States—and that spurred Congress to authorize court-authorized wiretapping in 1968.

So we see, over the course of time, how law enforcement strives to catch up with technology. And I think that's where we are right now with computer crime, with the encryption issues, with the telecommunication issues, and with the wireless communication issues—all of which need to be addressed and solved.

So I really salute all of you —and of the different countries and corporations that you represent—for putting together a conference which, for the first time, focuses internationally on this problem. Your lead is one that law enforcement must follow.

Today when new FBI agents graduate from our training academy in Virginia, they leave with their firearms and their badges, but they also leave with a lap top computer. It's an excellent symbol of the changing environment in which these young men and women will function over the next 20 years. It is also imperative for the way they must conduct investigations. When they serve law enforcement search warrants, they seize hard drives and disks instead of the boxes and boxes of records and books and ledgers that their predecessors, myself included, used to seize to support our cases.

Today, also, they chase fugitives over cyberspace as well as over fences. You may remember when we arrested Mr. Mitnik a year or so ago. He was found by the FBI, but he was found because we hired a 23-year-old computer specialist to locate exactly where he was and where he was transmitting from. That was the basis of effecting that arrest.

I though also I would mention, very briefly, some of the cases where the technology of computers and cyber crime is evidencing itself, and then talk generally and briefly about recent initiatives undertaken between the FBI and other government organizations, in partnership with the private sector, to deal with some of these problems.

Clearly these problems and issues cannot be solved unilaterally by law enforcement, no more than they could be solved unilaterally by the private sector. If we are to identify and respond to these various problems, we've got to unite the efforts of industry and law enforcement on an international scale.

Let me mention very quickly a couple of cases; these are all public cases so I can comment on them. The Citibank case, which you've heard Mary Jo White refer to, was a case where someone with a lap top computer, sitting in an apartment in St. Petersburg, Russia, intrudes into a bank and attempts to move millions of dollars out of accounts to a place where they can be exploited.

We had a similar case recently with a so-called "phone phreaker" in Sweden—and because of the assistance we received from Swedish authorities, we were able to solve that case. There a young man, sitting in his own apartment, hacked his way across the Atlantic Ocean into U.S. telephone switching systems and worked his way down to Northern Florida, where over the course of several weeks, he interfered with 911 systems and had the capability to disable the system. It could have been disastrous, because 911 systems not only affect the police but also affect fire and emergency services.

Now extrapolate that to imagine if he'd hit to larger systems—banking systems, stock exchanges, or power grids in the northeast or northwest in the middle of winter.

We had another recent and continuing case in Baltimore, which we call the Innocent Images case. Several speakers today have already referred to it. It goes back to 1993 when we began investigating a kidnapping case. When we began to focus on several subjects, it became clear that they were using computers—computer telecommunications networks—to contact, identify and, in some cases, arrange for meetings with children. In a sense, they were entering the homes of the children, not on the telephone or by a knock on the door, but through computer modems. That case, which has become a national initiative by the FBI, has resulted, so far, in approximately 88 arrests and 78 convictions. And the only people targeted in those cases are the individuals who are involved in large-scale distributions of pornography, and that's just, in our view, the tip of the iceberg.

We had a recent terrorism case where an individual maintained plans in his lap top computer to attack airliners and other targets. Part of the files contained in that lap top is still encrypted and is still in need of being deciphered by the law enforcement authorities.

Those are just several cases on the menu—again, not representative of thousands of others, but all of a very serious nature and with grave implications.

If you take the context of those cases and translate them to large scale industries, to infrastructure, and to informational systems then you can see that the potential is catastrophic.

Consider for example, a recent exercise by a government agency that is responsible for maintaining and transmitting secure information. This agency ran some computer attacks against its own very well defended systems, using people inside and outside the agency to perform them. The results of the test were that 88 percent of the attacks were successful. Again, the implications of that exercise, translated to all our informational systems, are sobering.

We have been trying to respond to the prospective issues involved in this issue in a number of different ways. In June last year the President signed an Executive Order that asked all government agencies, coordinated by the FBI, to do a critical infrastructure study over a 1-year period that would focus on the vulnerabilities of the systems—both physical and informational security—and that would compose and design protocols of plans and systems to protect key areas of government, as well as private industry infrastructure.

That process has been ongoing now for several months. We have enlisted the assistance of many agencies, particularly Department of Defense agencies, which have great expertise in this area. We have also heavily relied upon private industry and private consultants to supply some of the necessary expertise for analysis and planning.

In addition, pursuant to Executive Order and to a Presidential Directive on terrorism, we established an FBI Computer Investigations and Threat Assessment Center in our Headquarters, which we call CITAC. The purpose of that center is two-fold. One, to develop and provide expertise in computer investigations. Secondly, to do threat assessments with respect to computer crime infrastructure defenses. This ties in as best it can with the infrastructure analysis program which is ongoing at the same time.

We've established three FBI computer crime squads in the field now: one is there in New York; two, in other cities around the country. These are very different animals in terms of our FBI structure. Most FBI squads are programmatic squads, dealing with bank robbery or with theft from interstate shipment. These new computer crime squads, however, are disciplinary squads—nonprogrammatic and specifically designed and ordered to gather up, within a particular division, all of the computer investigative expertise that we have both from an analytic and an operational point of view. We then use them as a resource for all other programs in the field divisions, whether they be counterterrorism programs, criminal programs, or national security matters. We also use them to assist our partners in other law enforcement agencies.

Part of our CITAC program also requires the SACs in all our 56 field divisions to form working groups with local industry—banking, utilities, energy, whatever the framework may be for that particular location—and put together working groups that will serve both to advise and respond to a crisis that threatens infrastructure, whether it be a criminal or a national security matter. To date, those advisory working groups are working very well around the country.

Again, the real key to success here—and I don't think it can be repeated too much—is the critical partnership of government with the private sector and private industry. Beyond our best expertise, we need additional expertise based on system-design and system-engineering infrastructure protection. It is critical that, prior to an emergency, we develop the contacts, the associations, and the working groups to deal with some of those problems.

We have worked very hard, as you know, in the legislation area to obtain the authorities that not only enable us to continue our investigative programs and techniques, but also help us anticipate some of the emerging problems. Last year, for instance, the FBI worked very hard with private industry and with many distinguished academics, to propose and ultimately to see pass the economic espionage statute, which is really a trade secrets act.

The interesting thing behind that initiative, however, was the fact that it was computer crime—particularly computer intrusions into major companies to valuable trade secrets—that focused our efforts to protect commerce and industry here in the United States. We found, for instance, that the traditional theft statutes, like the transportation of stolen property, just didn't apply to the situations where an intruder into or an employee of a corporation quickly downloads an important trade secret and transports that information on a disk either locally or globally.

The courts had said in many cases that intellectual property or knowledge of a trade secret was not really a "good" or "ware" as intended by the Congress in the interstate transportation of stolen property act. So we found we had a large area of criminal activity legally exempted from the FBI's program.

A major impetus for this trade secrets act was thus the ability of computer criminals to steal valuable intellectual property that doesn't quite fit the 1930's definition of "goods, wares, and merchandise." This is just one example of our legislative initiatives directed towards those technology problems.

Encryption is another important one. We realize that the need for robust encryption is critical for the health of our national economy and for American competitiveness, here and overseas. As a law enforcement agency that wears a national security hat, however, we also realize that we need encryption with some exempted or court-authorized recovery mechanism for those very rare instances when encrypted channels are used to either transmit or store information, relative to a crime, an act of terror, or a national security matter. Of course such a mechanism would to be available to us only under court orders and very stringent requirements, as with the 1968 court-authorized wiretapping statute.

So there is a friction, as there always is, between technology issues and law enforcement, with respect to balancing public safety issues with first amendment issues regarding free trade. We have worked very hard to try to achieve that critical balance. We are not out of the woods yet. We cannot do, nor would we advocate doing, what some countries have done; Russia, Israel, and France, for example, have outlawed encryption. That's a solution that doesn't really deal with the problem in all of its dynamics and all of its dimensions. We are rather looking for a solution that balances the protection of robust encryption with the protection of national security and public safety—one that affords both sides the protections and restrictions that will keep our intrusiveness limited and will not affect competition or the economy.

Another FBI initiative that deals with cyber crime and global crime issues is the expansion of our Legal Attache program. As many of our international friends here know, the FBI has had, for many, many years, a "Legat" program, as we call it, where FBI agents are assigned to various embassies to engage in liaison functions with the host law enforcement authorities. They deal exclusively in law enforcement-related activities. They don't engage in any other non-criminal activities except as liaison.

We have Legats now in 30 countries. On my last trip, in fact, I dedicated an office in Tel Aviv, which will also serve as liaison with Jordan, with the Palestinian authority, and with the office in Cairo. Over the next two years, pursuant to a plan approved by the Congress, we'll open another 16 Legats which will take us to the places like Beijing, South Africa, and Buenos Aires—places where law enforcement needs cop-to-cop bridges and the police-to-police contacts that are necessary to deal with crimes like computer crime and others that have no boundaries and that are committed in the twinkling of a eye. These crimes absolutely require us to work with our partners in order to identify and solve them.

We are behind the eight ball, I think, in our efforts to deal both with cyber crime and global crime. But the initiatives I've mentioned—including infrastructure initiatives, training initiatives, the Legat programs—are all certainly moving in the right direction. My concern is that we are moving too slowly and that the pace of change is so rapid that, despite our best efforts and our resources, we will still remain a little bit behind the curve. If you think about what's happening now with respect to cyber crime and global crime, it's not unfair to compare it to the advent of the automobile back in the early part of the century. Easy automobile use then changed everything. It didn't just changed the economy, it also had an immense impact on law enforcement, which had been dealing with crime on a localized basis.

Today the change is from national to international borders. I remember when I was a new FBI agent here in New York in 1975. It was an anomaly to have a lead in your case which went overseas to a foreign bank account, or to need to speak to a witness who was outside our jurisdiction, or to need records from an off-shore location where we had no jurisdiction.

Now in 1997, that's all changed. It's probably rare when we don't have an international connection in a drug case or an economic crime case, or a fugitive case or a national security matter. Just like the automobile back in the 1920s and the 1930s, the computer is impacting the economy and the science of law enforcement today, except, probably, with a ten-fold greater impact.

It is affecting everything we do in law enforcement. It is changing the rules of the game with respect to how we prepare for and deal with national security issues. And it will continue to do that at an even more alarming rate.

Remember the old gangster movies where somebody robbed a bank, got in an old Model T, and raced away from the police to a state line—where the police had to stop because they didn't have jurisdiction to take the bank robbers over the state line? Congress intervened when that happened. It established interstate banking authority for the FBI with respect to the bank robbery jurisdiction. And everybody thought we had solved the problem. In today's world, though, we're dealing with global borders and economic borders that have ceased to exist. We now need to have authorities from Congress, and we also need technological means to deal with a problem that is getting increasingly more complex and global. We need to draw on your expertise and advice and partner our efforts.

So, let me just close by again thanking you all of your attendance and participation here. We certainly appreciate your interest. We thank your Chiefs and Director Generals for approving this. And we ask that you help us to combat these new crime phenomena. A critical part, of our success will come from industry and from you. Thank you very much.

CHAPTER 23

ENCRYPTION

STATEMENT OF LOUIS J. FREEH, DIRECTOR FEDERAL BUREAU OF INVESTIGATION

Before the Senate Judiciary Committee Hearing on Encryption United States Senate

Washington, D.C.
July 9, 1997

Mr. Chairman and members of the committee, I appreciate the opportunity to discuss the issue of encryption and I applaud your willingness to deal with this vital public safety issue.

The looming spectre of the widespread use of robust, virtually uncrackable encryption is one of the most difficult problems confronting law enforcement as the next century approaches. At stake are some of our most valuable and reliable investigative techniques, and the public safety of our citizens.

We believe that unless a balanced approach to encryption is adopted that includes a viable key management infrastructure, the ability of law enforcement to investigate and sometimes prevent the most serious crimes and terrorism will be severely impaired. Our national security will also be jeopardized.

For law enforcement, framing the issue is simple. In this time of dazzling telecommunications and computer technology where information can have extraordinary value, the ready availability of robust encryption is essential. No one in law enforcement disputes that. Clearly, in today's world and more so in the future, the ability to encrypt both contemporaneous communications and stored data is a vital component of information security.

As is so often the case, however, there is another aspect to the encryption issue that if left unaddressed will have severe public safety and national security ramifications. Law enforcement is in unanimous agreement that the widespread use of robust non-key recovery encryption will allow drug lords, spies, terrorists and even violent gangs to communicate about their crimes and their conspiracies with impunity. We will lose one of the few remaining vulnerabilities of the worst criminals and terrorists upon which law enforcement depends to successfully investigate and often prevent the worst crimes.

For this reason, the law enforcement community is unanimous in calling for a balanced solution to this problem. It is called "key recovery" encryption and, in our view, any legislative approach that does not achieve such a balanced approach seriously jeopardizes the long-term viability and usefulness of court-authorized access to transmitted as well as stored evidence and information. Electronic surveillance and search and seizure are techniques upon which law enforcement depends to ensure public safety and maintain national security.

Under one type of key recovery approach, a decryption "key" for a given encryption product is deposited with a trustworthy key recovery agent for safe keeping. The key recovery agent could be a private company, a bank, or other commercial or government entity that meets established trustworthiness criteria. Should encryption users need access to their encrypted information, they could obtain the decryption key from the key recovery agent. Additionally, when law enforcement needs to decrypt criminal-related communications or computer files lawfully seized under established legal authorities, they too, under conditions prescribed by law and with the presentation of proper legal process, could obtain the decryption key from the key recovery agent. This is the only viable way to permit the timely decryption of lawfully seized communications or computer files that are in furtherance of criminal activity. The key recovery information would be provided to the law enforcement agency under very strict controls and would be used only for its intended public safety purpose. Under this approach, the law-abiding would gain the benefits of strong, robust encryption with emergency access capabilities and public safety and national security would be maintained—as manufacturers produce and sell encryption products that provide key recovery.

This solution meets industry's information security and communications privacy needs for strong encryption while addressing law enforcement's public safety needs for timely decryption when such products are used to conceal crimes or impending acts of terrorism or espionage.

Some have argued that government policy makers should step aside and let market forces solely determine the direction of key recovery encryption, letting market forces determine the type of technologies that will be used and under what circumstances. They argue that most corporations that see the need for encryption will also recognize the need for, and even insist on, key recovery encryption products to secure their electronically stored information and to protect their corporate interests should an encryption key be lost, stolen or used by a rogue employee for extortion purposes.

We agree that rational thinking corporations will act in a prudent manner and will insist on using key recovery encryption for electronically stored information. However, law enforcement has a unique public safety requirement in the area of perishable communications which are in transit (telephone calls, e-mail, etc.). It is law enforcement, not corporations, that has a need for timely decryption of communications in transit. There is extraordinary risk in trusting public safety and national security to market forces that rightfully are protecting important but unrelated interests. Law enforcement's needs will not be adequately addressed by this type of an approach.

It is for this reason that government policy makers and Congress should play a direct role in shaping our national encryption policy and adopt a balanced approach that addresses both the commercial and the public safety needs. The adverse impact to public safety and national security associated with any type of "wait and see" or voluntary market force approach would be far too great of a price for the American public to pay.

Several bills have recently been introduced which address encryption. Language in some of the proposed bills makes it unlawful to use encryption in the furtherance of criminal activity and set out procedures for law enforcement access to stored keys in those instances where key recovery encryption was voluntarily used. One of these bills, S.909, takes significant strides in the direction of protecting public safety by encouraging the use of key recovery encryption through market based incentives and other inducements.

Unfortunately, these legislative proposals still do not contain adequate assurances that the impact on public safety and effective law enforcement of the widespread availability of encryption will be addressed. We look forward to working with you to develop legislative accommodations that adequately address the public safety needs of law enforcement and a balanced encryption policy.

Further, some argue the encryption "genie is out of the bottle," and that attempts to influence the future use of encryption are futile. I do not believe that to be the case. Key recovery encryption products can, with government and industry support, become a standard for use in the global information infrastructure.

No one contends that a key recovery-based encryption policy will prevent all criminals, spies and terrorists from using non-key recovery encryption. But if we, as a nation, act responsibly and build systems and products that support and rely upon key recovery, all facets of the public's interest can be served.

And as this committee knows, export controls on encryption products exist primarily to protect national security and foreign policy interests. However, law enforcement is more concerned about the significant and growing threat to public safety and effective law enforcement that would be caused by the proliferation and use within the United States of a communications infrastructure that supports strong encryption products but cannot support timely law enforcement decryption. Without question, such an infrastructure will be used by dangerous criminals and terrorists to conceal their illegal plans and activities from law enforcement, thus inhibiting our ability to enforce the laws and prevent terrorism.

Congress has on many occasions accepted the premise that the use of electronic surveillance is a tool of utmost importance in terrorism cases and in many criminal investigations, especially those involving serious and violent crime, terrorism, espionage, organized crime, drug-trafficking, corruption and fraud. There have been numerous cases where law enforcement, through the use of electronic surveillance, has not only solved and successfully prosecuted serious crimes and dangerous criminals, but has also been able to prevent serious and life-threatening criminal acts. For example, terrorists in New York were plotting to bomb the United Nations Building, the Lincoln and Holland Tunnels, and 26 Federal Plaza as well as conduct assassinations of political figures. Court-authorized electronic surveillance enabled the FBI to disrupt the plot as explosives were being mixed. Ultimately, the evidence obtained was used to convict the conspirators. In another example, electronic surveillance was used to prevent and then convict two men who intended to kidnap, molest and then kill a male child.

Most encryption products manufactured today do not provide for timely law enforcement decryption.

Widespread use of non-key recovery encryption or communications infrastructure that supports non-key recovery encryption use clearly will undermine law enforcement's ability to effectively carry out its public safety mission and to combat ultra-dangerous criminals and terrorists.

This is not a problem that will begin sometime in the future. Law enforcement is already encountering the harmful effects of encryption in many important investigations today. For example:

- Convicted spy Aldrich Ames was told by the Russian intelligence service to encrypt computer file information that was to be passed to them.
- An international terrorist was plotting to blow up 11 U.S. -owned commercial airliners in the far east. His laptop computer which was seized during his arrest in Manilla contained encrypted files concerning this terrorist plot.
- A subject in a child pornography case used encryption in transmitting obscene and pornographic images of children over the Internet.

- A major international drug trafficking subject recently used a telephone encryption device to frustrate court-approved electronic surveillance.

Requests for cryptographic support pertaining to electronic surveillance interceptions from FBI field offices and other law enforcement agencies have steadily risen over the past several years. For example, from 1995 to 1996, there was a two-fold increase (from 5 to 12) in the number of instances where the FBI's court-authorized electronic efforts were frustrated by the use of encryption that did not allow for law enforcement access.

Over the last three (3) years, the FBI has also seen the number of computer related cases utilizing encryption and/or password protection increase from 20 or two (2) percent of the cases involving electronically stored information to 140 or seven (7) percent. These included the use of 56 bit data encryption standard (DES) and 128 bit "pretty good privacy" (PGP) encryption.

Just as when this committee so boldly addressed digital telephone, the government and the nation are again at an historic crossroad on this issue. The International Association of Chiefs of Police, the National Sheriff's Association and the National District Attorneys Association have all enacted resolutions supporting a balanced encryption policy and opposing any legislation that undercuts or falls short such a balanced policy. If public policy makers act wisely, the safety of all Americans will be enhanced for decades to come. But if narrow interests prevail, then law enforcement will be unable to provide the level of protection that people in a democracy properly expect and deserve.

Conclusion

We are not asking that the magnificent advances in encryption technology be abandoned. We are the strongest proponents of robust, reliable encryption manufactured and sold by American companies all over the world. Our position is simple and, we believe, vital. Encryption is certainly a commercial interest of great importance to this great nation. But it's not merely a commercial or business issue. To those of us charged with the protection of public safety and national security, encryption technology and its application in the information age—here at the dawn of the 21st century and thereafter—will become a matter of life and death in many instances which will directly impact on our safety and freedoms. Good and sound public policy decisions about encryption must be made now by the Congress and not be left to private enterprise. Legislation which carefully balances public safety and private enterprise must be established with respect to encryption.

Would we allow a car to be driven with features which would evade and outrun police cars? Would we build houses or buildings which firefighters could not enter to save people?

Most importantly, we are not advocating that the privacy rights or personal security of any person or enterprise be compromised or threatened. You can't yell "fire" in a crowded theater. You can't with impunity commit libel or slander. You can't use common law honored privileges to commit crimes.

In support of our position for a rational encryption policy which balances public safety with the right to secure communications, we rely on the Fourth Amendment to the Constitution. There the framers established a delicate balance between "the right of the people to be secure in their persons, houses, papers, and effects (today we might add personal computers, modems, data streams, discs, etc.) against unreasonable searches and seizures." Those precious rights, however, were balanced against the legitimate right and necessity of the police, acting through strict legal process, to gain access by lawful search and seizure to the conversations and stored evidence of criminals, spies and terrorists.

The precepts and balance of the Fourth Amendment have not changed or altered. What has changed from the late eighteenth to the late twentieth century is technology and telecommunications well beyond the contemplation of the framers.

The unchecked proliferation of non key recovery encryption will drastically change the balance of the Fourth Amendment in a way which would shock its original proponents. Police soon may be unable through legal process and with sufficient probable cause to conduct a reasonable and lawful search or seizure, because they cannot gain access to evidence being channeled or stored by criminals, terrorists and spies. Significantly, their lack of future access may be in part due to policy decisions about encryption made or not made by the United States. This would be a terrible upset of the balance so wisely set forth in the Fourth Amendment on December 15, 1791. I urge you to maintain that balance and allow your police departments, district attorneys, sheriffs and federal law enforcement authorities to continue to use their most effective techniques to fight crime and terrorism—techniques well understood and authorized by the framers and Congress for over two hundred years.

I look forward to working with you on this matter and at this time would be pleased to answer any questions.

CHAPTER 24

THE IMPACT OF ENCRYPTION ON PUBLIC SAFETY

STATEMENT OF LOUIS J. FREEH, DIRECTOR FEDERAL BUREAU OF INVESTIGATION

Before the Permanent Select Committee on Intelligence United States House of Representatives

Washington, D.C.
September 9, 1997

Mr. Chairman and members of the committee, I appreciate the opportunity to discuss the issue of encryption and I applaud your willingness to deal with this vital public safety issue.

The looming spectre of the widespread use of robust, virtually unbreakable encryption is one of the most difficult problems confronting law enforcement as the next century approaches. At stake are some of our most valuable and reliable investigative techniques, and the public safety of our citizens. We believe that unless a balanced approach to encryption is adopted that includes a viable key management infrastructure that supports immediate decryption capabilities for lawful purposes, our ability to investigate and sometimes prevent the most serious crimes and terrorism will be severely impaired. Our national security will also be jeopardized.

For law enforcement, framing the issue is simple. In this time of dazzling telecommunications and computer technology where information can have extraordinary value, the ready availability of robust encryption is essential. No one in law enforcement disputes that. Clearly, in today's world and more so in the future, the ability to encrypt both contemporaneous communications and stored data is a vital component of information security.

As is so often the case, however, there is another aspect to the encryption issue that if left unaddressed will have severe public safety and national security ramifications. Law enforcement is in unanimous agreement that the widespread use of robust unbreakable encryption ultimately will devastate our ability to fight crime and prevent terrorism. Unbreakable encryption will allow drug lords, spies, terrorists and even violent gangs to communicate about their crimes and their conspiracies with impunity. We will lose one of the few remaining vulnerabilities of the worst criminals and terrorists upon which law enforcement depends to successfully investigate and often prevent the worst crimes.

For this reason, the law enforcement community is unanimous in calling for a balanced solution to this problem. Such a solution must satisfy both the commercial needs of industry for strong encryption and law enforcement's public safety decryption needs. In our view, any legislative approach that does not achieve such a balanced approach seriously jeopardizes the long-term viability and usefulness of court-authorized access to transmitted as well as stored evidence and information. Electronic surveillance and search and seizure are techniques upon which law enforcement depends to ensure public safety and maintain national security.

One such balanced solution to this problem is key recovery encryption. Under this approach, a decryption "key" for a given encryption product is deposited with a trustworthy key recovery agent for safe keeping. The key recovery agent could be a private company, a bank, or other commercial or government entity that meets established trustworthiness criteria. Should encryption users need access to their encrypted information, they could obtain the decryption key from the key recovery agent. Additionally, when law enforcement needs to decrypt criminal-related communications or computer files lawfully seized under established legal authorities, they too, under conditions prescribed by law and with the presentation of proper legal process, could obtain the decryption key from the key recovery agent. This is the only viable way to permit the timely decryption of lawfully seized communications or computer files that are in furtherance of criminal activity.

The decryption key or information would be provided to the law enforcement agency under very strict controls and would be used only for its intended public safety purpose. Under this approach, the law-abiding would gain the benefits of strong, robust encryption products and services with emergency decryption capabilities and public safety and national security would be maintained—as manufacturers produce and sell encryption products that include features that allow for the immediate decryption of criminal-related encrypted communications or electronic information.

This solution meets industry's information security and communications privacy needs for strong encryption while addressing law enforcement's public safety needs for immediate decryption when such products are used to conceal crimes or impending acts of terrorism or espionage.

Some have argued that government policy makers should step aside and let market forces solely determine the direction of key recovery encryption, letting market forces determine the type of technologies that will be used and under what circumstances. They argue that most corporations that see the need for encryption will also recognize the need for, and even insist on, key recovery encryption products to secure their electronically stored information and to protect their corporate interests should an encryption key be lost, stolen or used by a rogue employee for extortion purposes.

We agree that rational thinking corporations will act in a prudent manner and will insist on using key recovery encryption for electronically stored information. However, law enforcement has a unique public safety requirement in the area of perishable communications which are in transit (telephone calls, e-mail, etc.). It is law enforcement, not corporations, that has a need for the immediate decryption of communications in transit. There is extraordinary risk in trusting public safety and national security to market forces that rightfully are protecting important but unrelated interests. Law enforcement's needs will not be adequately addressed by this type of an approach.

It is for this reason that government policy makers and Congress should play a direct role in shaping our national encryption policy and adopt a balanced approach that addresses both the commercial and the public safety needs. The adverse impact to public safety and national security associated with any type of "wait and see" or voluntary market force approach would be far too great of a price for the American public to pay.

Several bills have recently been introduced which address encryption. Language in some of the proposed bills makes it unlawful to use encryption in the furtherance of criminal activity and set out procedures for law enforcement access to stored decryption keys in those instances where key recovery encryption was voluntarily used. **Only one of these bills, S. 909, comes close to meeting our core public safety, effective law enforcement, and national security needs.** S. 909 takes significant strides in the direction of protecting public safety by encouraging the use of key recovery encryption through market based incentives and other inducements. All of the other bills currently under consideration by the Congress, **to include S. 376, S. 377, and H.R. 695,** would have a significant **negative** impact on public safety and national security and would risk great harm to our ability to enforce the laws and protect our citizens if enacted.

Unfortunately, S. 909 still does not contain sufficient assurances that the impact on public safety and effective law enforcement caused by the widespread availability of encryption will be adequately addressed. We look forward to working with you to develop legislative accommodations that adequately address the public safety needs of law enforcement and a balanced encryption policy.

Further, some argue the encryption "Genie is out of the bottle," and that attempts to influence the future use of encryption are futile. I do not believe that to be the case. Strong encryption products that include decryption features for lawful purposes can, with government and industry support, become the standard for use in the global information infrastructure.

No one contends that the adoption of a balanced encryption policy will prevent all criminals, spies and terrorists from gaining access to and using unbreakable encryption. But if we, as a nation, act responsibly and only build systems and encryption products that support and include appropriate decryption features, all facets of the public's interest can be served.

And as this committee knows, export controls on encryption products exist primarily to protect national security and foreign policy interests. However, law enforcement is more concerned about the significant and growing threat to public safety and effective law enforcement that would be caused by the proliferation and use within the United States of a communications infrastructure that supports the use of strong encryption products but that does not support law enforcement's immediate decryption needs. Without question, such an infrastructure will be used by dangerous criminals and terrorists to conceal their illegal plans and activities from law enforcement, thus inhibiting our ability to enforce the laws and prevent terrorism.

Congress has on many occasions accepted the premise that the use of electronic surveillance is a tool of utmost importance in terrorism cases and in many criminal investigations, especially those involving serious and violent crime, terrorism, espionage, organized crime, drug-trafficking, corruption and fraud. There have been numerous cases where law enforcement, through the use of electronic surveillance, has not only solved and successfully prosecuted serious crimes and dangerous criminals, but has also been able to prevent serious and life-threatening criminal acts. For example, terrorists in New York were plotting to bomb the United Nations building, the Lincoln and Holland tunnels, and 26 Federal Plaza as well as conduct assassinations of political figures. Court-authorized electronic surveillance enabled the FBI to disrupt the plot as explosives were being mixed. Ultimately, the evidence obtained was used to convict the conspirators. In another example, electronic surveillance was used to prevent and then convict two men who intended to kidnap, molest and then kill a male child.

Most encryption products manufactured today do not contain features that provide for immediate law enforcement decryption. Widespread use of unbreakable encryption or communications infrastructure that supports the use of unbreakable encryption clearly will undermine law enforcement's ability to effectively carry out its public safety mission and to combat dangerous criminals and terrorists.

This is not a problem that will begin sometime in the future. Law enforcement is already encountering the harmful effects of encryption in many important investigations today. For example:

- convicted spy Aldrich Ames was told by the Russian Intelligence Service to encrypt computer file information that was to be passed to them.
- an international terrorist was plotting to blow up 11 U.S.-owned commercial airliners in the Far East. His laptop computer which was seized during his arrest in Manilla contained encrypted files concerning this terrorist plot.
- a subject in a child pornography case used encryption in transmitting obscene and pornographic images of children over the Internet.
- a major international dug trafficking subject recently used a telephone encryption device to frustrate court-approved electronic surveillance.

Requests for cryptographic support pertaining to electronic surveillance interceptions from FBI field offices and other law enforcement agencies have steadily risen over the past several years. For example, from 1995 to 1996, there was a two-fold increase (from 5 to 12) in the number of instances where the FBI's court-authorized electronic efforts were frustrated by the use of encryption products that did not allow for lawful law enforcement decryption.

Over the last three (3) years, the FBI has also seen the number of computer-related cases utilizing encryption and/or password protection increase from 20 or two (2) percent of the cases involving electronically stored information to 140 or seven (7) percent. These included the use of 56-bit data encryption standard (DES) and 128-bit "pretty good privacy" (PGP) encryption.

Just as when the Congress so boldly addressed the digital telephony issue in 1994, the government and the nation are again at an historic crossroad on this issue. The Attorney General and the heads of federal law enforcement agencies as well as the presidents of several state and local law enforcement associations recently sent letters to every member of Congress urging the adoption of a balanced encryption policy. In addition, the International Association of Chiefs of Police, the National Sheriff's Association and the National District Attorneys Association have all enacted resolutions supporting a balanced encryption policy and opposing any legislation that undercuts or falls short such a balanced policy.

If public policy makers act wisely, the safety of all Americans will be enhanced for decades to come. But if narrow interests prevail, then law enforcement will be unable to provide the level of protection that people in a democracy properly expect and deserve.

Conclusion

We are not asking that the magnificent advances in encryption technology be abandoned. We are the strongest proponents of robust, reliable encryption manufactured and sold by American companies all over the world. Our position is simple and, we believe, vital. Encryption is certainly a commercial interest of great importance to this great nation. But it's not merely a commercial or business issue. To those of us charged with the protection of public safety and national security, encryption technology and its application in the information age—here at the dawn of the 21st century and thereafter—will become a matter of life and death in many instances which will directly impact on our safety and freedoms. Good and sound public policy decisions about encryption must be made now by the Congress and not be left to private enterprise. Legislation which carefully balances public safety and private enterprise must be established with respect to encryption.

Would we allow a car to be driven with features which would evade and outrun police cars? Would we build houses or buildings which firefighters could not enter to save people?

Most importantly, we are not advocating that the privacy rights or personal security of any person or enterprise be compromised or threatened. You can't yell "fire" in a crowded theater. You can't with impunity commit libel or slander. You can't use common law honored privileges to commit crimes.

In support of our position for a rational encryption policy which balances public safety with the right to secure communications, we rely on the Fourth Amendment to the Constitution. There the framers established a delicate balance between "the right of the people to be secure in their persons, houses, papers, and effects (today we might add personal computers, modems, data streams, discs, etc.) against unreasonable searches and seizures." Those precious rights, however, were balanced against the legitimate right and necessity of the police, acting through strict legal process, to gain access by lawful search and seizure to the conversations and stored evidence of criminals, spies and terrorists.

The precepts and balance of the Fourth Amendment have not changed or altered. What has changed from the late eighteenth to the late twentieth century is technology and telecommunications well beyond the contemplation of the framers.

The unchecked proliferation of unbreakable encryption will drastically change the balance of the Fourth Amendment in a way which would shock its original proponents. Police soon may be unable through legal process and with sufficient probable cause to conduct a reasonable and lawful search or seizure, because they cannot gain access to evidence being channeled or stored by criminals, terrorists and spies. Significantly, their lack of future access may be in part due to policy decisions about encryption made or not made by the United States. This would be a terrible upset of the balance so wisely set forth in the Fourth Amendment on December 15, 1791. I urge you to maintain that balance and allow your police departments, district attorneys, sheriffs and federal law enforcement authorities to continue to use their most effective techniques to fight crime and terrorism—techniques well understood and authorized by the framers and Congress for over two hundred years.

I look forward to working with you on this matter and at this time would be pleased to answer any questions.

CHAPTER 25

THREATS TO U.S. NATIONAL SECURITY

Statement for the Record
Louis J. Freeh, Director
Federal Bureau of Investigation

Hearing on Current and Projected Threats to U.S. National Security, before the Senate Select Committee on Intelligence

Washington, D.C.
January 28, 1998

The New Global Threat Paradigm

The overriding concern now facing law enforcement is how rapidly the threats from terrorists and criminals are changing, particularly in terms of technology, and the resulting challenge to law enforcement's ability to keep pace with those who wish to do harm to our nation and our nation's citizens. This is why the encryption issue is one of the most important issues confronting law enforcement and potentially has catastrophic implications for our ability to combat every threat to national security that I am about to address in my statement here today. Law enforcement remains in unanimous agreement that the widespread use of robust non-recovery encryption ultimately will devastate our ability to fight crime and terrorism. Uncrackable encryption is now and will continue, with ever increasing regularity, allow drug lords, terrorists and even violent gangs to communicate about their criminal intentions with impunity and to maintain electronically stored evidence of their crimes impervious to lawful search and seizure. Other than some type of key-recoverable system, there is currently no viable technical solution to this problem for law enforcement.

This is not a problem that will begin sometime in the future with theoretical implications. In many important investigations effective law enforcement is being frustrated by criminals and terrorists using non-recoverable encryption. For example:

- Convicted spy Aldrich Ames was told by his Soviet handlers to encrypt computer file information that was to be passed to them.
- Ramzi Yousef and other international terrorists were plotting to blow up 11 U.S.-owned commercial airliners in the far east. Yousef's laptop computer, which was seized in Manila, contained encrypted files concerning this terrorist plot.

- A major international drug trafficking subject recently used a telephone encryption device to frustrate court-approved electronic surveillance.

Requests for cryptographic support pertaining to electronic surveillance interceptions from FBI field offices and other law enforcement agencies have steadily risen over the past several years. From 1995 to 1996, there was a two-fold increase (from 5 to 12) in the number of instances where the FBI's court-authorized electronic efforts were frustrated by the criminal's use of encryption that did not allow for law enforcement access.

Over the last two years, the FBI has also seen the number of computer-related cases utilizing encryption and/or password protection increase from two (2) percent to seven (7) percent, to include the use of 56 bit Data Encryption Standard (DES) and 128 bit "Pretty Good Privacy" (PGP) encryption.

It is for this reason that the law enforcement community is urgently calling for our Nation's policy makers to adopt a balanced public policy on encryption. In our view, any legislative approach to the encryption issue that does not achieve such a balanced approach seriously jeopardizes the utility of some of our most important and effective investigative techniques upon which law enforcement must depend to ensure public safety and to maintain national security.

Several bills have been introduced in this Congress that address certain aspects of the encryption issue. Unfortunately, most of these legislative proposals would largely remove existing export controls on encryption products, and would promote the widespread availability and use of uncrackable encryption products regardless of the impact on public safety and national security.

It is important to note that S.909, the "Secure Public Networks Act," introduced by Senators Kerrey, McCain, and Hollings, comes close to addressing law enforcement's public safety needs in the area of encryption. However, law enforcement believes that the bill does not contain sufficient legislative assurances to adequately address law enforcement's public safety needs regarding the use and availability of encryption products and service within the United States.

Conversely, the substitute bill adopted by the House Permanent Select Committee on Intelligence (HPSCI) on September 11, 1997 during their mark-up of H.R. 695 does effectively address all of law enforcement's public safety and national security concerns regarding encryption products and services manufactured for use in the United States or imported into the United States. The HPSCI substitute bill would require all such encryption products and services to contain features that would allow for the immediate access by law enforcement to the "plaintext" of encrypted criminal-related communications or electronically stored data pursuant to a court order.

We are now at an historic crossroad on this issue. If public policy makers act wisely, the safety of all Americans will be enhanced for decades to come. But if narrow interests prevail, law enforcement will be unable to provide the level of protection that people in a democracy properly expect and deserve. I do not believe it is too late to deal effectively with this issue and would encourage the Committee to look closely at the action taken by the HPSCI in their efforts to adopt a balanced encryption policy.

CRITICAL INFRASTRUCTURE PROTECTION

In a few short months, on July 26, 1998, the FBI will celebrate its 90th birthday. The FBI has been a remarkable institution for many reasons, not the least of which has been its ability to remake itself to address new challenges to U.S. national security and criminal justice. In the beginning, FBI agents were not authorized to carry firearms. In response to the gang-era of the 1920s and 1930s, agents were first issued handguns, and then the storied Thompson submachine gun. In what is much more than a symbolic shift, todays agents are issued laptop computers. This important advancement is the direct result of an evolution in national security vulnerabilities.

As one consequence of technological innovation, deregulation, and economic imperatives, critical infrastructure systems have become more complex and interdependent. Digital control systems-based on commercial off-the-shelf hardware and software are being used to streamline network operations and reduce personnel requirements. These control networks frequently are connected by publicly-accessible telecommunications systems and commercially available information technologies—the National Information Infrastructure (NII)—a trend that will accelerate as utility, transportation, and government activities eliminate antiquated, expensive private telecommunications networks. The result is a revolutionary and systemic improvement in industrial and commercial processes that has been widely recognized and exploited by both public and private sectors.

Public—and private—sector organizations that rely on information technologies are diverse. Within the government, information technologies provide leverage for performing traditional missions more efficiently, e.g., law enforcement, intelligence gathering and exploitation, and national defense. In the private sector information systems allow rapid, efficient transfers of information and capital, enable a new wave of electronic commerce, and enable far-flung, technically complex operations to exist over vast geographic distances.

However, as commercial information technologies create advantages, their increasingly indispensable nature transforms them into high-value targets. Moreover, in practice these developments have resulted in diminished systems redundancy and the consolidation of core assets, heightening the risk of catastrophic single-point failures. These vulnerabilities are accompanied by a more variegated threat picture. The range of potential adversaries that may seek to attack U.S. infrastructure systems is broad and growing. Disgruntled employees, disaffected individuals or groups, organized crime, domestic and international terrorists, and adversary nations are all potential sources of attack.

Vulnerabilities

Hundreds of information system vulnerabilities are discovered every day —many directly related to U.S. national security. Dozens of previously unknown computer system vulnerabilities are uncovered every day by the vast yet interconnected community of technical experts. Most of these vulnerabilities are subsequently posted publicly, usually on the Internet first. For example, Internet mailing lists routinely distribute vulnerability information and software that can be used to exploit vulnerabilities.

Vulnerability publicity usually follows through a succession of books, magazine and newspaper articles, electronic bulletin board messages, and a growing list of World Wide Web sites that are targeted at informing hackers, crackers, "phreakers," and, potentially, members of terrorist organizations and foreign intelligence services, about the latest methodology for staging successful cyber attacks. Lists of Frequently Asked Questions (FAQs) outlining the specifics of system vulnerabilities are widespread. "The Unofficial Web Hack FAQ," "The Hacker FAQ," and "How To Hack a Website" are popular, accessible, and easily downloaded from the Web. These vulnerabilities are present in the same commercial, off-the-shelf computer hardware and software used by both private industry and government.

These developing phenomena are associated with another, perhaps even more worrisome development. As information technologies and the physical infrastructure systems they control become increasingly complex, our ability to reliably anticipate system-wide behavior diminishes. This is because as systems increase in complexity, the effect of manipulating, degrading, or eliminating a single component is difficult to predict. Accidents, even seemingly minor ones, can have catastrophic effects. This creates a growing increase in the likelihood of catastrophic, single-point failures —accidental or induced.

The details of the nations infrastructure vulnerabilities are clearly illustrated in both classified and unclassified arenas. For instance, the myriad of government studies and reports on the matter include one recently published by the President's Commission on Critical Infrastructure Protection. The report was preceded by similar assessments by the Defense Science Board, the Office of Science and Technology Policy, and the Office of Management and Budget.

Threats

With very few exceptions, attacks against the nation's cyber assets can be aggregated into one of four categories: crime, terrorism, foreign intelligence, or war. Regardless of the category, any country or group can acquire the capability to conduct limited attacks against information systems from friendly nations, commercial vendors, arms dealers, hacker conferences, the Internet, and computer bulletin boards.

Software is one weapon of information-based attacks. Such software includes computer viruses, Trojan Horses, worms, logic bombs and eavesdropping sniffers. Advanced electronic hardware can also be useful in information attacks. Examples of such hardware are high-energy radio frequency (RF) weapons, electromagnetic pulse weapons, RF jamming equipment, or RF interception equipment. Such weapons can be used to destroy property and data; intercept communications or modify traffic; reduce productivity; degrade the integrity of data, communications, or navigation systems; and deny crucial services to users of information and telecommunications systems.

Where hackers formerly may have been motivated by the technical challenge of breaking into a computer system, the motivation may be shifting more toward hacking for profit. As more and more money is transferred through computer systems, as more fee-based computer services are introduced, as more sensitive proprietary economic and commercial information is exchanged electronically, and as the nation's defense and intelligence communities increasingly rely on commercially-available information technology, the tendency toward information threats emerging as national security threats will increase.

Terrorists, transnational criminals, and intelligence services are quickly becoming aware of and exploiting the power of information tools and weapons. This has been true in the past as new means of communication, transportation, and secrecy have been introduced to the public. For example, narcotics traffickers began using communications advances such as pagers and cellular phones soon after their introduction to the public.

Perhaps the most imminent threats today come from insiders. Insiders have the advantage of not needing to break into computer systems from the outside, but only to use, or abuse, their legitimate access. A large portion of the computer intrusion reports that the FBI and other law enforcement organizations receive have at their core an employee, or a former employee, who has exceeded his or her access, often in revenge for a perceived offense or wrong. These individuals have the knowledge of where the most sensitive information is stored, how to access the information, and, at times, how to steal or damage the data.

One such example involves a U.S. defense contractor firm that subcontracted with a foreign firm. The U.S. firm employed foreign contractors onsite, and allowed these employees access to certain areas of the premises that were necessary to their duties. However, the foreign contractors used their knowledge of the company's computer system to access other areas of the company's computer network that were off limits to non-U.S. employees. The foreign contractors were able to access proprietary and potentially classified information regarding the U.S. company's government contracts. Their activities jeopardized the competitiveness of the company and posed a potential threat to U.S. national security.

Another insider incident occurred in October 1997 when a former Pacific Gas & Electric Co. worker caused a widespread power outage in San Francisco. Moreover, security experts have repeatedly and publicly demonstrated the ease of compromising security at both private- and public-sector facilities through social engineering—posing as an insider to easily acquire information on internal security, passwords, and system configurations.

In terms of the maturity of the threat, the numbers tell the story. So far, in the month of January 1998 alone, there have been over forty reported hacked web sites according to hacked.net, a website that tracks such statistics. The FBI's load of computer intrusion-related cases is more than doubling every year. Because of the uncertainties associated with the evolutionary path of information technologies, the threat picture fifteen years hence is difficult to predict. However, some certainties apply: information technology is sure to proliferate, and those who would exploit these technologies for nefarious purposes are sure to multiply.

The FBI Response

The FBI was among the first to recognize the importance of predicting information-based attacks on critical infrastructures, preventing their occurrence, and mitigating damage in the event such attacks did occur. Since the 1992 creation of the National Computer Crime Squad in the FBI's Washington Field Office, additional regional computer squads in New York, San Francisco, Boston, Atlanta, Dallas, and Los Angeles. In addition to regional squads, the FBI has created computer investigative teams in each of its 56 field offices that will respond to computer incidents within their geographical area of responsibility.

The expansion of the National Computer Crime Squad was accompanied by the creation of a new National Security Threat List (NSTL) issue within the FBI's FCI program: Targeting the National Information Infrastructure, which the Attorney General approved in September 1995. The addition of this issue to the NSTL makes it possible for the FBI, working within its FCI authority, to investigate information infrastructure-related incidents perpetrated or coordinated by foreign intelligence services. These attacks might be directed against the U.S. Government or U.S. corporations, establishments, or persons and could target physical facilities, personnel, information, or computer, cable, satellite, or telecommunications systems. With the new NSTL issue, these teams have responsibilities over both the criminal investigative and the potential national security implications of computer intrusions.

The FBI is responding to these novel threat and vulnerability combinations through a coordinated interagency effort that includes, among other important participants, the Department of Defense, the National Security Agency, the Central Intelligence Agency, and the President's Commission on Critical Infrastructure Protection.

The Infrastructure Protection Center

One important interagency attempt to meet the emerging threat to the nation's critical infrastructures is the Infrastructure Protection Center (IPC). The IPC, a government-industry partnership hosted by the FBI, will provide a mechanism for assessing, warning, investigating, and responding to attacks on interconnected and interdependent infrastructures. The IPC units will be staffed with representatives from FBI and DOD, Intelligence Community, and agency detailees experienced in computer crimes and infrastructure protection. To build private confidence and information sharing, IPC will hire representatives of private industry or private sector computer emergency response teams (CERTs) making them an integral part of the center. Direct electronic connectivity is also being established with private industry and the CERTs.

Twenty-four-hour watch presence and connectivity maintained between the FBI, the National Security Agency, the Defense Intelligence Agency, and the Defense Information Systems Agency reinforce the respective strengths these parties bring to the infrastructure assurance mission. Future connectivity will include other government participants.

The IPC builds on and enhances close ties to the "first responders" to an attack on critical infrastructures—state and local law enforcement and government. Building on the FBI's long standing relationships with state and local law enforcement (through mechanisms like the Joint Terrorism Task Forces), the IPC will conduct outreach, provide training, share information, and coordinate interagency efforts during an attack. The IPC would also establish direct electronic connectivity to state and local governments building on existing FBI programs such as the Law Enforcement On-line (LEO) and Awareness of National Security Issues and Response (ANSIR) systems.

Though the national security threat from cyber-related issues is of concern, the FBI, with its private- and public-sector partners, is building a firewall of protection between malevolent actors and critical U.S. infrastructure systems. The threat is real and growing, but an effective response is underway. The IPC will embody the collected interagency expertise in the infrastructure protection mission, and exemplifies the depth of commitment the FBI has made to this important law enforcement and national security issue.

Economic Espionage:

- Boston: This case involved unauthorized intrusion into a voice-mail system by a disgruntled former employee. The victim was Standard Duplicating Machines Corporation (Standard), whose main competitor was the U.S. affiliate, Duplo Manufacturing Corporation of Japan (Duplo). John Hebel was employed by Standard as a field sales manager from 1990 to 1992, when he was terminated. Through an unsolicited phone call from a customer, Standard discovered Hebel had accessed Standard's phone mail system and had used the information to compete against Standard. Hebel was employed by Duplo at the time of the intrusions. On November 6, 1996, Hebel was charged with one count of violating Title 18, U.S.C. Section 1343 (Wire Fraud). On March 14, 1997, Hebel was sentenced to two years probation. Recent press accounts have highlighted the high financial risk of economic espionage to American businesses, communities, and jobs. The American Society of Industrial Security (ASIS) reported this month to the FBI that over $30 billion in American intellectual property were "placed at risk" from attempted theft in 1996 alone. Over 270 separate incidents were confirmed in the ASIS study. ASIS presently is finalizing its results for publication this year.

CHAPTER 26

CYBERCRIME, TRANSNATIONAL CRIME, AND INTELLECTUAL PROPERTY THEFT

Statement for the Record
Michael A. Vatis, Deputy Assistant Director
and
Chief, National Infrastructure Protection Center
Federal Bureau of Investigation
Before the Congressional Joint Economic Committee

Washington, DC
March 24, 1998

Chairman Saxton, Vice Chairman Mack, and Members of the Joint Economic Committee: Thank you for this opportunity to discuss cybercrime, the vulnerabilities of our Nation's critical infrastructures to increasing cyber threats, and what the Federal Bureau of Investigation (FBI) is doing to combat these problems.

As we continue to rush into the Information Age, our society is moving increasingly on-line. We use computers, the Internet, and other new "information technologies" to conduct business, perform scientific research, engage in personal communications, and do just about anything else that inventive minds can think of. But as society as a whole is moving on-line, so are criminals. Criminals use computers to facilitate crimes committed in the physical world. For example, they can use computers and the Internet to communicate with co-conspirators or to keep accounts of their illicit gains. Criminals also use these tools to engage in criminal activity on-line. For example, they use the Internet to defraud unsuspecting senior citizens, disseminate child pornography, steal credit card numbers, and rob banks by electronically shifting funds to their own off-shore accounts.

But the Internet and other advances in information technology do not merely give criminals new means to commit traditional crimes like theft or fraud. They also allow criminals and other malicious actors to cause new types of harm that go well beyond the potential loss to the individual victim and can affect our national economy and, indeed, our national security.

What type of harm am I talking about? The everyday functioning of our economy depends on the delivery of certain critical services. While we once got along fine without electrical power, think of the consequences if the power went out for a week—not just in one town or city, but across the whole Eastern Seaboard. And while plenty of people made their fortunes before the telephone, imagine what would happen to the Fortune 500 if they were deprived of telephone service for a few days.

There are several services whose availability we may take for granted, but which are truly critical to the smooth functioning of our society. We call these vital services our "critical infrastructures." Executive Order 13010, signed in 1996, lists the following eight infrastructures as "critical" to our economic health and our national security: telecommunications, banking and finance, transportation (including roads, railroads, airplanes and airports, mass transit, ports and harbors), electrical energy, gas and oil supply, water supply, emergency services (fire, health, police), and government operations. These infrastructures are defined as "critical" because their debilitation or destruction would have a significant adverse impact on our national economy or national security.

In the United States, we are able to expect things to work because our infrastructures are highly developed and efficient. Individuals and families can wake up in the morning confident that the lights will work, water will flow from the tap, and the trains will run. Businesses, too, can plan their activities and investments around the certainty that they will have ready access to telecommunications, that gas or oil will supply power to their factories, that their goods will be transported by truck, rail or airplane, and that funds can be safely deposited or withdrawn from their bank accounts. It is a given, in both our personal and professional lives, that essential goods and services will be available when needed.

Not so long ago, our dependence on these infrastructures did not pose a significant problem because there was little risk that these vital services would be knocked out. Only a rare and isolated occurrence, such as an earthquake or tornado or an accidental power outage could knock out a critical service over a broad area. The physical breadth of the infrastructures made it difficult for a potential malefactor to cause anything other than an isolated disturbance. And physical security measures adopted to prevent theft or vandalism generally also kept out those who would seek to destroy an infrastructure's ability to continue operating. A strong fence and a good security staff fended off not only thieves and vandals, but also terrorists. Moreover, our geographic isolation from other countries made it difficult for foreign adversaries to launch an attack on our infrastructures.

The Information Age, however, has changed things dramatically. For while information technologies create dramatic increases in efficiency and productivity, our dependence on them creates new vulnerabilities.

All critical infrastructures now rely on computers, advanced telecommunications, and, to an ever increasing degree, the Internet, for the control and management of their own systems, for their interaction with other infrastructures, and for communications with their suppliers and their customer base. For example, electric power grids and natural gas pipelines are controlled by computer systems, and those computers may be linked to each other and to the company headquarters by publicly-accessible telecommunications systems and commercially available information technologies to allow efficient management of power generation and smooth delivery to consumers. Billions of shares are traded each day over the telephone or Internet, and the stock exchanges could not function today without their vast networks of computers. Banks no longer rely on ledger books and safe deposit boxes to account for and secure their holdings, but depend on computerized accounting systems to manage depositors' accounts. The telecommunications system itself no longer uses operators to manually plug in calls to a switchboard but depends on computerized switching stations to handle the billions of calls placed each day. The government also relies on computers and publicly available communications systems to conduct the nation's business. Public and private networks and databases use the same technology, and vulnerabilities that affect one also affect the other.

But this reliance on new technologies comes with a price, and that price is a new vulnerability to those who would cause harm. For just as the new technologies make it easier for companies to communicate and control their businesses, they also make it easier for malicious actors to cause harm. The new vulnerability stems in part from the fact that the Internet and modern telecommunications systems are inherently open and accessible. That means that, with a certain amount of technical skill, one can use these communications media to get inside a company's or a government agency's computer system without ever physically penetrating its four walls. Moreover, the increased centralization of command and control systems afforded by the new technologies also means that, once inside that system, a potential malefactor can use those same technologies to cause harm over a much broader area than he ever could have hoped using physical weapons such as a bomb.

This vulnerability is exacerbated by several factors. First, most of our infrastructures rely on commercially available, off-the-shelf technology. This means that a vulnerability in hardware or software is not limited to one company, but is likely to be widespread, affecting every entity that uses the same equipment. A malefactor with knowledge of this one vulnerability can therefore attack multiple victims across the country, with just a few strokes on a keyboard.

Second, our infrastructures are increasingly interdependent and interconnected with one another. For example, the banking system depends on the availability and reliability of the telecommunications system and the Internet, which in turn rely on electrical power. Our transportation system depends on the availability of gas and oil supplies, which in turn are controlled through the use of new information technologies. The infrastructures are thus increasingly interdependent, so much so that it is difficult to predict the cascading effects that the disruption of one infrastructure would have on others.

Third, our telecommunications infrastructure is now truly global. Satellite communications, the Internet, and foreign ownership of telecommunications carriers in the U.S. have all combined to undermine the notion of a "National" Information Infrastructure. This means that our geographic isolation no longer acts as a moat to fend off foreign adversaries. Instead, it is now as easy to break into an infrastructure's network from St. Petersburg, Russia, as St. Petersburg, Florida. A personal computer and a telephone connection to an Internet Service Provider anywhere in the world are enough to conduct an attack.

Software is one weapon of cyber attacks. Such software includes, among others, computer viruses, Trojan Horses, worms, logic bombs, and eavesdropping "sniffers" that can be used to obtain passwords that allow hackers "root access" control of a computer system. Advanced electronic hardware also can be used in cyber attacks, including such items as high-energy radio frequency (RF) weapons, electromagnetic pulse weapons, RF jamming equipment, or RF interception equipment. These weapons can be used to destroy property and data; intercept communications or modify traffic; degrade the integrity of data, communications, or navigation systems; and deny crucial services to users of information and telecommunications systems.

So that's the vulnerability picture in the cyber world. But what about the corresponding threat? In the physical world, the range of people or groups that would have the means and motive to cause widespread destruction of an infrastructure are relatively limited—terrorist groups and hostile nations are the most likely actors. But the accessibility of the information infrastructure, global connectivity, and the rapid growth of a computer-literate population combine to ensure that millions of people around the world possess the means to engage in a cyber attack. The spectrum of threats in this new cyber world is staggeringly broad and varied, including: the disgruntled insider seeking revenge against his employer; the recreational hacker out to test his "cracking" skills; organized crime groups seeking illicit financial gain; domestic or international terrorist groups bent on causing harm to send a political message; foreign intelligence services seeking companies' proprietary data or sensitive government information; and hostile nation states utilizing information warfare as part, or instead, of a strategic military attack. Let me discuss each of these threats in a little more detail.

Perhaps the most imminent threat today comes from insiders. Insiders have the advantage of not needing to break into computer systems from the outside, but only to use, or abuse, their ligitimate access. Many of the computer intrusion reports the FBI and other law enforcement organizations receive have at their core an employee, former employee, consultant, or temporary employee who has exceeded his or her access, often in revenge for some perceived wrong. These individuals often have intimate knowledge of where the most sensitive information is stored, how to access the information, and how to steal or damage the data.

Recreational hackers are also increasingly dangerous, in part because of the widespread availability of "cracking" tools on hacker websites. One no longer needs to have a sophisticated understanding of computers and the Internet to successfully crack into a company's systems. Rather, one needs only to download an automated hacking tool from a website, compile the source code using a program readily available on the Internet, and click on a button to launch an attack on any number of target sites.

Moreover, the problem is exacerbated by our continued romanticization of hackers as technical whizzes who are not really doing anything wrong but are actually providing a service by pointing out the vulnerabilities in an individual's or a company's or government agency's system. But do we praise the burglar for demonstrating the vulnerability of our home security by breaking in and stealing our cash or jewelry? Even if he does not steal or break anything, the simple invasion of our private property causes a feeling of violation and vulnerability that would send chills down all our spines. Or do we thank the vandal who breaks into the corner store and defaces or destroys someone else's property? Of course not. But, similarly, we should not tolerate or condone analogous acts committed with computers. These are not acts that occur in some ethereal "cyberspace" that is somehow divorced from the real world. These are acts that are very real, and can cause serious harm. It is no joke when an individual's private E-mail communications are intercepted, or when a company's proprietary data is stolen or destroyed, or when a government agency's sensitive data is compromised. And these acts can have serious physical consequences. No one would laugh if a hacker caused air traffic control to go down at an airport, as happened in a case in Massachusetts that recently resulted in a plea bargain. Or if a hacker tied up 911 emergency phone services, potentially denying critical aid to people with true emergencies, as happened in a recent case in Florida. Our society has to do a better job of educating our children and young adults that breaking into someone else's computer system has serious real-world consequences, and is a serious crime.

Where hackers formerly may have been motivated by the technical challenge of breaking into a computer system, the motivation may now be shifting more toward hacking for profit. As more and more money is transferred through computer systems, as more fee-based computer services are introduced, and as more sensitive proprietary economic and commercial information is stored and exchanged electronically, we will see criminal hackers use their computer skills for illicit gain.

Terrorists and transnational criminals also rapidly are becoming aware of and exploiting the power of cyber tools. This has been true in the past as new means of communication and secrecy have been introduced to the public. For example, narcotics traffickers began using communications advances such as pagers, cellular phones, and unbreakable encryption soon after their introduction to the public. The fantastic growth of the Internet and other global information networks grants increasing numbers of users with hostile intentions access to global networks—and to those United States networks upon which critical infrastructures depend.

Finally, as our nation's defense and intelligence agencies increasingly rely on commercially available information technologies and publicly accessible communications systems for their everyday work, foreign intelligence services and hostile nation states will increasingly seek to acquire and use cyber tools to conduct espionage or engage in "information warfare" against us. Several different commissions, including the President's Commission on Critical Infrastructure Protection and the National Defense Panel, have recognized that no nation or group hostile to the United States can match us in traditional military firepower. Because of this, they would not be expected to take us on in a frontal or "symmetrical" attack. Rather, they would utilize irregular, "asymmetrical" attacks that hit us where we are most vulnerable. And one of those vulnerabilities is our reliance on information technologies for command and control of our national security activities as well as for the daily functioning of our privately-owned critical infrastructures. This vulnerability is particularly attractive to foreign enemies in that it is just as easy to crash a system from a computer terminal overseas as it is from one in the United States.

Some would say that this vulnerability is overstated, that there are sufficient technological security tools to protect against malicious hackers and crackers, and that infrastructures have built in redundancies to their systems to prevent catastrophic system failures in the event of a successful intrusion. I'm afraid that the facts prove otherwise. Although we have not experienced the electronic equivalent of a Pearl Harbor or Oklahoma City as some have foretold, the statistics and our cases demonstrate our dangerous vulnerabilities to cyber attacks.

A 1998 study by the Computer Security Institute shows that 64% of companies polled reported information system security breaches—an increase of 16% over last year. The total financial losses from the 241 organizations that could put a dollar figure on them adds up to $136,822,000. This figure represents a 36% increase in reported losses over the 1997 figure of $100,115,555 in losses.

While the Carnegie Melon CERT/Coordination Center reported a small reduction in security incidents (2,134 in 1997, down from 2,573 in 1996), the type and scope of attacks indicates a disturbing increase in the use of automated scripts, enabling malevolent network users to attack very large numbers of systems with much greater efficiency.

A study of 300 Australian companies by Deloitte Touche Tohmatsu found that over 37 percent of the companies experienced some form of security compromise in 1997, with the highest percentage of intrusions (57%) occurring in the banking and finance industry.

A 1996 survey by the American Bar Association of 1,000 companies showed that 48 percent had experienced computer fraud in the last five years. Company losses were reported to have ranged from $2-10 million.

In 1996 the Defense Information Systems Agency (DISA) estimated that as many as 250,000 attacks on DOD systems may have occurred in 1995. DISA indicates that the number of attacks has been increasing each year for the past few years, and that trend is expected to continue.

Finally, we at the FBI have seen significant increase in the number of pending computer intrusion investigations and in the number of successful prosecutions. Pending cases have increased 133% from the beginning of FY 1997, from 206 to 480. In FY 1997, there was a 110% increase in informations and indictments (from 10 to 21), a 950% increase in arrests (from 4 to 42), and an 88% increase in convictions (from 16 to 30).

As a caveat, let me state that it is not clear what accounts for these increases in our own case statistics or in the numbers reported by the private studies. It may be that systems administrators have simply gotten better at detecting intrusions, or that companies have become more willing to share information about their own exploited vulnerabilities. Or, it may be that the number of intrusions has risen significantly. Most likely, in my view, all three things are occurring. Regardless of the cause, however, these numbers clearly indicate significant vulnerabilities to cyber attacks.

Let me now give you a few examples of the types of computer crimes we have seen in recent years to further illustrate the problem:

You are undoubtedly aware of the recent series of intrusions into Department of Defense and other government agency computers across the country. This case involved widespread illegal intrusions into government systems using holes in the systems' software. I cannot go into detail on this matter because it is a pending case, but the FBI recently identified two juveniles in California who appear to have been responsible for many of the intrusions. And the Israeli National Police, working with FBI, Air Force, and NASA investigators, this week placed under house arrest one individual who also appears responsible for many of the intrusions. While we are still determining the extent of harm caused by these intrusions, the potential harm was obviously enormous. Even the unclassified systems used by DoD and other government agencies contain an enormous amount of important and sensitive data, the loss or alteration of which would have serious adverse consequences for our national security.

Many of you have also probably read about the plea bargain in Massachusetts this week of a teenage hacker who was able to break into the former NYNEX (now Bell Atlantic) system and, through it, disable telecommunications at a regional airport, cut off services to the airport's control tower, and prevent incoming planes from turning on the runway lights. This case is a wake-up call for those who would argue that hacking is simply harmless fun.

In 1994, foreign crime groups operating in several different countries were able to hack into the Citibank Cash Management System, which is used for banking functions such as wire transfers. The criminals compromised passwords to impersonate account holders worldwide, and attempted 40 transfers totaling $10 million. As a result of early detection by Citibank officials, and close cooperation between Citibank investigators, payee banks, foreign police, and the FBI, the perpetrators were tracked down and arrested, and actual losses were limited to $400,000. But imagine if the hackers had been intent not simply on stealing funds, but on destroying Citibank's account records or denying service to Citibank customers. The effects in such a scenario would have had much more serious and widespread consequences.

In another case, hackers from Germany recently captured the customer credit card files of a Miami company. The hackers threatened to distribute all the credit card numbers unless they were paid ransom. When one of the hackers tried to pick up the money, he was arrested by German authorities. If the hackers had chosen to use the numbers instead of trying extortion, law enforcement may not have been able to stop them before they had caused significant financial loss.

An international computer hacker organization headquartered in Dallas, Texas successfully penetrated the networks of several telecommunications providers and acquired unlisted telephone numbers, personal addresses, credit information, and National Crime Information Center data, causing losses in excess of $500,000. The hackers installed a sniffer which compromised at least 15 telephone company systems including records, maintenance, and operational control system, and also illegally wiretapped the phone lines. The advanced level of expertise of the hackers was comparable to telephone company experts, and suggests that they could have disrupted telecommunications on a national basis if they had wanted to.

In July, 1997, the owner of a computer communications company sent, or caused to be sent, malicious computer code which resulted in the redirection of computer communications away from the computers of one of his competitors. This redirection of computer communications resulted in a direct loss to the victim company of at least $1,500,000. Additionally, millions of Internet users were denied access to various affected Internet sites.

These are just a few examples of the computer crime problem that we are seeing. But they illustrate the growing problem of cybercrime, the international dimension of the problem, and the increasing threat to our critical infrastructures. And, as I stated earlier, they demonstrate that this is not simply a problem of enforcing the law against imaginative criminals, but of protecting our economic health and national security.

Now let me tell you what the FBI is doing about it. On February 26 of this year, the FBI created the National Infrastructure Protection Center (NIPC). The NIPC's mission is to detect, deter, prevent, assess, warn, respond to, and investigate unlawful acts involving computer and information technologies and unlawful acts, both physical and cyber, that threaten or target our critical infrastructures. This means we do not simply investigate and respond to attacks after they occur, but we try to learn about them and prevent them beforehand. This requires the collection and analysis of information gathered from all available sources, and the dissemination of our analyses and of warnings of possible attacks to potential victims, whether in the government or private sector.

This broader mission also means that we in the FBI, and indeed law enforcement as a whole, cannot do this alone. Rather, this mission requires the combined efforts of many different agencies. The Defense Department has a critical role to play because its reliance on information technologies makes it a prime target for our adversaries and because it holds much of the government's expertise in defending against cyber attacks. Our intelligence agencies have an important role because of their responsibility for gathering information about threats from abroad. And other civilian agencies with jurisdiction over critical infrastructures, such as the Departments of Treasury, Energy, and Transportation, have similarly significant roles.

But his is also not just a role for the federal government. State governments must be involved because they own and operate some of the critical infrastructures and because their agencies are often the first responders in the event of a crisis.

And, perhaps most importantly, this mission requires the intensive involvement of the private sector. Private industry owns and operates most of the infrastructures, so it must be involved in helping us defend them. And it also has the greatest expertise in the technical problems and solutions.

In recognition of the vital roles all of these entities must play, the NIPC is founded on the notion of a partnership. It creates a partnership by including representatives from the other critical federal agencies, from state and local law enforcement, and from private industry. This will foster the sharing of information and expertise, and improve coordination among all the relevant actors in the event of a crisis. And it will augment the physical presence of these representatives by establishing electronic connectivity to the many different entities in government and industry who might have, and need, information about threats to our infrastructures.

Let me say at this point something about what we are not. We are not the Nation's super-systems administrator, responsible for physically securing everyone's systems against intruders or advising on the latest security software or patches to fix vulnerabilities. That role clearly must be filled by systems administrators in each company, by chief information officers in government agencies, and by industry groups and other entities with expertise in reducing vulnerabilities and restoring service. Rather, our role is to help prevent intrusions and attacks by gathering information about threats from sources that are uniquely available to the government (such as from law enforcement and intelligence sources), combining it with information voluntarily provided by the private sector or obtained from open sources, conducting analysis, and disseminating our analyses and warnings to all relevant consumers. And if an attack does occur, our role is to serve as the federal government's focal point for crisis response and investigation. That job is big and difficult enough, so I don't want to create any unwarranted expectations about what else we might do.

The NIPC incorporates and expands the mission and personnel of the FBI's former Computer Investigations and Infrastructure Threat Assessment Center (CITAC) which was created in 1996 to coordinate the FBI's investigations and response to the increasing problem of computer crime. The NIPC, located at FBI Headquarters in Washington, D.C., consists of three sections. The Computer Investigations and Operations Section (CIOS) is responsible for managing support to computer intrusion investigations conducted by our Field Offices, providing and coordinating technological support to all FBI investigations involving computers and information technologies, and for developing and managing an interagency Cyber Emergency Support Team (CEST) analogous to the Domestic Emergency Support Team and Foreign Emergency Support Teams that are responsible for responding to terrorist acts in the U.S. or abroad. In addition, CIOS provides and coordinates subject matter experts, equipment, and technological support to cyber investigators from our Field Offices and other federal, state or local government agencies.

The Analysis and Warning Section (AWS) provides analytical support for computer investigations, and serves as the information clearing-house for research and analysis about physical and cyber threats and unlawful acts that target the critical infrastructures of the United States. It is charged with obtaining relevant information from all sources —law enforcement investigations, intelligence sources, open sources, and voluntarily provided industry data—analyzing it, and disseminating its analyses and tactical warnings to relevant consumers.

The Training, Administration, and Outreach Section (TAOS) has at its core the responsibility for coordinating the training and continuing education of cyber investigators in the FBI Field Offices, in other federal agencies, and in state and local law enforcement; and of personnel in the public and private sector involved in infrastructure protection. It also will direct our extensive outreach efforts to FBI Field Offices, other government agencies, industry, and academia, which are necessary to encourage the sharing of information about threats, vulnerabilities, and technological developments. In addition, the TAOS provides the administrative support that underlies and is necessary to all of the other activities of the Center.

Let me note, finally, that we have been in existence less than a month, so we are still very much in the early stages of building the Center. We have a lot of work to do in order to establish the necessary liaison with other agencies and the private sector, and to put in place our personnel and equipment. This will take time. But the Department of Justice and the FBI have taken an important first step in establishing this Center, in recognizing the need for an interagency and public-private partnership, and in realizing that the new challenges of the next century require new ways of thinking and creative solutions.

Thank you.

CHAPTER 27

NATIONAL INFRASTRUCTURE PROTECTION CENTER CYBERNOTES

January 5, 1999

CyberNotes is published every two weeks by the National Infrastructure Protection Center (NIPC). Its mission is to support security and information system professionals with timely information on cyber vulnerabilities, hacker exploit scripts, hacker trends, virus information, and other critical infrastructure-related best practices.

You are encouraged to share this publication with colleagues in the information and infrastructure protection field. Electronic copies are available on the NIPC Web site at www.fbi/nipc/index.htm.

Please direct any inquiries regarding this publication to the Editor-CyberNotes, National Infrastructure Protection Center, FBI Building, Room 11719, 935 Pennsylvania Avenue, NW, Washington, DC, 20535.

Bugs, Holes & Patches

The following table provides a summary of software vulnerabilities identified between December 10 and December 31, 1998. The table provides the operating system, software name, potential vulnerability/impact, identified patches/workarounds/alerts, common name of the vulnerability, potential risk, and an indication of whether attacks have utilized this vulnerability or an exploit script is known to exist. Software versions are identified if known. This information is presented only as a summary; complete details are available from the source of the patch/workaround/alert, indicated in the footnote or linked site. Please note that even if the method of attack has not been utilized or an exploit script is not currently widely available on the Internet, a potential vulnerability has been identified.

Operating System	Software Name	Vulnerability Impact	Patches/ Workarounds/ Alerts	Common Name	Risk*	Attacks/ Scripts
3Com Network Management Card (NMC) for Total Control Hubs (TCHs)[1]	HARDWARE	Remote user can access remotely using a default account.	The adm account should be disabled, not deleted. If deleted, the account appears to reinstall on hardware reboot. Note that any hardware reboot over time will reactivate the account.	3Com default adm account on NMCs	Medium/ High	Default passwords available on Web sites and newsgroups.
3Com HiPer[2]	HiPer ARCs running 4.1.11	Nestea Denial-of-Service (DOS) exploit script will cause card to fail and reboot.	Total Control NetServer card fixed problem but the HiPer ARC appears to have reintroduced the problem.	Nestea DOS	Medium	Exploit scripts posted to newsgroups and Web sites.
BreezeCOM[3] (SA 10, SA 40, AP 10)	HARDWARE	Local user can access adapter through default passwords.	No workaround available.	BreezeCOM default password	Low/ Medium	Default passwords available on Web sites and newsgroups.
Cisco IOS[4] 12.X,11.3.AA, and 11.3DB,	Operating System	NMAP User Datagram Protocol (UDP) scans crashes routers running listed version of the software.	A tested workaround is to block incoming syslog traffic using the access list.	Cisco IOS Nmap crash	High	Exploit script (nmap) posted to newsgroups and Web sites. Hackers known to be using.

Operating System	Software Name	Vulnerability Impact	Patches/ Workarounds/ Alerts	Common Name	Risk*	Attacks/ Scripts
Embedded intelligent controllers or Programmed logic Controllers (PLCs)[5]	OS-9	Internet Control Message Protocol (ICMP) redirect attacks leading to system shutdown or hanging.	Workaround is to remove, where possible, these systems from the TCP/IP network.	Embedded intelligent controllers ICMP shutdown	High (individual site risk is Low/ Medium)	Current hacker tools will initiate this condition.
HP JetDirect[6]	JetDirect interface card	DOS attack against print services and printers.	Newer JetDirect cards are not affected.	JetDirect Denial of Service	Low	No scripts or exploits identified at time of publishing.
HP9000 series 700/800[7] 9.x to 11.X	Operating system (SNMP)	Remote user can gain root access using a buffer overflow (previous advisory issued).	Patches now available from Hewlett-Packard (HP).	SNMP buffer overflow	High	Exploit scripts posted to newsgroups and Web sites.
Iomega Zip Disk[8]		Unauthorized user can gain access to a password protected Zip disk.	Workaround is to place all password protected disks in a locked container when not in use.	ZIP disk password removal	Low/ Medium	Unauthorized user must have physical access to the Zip disk that is protected.
Irix[9] 6.4 or higher	Fcagent daemon	DOS against FiberVault.	Patches are available at: http://www.sgi.com/Support/security/security.html	Fcagent DOS	Low	No scripts or exploits identified at time of publishing.

Operating System	Software Name	Vulnerability Impact	Patches/ Workarounds/ Alerts	Common Name	Risk*	Attacks/ Scripts
Irix[10] V 6.5	Operating system (inetd)	Nmap scan will cause inetd to fail if multiple server scans occur.	No workarounds known at time of publishing	Nmap scanning	Medium	Exploit script (nmap) posted to newsgroups and Web sites. Hackers known to be using.
Linux[11]	Pluggable Authentication Modules (PAMs)	Local user can read and write to/etc/shadow and gain superuser privileges. This vulnerability exists for a limited time.	Patches available on some Linux software support sites. A temporary fix is to chmod -s/usr/bin/passwd.	Pam default exploit	Low/ Medium	No scripts or exploits identified at time of publishing.
Linux[12] (all versions)	WordPerfect 8	If a privileged user installs WordPerfect, root compromise is possible.	Set $TMPDIR before running the install script and do not run the install script as superuser or other highly privileged account.	WordPerfect 8 directory permission problem	High	No scripts identified at time of publishing. Explanation of exploit available in newsgroups.
Linux Red Hat[13] (lpr 0.31-1 or 0.33-1)	Lpd	DOS caused by remote user occurs if a remote user attempts to print to a printer server they do not have an account on.	No tested patches or workarounds available.	Linux Red Hat lpd remote user Denial of Service	Low/ Medium	Properly configured firewall will stop this attack from outside.

Operating System	Software Name	Vulnerability Impact	Patches/ Workarounds/ Alerts	Common Name	Risk*	Attacks/ Scripts
Microsoft[14]	Excel	Using the call function in Excel some executables can be run without the user's knowledge.	Patch is available at: http://officeup-date.microsoft.com/download details/x197cfp.htm	Excel call function vulner-ability (aka. Russian New Year exploit-ation)	Low/ Medium	No scripts identified at time of publishing.
Microsoft[15]	Internet Explorer	Malicious Web site can impersonate a legitimate Web site.	Patch is available at: http://www.microsoft.com/windows/ie/security/spoof.asp	Internet Explorer Frame Spoofing or Untrusted Scripted Paste	Low/ Medium	No scripts identified at time of publishing.
Microsoft Window NT[16]	Operat-ing System Server 4	The network monitor process netmon.exe will fail when attempting to view a session request from a machine with a NetBIOS Scope Identifier of 190 or more characters.	No workarounds known at time of publishing.	Windows NetBIOS Scope ID problem	Low/ Medium	No scripts identified at time of publishing.
Microsoft Windows 98[17]	Operat-ing System	Nmap scan will cause a critical error to develop (Blue Screen) and network connectivity is lost.	No workarounds known at time of publishing.	Nmap scanning	Medium	Exploit script (nmap) posted to newsgroups and Web sites. Hackers known to be using.

Operating System	Software Name	Vulnerability Impact	Patches/ Workarounds/ Alerts	Common Name	Risk*	Attacks/ Scripts
Yahoo—Multiple systems[18]	Yahoo pager	DOS by deactivating an identity.	No workarounds known at time of publishing.	Yahoo paper identity DOS	Medium	Exploit script posted to newsgroups.
NeXTStep[19] V3.3	Operating System	Nmap scan will cause the system to panic and reboot.	No workarounds known at time of publishing.	Nmap scanning	Medium	Exploit script (nmap) posted to newsgroups and Web sites. Hackers known to be using.
SCO UNIX OpenServer[20]	Operating system (calserver)	Remote user can gain root access using a buffer overflow.	No workarounds known at time of publishing.	Calserver	High	Exploit script posted to newsgroups and Web sites.
Sun Internet Mail Server (SIMS) 3.x[21]	SIMS	Remote user can read and write to the slapd.log file. This file has clear text passwords and uids of everyone checking mail.	No workarounds known at time of publishing.	SIMS slapd.log world read/ write vulnerability	High	Limited Unix knowledge required to exploit this vulnerability. Hackers known to be exploiting this hole.
Sun Lightweight Directory Access Protocol (LDAP) Directory Services (SDS)[22] 1.x and 3.1	LDAP SDS	Remote user can read and write to the slapd.log file. This file has clear text passwords and uids of everyone checking mail.	No workarounds known at time of publishing.	slapd.log world read/ write vulnerability	High	Limited Unix knowledge required to exploit this vulnerability. Hackers known to be exploiting this hole.

Operating System	Software Name	Vulnerability Impact	Patches/ Workarounds/ Alerts	Common Name	Risk*	Attacks/ Scripts
Sun Solaris[23] (multiple ersions)	Operating system (inetd)	Nmap scan will cause inetd to fail if server scans occur.	No workarounds known at time of publishing.	Nmap scanning	Medium	Exploit script (nmap) posted to newsgroups and Web sites. Hackers known to be using.
Sun Solaris[24] 2.3 to 2.6	Operating system (Bind)	Remote user can gain root access using a buffer overflow (previous advisory issued).	Patches available at: http://sunsolve.com/sunsolve/pubpatches/patches.html	Bind	High	Exploit scripts posted to newsgroups and Web sites.
Sun Solaris[25] 2.4 to 2.6	Common Desktop Environment (CDE) dtmail	Remote user can execute programs with the privileges of mail or user reading mail.	Patches available at: http://sunsolve.com/sunsolve/pubpatches/patches.html.	Dtmail buffer overflows	High	No scripts or exploits identified at time of publishing.
Unix (many Berkeley Software Distributions [BSD] derived sytems)[26] BSD/OS 3.1 Free BSD prior to 2.2.8 FreeBSD 3.0 prior to 11/12/1998	Operating system (TCP/IP stack)	Remote user can cause a DOS condition in many implementations.	Patches available at: ftp://ftp.freebsd.org/pub/FreeBSD/CERT/patches/CA-98-13/patch	BSD TCP/IP stack weakness	Low/ Medium	Current hacker tools will initiate this condition.

*Risk is defined in the following manner:

High – A vulnerability that will allow an intruder to immediately gain privileged access (e.g., Sysadmin, root) to the system. An example of this would be a vulnerability in which a sequence of instructions is sent to a machine by an unauthorized user and the machine responds with a command prompt.

Medium – Any vulnerability that will allow an intruder immediate access to the system that is not privileged access. This allows the intruder the opportunity to continue the attempt to gain root access. An example would be a configuration error that allows an intruder to capture the password file.

Low - A vulnerability that provides information to an intruder that could lead to further compromise attempts or a Denial-of-Service attack. The reader should note that while the Denial-of-Service attack is deemed low from a threat potential, the frequency of this type of attack is very high. Denial-of-Service attacks against mission-critical nodes are not included in this rating and any attack of this nature should instead be considered as a "High" threat.

---

[1] Bugtraq, December 21, 1998.

[2] Bugtraq, December 21, 1998.

[3] Bugtraq, December 10, 1998.

[4] Bugtraq, December 22, 1998.

[5] ISS Security Advisory, ICMP redirects against embedded controllers.

[6] ISS Security Advisory, HP Jetdirect TCP/IP problems.

[7] HP Daily Securoty Bulletin , HPSBUX9811-088 updated.

[8] Counterpane System's Ctypto-gram, December 15, 1998.

[9] Silicon Graphics Inc. Security Advisory, 19981201-01-PX.

[10] Bugtraq, December 22, 1998.

[11] Bugtraq, December 23, 1998.

[12] Bugtraq, December 18, 1998.

[13] Bugtraq, December 18, 1998.

[14] Microsoft Security Bulletin (MS98-018).

[15] NTSHOP, December 23, 1998.

[16] NTBugtraq, Decemver 12, 1998.

[17] SecureXpert labs Advisory SC-98.12.23-01.

[18] BugTraq, December 25, 1998.

[19] SecureXpert labs Advisory SC-98.12.23-01.

[20] BugTraq, December 29, 1998.

[21] BugTraq, December 25, 1998.

[22] Ibid.

[23] SecureXpert labs Advisory SC-98.12.23-01.

[24] Sun Microsystems, Inc. Security Bulletin, #00180.

[25] Sun Microsystems, Inc. Security Bulletin, #00181.

[26] CERT Advisory CA-98.13.

Recent Exploit Scripts

The table below contains a representative sample of exploit scripts, identified between December 10 and December 31, 1998, listed by date of script, script name, script description, and testing conducted. Items listed in boldface (if any) are attack scripts for which vendors, security vulnerability listservs, or Computer Emergency Response Teams (CERTs) have not published workarounds or patches. During this time period, 50 scripts, programs, and net-news messages containing holes or exploits were identified.

Date of Script (Reverse Chronological Order)	Script Name	Script Description	Testing Conducted
Dec 31, 1998	Flushot.c	**Denial-of-Service (DOS) C code that sends Central Processing Unit (CPU) usage to 100% on Windows 95/98 machines.**	
Dec 31, 1998	Net-RawIP v0.03d	**Perl module that manipulates raw ip packets and Ethernet headers.**	
Dec 30, 1998	Net-RawIP v0.03c	**Perl module that manipulates raw ip packets and Ethernet headers.**	
Dec 29, 1998	Aggressor PRO v1.0	Network vulnerability testing tool.	
Dec 29, 1998	Sco-calserver-bof.c	Exploit script in C code for the SCO Unix calserver buffer overflow.	
Dec 29, 1998	Sco-calserver-bof.sh	Exploit script in shell code for the SCO Unix calserver buffer overflow.	
Dec 29, 1998	Sbouncer.c	Bouncer that is used with wingate and socks proxies.	
Dec 29, 1998	Nmap v 2.02	Network scanning tool that has a variety of scanning modes, including stealth, Xmas, and Null stealth scans. Note: This tool is being used by hackers and may cause systems to become unstable.	
Dec 29, 1998	Lk2.pl	Perl script that scans for AFS file systems running Hesiod name servers checking for user directory permissions.	
Dec 28, 1998	Net-RawIP v 0.03b	**Perl module that manipulates raw ip packets and Ethernet headers.**	
Dec 28, 1998	Freebsd-mbuf-crash.c	DOS script which causes FreeBSD to fail.	

Date of Script (Reverse Chronological Order)	Script Name	Script Description	Testing Conducted
Dec 27, 1998	Pschack.c	Program that attempts to scavenge connections of an Internet Service Provide (ISP). Scavenging attempts to capture and use a connection that a user believes is closed. This is an effective method for hiding an attacker's identity.	
Dec 27, 1998	SDI-bnc.c	Buffer overflow exploit script for bnc.	
Dec 27, 1998	THC-SCAN v2.0	War dialer with many auto-identification features.	
Dec 27, 1998	Vanity.c	BNC remote buffer overflow for Linux x86 (without the stack-non-exec patch). Works on versions below 2.4.4.	
Dec 24, 1998	Kcmsex.c	Exploit script for the Solaris i386/usr/openwin/bin/kcms_configure hole.	
Dec 22, 1998	Bootpd-bb.tar.gz	Bootp exploit against Debian Linux 1.3 and 2.0.	
Dec 22, 1998	Cheops v0.58	Tool for network scanning and discovery of operating system. Note: The author of the tool has posted warnings that the tool has the potential of being misused and there have been indications that hackers are currently using this tool.	
Dec 22, 1998	Lamescan v1.4	Port scanner that will perform random scans and multithread scans. It will also scan host names as they are typed on stdin, ansi, and identd lookups. It will send a string "user XXXX" in an effort to defeat automated port scan detectors.	
Dec 22, 1998	Mdmrst.c	Implements the +++ATH0 modem bug and is capable of spoofing.	
Dec 22,1998	Miffo-check.c	Port scanner.	
Dec 22,1998	Rely v1.0	Forwards traffic for a port to any other host and port designated by the user.	
Dec 22, 1998	Resetter.c	DOS tool that attempts to reset all connections of a network segment by sending spoofed RST and ICMP UNREACHABLE packets.	
Dec 22, 1998	Snort v0.96	Fairly portable packet (Solaris, Linux, and FreeBSD) sniffer.	

Date of Script (Reverse Chronological Order)	Script Name	Script Description	Testing Conducted
Dec 22, 1998	Stuffit.c	Modified ping flood that contains an illegal character for Point-to-Point Protocol (PPP).	
Dec 22, 1998	Tgk-log v2.1	Designed to record contents of packets as they pass through an ipmasq gateway.	
Dec 22, 1998	Usr-totalswitch.txt	Backdoor passwords for all versions of US Robotics' Totalswitch.	
Dec 21, 1998	Nessus WIP 122198	Security audit tool that checks for over 150 known holes. (Trojan horse versions found.)	
Dec 19, 1998	dcc-hijack.c	Hijacking tool for the DCC chat/send/receive sessions of many Internet Relay Chat (IRC) clients.	
Dec 19, 1998	Exscan v0.2	**Port scanner with a strobe scanning capability.**	
Dec 19, 1998	Valueclick-cgi.txt	**Exploit for Valueclick online advertising agency Web interface that captures usernames/passwords.**	
Dec 18, 1998	Bootp_exploit.c	Remote buffer overflow (bootpd) exploit code for many Unix systems.	
Dec 18, 1998	PortScanner v1.2	A port scanner with strobe scanning capability.	
Dec 18, 1998	Squid_connect.tar.gz	Exploit on squid proxy server used to hide an intruder's identity.	
Dec 18, 1998	Suck.c	Sample sniffer code.	
Dec 18, 1998	Tellme.tar.gz	Script that retrieves netbios name over the network.	
Dec 18, 1998	Udplstn.tar.gz	Daemon that hooks User Datagram Protocol (UDP) ports and then records all incoming packets.	
Dec 18, 1998	Webspf.pl	Program that spoofs referer page and will bypass most reference checking.	
Dec 17, 1998	Nmap v2.01	Network scanning tool that has a variety of scanning modes, including stealth, Xmas, and Null stealth scans. Note: This tool is being used by hackers and may cause systems to become unstable.	
Dec 17, 1998	Scancd.pl.	New port scanner written in Perl.	
Dec 17, 1998	Syn.pl.	Script designed to detect stealth scans performed by nmap.	
Dec 14, 1998	Cheops v0.57	See description above.	

Date of Script (Reverse Chronological Order)	Script Name	Script Description	Testing Conducted
Dec 14, 1998	Game.exe	Trojan horse program with the NetBus exploit attached.	
Dec 14, 1998	Hping v0.66	Ping-based program for network mapping. Note: A number of hacker sites are recommending that this tool be used.	
Dec 14, 1998	Kkill.c	Opens hundreds of connections to a single host in a DOS effort.	
Dec 14, 1998	Nmap v2.00	See description above.	
Dec 14, 1998	Sxploit.sh	Tool that scans local system for a number of old vulnerabilities and then attempts to exploit vulnerabilities discovered.	
Dec 13, 1998	Lportfuck.c	DOS attack script.	
Dec 13, 1998	Nessus WIP 121398	See description above.	
Dec 11, 1998	Murderkill Deluxe v2.0	Package of many DOS scripts that automate the launching of multiple attacks.	

Note: A number of scripts have been updated during this two-week period (e.g., Nmap, Nessus WIP, Cheops). The hacker community is constantly improving the tools they use.

Trends

1. Several hackers appear to be using coordinated scans and probes from different sites.

2. More phreaker tools are appearing on hacker Web sites.

3. Scanning for IMAP and POP continues.

4. A significant increase in reports of NetBus and BackOrifice scanning is occurring.

5. A set of scripts has appeared that identifies hosts running Linux, responds on TCP port 111, and then attempts exploitation of the mountd vulnerability.

6. A new TCP port scan method has appeared that uses spoofed addresses of "zero traffic" hosts to identify open ports.

7. A rapid increase is noted in the use of hacker/Phreaker tools developed for the PalmPilot.

Viruses

Remote Explorer - In the last few weeks, a virus known as Remote Explorer has received a great deal of press. Network Associates Inc. (NAI) issued warnings to the Microsoft Windows computing community. This virus was first discovered on December 17, 1998, at an MCI WorldCom facility. This virus compresses executables and encrypts some file types, hampering recovery of data. Detection of the current version can be accomplished by going to the control panel and checking services for a file titled "Remote Explorer." If this file is discovered, then the machine is infected. The virus is unique in that it is the first virus discovered that uses network services to allow it to spread. Once on a network server, the virus spread rapidly to workstations and other trusted servers. Most major anti-virus vendors have upgraded their products to detect the virus and some have the capability to remove the virus and recover any affected files. Anti-virus vendors recommend that their users update the anti-virus packages on a regular basis (every two weeks).

CHAPTER 28

ENCRYPTION: IMPACT ON LAW ENFORCEMENT

Laboratory Division
Engineering Research Facility
Quantico, Virginia
January 8, 1999 and June 3, 1999

EXECUTIVE SUMMARY

Encryption is extremely beneficial when used legitimately to protect commercially sensitive information and communications. The law enforcement community, both domestically and abroad, is extremely concerned about the serious threat posed by the proliferation and use of robust encryption products that do not allow for the immediate, lawful access to the plaintext of encrypted, criminally-related communications and electronically stored data in accordance with strict legal requirements and procedures.

The potential use of such commercially-available encryption products by a vast array of criminals and terrorists to conceal their criminal communications and information poses an extremely serious threat to public safety and national security. Law enforcement fully supports a balanced encryption policy that satisfies both the commercial needs of industry for robust encryption while at the same time satisfying law enforcement's public safety and national security needs. Robust, commercially-available encryption products, which include some type of recoverable capability that allows for immediate, lawful access to plaintext is clearly the best method to achieve the goals of both industry and law enforcement.

Since April of 1993, the Clinton Administration has expressed support for the adoption of a balanced encryption policy. In lieu of legislation, the Clinton Administration continues to favor a **voluntary** approach to address law enforcement's public safety concerns regarding encryption for domestic use. The Administration has been attempting to work with industry, through "good faith dialogue," and by allowing "market forces," influence and inducements (mainly changes to existing export regulations) to bring about the development, sale and use of recoverable encryption products within the U.S.

During the 105th Congress, several encryption-related bills were introduced; however, none were enacted. The main focus of these bills was the relaxation of existing export controls on encryption, regardless of the impact on national security and foreign policy.

During the 106th Congress, three encryption-related bills have been introduced. Like last Congress' encryption related bills, the main focus of these bills is to either relax existing export controls on encryption products and/or prevent the government (federal or state) from imposing domestic requirements on encryption products to ensure that such domestic encryption products include some type of plaintext access for law enforcement should these products be used in the furtherance of serious criminal activity. These bills included: **H.R. 850, S. 798,** and **S. 854.**

THE PROLIFERATION OF SECURE OR ENCRYPTED COMMUNICATIONS AND ELECTRONICALLY STORED INFORMATION WILL MAKE IT INCREASINGLY DIFFICULT FOR LAW ENFORCEMENT TO OBTAIN AND DECIPHER THE ENCRYPTED CONTENT OF LAWFULLY INTERCEPTED COMMUNICATIONS AND LAWFULLY OBTAINED ELECTRONICALLY STORED INFORMATION THAT IS NECESSARY TO PROVIDE FOR EFFECTIVE LAW ENFORCEMENT, PUBLIC SAFETY, AND NATIONAL SECURITY.

WHAT IS ENCRYPTION?

Encryption is the method of hiding the content of a message. In broad terms, any system or technique that renders a message unintelligible by anyone other than the intended recipient of the message is utilizing encryption. A message which has not been encrypted is often referred to as "plaintext". After a message has been encrypted, it is referred to as "ciphertext". Whereas encryption is used to secure a message, decryption is the method for converting ciphertext back to its original plaintext.

Many encryption systems use a mathematical function, known as a cryptographic algorithm, to encrypt and decrypt messages. Just as a lock box requires a key to lock or unlock it, a cryptographic algorithm requires a key to encrypt and decrypt a message.

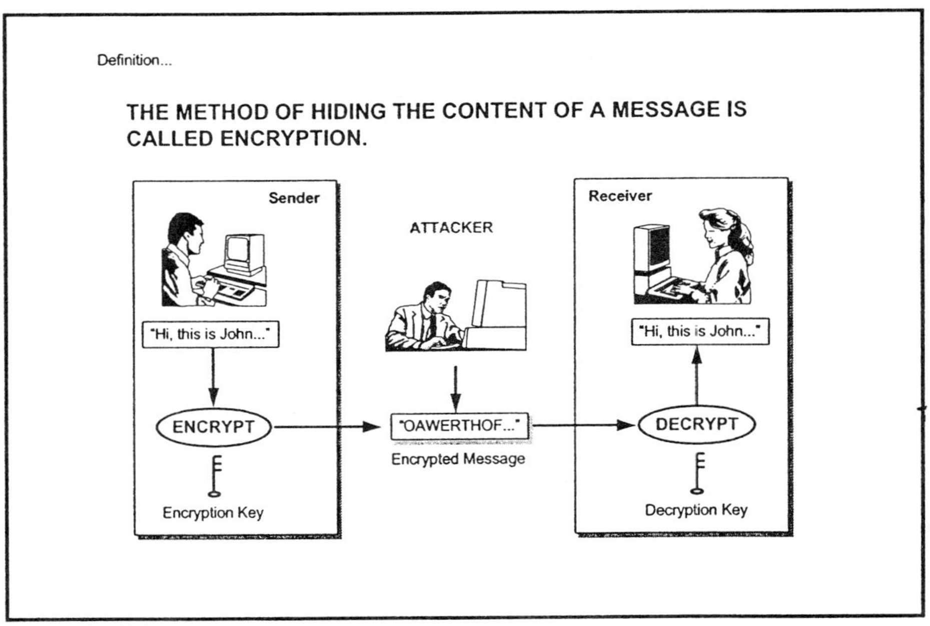

USES AND BENEFITS OF ENCRYPTION

Governments have always been very concerned with the secrecy of information related to military, economic and foreign policy issues. For many years, military and government missions drove the development and use of applications for encryption. Protecting one's intentions from an opposing party is critical and for that reason information security is very important.

Although encryption software and hardware devices have been commercially available for years, their cost, degradation of voice quality, and user "friendliness" have, in the past made these devices unattractive to the general public. The introduction of digitally-based technologies as well as the widespread use of computers and computer networks which may incorporate privacy features/capabilities through the use of encryption are facilitating the development, production, and use of affordable and robust commercially-available encryption products and services for use by the general public. These encryption systems provide robust security for conventional and cellular telephone conversations, facsimile transmissions, local and wide area networks, communications transmitted over the Internet (E-mail, etc.), personal computers, wireless communications systems, electronically stored information, remote keyless entry systems, advanced messaging systems, and radio frequency communications systems.

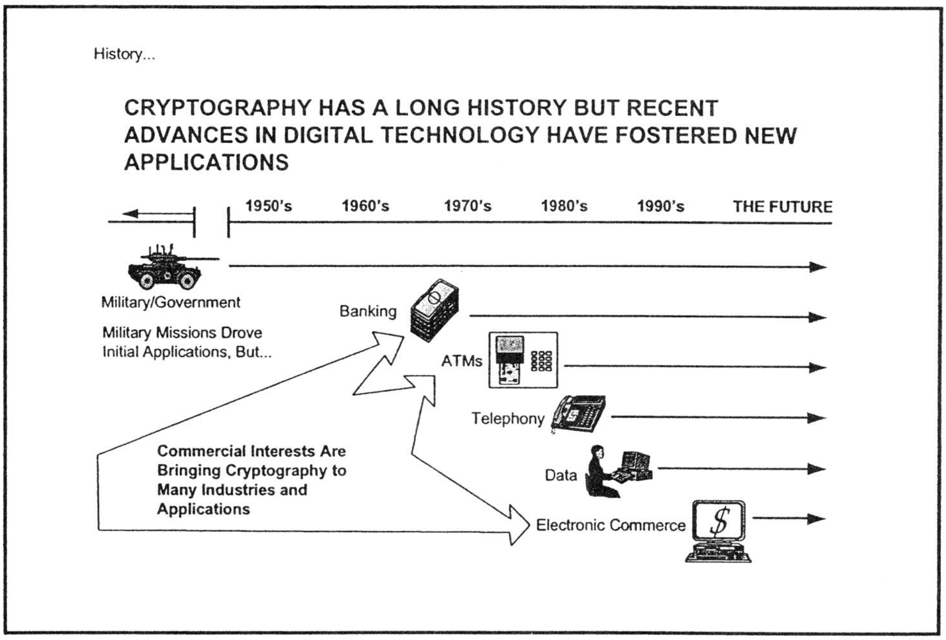

Various applications will use encryption to provide privacy, information integrity, authentication and non-repudiation. Privacy, or confidentiality, is probably the best known application of encryption. Unauthorized individuals are prevented from listening in or viewing electronic information. Information integrity protects against unauthorized changes to information after it is sent. This is important for the validation of legal electronic documents. Authentication techniques verify the identity of a sender of a message. This provides assurance that the claimed sender (e.g., return address on a letter envelope) of information is the actual sender and vice versa for destination authentication. Non-repudiation ensures that a sender is not able to deny that he or she sent a particular message. This verification is important when auditing or when litigation is being considered.

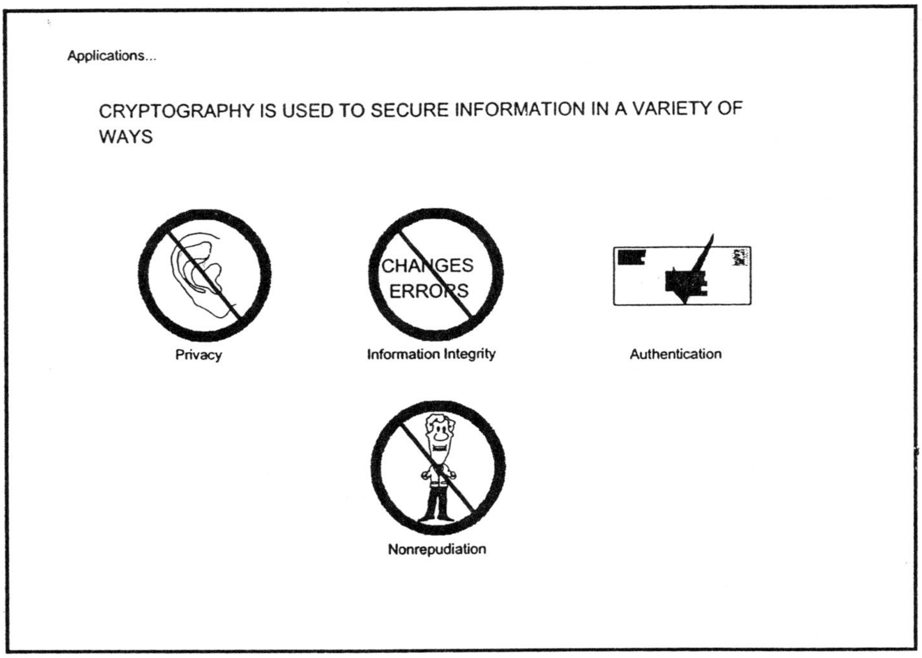

ADVERSE IMPACTS OF ENCRYPTION

The ability of encryption to ensure the confidentiality and the content of important messages, files or communications of corporations and private citizens can also prevent those same entities from gaining plaintext access to that critical information should the keys needed for decryption become lost or corrupted. Unless there is an alternative plaintext access method, such as a recovery feature incorporated in the encryption product to allow such plaintext access, this important information could be lost forever.

The use of encryption can effectively prevent plaintext access not only to law enforcement acting under proper legal authority, but also to corporations in situations where an employee could potentially use encryption to commit illegal acts, including acts against the corporation. A report from Congress's Office of Technology Assessment entitled, "Information Security and Privacy in Network Environments," cited the following: "There is also growing recognition of the potential misuses of encryption, such as by disgruntled employees as a means to sabotage an employer's database."

Encryption can also be used to conceal criminal activity and thwart law enforcement efforts to collect critical evidence needed to prevent, solve and prosecute serious and often violent criminal activities, including illegal drug trafficking, organized crime, child pornography and terrorism. In these instances, the use of encryption to secure the content or confidentiality of information poses substantial threats to law enforcement's abilities to: 1) interpret and analyze stored electronic records and files which have been obtained through court-order or other lawful procedures; and 2) perform court-ordered electronic surveillance. Encrypted information obtained through the use of lawfully intercepted communications and/or lawfully accessed electronic records or files will be useless in solving crimes and preventing criminal activity unless law enforcement, pursuant to a court order, has immediate access to the plaintext of such encrypted, criminally-related communications and electronically stored data.

In addition to the following notable cases which have been previously highlighted:

1) the Aldrich Ames spy case where Ames was told by his Soviet handlers to encrypt computer file information to be passed to them;

2) the Ramzi Yousef (mastermind of the World Trade Center) /Manilla Air terrorist case where Yousef and other international terrorists were plotting to blow up 11 U.S. owned airliners in the Far East in which data regarding this terrorist plan was found in encrypted computer files discovered in Manilla after Yousef's arrest; and

3) a child pornography case where the subject used commercially-available encryption to encrypt pornographic images of children that were transmitted to other subjects of the investigation,

law enforcement continues to experience an increase in the number of encounters with, and the subsequent damaging and detrimental effects of, the use of commercially-available encryption by criminals, terrorists and in hostile intelligence activities throughout the United States and across international borders. In particular, the FBI is currently experiencing the greatest impact from encryption in Foreign Counter-Intelligence (FCI), Violent Crime (which includes sexual exploitation of children and child pornography cases), and White Collar Crime/Computer Crimes and Threat Assessments investigations. However, the use of encryption in Organized Crime, Drug, as well as Domestic and International Terrorism investigations have also been encountered.

In fact, at least 50 percent of the FBI's 56 field offices have recently encountered encryption in cases involving one or more of the aforementioned investigative program areas. Given the ongoing nature of many of these cases, it is extremely difficult to publicly discuss, with any degree of specificity, the details of these specific case examples, which would further illustrate the adverse impact that the use of encryption is having on law enforcement's ability to effectively conduct investigations and carry out its public safety mission, because such public disclouser could compromise many of these ongoing investigations. The following represents some of the general information which can be publicly discussed concerning ongoing FBI investigations that have been adversly impacted by the use of encryption:

1) At least eight major FBI field offices have identified the use of encryption by the subjects of foreign counter-intelligence and international terrorism investigations in a deliberate attempt by these subjects to protect their suspected criminally-related communications and prevent detection by law enforcement of their suspected criminal activities;

2) In addition to the previously stated example of a child pornography case where pornographic images of children were encrypted and transmitted between subjects, there continues to be evidence of the increased use of encryption by a number of subjects involved in the sexual exploitation of children and child pornography who are utilizing various commercially available encryption products to protect suspected criminally-related information and communications as well as to protect their illegal pornographic images of children from discovery and detection; and

3) There are also numerous ongoing FBI investigations involving illegal hacker intrusions into several university and government computer systems as well as software piracy in which the subjects of these investigations have encrypted potentially incriminating information which, if decrypted, could indicate the level of their involvement in these illegal activities as well as disclose the specific information stolen from these various computer systems.

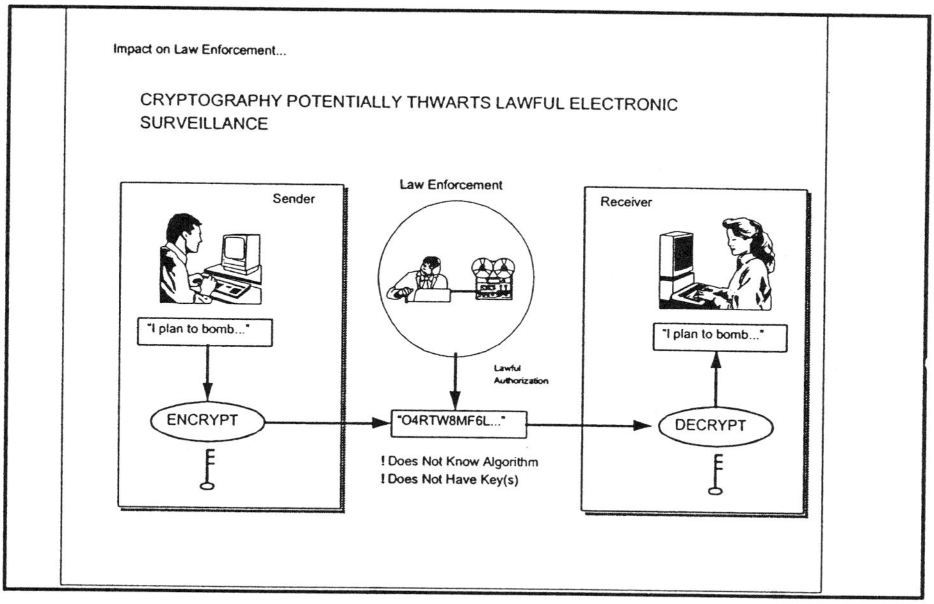

The FBI Laboratory Division's Computer Analysis and Response Team (CART) is responsible for providing assistance in law enforcement investigations where computer generated and/or electronically stored information has been obtained pursuant to court authorized search and seizure. The CART has seen the number of cases utilizing encryption and/or password protection increase from **two (2)** percent to approximately **twenty (20)** percent over the past four years, to include the use of 56-bit Data Encryption Standard and 128-bit Pretty Good Privacy encryption. These totals are expected to increase significantly with the introduction of Microsoft's newest operating system, Windows 2000. This new operating system will allow users to employ an Encrypted File System (EFS) which will provide the individual computer users with easy to use "point and click" encryption thereby enabling the user, including criminals and terrorists, to easily encrypt all of the files stored on their computer.

It is important to understand that this is not an issue unique only to U.S. law enforcement. Included in a recently released paper by the British Government entitled "Encryption and Law Enforcement", several case examples were sited where encryption had been used in furtherance of serious criminal activity in the United Kingdom. The paper noted that the "development of encryption technology gives rise to a number of challenges to law enforcement, security and intelligence agencies. In particular, its widespread use will have an effect on the ability of these agencies to make use of lawfully intercepted communications and retrieved data for law enforcement purposes." Additionally, the paper emphasized the important role the use of lawful communication interceptions play in protecting society from crime and terrorism and concludes "that the development of electronic communications, which promises many benefits to businesses and individuals, should not also give assistance to those who are engaged in serious crime." The case examples set forth in the paper include the following:

1) a 1995 child pornography investigation involving two suspected paedophiles who used the Internet to distribute pornographic images of children that was severely hampered by the use of encryption by the primary suspect of the investigation;

2) a 1996 investigation of a Northern Irish terrorist group in which a computer was seized containing encrypted files concerning potential terrorist targets, including police officers and politicians; and

3) a 1998 attempted murder and sexual assault investigation that was impeded by the discovery of relevant encrypted files on a suspect's computer.

THE CONCEPT OF RECOVERABLE ENCRYPTION

Technical solutions that provide robust encryption, combined with some type of recoverable feature which allows for the immediate, lawful access to plaintext of encrypted, criminally-related communications and electronically stored data, is clearly the best way to achieve the goals of both industry and law enforcement. Law enforcement's needs in dealing with its responsibility for protecting public safety and national security are best met by ensuring that commercially-available encryption products manufactured or imported into the U.S. include some type of capability that allows for the immediate access to the plaintext of encrypted, criminal-related data (both transmitted and stored), pursuant to lawful authority (court order).

The concept of recoverable encryption:

- Provides a means for corporations to address the misuse of encryption by disgruntled employees;
- Ensures the integrity of the investigation through the obtainment of the recovery information or plaintext from a trusted third party (this would also provide the assurance to commercial and individual users of encryption that their encrypted communications and electronically stored information are secure against unauthorized disclosure and illegal "hacker-type" attacks);
- Allows for an overt process for legally obtaining recovery information or plaintext that is subject to public scrutiny and accountability;
- Provides confidentiality of law enforcement's request for recovery information or plaintext access;

- Provides an immediate decryption capability which is available to law enforcement upon presentation of proper legal authority (to include the state and local levels) of encrypted, criminally-related communications or electronically stored information.

Law enforcement's public safety needs can either be achieved through a voluntary approach, should their be a willingness on the part of industry as a whole (inclusive of all domestic manufacturers and all importers of foreign made products into the U.S.) to effectively address law enforcement's needs voluntarily; or through a legislative approach, should there be the will on the part of public policy makers to address the issue legislatively; or through a combination of both. Failure to effectively address law enforcement's needs regarding domestic encryption products will ultimately have a devastatingly adverse impact on the safety and security of the American public.

CLINTON ADMINISTRATION'S POSITION ON ENCRYPTION

Since April of 1993, the Clinton Administration has expressed support for the adoption of a balanced encryption policy that meets the commercial needs of industry for robust encryption while at the same time meeting the public safety needs of law enforcement. Administration representatives have been attempting to working with representatives of industry to encourage the **voluntary** development, sale, and use of recoverable encryption products within the U.S. The Clinton Administration has steadfastly opposed any legislative effort to impose domestic controls on encryption, favoring a voluntary approach to address law enforcement's public safety needs through the use of "good faith dialogue," "market-forces," influence and inducements (mainly regulatory change to exiting export controls on encryption products). Law enforcement remains optimistic that such a voluntary approach will be successful in addressing its public safety needs; however, this approach's ultimate success remains uncertain at this time.

On March 4, 1998, the Vice President announced a new Administration initiative to try and bring about the voluntary development of technical solutions that address law enforcement's public safety needs by calling for "good faith dialogue" between industry and law enforcement rather than seeking to legislate domestic controls at that time. Such "good faith dialogue" efforts remain ongoing. Additionally and of significance to law enforcement, on September 16, 1998, the Clinton Administration formally express support for the creation of a centralized law enforcement resource within the FBI to provide law enforcement with urgently needed technical capablities to fulfill its investigative responsibilities in light of the ever increasing proliferation and use of strong, commercially-available encryption products within the U.S.

Conversely, the Clinton Administration has steadfastly opposed any legislative effort to relax existing export controls on encryption. Such encryption export controls have existed for years to protect national security and foreign policy interest.

ENCRYPTION LEGISLATION

ENCRYPTION-RELATED BILLS INTRODUCED DURING THE 106TH CONGRESS:

H.R. 850, the **"Security and Freedom Through Encryption (SAFE) Act,"** introduced by Congressman Goodlatte (R-6th-VA) on February 25, 1999.

S. 798, the **"Promote Reliable On-Line Transactions to Encourage Commerce and Trade (PROTECT) Act of 1999,"** introduced by Senator McCain (R-AZ) on April 14, 1999.

S. 854, the **"Electronic Rights (E-Rights) for the 21st Century Act"** introduced by Senator Leahy (D-VT) on April 21, 1999.

H.R.850

H.R.850 is almost identical to the bill Congressman Goodlatte introduced during the 105th Congress (H.R. 695). H.R. 850 would largely remove existing export controls on hardware and software encryption products of comparable strength to those that are commercially available from a foreign supplier, regardless of the adverse impact to national security. The bill would also place a prohibition on mandatory key recovery encryption by the government and includes a provision making it a crime to use encryption in furtherance of a criminal act. At the time of the bill's introduction, it enjoyed over 200 bi-partisan co-sponsors. Since that time, the number of co-sponsors has grown to over 250.

The following is an overview of the bill's progress thus far during the 106th Congress:

On March 4, 1999, the House Judiciary Committee's Subcommittee on Courts and Intellectual Property held a hearing concerning the bill with witnesses from NSA, the Justice Department and the Commerce Department testifying in opposition to the bill. On March 11, 1999, despite the aforementioned testimony, the Subcommittee held a mark-up concerning the bill and favorably reported the bill out of Subcommittee (Congressman Goodlatte is a member of that Subcommittee).

On March 24, 1999, the full House Judiciary Committee held a mark-up concerning H.R. 850 and, by a voice vote, favorably reported the bill out of committee without amendments. During the mark-up Congressman McCollum (R-8th-FL) expressed concerns about the bill's adverse impact on law enforcement, national security and intelligence interest, should encryption products that do not allow for immediate plaintext access proliferate within the U.S. and aboard. Congressman McCollum then offered an amendment to the bill's export provisions that would require all hardware and software encryption products for export to include features that allow for immediate plaintext access capabilities for use when there is lawful authorization to obtain such plaintext. Congressman Goodlatte immediately raised a point of order objection, asserting that the amendment was not germane and, without debate, Chairman Hyde ruled that Congressman McCollum's amendment did not fall under the jurisdiction of the Judiciary Committee.

The bill has been referred to the House International Relations, Armed Services, Intelligence and Commerce Committees for consideration. The House International Relations and Commerce Committees held hearings on the bill in May, with the House Intelligence Committee planning to hold several hearings on the bill in June.

S. 798

S. 798, introduced by Senator McCain who is the Chairman of the Senate Commerce Committee, is somewhat of a pro-industry bill even though the bill's expressed purpose is "to promote electronic commerce by encouraging and facilitating the use of encryption in interstate commerce consistent with the protections of national security and other purposes."

The bill seeks to address both domestic and export related issues regarding encryption and would basically set in law many of the Clinton Administration's official positions on encryption [market driven/voluntary approach regarding domestic encryption, opposition to mandatory domestic controls (plaintext access requirement), and the appropriate relaxation of export controls on encryption for certain sectors and "responsible" governments while maintaining national security interests. [In Senator McCain's bill, the responsible governments include NATO, Association of Southeast Asian Nations (ASEAN) & Organization for Economic Cooperation and Development (OECD)].

The bill would also create an "Encryption Export Advisory Board" to make recommendations to the Secretary of Commerce regarding those encryption products that are generally available, publicly available or are comparable or will be comparable within 12 months from a foreign supplier thus making these encryption products eligible for export from the United States. In an attempt to address NSA's national security concerns with regard to the export relaxation of such "publicly available" foreign encryption products, the bill includes a provision that would allow the President to override any decision by the Advisory Board for purposes of national security and maintains current Presidential Authorities to prohibit the export of encryption to countries that support acts of terrorism or who pose a threat to United States national security.

From strictly a law enforcement prospective, the only provision of Senator McCain's bill that is not in sync with the Administration's current **voluntary** efforts in support of law enforcement is Title II of the bill, entitled "Government Procurement". This provision requires the federal government to purchase *only encryption products that are interoperable with any and all types of encryption products (recoverable and non-recoverable)* thus eliminating one of law enforcement's potentially most effective influencers in the market-place in our effort to encourage industry to manufacture and sale encryption products that meet law enforcement's plaintext access needs. Additionally, Title I of the bill, entitled "Domestic Encryption Provisions", would prohibit the *federal government or any state government* from requiring or mandating key escrow, key recovery or plaintext access capabilities be included in domestic encryption products.

The following is an overview of the bill's progress thus far during the 106th Congress:

> Introduced on April 14, 1999, and referred to the Senate Commerce Committee for consideration.

S. 854

S. 854 was introduced by Senator Leahy, who is the Ranking Minority Member of the Senate Judiciary Committee. The bill's expressed purpose is "to protect the privacy and constitutional rights of Americans, to establish standards and procedures regarding law enforcement access to location information, decryption assistance for encrypted communication and stored electronic information, and other private information, to affirm the rights of Americans to use and sell encryption as a tool for protecting their online privacy and for other purposes."

The bill is generally not in the best interest of law enforcement. Like the Goodlatte and McCain bills, S. 854 would also prohibit the *federal government or any state government* from requiring or mandating that key escrow, key recovery or plaintext access capabilities be included in domestic encryption products. The bill also seeks to significantly raise the current legal standard needed to obtain a court order to acquire electronically stored information held by a third party, decryption information held by a third party, and the location information of a wireless telecommunication instrument within a carrier's system to a probable cause standard; repeal last Congress' changes to the roving wiretap statute; institute minimization requirements on pen register intercepts with regard to "post cut through digits"; and institute reporting requirements on the Department of Justice with regards to Section 2703 court orders for the obtainment of electronically stored information.

The following is an overview of the bill's progress thus far during the 106th Congress:

> Introduced on April 21, 1999 and referred to the Senate Judiciary Committee for consideration.

ENCRYPTION-RELATED BILLS INTRODUCED DURING THE 105TH CONGRESS:*

H.R. 695, the **"Security and Freedom Through Encryption (SAFE) Act,"** introduced by Congressman Goodlatte (R-6th-VA) on February 12, 1997;

S. 376, the **"Encryption Communications Privacy Act of 1997,"** introduced by Senator Leahy (D-VT) on February 27, 1997;

S. 377, the **"Promotion of Commerce On-Line in the Digital Era (Pro-CODE) Act of 1997,"** introduced by Senator Burns (R-MT) on February 27, 1997;

S. 909, the **"Secure Public Networks Act,"** introduced by Senators McCain (R-AZ), Kerrey (D-NE), Hollings (D-SC) on June 16, 1997.

S. 2067, the **"Encryption Protects the Rights of Individuals from Violation and Abuse in Cyberspace (E-Privacy) Act,"** introduced by Senators Ashcroft (R-MO) and Leahy (D-VT) on May 12, 1998.

The following is an overview of each to the aforementioned bills' progress during the 105th Congress:

H.R. 695

Reported favorably out of the House Judiciary Committee on May 14, 1997 with three amendments. (Congressman McCollum's amendment—members of the Intelligence Community could obtain key recovery information if escrowed, Congressman Asa Hutchinson's amendment—AG is to maintain records regarding the number of cases where encryption prevented law enforcement from enforcing the law, and Congressman Delahunt's amendment—would make it a felony to encrypt information of a criminal nature). The bill was then referred to the House International Relations Committee for consideration and appropriate action.

---

* **NONE OF THE AFOREMENTIONED ENCRYPTION BILLS WERE ENACTED DURING THE 105TH CONGRESS.** Four of the aforementioned encryption-related bills [Goodlatte (H.R. 695), Leahy (S. 376), Burns (S. 377), and Ashcroft/Leahy (S. 2067)] would have largely removed existing export controls on hardware and software encryption products of comparable strength to those that were commercially available from a foreign supplier, regardless of the adverse impact to national security. All five bills would have placed a prohibition on *mandatory* key recovery encryption by the government and included provisions making it a crime to use encryption in furtherance of a criminal act. The McCain/Kerrey, Leahy, and Ashcroft/Leahy bills would have allowed for the *voluntary* use of key recovery encryption and would have established in law requirements for the release of decryption keys to law enforcement (Leahy and Ashcroft/Leahy bills by court order, McCain/Kerrey bill by subpoena).

On May 24, 1997, the House International Relations Committee's Subcommittee on International Economic Policy and Trade held a mark-up concerning the bill and favorably reported the bill out of subcommittee by a fourteen (14) to one (1) vote.

On 7/22/97, the House International Relations Committee held a mark-up concerning the bill. The Committee voted to report H.R. 695 out of Committee with no amendments. The bill was then referred to the House National Security Committee, the House Permanent Select Committee on Intelligence and the House Commerce Committee for appropriate action.

Hearings were also held concerning H.R. 695 before the House National Security Committee on July 30, 1997, before the House Commerce Committee's Subcommittee on Telecommunications, Trade and Consumer Protection on September 4, 1997 and before the House Permanent Select committee on Intelligence on September 9, 1997.

The House National Security Committee held a mark-up of H.R. 695 on September 9, 1997 and adopted an amendment which continues to require a "one time review" and export license for export of encryption products. This action effectively addressed the national security concerns associated with the bill.

The House Permanent Select Committee on Intelligence held a mark-up of H.R. 695 on September 11, 1997 and adopted an amendment by way of a substitute bill that effectively addressed all of the law enforcement and national security concerns associated with commercially-available encryption products and services manufactured for use in the U.S. as well as for export. Highlights include: requirements for immediate access to plaintext features to be included in all encryption products and services manufactured for use in the United States or imported for use in the United States by 1/31/2000; "one time review" by NSA of all encryption products for export and voluntary enabling of any decryption feature included in encryption products for export by the destination country; provide for criminal and civil penalties for unauthorized access to plaintext or decryption information; and, require the U.S. government to only purchase encryption products which include such immediate access to plaintext features.

On September 24, 1997, the House Commerce Committee held a mark-up of H.R. 695. Two competing amendments were offered: Congressmen Oxley and Manton offered an amendment to require all encryption products manufactured for use in the U.S. or imported into the U.S. to contain an immediate access to plaintext feature which would have effectively address law enforcement's domestic encryption needs and would be supported by law enforcement; Congressmen Markey and

White offered an amendment to establish a "National Electronic Technologies Center" to foster the "exchange of information and expertise" between government and industry. However, the Markey/White amendment provided no funding for this center. It did not mandate industry participation, nor is it the goal of the "Center" to provide law enforcement with immediate decryption technical capabilities. Markey/White was supported by industry but was opposed by law enforcement. The Commerce Committee defeated the Oxley/Manton proposal and adopted the Markey/White Amendment, agreeing to favorably report H.R. 695 out of committee as amended.

H.R. 695, as amended by the five committees, was then sent to the House Rules Committee. The Rules Committee, at the discretion of its Chairman, was to consider the different versions of the bill adopted by the five House Committees (Judiciary, International Relations, National Security, Intelligence and Commerce) to determine if a workable compromise bill could be developed and forwarded to the House floor for action. At the insistence of Chairman Solomon (R-NY-22nd), no action was taken by the Rules Committee concerning H.R. 695 as to ensure that the House did not pass an encryption bill that failed to meet all of the law enforcement and national security needs concerning encryption.

S. 909

Reported favorably out of the Senate Commerce Committee on June 19, 1997 with five amendments: one amendment to section 106 regarding the strength of the subpoena used to obtain recovery information; one amendment to section 201 requiring NIST to release a public reference plan regarding key recovery systems prior to the policy provisions of this section being enforced; one amendment to section 205 to clarify that this section only covers networks for the transaction of government business; and one amendment to section 1005 to define what key recovery means. Another amendment was introduced that would create an export advisory board consisting of a chairman appointed by the President, four (4) industry representatives and four (4) government representatives-one each from the CIA, NSA, FBI and Commerce. The bill was scheduled to be referred to the Senate Judiciary and Intelligence Committees for appropriate action but was never officially reported out of the Senate Commerce Committee.

S. 377

Introduced. Failed to be favorably reported out of the Commerce Committee by a 12 to 8 vote on June 19, 1997 as a substitute to S. 909. Senators Burns, Gorton, Lott, Ashcroft**, Abraham**, Brownback, Dorgan and Wyden voted in favor of S. 377; Senators McCain, Stevens, Hutchison, Snowe, Frist, Hollings, Inouye, Ford, Rockefeller, Kerry, Breaux and Bryan voted against S. 377.

S. 376

Only introduced.

S. 2067

Only introduced.

**denotes member of Senate Judiciary Committee)

CHAPTER 29

NATIONAL INFRASTRUCTURE PROTECTION CENTER CYBERNOTES

January 20, 1999

CyberNotes is published every two weeks by the National Infrastructure Protection Center (NIPC). Its mission is to support security and information system professionals with timely information on cyber vulnerabilities, hacker exploit scripts, hacker trends, virus information, and other critical infrastructure-related best practices.

You are encouraged to share this publication with colleagues in the information and infrastructure protection field. Electronic copies are available on the NIPC Web site at www.fbi/nipc/index.htm.

Please direct any inquiries regarding this publication to the Editor-CyberNotes, National Infrastructure Protection Center, FBI Building, Room 11719, 935 Pennsylvania Avenue, NW, Washington, DC, 20535.

Bugs, Holes & Patches

The following table provides a summary of software vulnerabilities identified between January 1 and January 15, 1999. The table provides the operating system, software name, potential vulnerability/impact, identified patches/workarounds/alerts, common name of the vulnerability, potential risk, and an indication of whether attacks have utilized this vulnerability or an exploit script is known to exist Software versions are identified if known. This information is presented only as a summary; complete details are available from the source of the patch/workaround/alert, indicated in the footnote or linked site. Please note that even if the method of attack has not been utilized or an exploit script is not currently widely available on the Internet, a potential vulnerability has been identified.

Operating System	Software Name	Vulnerability /Impact	Patches/ Workarounds /Alerts	Common Name	Risk*	Attacks/ Scripts
ACC's Tigris Access Terminal[1]	HARD-WARE	Unauthorized individuals can gain access to sensitive data without logging in.	No workarounds or patches known at time of publishing.	Tigris Access Terminal Display	Medium	No scripts identified at time of publishing. Explanation of exploit available in news-groups.

Operating System	Software Name	Vulnerability /Impact	Patches/ Workarounds /Alerts	Common Name	Risk*	Attacks/ Scripts
ACC's Tigris Access Terminal Operating System[2] Up to 10.5.8	Operating System	Unauthorized individuals can use the box as a telnet hopping point. The terminal also contains an apparently undocumented default account of public.	Workaround is to restrict telnet access to specific host and to disable source routing.	Tigris default password	Medium/ High	No scripts identified at time of publishing. Explanation of exploit available in newsgroups.
Fore PowerHUB[3]	HARDWARE	Nmap scan will cause the hub to halt for approximately 80 seconds.	Fore has a fix in the form of Powerhub Software 5.0.1. Non-accelerated hardware has a new software release (2.6.4.3) that corrects this.	Nmap scan	Medium	Exploit script (nmap) posted to newsgroups and Web sites. Hacker known to be using.
FreeBSD[4] 2.2.5-r	Operating system (syslog)	Default configuration of FreeBSD is to allow 10 log-in attempts before a log entry is entered. Times of 30, 120 and 600 seconds are also used to log entries. Unusual system failure during this window may result in log loss.	The number of log-in attempts allowed before logging can be modified. The source is located in /usr/src/usr.bin/login/login.c.	FreeBSD syslog timing	Low	This default condition has been discussed in newsgroups.

Operating System	Software Name	Vulnerability /Impact	Patches/ Workarounds /Alerts	Common Name	Risk*	Attacks/ Scripts
Linux[5] (old bug that has resurfaced)	Tripwire 1.2 or earlier (part of RedHat's Linux powertools CD-ROM)	Tripwire will stop functioning and cause a core dump if it receives a filename of 128-255 characters.	An earlier published patch contained an error. Commercial version of tripwire has problem corrected. Support possible from http:/www.tripwiresecurity.com	Tripwire long filename crash	Low	No scripts identified at time of publishing. Explanation of exploit available in newsgroup.
Linux - Red Hat[6] 4.2 and 5.X	Operating System (Dosemu + S-Lang)	Local user can execute a buffer overflow and gain root access.	Red Hat posted a patch in June 1998. This patch does not appear to be incorporated in the Dosemu routine of the recently released Version 5.2.	Dosemu + S-Lang	Low	Exploit scripts posted to newsgroups and Web sites.
Microsoft[7]	Excel	When using the call function in Excel some executables can be run without the user's knowledge.	Patch is available at: http://officeupate.microsoft.com/downloaddetails/x197cfp.htm	Excel call function vulnerability (a.k.a. Russian New Year exploitation)	Medium/ High (Risk increase from the Jan 6 CyberNotes due to publicity)	Attack scripts have been discussed on hacker IRC channels.

Operating System	Software Name	Vulnerability /Impact	Patches/ Workarounds /Alerts	Common Name	Risk*	Attacks/ Scripts
Microsoft Windows[8] 95, 98, and NT	Internet Explorer (IE) 4.0, 4.1, and 5 beta	Latest Java Virtual Machine (JVM) has reintroduced a hostile applet attack. These applets will cause 95 and 98 to crash. On NT machines, IE will crash.	Patches are available at: http://www.microsoft.com/java/vm/dl_vm31.htm. or http://www.microsoft.com/windows/ie/download/jvm.htm.	Java hostile applet	Medium/ High	Exploit scripts posted to newsgroups and Web sites. Attacks have occurred in the past.
Microsoft Windows[9] 95, 98, and NT	Operating System	Between April 1, 2001, and April 8, 2001, Microsoft Windows applications will believe that it is one hour earlier.	Microsoft has stated that a release date for the fix is not available at this time.	April fool's 2001 bug	Low (unless you have time critical functions)	Bug has been discussed on various newgroups.
Microsoft Windows 98[10]	Novell Intranet-ware Client	Nmap scan on port 427 will cause a critical error to develop (Blue Screen) and network connectivity to be lost.	No workarounds or patches known at time of publishing.	Nmap Scanning	Medium/ High	Exploit script (nmap) posted to newsgroups and Web sites. Hackers known to be using nmap with increased frequency.

Operating System	Software Name	Vulnerability /Impact	Patches/ Workarounds /Alerts	Common Name	Risk*	Attacks/ Scripts
Microsoft Windows 98[11] (CORRECTIONS[12])	Operating System	Nmap scan will cause a critical error to develop (Blue Screen) and network connectivity to be lost.	Organization that published this advisory has updated it. The new advisory states that default installation of Windows 98 is not vulnerable (see new listing Microsoft Windows 98. -Novell Intranetware Client).	Nmap scanning	Medium/ High	Exploit script (nmap) posted to newsgroups and Web sites. Hackers known to be using.
Microsoft Windows[13] 9X	Operating System (system acting as file servers)	It is possible to sniff the Challenge Response sequence and replay it for 15 minutes. This allows an unauthorized user to assume the identity of an authorized user.	Suggested workaround disables LM Authenticationresulting in the inability to use 9X boxes as file servers.	Window 9X Challenge Response Timeout	Medium/ High	Bug has been discussed on various newgroups. Hackers typically install sniffers on networks. Sniffers will capture the Challenge Response sequence.

Operating System	Software Name	Vulnerability /Impact	Patches/ Workarounds /Alerts	Common Name	Risk*	Attacks/ Scripts
Sun Solaris[14] 2.5, 2.5.1, 2.6, and, 2.7 (Note: advisories have been previously issued regarding this vulnerability)	Operating System (Automount and rpc.statd)	Unauthorized user can gain root by sending specific remote procedure cail (rpc) packets.	Patch is available that partially corrects this problem. Patch is available from: http://sunsolve.com/sunsolve/pubpatches/patches.html	Solaris Automount	High	Exploit script posted to newsgroups and Web sites. Hackers known to be exploiting.
Sun Solaris[15] 2.7 (See Note at end of table)	Operating System (ufsdump)	Buffer overflow conditions exists in ufsdump that can allow an unauthorized user to gain root access.	Patch is available from: http://sunsolve.com/sunsolve/pubpatches/patches.html	Solaris ufsdump	High	Exploit script posted to newsgroups and Web sites. Hackers known to be exploiting this overflow.
Sun Solaris[16] 2.5.1 and 2.6	Operating System	Unauthorized user can rename files. This can result in manipulation of the file system to provide a root log-in.	Workaround is to chmod ug-s /usr/openwin/bin/ff.core. Note that this exploit is very dependent on system configuration.	Sun Solaris ff.core	Medium	Exploit script posted to newsgroups and Web sites.
Unix[17]	L0phtcrack 2.5	Creates temporary files in the system TEMP directory. These files contain password hashes. If auto-save is used	cracked passwords will be stored in the TEMP directory.	New version has been made available at: http://www.l0pht.com/l0phtcrack/dist/l0phtcrack251.exe L0phtcrack temporary files	Medium/ High	No scripts identified at time of publishing. Explanation of exploit available in newsgroups.

Operating System	Software Name	Vulnerability /Impact	Patches/ Workarounds /Alerts	Common Name	Risk*	Attacks/ Scripts
Unix[18]	suGuard	Any user can gain root access to the system.	No workarounds or patches known at time of publishing.	SuGuard pathing problem	High	Exploit scripts posted to newsgroups and Web sites.
Unix and Microsoft Windows[19]	mSQL	Multiple buffer overflows exist in the program that may allow unauthorized users to execute commands.	No workarounds or patches known at time of publishing.	MSQL buffer overflows	High	Exploit scripts posted to newsgroups and Web sites.

*Risk is defined in the following manner:

High - A vulnerability that will allow an intruder to immediately gain privileged access (e.g., Sysadmin, root) to the system. An example of this would be a vulnerability in which a sequence of instructions is sent to a machine by an unauthorized user and the machine responds with a command prompt.

Medium - Any vulnerability that will allow an intruder immediate access to the system that is not privileged access. This allows the intruder the opportunity to continue the attempt to gain root access. An example would be a configuration error that allows an intruder to capture the password file.

Low - A vulnerability that provides information to an intruder that could load to further compromise attempts or a Denial-of-Service (DoS) attack. The reader should note that while the DoS attack is deemed low from a threat potential, the frequency of this type of attack is very high. DoS attacks against mission-critical nodes are not included in this rating and any attack of this nature should instead be considered as a "High" threat.

1BUGTRAQ, January 4, 1999.

[2]BUGTRAQ, January 4, 1999.

[3]BUGTRAQ, January 5, 1999.

[4]BUGTRAQ, January 2, 1999.

[5]BUGTRAQ, January 4,1999.

[6]BUGTRAQ, January 4, 1999.

[7]Microsoft Security Bulletin (MS98-018).

[8]Newsbytes, January 4, 1999.

[9]NTBUGTRAQ, January 7, 1999.

[10]SecureXpert labs Advisory SX-98.12.30-01.

[11]SecureXpert labs Advisory SC-98.12.23-01.

[12]SecureXpert labs Advisory SX-98.12.30-01.

[13]L0pht Security Advisory, January 5,1999.

[14]BUGTRAQ, January 3, 1999.

[15]BUGTRAQ, January 2,1999.

[16]BUGTRAQ, January 7, 1999.

[17]NTBUGTRAQ, January 6, 1999.

[18]L0pht Security Advisory, January 3, 1999.

[19]Brazilian Information Security Team, 01/99.

Note: If you have recently purchased a CD-ROM, it is possible that patches released several weeks before your purchase are not included on the CD-ROM. CD-ROM's generally have cut off dates for software additions weeks before they are available for purchase.

Recent Exploit Scripts

The table below contains a representative sample of exploit scripts, identified between January 1 and January 15, 1999, listed by date of script, script name, script description, and testing conducted. Items listed in boldface (if any) are attack scripts for which vendors, security vulnerability listservs, or Computer Emergency Response Teams (CERTs) have not published workarounds or patches. Those items in red represent scripts that hackers/crackers are utilizing. During this time period, 50 scripts, programs, and net-news messages containing holes or exploits were identified.

Date of Script (Reverse Chronological Order)	Script Name	Script Description	Testing Conducted
Jan 14, 1999	Ath0-2.sh	Shell script to exploit "+++ATH0" DoS attack against Unix boxes running modems.	
Jan 14, 1999	Bogtk v1.0.1	Linux/Unix Back Orifice client graphical user interface.	
Jan 13, 1999	IPTraf v1.4.2	A monitor that collects Local Area Network (LAN) statistics, checksums, and packet information.	
Jan 13, 1999	Namp	Network scanning tool that has a variety of scanning modes, including stealth, Xmas, and Null stealth. Note: This tool is being used by hackers and may cause systems to become unstable.	

Date of Script (Reverse Chronological Order)	Script Name	Script Description	Testing Conducted
Jan 12, 1999	Proxys-4-all	Explanation of how to use proxies for anonymous Web surfing. Includes a list of sites that allow anonymous Web surfing.	
Jan 11, 1999	Domscan 2.0pl	A domain scanner written in Perl.	
Jan 11, 1999	Gammaprog 1.4	Bruteforce password cracker for Web-based and Point-of-Presence 3 (POP3) e-mail addresses.	
Jan 11, 1999	GetadmforSop	Exploit that elevates privileges to both local and global Administrator level.	
Jan 11, 1999	K2v1017	Program that allows remote user access to current and old dial-up networking passwords, also shows passwords hidden behind asterisks (*).	
Jan 11, 1999	MSQL-DoS.txt	Explanation of several buffer overflows that exist in MSQL.	
Jan 11, 1999	Netboot v0.9.0b	Allows booting of computer with Intel processor via an Internet Protocol (IP) network.	
Jan 11, 1999	Pass10	Program that shows passwords hidden behind asterisks (*).	
Jan 11, 1999	S10scan v0.1	Sends large volume of "fake scans" with "real scans" embedded within.	
Jan 11, 1999	SDI-msql	Exploit code for the mSQL vulnerabilities (several buffer overflows.)	
Jan 11, 1999	SilentBot	Silent bot for IRCs.	
Jan 11, 1999	Trojanit	Trojan horse/root kit for Linux. Some code included for Berkeley Software Distribution (BSD) root kit.	
Jan 10, 1999	Arptool	Program written in C that can map the IP of machines on the same Ethernet segment and can be used for spoofing hosts.	
Jan 10, 1999	BoClient 1.4.1	Back Orifice with built-in Transmission Control Protocol (TCP) send/receive and the ability to remember most parameters.	

Date of Script (Reverse Chronological Order)	Script Name	Script Description	Testing Conducted
Jan 10, 1999	ff.core.sh	Exploit script for the ff.core vulnerability listed in "Bugs, Holes, & Patches" table above.	
Jan 10, 1999	Mother2	Shell script that finds all setuid and setgid programs.	
Jan 10, 1999	Nessus WIP 010999	Security audit tool that has 180 plug-ins for checking known holes. (Earlier version found to contain a Trojan horse.)	
Jan 10, 1999	Sucker v0.1	Shell script that checks for various vulnerabilities in a TCP/IP-based network.	
Jan 10, 1999	Tcpscan v1.02	TCP port scanner.	
Jan 7, 1999	Exploits	A 3.1megabyte (MB) file that contains 1000 exploits that have been sorted and categorized.	
Jan 7, 1999	ICMP	Program that uses Internet Control Message Protocol (ICMP) packets to obtain information from remote hosts. This program supports spoofing and broadcasting.	
Jan 7, 1999	Lanlord v0.1.1	Program designed to discover what machine has which Dynamic Host Configuration Protocol (DHCP) address.	
Jan 6, 1999	Cheops v0.59a	Tool for network scanning and discovery of operating system. Note: The author of the tool has posted warnings that the tool has the potential of being misused and there have been indications that hackers are currently using this tool.	
Jan 6, 1999	Cyberanon v1.1	Common Gateway Interface (CGI) script anonymizer that attempt to allow users to surf the web anonymously.	
Jan 6, 1999	L0phtCrack 2.51 for Win95/NT	Password cracker for Microsoft Windows 95/NT.	
Jan 6, 1999	Net RawIP v 0.03f	Perl module that manipulates raw IP packets and Ethernet headers.	

Date of Script (Reverse Chronological Order)	Script Name	Script Description	Testing Conducted
Jan 6, 1999	Novell-iwc-DoS	DoS explanation for the Novell Intranet Client vulnerability (see "Bugs, Holes, & Patches" above.)	
Jan 6, 1999	Oracle8-tnslsnr-DoS	DoS explanation for the Oracle version 8 DoS vulnerability (see January 6, 1999, edition of CyberNotes.)	
Jan 6, 1999	Pwdump2	Program that dumps the password hashes from NT's SAM database without SYSKEY enabled.	
Jan 6, 1999	Qmail-DoS-anonymous	A number of DoS attack scripts against Qmail.	
Jan 6, 1999	Windows File Sharing hole	Explanation of the authentication timing problem with Microsoft Windows 95 (see "Bugs, Holes & Patches" above.)	
Jan 4, 1999	Automountd exploit	Program to exploit the Automountd vulnerability in Sun Solaris (see "Bugs, Holes & Patches" above.)	
Jan 4, 1999	Bus Conquerer v1.3	Scans for the NetBus program, cracks the password and then reassings a new password.	
Jan 4, 1999	Either	Program for modification of MAC addresses on Microsoft Windows machines.	
Jan 4, 1999	EliteSys Entry v2.05	Brute force password cracker for File Transfer Protocol (FTP), Web, and POP3 machines.	
Jan 4, 1999	GnuSniff v0.0.5	Packet sniffer.	
Jan 4, 1999	Mirc-hidden-files	Exploit program for ICQ "hidden file" security hole.	
Jan 4, 1999	Net-RawIP v 0.03e	See entry above.	
Jan 4, 1999	Suguard	Exploit code for DataLynx's SuGuard (see "Bugs, Holes & Patches" above.)	
Jan 4, 1999	Termz.c	Exploit code for the S-Lang buffer overflow in Red Hat Linux (see "Bugs, Holes & Patches" above.)	
Jan 4, 1999	Web Cracker v2.0	Brute force password cracker for Web sites.	

Date of Script (Reverse Chronological Order)	Script Name	Script Description	Testing Conducted
Jan 3, 1999	Bogtk v1.0	See entry above.	
Jan 3, 1999	Exscan v0.3	Port scanner with a strobe scanning capability.	
Jan 2, 1999	Electronic Civil Disobedience Disturbance Developer's Kit	Explanation and code for flooding a host server with requests.	
Jan 2, 1999	Gammaprog 1.3	See description above (Gammaprog 1.4).	
Jan 1, 1999	Ufodump	Local root exploit for the ufsdump vulnerability in Solaris 2.6	

Trends

1. Several hackers/hacker groups appear to be using coordinated scans and probes from different sites.

2. More phreaker tools are appearing on hacker Web sites.

3. Increase in organized crime using compromised Private Automatic Branch Exchanges (PABXs.)

4. Scanning for Internet Map Access Protocol (IMAP) and POP continues.

5. Significant increase in reports of NetBus and BackOrifice scanning.

6. Large increase in the number of scans directed specifically against Domain Name Servers.

Viruses

During the last quarter, anti-virus vendors have reported that there has been a sharp increase (5 - time increase over the same period last year) in the number of customers submitting files suspected of containing viruses. The contaminated files during the quarter were classified in the following manner: 45 to 50 percent contained Trojan horses, 25 to 30 percent contained a macro virus, approximately 20 percent contained a file virus, and 3 to 5 percent contained multi-partite viruses.

In the last two weeks, there has been an increase in reports of many people receiving e-mail messages with Trojan horse programs attached. The most prevalent of these appears to be the picture.exe (aka. URLsnoop) Trojan horse program. Most major anti-virus vendors have included code in their latest data file updates to detect this Trojan horse program.

CHAPTER 30

NATIONAL INFRASTRUCTURE PROTECTION CENTER CYBERNOTES

February 3, 1999

CyberNotes is published every two weeks by the National Infrastructure Protection Center (NIPC). Its mission is to support security and information system professionals with timely information on cyber vulnerabilities, hacker exploit scripts, hacker trends, virus information, and other critical infrastructure-related best practices.

You are encouraged to share this publication with colleagues in the information and infrastructure protection field.

Please direct any inquiries regarding this publication to the Editor-CyberNotes, National Infrastructure Protection Center, FBI Building, Room 11719, 935 Pennsylvania Avenue, NW, Washington, DC, 20535.

Bugs, Holes & Patches

The following table provides a summary of software vulnerabilities identified between January 16 and January 27, 1999. The table provides the operating system, software name, potential vulnerability/impact, identified patches/workarounds/alerts, common name of the vulnerability, potential risk, and an indication of whether attacks have utilized this vulnerability or an exploit script is known to exist. Software versions are identified if known. This information is presented only as a summary; complete details are available from the source of the patch/workaround/alert, indicated in the footnote or linked site. Please note that even if the method of attack has not been utilized or an exploit script is not currently widely available on the Internet, a potential vulnerability has been identified.

Operating System	Software Name	Vulnerability/ Impact	Patches/ Workarounds/ Alerts	Common Name	Risk*	Attacks/ Scripts
Cisco IOS[1]	Operating System	Any Cisco device running IOS code will identify the fact that it is a Cisco product when a SYN packet is sent to port 1999.	Workaround is to deny incoming Transmission Control Protocol (TCP) connections to port 1999 where possible.	Remote Cisco Identification	Low / Medium	Bug discussed in newsgroups. Exploit script posted to newsgroups and Web sites.
Linux[2] 2.0.36	Operating System	Local user can cause a Denial-of-Service (DoS) on ports above 1024.	Versions 2.1.X and 2.2.X-prebuild are not affected.	2.0.36 local user DoS	Low	Exploit script posted to newsgroups and Web sites.
Linux[3] (may also affect FreeBSD prior to Perl 5 implementation)	Operating system and Perl	It is possible for an unauthorized local user to gain root access using a setuid Perl script on a CD-ROM or floppy disk.	No workarounds or patches known at time of publishing.	Linux suidperl	Medium/ High	Bug discussed in newsgroups.
Linux[4] – Debian GNU 1.3	FTPwatch	Unauthorized users can gain root access.	Authors recommend removing FTPwatch immediately.	Debian Linux FTPwatch	High	Bug discussed in newsgroups.
Microsoft[5]	Internet Explorer 4 (IE4)	IE4, under certain circumstances, fails to close connections to a proxy server. This can open the session to man-in-the-middle attacks/spoofs.	No workarounds or patches known at time of publishing.	IE4 Persistent Keepalive Bug	Low	Condition discussed in newsgroups.

Operating System	Software Name	Vulnerability/ Impact	Patches/ Workarounds/ Alerts	Common Name	Risk*	Attacks/ Scripts
Microsoft[6]	Applications using Forms 2.0 ActiveX control (any application that includes Visual Basic for Applications 5.0)	Unauthorized Web site or maliciously constructed HyperText Markup Language (HTML) email can reveal items contained in the clipboard.	Patch is available at: http://support.microsoft.com/support/kb/articles/q214/7/57.asp	Forms 2.0 ActiveX Clipboard	Low	Bug discussed in newsgroups. Exploit script posted to newsgroups and Web sites.
Microsoft Windows[7]	Internet Information Server 4 (IIS 4) with sample site ExAir loaded	Direct request to three of the ExAir files will cause a DoS condition for 90 seconds even if ExAir was not loaded.	Workaround is to remove: ExAir – root/search/advsearch.asp, ExAir – root/search/query.asp, and ExAir – root/search/search.asp	ExAir sample site DoS	Low	Bug discussed in newsgroups and Web sites. Exploit script is not required for the exploit.
Microsoft Windows[8]	IIS	If upgrading from IIS 2 or 3 to 4, a dll called ism.dll is left in/scripts/iisadmin. An unauthorized user can gain access to sensitive server information, including potentially the Administrator's password.	Remove ism.dll from /scripts/issadmin.	IIS upgrade ism.dll	Medium	Bug discussed in newsgroups and Web sites. Exploit script not required for the exploit.
Microsoft Windows[9]	IIS	IIS will not log requests if the path is over approximately 10,100 bytes. The request will still be honored.	Workaround is to use additional logging methods.	IIS long path request	Low	Bug discussed in newsgroups and Web sites. Exploit script not required for the exploit.

Operating System	Software Name	Vulnerability/ Impact	Patches/ Workarounds/ Alerts	Common Name	Risk*	Attacks/ Scripts
Microsoft Windows[10]	IIS and perl.exe	On all versions of ISS, a request for a non-existent file will return the physical location of the Web directory on the machine.	Workaround is to use the perlis.dll instead of perl.exe. You can use perl.exe and have IIS check for the file's existence.	perl.exe and IIS	Low	Bug discussed in newsgroups and Web sites.
Microsoft Windows[11] 95, 98, and NT[12]	IIS and File Transfer Protocol (FTP)	Unauthorized remote user can log in (anonymous access included) and create a DoS condition. This may be a result of a stack overflow or bss (static pointers) overflow.	Specific configurations appear to be necessary for the vulnerability to occur. NT + Option pack 4 + SP 4 or NT + IIS3 + SP 4 or PWS 1.0 No workarounds or patches known at time of publishing.	IIS DoS FTP	Low / Medium	Script identified at time of publishing but does not appear to be posted. Explanation of exploit available in newgroups.
Microsoft Windows[13] NT	IIS (installed from NT Option Pack and FrontPage Server Extensions)	A buffer overflow that exists in the program that may allow unauthorized users to execute commands.	Obtain FrontPage Server Extensions 98 from: http://www.microsoft.com/frontpage.	IIS on NT with FrontPage buffer overflow	Medium/ High	Bug discussed in newsgroups and Web sites. References to exploit script found.
Microsoft Windows[14] NT	Quakenbush Windows NT Password Appraiser	When Internet Query option is selected, the standard and professional versions send LANMAN hashes and the decoded password in the clear across the Internet.	Update issued to the Password Appraiser program (standard and professional).	Quakenbush password problem	Medium/ High	Default condition discussed in newsgroups. Hacker tools exist that can decrypt password hashes.

Operating System	Software Name	Vulnerability/ Impact	Patches/ Workarounds/ Alerts	Common Name	Risk*	Attacks/ Scripts
Multiple[15]	Computer Associates' (CA) ControlIT enterprise management software	Use of weak password encryption when transmitting passwords over the Internet could enable a hacker to obtain user and administrator passwords on Microsoft Windows NT machines.	CA recommends using ControlIT's "built-in" security, because NT usernames/ passwords are not required.	ControlIT weak password encryption	Medium/ High	Bug discussed in newsgroups. Hackers typically install sniffers on networks. Sniffers will capture the NT username/ password if used.
Multiple[16]	CA's ControlIT enterprise management software	When "reboot on disconnect" is enabled, any invalid username/ password will reboot the system.	CA has a patch available at http://www.cai.com or by calling 1-800-DIALCAI.	ControlIT reboot on invalid username/ password	Medium (High -if you have machine that performs critical functions)	Bug discussed in newsgroups and Web sites. Exploit script not required for the exploit.
Multiple[17]	BackWeb Technologies' BackWeb Polite Agent Protocol	Unauthorized individual may be able to hijack sessions between client and server. This hijacking could result in the delivery of unwanted software.	Upgrade to V5.0.	BackWeb Polite Agent Spoofing	Low/ Medium	Bug discussed in newsgroups and Web sites. Exploit script not required for the exploit.
Multiple[18]	DPEC Online Courseware	Unauthorized user can gain access to and change the password of any user, including admin.	No workarounds or patches known at time of publishing.	DPEC Online Courseware password problem	Low	Bug discussed in newsgroups and Web sites.

Operating System	Software Name	Vulnerability/ Impact	Patches/ Workarounds/ Alerts	Common Name	Risk*	Attacks/ Scripts
Multiple[19]	Lotus Notes (Windows and Sun Solaris versions)	Multiple long strings (2,048 characters) establishing connections (helo A and then helo B) cause a DoS condition.	No workarounds or patches known at time of publishing.	Lotus Notes long string DoS	Low	Bug discussed in news-groups.
Multiple[20]	Most Web servers	It is possible to confuse Hypertext Transfer Protocol (HTTP) logging with the use of special characters in REQUEST_ METHOD.	Workaround is to use additional logging methods.	HTTP REQUEST_ METHOD flaw	Low	Bug discussed in news-groups and Web sites.
Multiple[21]	Multi-active's Maximizer	Unauthorized individual has the ability to alter data.	No workarounds or patches known at time of publishing.	Maximizer data change	Low	Bug discussed in news-groups and Web sites.
Multiple[22]	Sscan (0.1 alpha)	Unauthorized user can execute buffer overflow conditions.	Update to new version (See "Recent Exploit Scripts" below for explanation).	sscan	Low	Default condition discussed in news-groups.

Operating System	Software Name	Vulnerability/ Impact	Patches/ Workarounds/ Alerts	Common Name	Risk*	Attacks/ Scripts
Sun Solaris[23] 2.6	Operating System	On some installations using Network Information Services (NIS) and NIS+, an unauthorized individual can directly connect without using portmapper. This connection will provide hashed passwords.	Workaround is to use /var/yp/ securenets, which pro-- vides a secure distributed name service. Further details are available at: http://www. sunworld. com/sun- worldonline/ common/ security- faq.html.	NIS and NIS+ ephemeral ports	Medium/ High	Bug discussed in news- groups and Web sites. Hackers known to be attempting to exploit.
Sun Solaris[24] 2.5.1 (unpatched)	Operating System	Any process that runs with a user's privilege is susceptible to tracing by the user.	Patch for version 2.5.1 is kemel update 103640.	Solaris 2.5.1 user tracing problem	Low/ Medium	Bug discussed in news- groups.
Unix[25]	TCP Wrappers 7.6	A Trojan horse version of the program (tcp_wrappers_ 7.6.tar.gz) was placed on numerous sites. Unauthorized user can gain root access.	Obtain new version of TCP Wrappers if you downloaded the program since January 20, 1999. Workaround is to block port 421 source, not destination, if possible.	TCP Wrappers Trojan horse (Note: some sources have called this the Unix version of Back Orifice).	High	Hackers known to be attempting to exploit this Trojan horse.

Operating System	Software Name	Vulnerability/ Impact	Patches/ Workarounds/ Alerts	Common Name	Risk*	Attacks/ Scripts
Unix – Digital[26] 4.0 prior to 4.0D	Operating System (at)	Buffer overflow condition exists in "at" that can allow an unauthorized user to gain root access.	Upgrade to Digital Unix 4.0D or obtain appropriate patch from: ftp://ftp. service.digital. com/public/ dunix.	Digital Unix at buffer overflow	High	Script identified at time of publishing but does not appear to be posted. Explanation of exploit available in news-groups.
Unix – Digital[27] 4.0	Operating System (inc)	Buffer overflow condition exists in /usr/bin/mh/ inc that can allow an unauthorized user to gain root access.	No workarounds or patches known at time of publishing.	Digital Unix inc buffer overflow	High	Script identified at time of publishing but does not appear to be posted. Explanation of exploit available in news-groups.
Unix[28] _ NetBSD	Operating System	Unauthorized user may cause a DoS condition in TCP services due to select(2)/accept (2) race condition.	Patch available at: ftp://ftp. NetBSD.ORG. under/pub/ NetBSD/misc/ security/ patches/199901 20-accept.	Net BSD Select (2)/Accept (2) race condition in TCP servers	Medium/ High	Exploit script (nmap) posted to newsgroups and Web sites. Hackers known to be using nmap with increased frequency.
Sun So-laris[29] 2.6, 2.6x86, 2.7 and 2.7x86	Operating System	Buffer overflow condition exists in /usr/bin/ lpstat that can allow an unauthorized user to gain root access.	Workaround it to: chmod -s /usr/bin/ lpstat chmod -s/usr/bin/lpq	Buffer overflow in Solaris /usr/bin/ lpstat	High	Bug discussed in news-groups and Web sites. Hackers known to be attempting to exploit.

Operating System	Software Name	Vulnerability/ Impact	Patches/ Workarounds/ Alerts	Common Name	Risk*	Attacks/ Scripts
WE-BRAMP[30]	Operating System	Default password allows full access to device.	Change default admin password.	WEBRAMP default admin password	High	Bug discussed in news-groups. Hackers known to be attempting to exploit.

*Risk is defined in the following manner:

High – A vulnerability that will allow an intruder to immediately gain privileged access (e.g., Sysadmin, root) to the system. An example of this would be a vulnerability in which a sequence of instructions is sent to a machine by an unauthorized user and the machine responds with a command prompt.

Medium – Any vulnerability that will allow an intruder immediate access to the system that is not privileged access. This allows the intruder the opportunity to continue the attempt to gain root access. An example would be a configuration error that allows an intruder to capture the password file.

Low – A vulnerability that provides information to an intruder that could lead to further compromise attempts or a Denial-of-Service (DoS) attack. The reader should note that while the DoS attack is deemed low from a threat potential, the frequency of this type of attack is very high. DoS attacks against mission-critical nodes are not included in this rating and any attack of this nature should instead be considered as a "High" threat.

1BUGTRAQ, January 18, 1999.

[2]BUGTRAQ, January 19, 1999.

[3]BUGTRAQ, January 14, 1999.

[4]Debian security announcement, January 17, 1999.

[5]BUGTRAQ, January 22, 1999.

[6]Microsoft Security Bulletin, MS99-001

[7]NTBUGTRAQ, January 26, 1999.

[8]BUGTRAQ, January 14, 1999.

[9]Provider requested anonymity.

[10]NTBUGTRAG, January 26, 1999.

[11]eEye Digital Security Team, Advisory Code - IISE01.

[12]BUGTRAQ, January 25, 1999.

[13]BUGTRAQ, January 14, 1999.

[14]L0pht Security Advisory, January 21, 1999.

[15]IIS Security Advisory, Multiple vulnerabilities in ControlIT.

[16]Ibid.

[17]ISS Security Advisory, Vulnerability in the BackWeb Polite Agent 18Protocol.

[18]BUGTRAQ, January 15, 1999.

[19]BUGTRAQ, January 15, 1999.

[20]BUGTRAQ, January 6, 1999.

[21]BUGTRAQ, January 14, 1999.

[22]BUGTRAQ, January 20, 1999.

[23]BUGTRAQ, January 13, 1999.

[24]BUGTRAQ, January 13, 1999.

[25]CERT advisory CA-99-01-Trojan-TCP-Wrappers, revised January 22, 1999.

[26]BUGTRAQ, January 25, 1999.

[27]BUGTRAQ, January 25, 1999.

[28]BUGTRAQ, January 20, 1999.

[29]HERT advisory #001.

[30]BUGTRAQ, January 21, 1999.

Note: If you have recently purchased a CD-ROM, it is possible that patches released several weeks before your purchase are not included on the CD-ROM. CD-ROMs generally have cut-off dates for software additions weeks before they are available for purchase.

Recent Exploit Scripts

The table below contains a representative sample of exploit scripts, identified between January 16 and January 27, 1999, listed by date of script, script name, script description, and comments. Items listed in boldface (if any) are attack scripts for which vendors, security vulnerability listservs, or Computer Emergency Response Teams (CERTs) have not published workarounds or patches. Those items in red represent scripts that hackers/crackers are utilizing. During this time period, 48 scripts, programs, and net-news messages containing holes or exploits were identified.

Date of Script (Reverse Chronological Order)	Script Name	Script Description	Comments
Jan 27, 1999	Bandwidthrape	Explanation and sample exploit code that causes a remote server to consume large amounts of bandwidth.	
Jan 27, 1999	Cisco.c	Code segment that identifies Cisco equipment. See Remote Cisco Identification in "Bug Holes & Patches" above.	

Date of Script (Reverse Chronological Order)	Script Name	Script Description	Comments
Jan 27, 1999	Digital-Unix4.0-asm-shell	Exploit code that uses a buffer overflow, which results in a root shell. This is a local exploit.	
Jan 27, 1999	Lpstat x86.c	Exploit code for the lpstat vulnerability in Sun Solaris 2.6/2.7 x.86.	
Jan 27, 1999	Spoof.c	Demonstration code for creating spoofed packets on Linux.	
Jan 27, 1999	Warp Scanner 2	Windows-based TCP port scanner.	
Jan 26, 1999	Apc.c	Unix password cracker.	
Jan 26, 1999	CGIchk2.c	Program that checks for Common Gateway Interface (CGI) vulnerabilities and, if any are found, will attempt to exploit the most common.	
Jan 26, 1999	Exploit Scanner v2.0	Scans subnet for many well-known Trojan horses and exploits.	
Jan 26, 1999	Mailwatch.c	Program that watches a user mailbox and notifies another user when mail is received.	
Jan 26, 1999	Ppscan.c	Program that allows the use of a proxy server to hide scan attempts when checking for Web servers.	
Jan 26, 1999	Ps.s	Script that bind a shell to port 46256.	
Jan 25, 1999	Cgiscan.c	Scanner that checks for various common vulnerabilities in CGI scripts.	
Jan 25, 1999	Manipulate_data v1.0	Code that allows the user the search a hard drive for data and then write it back after modification.	
Jan 25, 1999	Net-RawIP v0.04	Perl module that manipulates raw Internet Protocol (IP) packets and Ethernet headers.	
Jan 25, 1999	Oshare_1_gou.c	Exploit code the cause a DoS condition on Microsoft Windows 98 machines by issuing malformed packets.	
Jan 25, 1999	Page.sh	Unix shell script that checks for the FrontPage_vti_pvt/service. pwd exploit.	

Date of Script (Reverse Chronological Order)	Script Name	Script Description	Comments
Jan 25, 1999	RiP FTP server v1.0	Win32 program that captures File Transfer Protocol (FTP) passwords from server client software.	
Jan 24, 1999	Balu.pl	Program that exploits a vulnerability in mIRC 5.5's new dcc server feature. Places program in any destination directory on the same hard drive in which mIRC resides.	
Jan 24, 1999	Tgk-log v2.3	Recoded version of linsniffer that is designed to record contents of packets as they pass through an ipmasq gateway.	
Jan 24, 1999	UTIL-Linux-29g	Trojan horse version of util-Linux2.9g.	
Jan 24, 1999	Wsftp-hit	Exploit for ws-ftp design weakness that allows hackers to gain access to cached passwords.	
Jan 23, 1999	Icmpmon.c	Program that shows all ICMP packets received by the machine on which the program is running.	
Jan 23, 1999	M3-entre	Explanation of how to conduct attacks on the WEBRAMP default admin password and the default login/password. This script includes a text file with several suggestions on malicious uses. See "Bugs Holes & Patches" section above.	
Jan 23, 1999	NetGuard v0.0.2	Combination of two programs that intercept and log TCP/IP and User Datagram Protocol (UDP)/IP connections. Allows limited logging of the TCP connections to syslog.	
Jan 22, 1999	ftp-spoof.pl	Program and instructions for altering site name of an FTP server, allowing any user to spoof its address.	
Jan 22, 1999	L0phtCrack 2.52 for Win95/NT	Password cracker for Microsoft Windows 95/NT. Improvements include 450% increase in speed, combination dictionary and numeric/symbol passwords.	

Date of Script (Reverse Chronological Order)	Script Name	Script Description	Comments
Jan 22, 1999	PoP-spoof.pl	Allows user to spoof a Post office Protocol (PoP) server and obtain passwords as clear text.	
Jan 22, 1999	TellME v0.1	Retrieves NetBIOS name from remote Microsoft Windows machines.	
Jan 21, 1999	Avoid	Program that tests for the IIS long path request.	
Jan 21, 1999	M3	Explanation of the WEBRAMP default admin password vulnerability. See "Bugs, Holes & Patches" section above.	
Jan 21, 1999	Quake2-bof-DOS	Explanation and exploit code for Quake2 buffer overflow DoS attack.	
Jan 20, 1999	Backwork v2.0	Trojan horse cleaner that states it is capable of identifying and removing 45 different Trojan horses.	
Jan 20, 1999	CGIc_DoS	Explanation and sample exploit code for a buffer overflow in the CGIc Library.	
Jan 20, 1999	Killport.c	Exploit code for the Linux local port/memory DoS attack. See Linux 2.0.36 local user DoS in "Bugs, Holes & Patches" section above.	
Jan 20, 1999	Resetter.c V2.1	DoS tool that attempts to reset all connections of a network segment by sending spoofed RST and ICMP UNREACHABLE packets.	
Jan 19, 1999	Against.c	Exploit code for the Sendmail 8.9.2 DoS problem.	
Jan 19, 1999	Backweb-spoof	Explanation of how to conduct an attack on Back Web. See Back Web Polite Agent Spoofing in "Bugs, Holes & Patches" section above.	
Jan 19, 1999	BusJack v1.0	Program that identifies and cleans the NetBus Trojan horse program.	

Date of Script (Reverse Chronological Order)	Script Name	Script Description	Comments
Jan 19, 1999	CISCO-IOS-ID	Code to automatically identify CISCO equipment. See Remote Cisco Identification in "Bugs, Holes & Patches" section above.	
Jan 19, 1999	Directory Snoop V2.1	Program that allows low-level review of disk drives on Microsoft Windows 95/98 machines.	
Jan 19, 1999	NetBus v2.0 Pro Beta	Trojan horse program with an improved Graphical User Interface (GUI), proxy support, file manager, web-cam capture, registry manager, hosts scheduler, and chat features.	
Jan 19, 1999	Net-RawIP v0.03	Perl module that manipulates raw IP packets and Ethernet headers.	
Jan 19, 1999	Sscan v0.1 alpha	Improved version of mscan that is now configurable by "anyone with no programming knowledge."	
Jan 19, 1999	TellMe v0.1	See explanation above.	
Jan 19, 1999	Vnmap	Tcl/Tk/wish-based front-end for nmap.	
Jan 19, 1999	Wyjeb.c	Local DoS that uses syslogd for the attack.	
Jan 19, 1999	xcrack	Unix/Linux Password cracking program written in Perl.	

Note: In the last edition of CyberNotes, in the "Recent Exploit Scripts" section under the name Gammaprog 1.4, the acronym POP was incorrectly referenced. The correct reference should read Post Office Protocol.

Trends

1. Several hackers/hacker groups appear to be using coordinated scans and probes from different sites.

2. Large numbers of scans and attacks continue to be directed at machines running the Linux operating system.

3. More phreaker tools are appearing on hacker Web sites.

4. Increase in organized crime using compromised Private Automatic Branch Exchanges (PABXs.)

5. Scanning for Internet Map Access Protocol (IMAP) and POP continues.

6. Significant increase in reports of NetBus and Back Orifice scanning.

7. Significant increase in the number of scans directed specifically against Domain Name Servers.

Viruses

A list of the top ten viruses infecting two or more sites as reported to various anti-virus vendors has been categorized into the two tables below. The first table list macro viruses, and the second table lists other viruses. Macro viruses have, historically, spread fastest due to their ability to be transferred by e-mail.

For the purposes of collecting and collating data, infections involving multiple systems at a single location are considered a single infection. The tables list the viruses by: ranking (number of sites affected), common virus name, type of virus (i.e., boot, file, macro, multi-partite), trends (based on number of infections during the last three months reported), and approximate date first found.

Note: Virus reporting is normally 6 to 8 weeks behind the first discovery of infection. With the number of viruses that appear each month, it is possible that a new virus will become widely distributed before the next edition of this publication. To limit the possibility of infection, readers are reminded to update their anti-virus packages, as updates become available.

The viruses listed in the virus table have infected over 510 sites in December, which is a slight increase in the number of reported infections from the last prevalence table. The number 1 ranked virus for December/January accounted for 35 infected sites, and the last virus listed in the tables infected 17 sites. A total of 297 distinct viruses were reported this month, infecting over 2,000 sites.

Table 1 – Macro viruses:

Ranking	Common Virus Name	Type of Virus	Trends	Date
1	Concept	Macro	Steady	December 1996
2	CAP	Macro	Steady	April 1997
3	Class	Macro	Increasing	September 1998
4	Npad	Macro	Steady	December 1996
5	Wazzu	Macro	Steady	December 1996
6	Laroux	Macro	Steady	July 1997
7	MDMA	Macro	Steady	December 1996
8	ColdApe	Macro	NEW - No Trend	December 1998
9	Hark	Macro	NEW - No Trend	December 1998
10	Temple	Macro	NEW - No Trend	December 1998
Not Ranked	W2KM_PSD	Macro		January 1999

Table 2 – Other viruses:

Ranking	Common Virus Name	Type of Virus	Trends	Date
1	Form	Boot	Steady	September 1991
2	One half	Multi	Decreasing	October 1995
3	Junkie	Multi	Steady	July 1994
4	AntiCMOS	Boot	Steady	October 1995
5	AntiEXE	Boot	Decreasing	September 1994
6	Empire.Monkey	Boot	Steady	July 1994
7	Parity_Boot	Boot	Steady	September 1993
8	Ripper	Boot	Steady	March 1994,
9	Natas	Multi	Steady	October 1995
10	NYB	Boot	Decreasing	July 1994

CHAPTER 31

NATIONAL INFRASTRUCTURE PROTECTION CENTER CYBERNOTES

February 17, 1999

CyberNotes is published every two weeks by the National Infrastructure Protection Center (NIPC). Its mission is to support security and information system professionals with timely information on cyber vulnerabilities, hacker exploit scripts, hacker trends, virus information, and other critical infrastructure-related best practices.

You are encouraged to share this publication with colleagues in the information and infrastructure protection field. Electronic copies are available on the NIPC Web site at http://www.nipc.gov.

Please direct any inquiries regarding this publication to the Editor-CyberNotes, National Infrastructure Protection Center, FBI Building, Room 11719, 935 Pennsylvania Avenue, NW, Washington, DC, 20535.

Bugs, Holes & Patches

The following table provides a summary of software vulnerabilities identified between January 27 and February 12, 1999. The table provides the operating system, software name, potential vulnerability/impact, identified patches/workarounds/alerts, common name of the vulnerability, potential risk, and an indication of whether attacks have utilized this vulnerability or an exploit script is known to exist. Software versions are identified if known. **This information is presented only as a summary; complete details are available from the source of the patch/workaround/alert, indicated in the footnote or linked site.** Please note that even if the method of attack has not been utilized or an exploit script is not currently widely available on the Internet, a potential vulnerability has been identified.

Operating System	Software Name	Vulnerability/ Impact	Patches/ Workarounds/ Alerts	Common Name	Risk*	Attacks/ Scripts
Allaire[1]	ColdFusion 4.0	Code snippets from sample applications make it possible to make HTTP calls from the machine and to execute Denial-of-Service (DoS) attacks.	Remove the CFDOCS directory or restrict access to the directory. A maintenance release is scheduled for April 1999.	ColdFusion Sample Code vulnerability	**High**	Bug discussed in news-groups and Web sites. Exploit script not required for the exploit.
Allaire[2]	ColdFusion Application Server 2.X, 3.X, 3.1.X and ColdFusion Server 4.X	Machines loaded with the Expression Evaluator (sample application) will allow authorized individuals to read and delete files on the server.	Remove all sample applications and code from production servers and obtain patch for Expression Evaluator at: http://www.allaire.com	ColdFusion Expression Evaluator unauthorized access	**High**	Bug discussed in news-groups and Web sites. Exploit script not required for the exploit. Other related hacker tools can be utilized to gain privileged access.

Operating System	Software Name	Vulnerability/ Impact	Patches/ Workarounds/ Alerts	Common Name	Risk*	Attacks/ Scripts
Allaire[3]	ColdFusion Server - all versions with Microsoft SQL and ColdFusion Server 4.0 Enterprise with Sybase SQL	It is possible to send malicious Structured Query Language (SQL) statements to a database through a dynamic query. ColdFusion may mishandle numbers, if not quoted, or a string that is processed with the Preserve Single Quotes() function.	Allaire has prepared a technical brief that provided guidance on this subject. The brief is titled "Securing Databases for ColdFusion Applications" and is available at http://www. allaire. com.	ColdFusion SQL statements in dynamic queries.	**High**	Bug discussed in news-groups and Web sites.
Cisco with Living-ston[4]	ComOS 3.8.2-PM3 ComOS 3.7L-Or-HS	DoS will occur if a large number of bytes are directed at the access port such as port 23.	Restrict telnet access to only those machines requiring access. Configure "service tcp-keepalives-in" or configure timeouts on TTY lines.	Router Access Port DoS	Medium (Note: The risk may be higher under certain condi-tions.)	Bug discussed in news-groups. Exploit script posted to news-groups and Web sites.
Cyrix[5]	HARD-WARE CPU	No privileged users can cause systems to lock up by issuing three opcodes. Testing has not been done to determine if bug may be invoked remotely.	No work-arounds or patches known at time of publishing.	Cyrix bug	Low/ Medium	Bug discussed in news-groups. Exploit script posted to news-groups and Web sites.

Operating System	Software Name	Vulnerability/ Impact	Patches/ Workarounds/ Alerts	Common Name	Risk*	Attacks/ Scripts
HP-UX[6] 11.0/800	Operating System	Patch PHCO_13214 will change the file usr/bin/newgrp to suid root.	Informational comment only.	PHCO_ 13214 suid root	Low	Bug discussed in news-groups.
IPSWITCH	IMail[7]	Local user can gain unauthorized privileged access to IMail including account creation/ deletion and reading.	No workarounds or patches known at time of publishing.	IMail 1920 key attack	Medium	Bug discussed in news-groups and Web sites. Exploit script not required for the exploit.
Linux[8]	Operating System - PLP Line Printer Control (lpc) program under SuSE 5.2	Local user can gain root access through a buffer overflow.	Patch code available along with alternative line printer suites.	Linux /usr/bin/ lpc buffer overflow	Low/ Medium	Bug discussed in news-group. Exploit script posted to news-groups and Web sites.

Operating System	Software Name	Vulnerability/ Impact	Patches/ Workarounds/ Alerts	Common Name	Risk*	Attacks/ Scripts
Linux - Multiple vendors	wu-ftpd	Unauthorized user can gain root access through a buffer overflow.	Upgrade available for wu-ftpd. Red Hat updates available at: ftp://updates. redhat.com/ 5.2 Slackware updates available at: ftp//ftp.cdrom .com/pub/ linux/ slack-ware-current/ slakware/n8 SCO updates available at: ftp://ftp.sco. com/SSE/	Palmetto. ftpd - remote root overflow	**High**	Bug discussed in news-groups. Exploit script posted to news-groups and Web sites.
Linux[9] - RedHat 5.x and below	MILO/ Alpha	A local user running MILO can cause an Alpha based Linux machine to become unstable lock up or reboot.	New version available at: ftp://genie. ucd.ie/pub/ apha/milo/ milo-latest	Non-Privileged Halt of Linux/ Alpha	Low	Bug discussed in news-groups. Exploit script posted to news-groups and Web sites.
Microsoft[10]	BackOffice 4.0 Setup utility	The REBOOT.INI file that is created by the Setup utility contains plaintext passwords for the SQL Executive logon, Exchange Services Account, and possibly others.	No workarounds or patches known at time of publishing.	Microsoft REBOOT. INI	Me-dium/ High	Bug discussed in news-groups and Web sites.

Operating System	Software Name	Vulnerability/ Impact	Patches/ Workarounds/ Alerts	Common Name	Risk*	Attacks/ Scripts
Microsoft[11]	Internet Information Server (IIS)	Virtual servers in the same physical directory can leak information through the cache.	Disable the cache.	Active Server Pages (ASP) Caching BUG	Medium	Bug discussed in news-groups.
Microsoft[12]	IIS with Site Server 2.0	If an administrator does not create a directory called "users" then on the first successful upload a "users" directory will be created. This directory will default to the EVERYBODY group and this group will have change access (including write permission).	If not needed remove Site Server including the files: cpshost.dll uploadn.asp uploadx.asp upload.asp repost.asp po stinfo.asp If needed check that no directory accessible from the web has write permissions.	IIS and Site Server users directory write	Me-dium/ High	Bug discussed in news-groups. Exploit script posted to news-groups and Web sites.

Operating System	Software Name	Vulnerability/ Impact	Patches/ Workarounds/ Alerts	Common Name	Risk*	Attacks/ Scripts
Microsoft[13]	Word	Microsoft Word will issue a warning if a document containing a macro is opened. If a template contains a macro the user receives no such warning. The template can be referenced by URL and thus load the contents from the Internet with out the user's knowledge.	Patch available at: http://officeupdate.microsoft.com/nonIE4/Download Details/wd97spnonie4.htm Note: If you load Service Release 2 (SR-2), update this patch will be removed.	Word 97 template Vulnerability	**High**	Exploit Script now available. **Virus Bulletin has reported that at least two viruses exploit or use this hole as a means of distribution.**
Microsoft Windows[14] 95, 98, and NT[15]	IIS and File Transfer Protocol (FTP)	Refer to CyberNotes #3-99	Patch available for various versions at: ftp://ftp.microsoft.com/bussys/iss	IIS DoS FTP	Low/ Medium	Script identified at time of publishing but does not appear to be posted. Explanation of exploit available in newsgroups.

Operating System	Software Name	Vulnerability/ Impact	Patches/ Workarounds/ Alerts	Common Name	Risk*	Attacks/ Scripts
Microsoft Windows NT[16] 4.0	NT 4.0 with Service Pack 4 (SP4)	If a user password is changed from a DOS, Windows 3.1, Windows for Macintosh client then, it is possible to log in from a Windows NT machine to the user account without a password.	Patch available at: ftp://ftp.microsoft.com/bussys/winnt/winnt-public/fixes/usa/nt40/hotfixes-postSP 4/Msv1-fix/	Authentication Processing Error in Windows NT 4.0	Medium/ High	Bug discussed in news-groups and Web sites. Exploit script not required for the exploit.
Microsoft Windows NT[17] 4.0 with SP3 and SP4	Operating System	If a user changes his/her password and then locks the workstation, the new password will not be accepted.	The user will discover the problem the first time he/she attempts to unlock the workstation.	NT4 locking	Low	**No attacks using this bug have been reported.**
Multiple[18]	FTP PASV mode	Unauthorized user may use the PASV command of FTP to gain access to information or cause a DoS condition.	Various solutions suggested, including updating the Request for Comment (RFC).	FTP PASV "Pizza Thief"	Low	Bug discussed in news-groups. Exploit script posted to news-groups and Web sites.
Multiple[19]	Nobo (Back Orifice program detector)	If sufficient information is received in a short period of time, Nobo may not be able to process the information resulting in a DoS condition.	Nobo V1.3 corrects this problems.	Nobo buffer overflow	Low	Bug discussed in news-groups and Web sites.

Operating System	Software Name	Vulnerability/ Impact	Patches/ Workarounds/ Alerts	Common Name	Risk*	Attacks/ Scripts
Norton[20]	For Your Eyes Only[21]	Product contains backdoor password.	No workarounds or patches known at time of publishing.	Backdoor password for Norton For Your Eyes Only	Medium	Bug discussed in news-groups and Web sites. Password discussed on IRC channel.
Rational Software[22]	Clear Case v3.2	Unauthorized local user can gain root access. This condition allows a user to suid any file.	No workarounds or patches known at time of publishing.	Clear Case root exploitable race condition	High	Bug discussed in news-groups. Exploit script posted to news-groups and Web sites.
Seat-tleLabs[23]	SLMail 3.1	Sending the "vrfy" or "expn" commands composed of 2,041 characters will cause the program to terminate.	No workarounds or patches known at time of publishing.	SLMail Vulner-abilities	Low	Bug discussed in news-groups. Exploit script not required for the exploit.
Seat-tleLabs[24]	SLMail 3.1	A DoS and a buffer overflow condition exist. If the "helo" command is sent to port 27 followed by 819 to 849 characters, a potential DoS condition occurs. If 855 to 2,041 characters are sent, a buffer overflow occurs.	No workarounds or patches known at time of publishing. SeattleLabs representatives report that they are working on a fix and it will be included in Slmail 3.2 maintenance release.	SLMail Vulner-abilities	Low / Medium	Bug discussed in news-groups. Exploit script not required for the exploit.

Operating System	Software Name	Vulnerability/ Impact	Patches/ Workarounds/ Alerts	Common Name	Risk*	Attacks/ Scripts
Sun Solaris[25] 2.4 to 2.6	Common Desktop Environment (CDE)	Unauthorized user can gain root access through several vulnerabilities.	Patches available at: http://sunsolve.com/sunsolve/pubpatches/patches.html	Sun Solaris CDE	**High**	Bug discussed in newsgroups and Web sites.
Sun Solaris[26] 7, 7_x86, 2.6, 2.5.1, many others	Operating system (sdtcm_convert)	Unauthorized user can gain root access through a buffer overflow.	Patches available at: http://sunsolve.sun.com/sunsolve/pubpatches/patches.html.	sdtcm_convert buffer overflow	**High**	Bug discussed in newsgroups and Web sites.
Sun Solaris[27] 7, 7_x86, 2.6, 2.5.1, many others	Operating system (man/catman)	When run as root, the man or catman commands may overwrite arbitrary files.	Patches available at: http://sunsolve.sun.com/sunsolve/pubpatches/patches.html	Sun Solaris man/catman overwrite	Low/ Medium (Note: The full extent of this bug has not been reviewed.)	Bug discussed in newsgroups and Web sites.
Unix	Pine 4.04	If not truncated, a message with a From line containing 10,000 characters causes Pine to fail. It may be possible to place data in memory.	No workarounds or patches known at time of publishing.	Pine 10,000 character From line	Low/ Medium	Bug discussed in newsgroups and Web sites.

Operating System	Software Name	Vulnerability/ Impact	Patches/ Workarounds/ Alerts	Common Name	Risk*	Attacks/ Scripts
Unix[28]	HARD-WARE Modem	If a user has access to the modem tty when dialing into the system, he/she an get passwords or cause a DoS condition.	Workarounds are (1) provide login access through modems connected to terminal servers, (2) use the mgetty-1.1.20 programs provided "ptylogin" as the login program, or (3) use the rlogin program.	Unix Modem problem	Low	Script identified at time of publish-ing but does not appear to be posted. Explana-tion of exploit available in news-groups.
Unix[29]	Pine up to 4.1	The mailcap file distributed with the metamail MIME-support package, while implementing RFC-1524 correctly, introduces the possibility that an unauthorized user can manipulate the system to execute arbitrary commands.	Remove the metamail package from Pine. The metamail package is not required by Pine. If this is not possible, set the "mail-search-path" to a personal mailcap file path. Note - A number of patches have been provided by individu-als, but many of these patches only serve to move the vulner-ability down one nesting level.[30]	Remote exploit on Pine 4.1	Medium / High	Bug discussed in news-groups.

Operating System	Software Name	Vulnerability/ Impact	Patches/ Workarounds/ Alerts	Common Name	Risk*	Attacks/ Scripts
Unix[31]	Wget 1.5.3	When invoked with the -N option, it attempts to change the permissions of symlinks but changes permissions at the target files.	Workaround has been posted in several newsgroups.	Wget-1.5.3 chmod + symlinks	Low	Bug discussed in news-groups.
Unix[32]	WS_FTP Server 1.0.1.E and 1.0.2.E	A DoS condition occurs if a "cwd" command is composed of over 875 characters.	No workarounds or patches known at time of publishing.	WS_FTP Server Remote DoS attack	Low/ Medium	Bug discussed in news-groups and Web sites. Exploit script not required for the exploit.
Unix - HP-UX[33]	rpc.pcnfsd	Unauthorized remote user can gain root access. This vulnerability allows the printer spool directory to be made world writeable.	Patch available for this problem. User should consult the patch matrix from Hewlett-Packard.	HP-UX rpc.pcnfsd	**High**	Bug discussed in news-groups and Web sites.
Unix - Digital Unix[34] 4.0 Updated since issu-ance of Cy-berNotes #3-99	Operating System (inc)	Buffer overflow condition exists in the /usr/bin/mh/ inc file. This condition can allow an unauthorized user to gain root access.	No workarounds or patches known at time of publishing.	Digital Unix inc buffer overflow	**High**	Exploit Script now available. Explana-tion of exploit available in news-groups.

Operating System	Software Name	Vulnerability/ Impact	Patches/ Workarounds/ Alerts	Common Name	Risk*	Attacks/ Scripts
Unix - Digital Unix[35] 4.0 prior to 4.0D Updated since issuance of CyberNotes #3-99	Operating System (at)	Buffer overflow condition exists in the "at" program that can allow an unauthorized user to gain root access.	Upgrade to Digital Unix 4.0D or obtain appropriate patch at: ftp://ftp.servce.digital.com/public/dunix	Digital Unix at buffer overflow	**High**	Exploit Script now available. Explanation of exploit available in newsgroups.
Unix - NetBSD[36] 19980603 to 19990208	netstat	Non-privileged user can read any kernel memory location.	Update to a source newer than 19990208. Disable netstat for non-root users: chmod 555/usr/bin/netstat Patch available at: ftp://ftp.NetBSD.ORG/pub/NetBSD/misc/security/patches/19990208-netstat	NetBSD netstat kernel viewing	Medium	Bug discussed in newsgroups and Web sites.

Operating System	Software Name	Vulnerability/ Impact	Patches/ Workarounds/ Alerts	Common Name	Risk*	Attacks/ Scripts
Unix -IBM AIX[37]	Navio NC browser install	One of the configuration scripts for the Navio NC browser sets permission of the /tmp directory to NFS exported and world read/writeable.	Users should check to see if /tmp has been exported and change NFS and permissions where appropriate.	Navio NC /tmp permission change	Medium	Bug discussed in news-groups and Web sites. Exploit script not required for the exploit. Other related hacker tools can be utilized to gain privileged access.

*Risk is defined in the following manner:

High—A vulnerability that will allow an intruder to immediately gain privileged access (e.g., sysadmin, root) to the system. An example of this would be a vulnerability in which a sequence of instructions is sent to a machine by an unauthorized user and the machine responds with a command prompt.

Medium—Any vulnerability that will allow an intruder immediate access to the system that is not privileged access. This allows the intruder the opportunity to continue the attempt to gain root access. An example would be a configuration error that allows an intruder to capture the password file.

Low—A vulnerability that provides information to an intruder that could lead to further compromise attempts of a Denial-of-Service (DoS) attack. The reader should note that while the DoS attack is deemed low from a threat potential, the frequency of this type of attack is very high. DoS attacks against mission-critical nodes are not included in this rating and any attack of this nature should instead be considered as a "High" threat.

1Allaire Security Bulletin, ASB99-02.

2Allaire Security Bulletin, ASB99-01.

3Allaire Security Bulletin, ASB99-04.

4BUGTRAQ, February 5, 1999.

5BUGTRAQ, February 4, 1999.

6BUGTRAQ, February 5, 1999.

7ibid.

8BUGTRAQ, February 3, 1999.

[9]KSR[T] Advisory #009

[10]NTBUGTRAQ, February 9, 1999.

[11]NTBUGTRAQ, January 27, 1999.

[12]BUGTRAQ, January 30, 1999.

[13]Microsoft Security Bulletin, MS99-002.

[14]eEye Digital Security Team, Advisory Code IISE01.

[15]BUGTRAQ, January 25, 1999.

[16]Microsoft Security Bulletin (MS99-004).

[17]BUGTRAQ, February 2, 1999.

[18]Info War Security Advisory #01

[19]BUGTRAQ, February 4, 1999.

[20]IRC channel discussion, February 4, 1999.

[21]NTSECURITY, February 6, 1999.

[22]L0pht Advisory - Rational Software ClearCase root exploitable race conditions.

[23]eEye Digital Security Team, Advisory Code AD02041999.

24idid.

[25]Sun Microsystems, Inc. Security Bulletin #00185.

[26]Sun Microsystems, Inc. Security Bulletin #00183.

27Sun Microsystems, Inc. Security Bulletin #00184.

[28]BUGTRAQ, January 27, 1999.

29BUGTRAQ, February 8, 1999.

[30]BUGTRAQ, February 10, 1999.

[31]BUGTRAQ, February 2, 1999.

[32]eEye Digital Security Team, Advisory Code AD02021999.

[33]HP Daily Security Bulletins Digest, HPSBUX9902-091.

[34]BUGTRAQ, January 25, 1999.

[35]BUGTRAQ, January 25, 1999.

[36]NetBSD Security Advisory 1999-002.

[37]BUGTRAQ, January 29, 1999.

Recent Exploit Scripts

The table below contains a representative sample of exploit scripts, identified between January 27 and February 12, 1999, listed by date of script, script name, script description, and comments. **Items listed in boldface/red (if any) are attack scripts for which vendors, security vulnerability listservs, or Computer Emergency Response Teams (CERTs) have not published workarounds or patches, or represent scripts that hackers/crackers are utilizing.** During this period, 31 scripts, programs, and net-news messages containing holes or exploits were identified.

Date of Script (Reverse Chronological Order)	Script Name	Script Description	Comments
Feb. 11, 1999	All-access.c	Unix source code that will retrieve and decrypt Microsoft Access 97 database passwords	
Feb 11, 1999	All-access.exe	Win32 executable that will retrieve and decrypt Microsoft Access 97 database passwords.	
Feb 11, 1999	Ftpchk.pl	Denial-of-Servers (DoS) code used against ProFTPD servers	
Feb 11, 1999	Net-RawIP v0.05c	Perl module that manipulates raw Internet Protocol (IP) packets and Ethernet headers. The version is ported to Perl 5.005 and BSD, and includes the oshare script (causes Microsoft Windows 98 machines to lock).	
Feb 11, 1999	Novell Admin Exploit	Exploit code and description of vulnerability that allows a user to gain supervisor access to Novell NetWare 2.x, 3.x, and 4.x.	

Date of Script (Reverse Chronological Order)	Script Name	Script Description	Comments
Feb 10, 1999	Ms-access97-passwds.c	Program written in C that checks for Microsoft Access 97 database password weaknesses.	
Feb 10, 1999	Pepsi5.c	Random source host User Datagram Protocol (UDP) flooder.	
Feb 10, 1999	Pwrspoof.tgz	Domain Name Server (DNS) spoofer.	
Feb 8, 1999	Net-RawIP v0.05b	Perl module that manipulates raw IP packets and Ethernet headers.	
Feb 8, 1999	Nmap v2.06	Network-scanning tool that has a variety of scanning modes, including stealth, Xmas, and Null stealth. This release adds more operating system fingerprinting and several bug fixes. Note: This tool continues to be used by hackers. A number of systems become unstable when scanned if patches are not applied.	
Feb 8, 1999	RCR Bot v1.1	A plug-in for Back Orifice that is an IRC client. Once logged into a server, the infected machine can be remotely administered through IRC /msg or /query commands.	
Feb 6, 1999	Router-tcpaccess-DoS	Code for conducting DoS attacks against a wide variety of routers.	

Date of Script (Reverse Chronological Order)	Script Name	Script Description	Comments
Feb 6, 1999	Slmail	Text that describes the procedure for exploiting the SLMail vulnerability. See SLMail Vulnerabilities in "Bugs, Holes & Patches" above.	
Feb 6, 1999	Spoof.0.1.tar.gz	IP/UDP/ Transmission Control Protocol (TCP) spoofing.	
Feb 6, 1999	Thermoprog	Brute force password cracker.	
Feb 6, 1999	Windows-FAT-recursion	Explanation of how to perform a DoS attack against Microsoft Windows using FAT and allowing FTP uploads.	
Feb 4, 1999	Cprobe.sh	Scanning program that checks for 23 known Common Gateway Interface (CGI) script vulnerabilities.	
Feb 4, 1999	How to Break out	Detailed description of how to break out (user exceeds allowed access level) of restricted shells and menus.	
Feb 4, 1999	Net-RawIP v0.04e	Perl module that manipulates raw IP packets and Ethernet headers.	
Feb 4, 1999	NT4ALL	Code that allows any user (even Guest user) with write access to \WINNT\SYSTEM32 to log in with any password.	
Feb 4, 1999	pcapture	Tool for capturing packets on a network (network sniffer).	

Date of Script (Reverse Chronological Order)	Script Name	Script Description	Comments
Feb 4, 1999	Smashdu.c	Code that exploits several buffer overflow conditions in Digital Unix. See Unix - Digital Unix in "Bugs, Holes & Patches" above.	
Feb 3, 1999	lanlord v0.2-1	Identifies who owns machines at specific IP addresses. It runs on the Dynamic Host Configuration Protocol (DHCP) Server as a CGI script and uses CSS to modify output.	
Feb 3, 1999	Net-RawIP v0.05a	Perl module that manipulates raw IP packets and Ethernet headers. The version is ported to Perl 5.005 and BSD.	
Feb 1, 1999	Chronicle Remote Registry Query Tool v1.0b	Determines the current service pack/hotfix level of all Microsoft Windows NT machines within a domain.	
Feb 1, 1999	Nessus 990201	Security audit tool that has 180 plug-ins for checking known security holes.	
Feb 1, 1999	Slurpie v2.0b	Password cracker designed to run in a distributed environment.	
Feb 1, 1999	Spoofscan.c	Program that conducts portscans from a spoofed IP. It spoofs the address of a machine located on the same Ethernet segments and then sniffs the responses.	

Date of Script (Reverse Chronological Order)	Script Name	Script Description	Comments
Jan 30, 1999	Nmapstub,pl	Code that reads the nmap output and puts it into port and host structures. It then calls any routine that is requested, on a host by host basis.	
Jan 30, 1999	Pmap_tools	Collection of tools that checks for portmap/rpc/rpcbind vulnerabilities.	
Jan 28, 1999	GnuSniff	Network packet sniffer.	

Script Analysis

This section will supply a short description of scripts that have been analyzed by various security professionals and organizations. If you or your organization wish to contribute, please send e-mail to nipc@fbi.gov with the subject line CyberNotes Script Analysis. While this section is intended only for short descriptions, contributions should include a full technical analysis of the script along with release instructions. The release options are: releasable to everyone, limited (originator-defined list of organizations), or provided for NIPC only. If you would like to receive a copy of the full technical analysis version of any summarized analysis, please send an e-mail requesting the full technical analysis of "X" where "X" is the script name. A member of the CyberNotes editorial team will contact you. All contributions will be credited to the individual or organization unless otherwise requested.

Nmap—The output of a Nmap scan provides crucial information to the hacker. First, it provides a list of services that are active on the remote host. Second, by sending invalid tcp packets, Nmap performs a TCP stack analysis of the remote system. Since these anomalous packets are not covered by the RFCs, each operating system handles them differently. Nmap compares the responses to these packets against an internal database and provides a Ôbest guess' as to the operating system and version number running there. This combination allows the hacker to target the specific vulnerabilities on a given host, providing a higher success rate and a much lower attack signature. Finally, Nmap tells the user how difficult tcp sequence number prediction is for the remote host. This information can be used to target hosts that have a high potential for session hijacking. Such measures might be employed when a remote system has no vulnerable services running, or when it is shielded behind a firewall.[1]

Nmap was a winner of Infoworld's first annual Golden Guardian award and is described as:

> "the port scanner extraordinaire that we rely on regularly to get a quick birds-eye view of a network. Besides identifying open ports in every shape and form, nmap can identify OSes via TCP fingerprinting. The capability to send non-Request for Comment-compliant packets to an IP stack does have its downsides, however. It can hang some kernels, so use it carefully. Despite this rare condition, the capability will forever change how risk assessments are performed. With the exception of commercial vulnerability-detection tools, nothing else comes close to nmap for rapid network-security assessments."[2]

Trends

1. Several hackers/hacker groups appear to be using coordinated scans and probes from different sites.

2. Large numbers of scans and attacks continue to be directed at machines running the Linux operating system.

3. Scanning for Internet Message Access Protocol (IMAP) and POP continues.

4. Significant increase in reports of NetBus and Back Orifice scanning.

5. Significant increase in the number of scans directed specifically against Domain Name Servers.

6. Viruses are now being written to capture and transmit information.

Note: In the last edition of CyberNotes, in the "Trends" section #5, the acronym IMAP should have been defined as Internet Message Access Protocol.Excerpt from a report by the US Navy's SHADOW team, Naval

Viruses

In the last few weeks, there has been an increase in reports of many people receiving e-mail messages containing Trojan horse programs. Along with the increase in Trojan horse programs, several viruses have either been reported in the press or exhibit new techniques/trends that the reader should made

---

1 Surface Warfare Center - Dahlgren Division.

2 Infoworld, The more the merrier: 1998's Golden Guardian award given to three top security solutions, February 15, 1999.

be aware of. Anti-virus vendors recommend that their users update the anti-virus definition packages on a regular basis (every two weeks).

IE0199.exe Trojan horse - This message, which appears to be sent from IEsupport@microsoft.com, informs the recipient that the attached program fixes a number of bugs in Internet Explorer. The program, if run, deletes sndvol32.exe from the System32 directory and installs a file called sndvol.exe in the System directory. A Registry key %SystemRoot%\System\sndvol.exe is added to the infected machine. Once installed, the Trojan horse launches a DoS attack against a Web site in Bulgaria.

To remove the program manually, you must remove the sndvol.exe file from the System directory, remove the Registry key, and reinstall a clean copy of sndvol32.exe into the System32 directory. Most major anti-virus vendors have included code in their latest data file updates to detect this Trojan horse program.

Happy99.exe Trojan horse - This program has been distributed via e-mail, newsgroups and Web software distribution sites. Happy99.exe is the most common name of this Trojan horse, but reports have identified the use of other files names. When run, the program displays fireworks graphics and installs two files (ska.exe and ska.dll) in the System directory. The program then attempts to spam e-mail recipients and newsgroups to which the infected machine has sent or received messages. Sites with an e-mail filtering capability can filter messages with the header "X-Spanska: Yes." Most major anti-virus vendors have included code in their latest data file updates to detect this Trojan horse program.

WM97.Caligula (aka.WM97.Cali) - This virus has received a great deal of press in the last two weeks for its attempt to steal PGP keys and then send the keys via FTP back to a site on the Internet. On the 31st of each month, this virus displays the following:

"WM97/Caligulq ©Opic [CodeBreakers 1998]

"No cia,"

"No nsa,"

"No satellite,"

"Could map our veins."

This virus also disables the macro warning prompt to save the normal.dot Word template, disables ToolsMacro - VB Editor, and disables the menu items: ToolsMacro, ToolsCustomize, ViewToolbars. This virus has been posted on a number of virus writers' Web sites. Most major anti-virus vendors have included code in their latest data file updates to detect this virus program.

W97M.Footprint[3] **-** This virus overwrites the footers of all open documents and attempts to overwrite the macros in open documents. One of the indicators of infection is the placement of two files (footprint.$$$ and footprint.$$1) on the C drive. This virus has infected several US Government sites. Most major anti-virus vendors have included code in their latest data file updates to detect this virus program.

O97M/Triplicate - This is a new virus that, at the time of publication, **could not be detected by many anti-virus packages**. This virus is unique in that it is the first virus known to infect Microsoft Word, Excel, and PowerPoint files. Triplicate disables PowerPoint's macro protection, but this ability appears only to work from an infected Word file. Several anti-virus vendors' technical support centers were contacted and reported that detection for this new virus should be available in the near future.

---

3 CIAC advisory J-025.

CHAPTER 32

NATIONAL INFRASTRUCTURE PROTECTION CENTER CYBERNOTES

March 3, 1999

CyberNotes is published every two weeks by the National Infrastructure Protection Center (NIPC). Its mission is to support security and information system professionals with timely information on cyber vulnerabilities, hacker exploit scripts, hacker trends, virus information, and other critical infrastructure-related best practices.

You are encouraged to share this publication with colleagues in the information and infrastructure protection field. Electronic copies are available on the NIPC Web site at http://www.nipc.gov.

Please direct any inquiries regarding this publication to the Editor-CyberNotes, National Infrastructure Protection Center, FBI Building, Room 11719, 935 Pennsylvania Avenue, NW, Washington, DC, 20535.

Bugs, Holes & Patches

The following table provides a summary of software vulnerabilities identified between February 12 and February 26, 1999. The table provides the operating system, software name, potential vulnerability/impact, identified patches/-workarounds/alerts, common name of the vulnerability, potential risk, and an indication of whether attacks have utilized this vulnerability or an exploit script is known to exist. Software versions are identified if known. **This information is presented only as a summary; complete details are available from the source of the patch/workaround/alert, indicated in the footnote or linked site.** Please note that even if the method of attack has not been utilized or an exploit script is not currently widely available on the Internet, a potential vulnerability has been identified.

Operating System	Software Name	Vulnerability/ Impact	Patches/ Workarounds/Alerts	Common Name	Risk*	Attacks/ Scripts
Cobalt Networks, Inc.[1]	Cobalt RaQ	The web server can read the shell history file of any user. Under certain circumstances this may expose passwords to unauthorized remote users.	Patch is available at: ftp://ftp. cobaltnet.com/ pub/security	Cobalt RaQ shell history vulner-ability	Medium	Bug discussed in news-groups.
Computer Associates Interna-tional Inc.[2]	ARCserve NT agents	The transfer of username and password between the ARCserve server and the client machine can be sniffed. Username is sent in the clear and the password is hidden with a simple XOR.	Updated files available at: http:// support.cai. com/ Download/ patches	Username/ Password vulnerability in ARCserve	Med-ium/ High	Bug discussed in news-groups and Web sites. The XOR value has also been provided in news-group.
Intel[3]	Pentium III	The identifying processor serial number contained on the Pentium III chip can be turned off and on without the user's knowledge.	Intel recommends that computer manufactures place on-off switch in the system BIOS.	Pentium III serial number track	**Special**	**Script to turn serial number on/off and track user has been demon-strated.**
Internet Security Systems Inc.[4] (ISS)	Internet Scanner for Linux (trial version 5.3)	Local user can create a Denial-of-Service condition by placing certain files in the /tmp directory.	No workarounds or patches known at time of publishing.	ISS install.iss security hole	Low	Explanation of exploit available in news-groups.

Operating System	Software Name	Vulnerability/ Impact	Patches/ Workarounds/Alerts	Common Name	Risk*	Attacks/ Scripts
Irix[5] 66.2, 6.5.3 and other versions	Operating System (rpcbind)	Unauthorized remote user can insert and delete files by spoofing source address.	Filter 127.0.0.1 and local nets at border routers.	Rpcbind	Medium	Bug discussed in newsgroups. Exploit script posted to newsgroups and Web sites.
Kabsoftware[6]	Lydia	User password is stored in the file Lydia.ini. This file is usually found in C:\windows\lydia.ini. This file is encrypted.	No workarounds or patches known at time of publishing.	Lydia password file	Medium	Bug discussed in newsgroups and Web sites Explanation of encryption and decrypting code posted to newsgroups and Web sites.
Linux[7]	Operating System (autofs)	Unauthorized user can potentially gain root access through a buffer overflow. At a minimum this overflow will cause kernel errors.	Various temporary solutions are available in newsgroup. The author has not presented his fix at time of publishing	Linux autofs buffer overflow	**High**	Bug discussed in newsgroups. Exploit script posted to newsgroups and Web sites.

Operating System	Software Name	Vulnerability/ Impact	Patches/ Workarounds/Alerts	Common Name	Risk*	Attacks/ Scripts
Linux[8]	System Administration utility (Super)	Buffer overflow condition exists that will allow a local user to gain root access.	Patch located at: frp://ftp. ucoiik.org/ pub/users/ will/ To disable super you can execute the following as root. This will disable the setuid bit: Chmod 755/usr/bin/ super	Super local user buffer overflow	**High**	Bug discussed in news-groups. Exploit script posted to newsgroups and Web sites.
Linux[9]	TetriNet daemon "Tetrix"	Buffer overflow condition exists that will allow an unauthorized user to execute code resulting in possible root compromise.	No workarounds or patches known at time of publishing. Post patches as of February 25, 1999 do not correct the problem.	Tetrix 1.13.16 buffer overflow	**High**	Bug discussed in news-groups. Exploit script posted to newsgroups and Web sites.
Linux[10]	zgv	A privilege leaked to a child process can lead to a root compromise. The process is a processor privilege on Intel [iopl(3)] access to i/o ports.	A workaround is to disable the "-a" option.	Zgv privilege leak to child process	Medium (this attack cur-rently requires a high degree of skill to obtain root access)	Bug discussed in news-groups.

Operating System	Software Name	Vulnerability/ Impact	Patches/ Workarounds/Alerts	Common Name	Risk*	Attacks/ Scripts
Linux[11] – Debian 3.9.6 to 3.11.6	System Administration utility (Super)	Two buffer overflow conditions exist that will allow an unauthorized user to gain root access.	Patch located at: ftp://ftp.ucolick.org/pub/users/will/	Debian Linux "Super" buffer overflow	**High**	Bug discussed in newsgroups. Exploit script posted to newsgroups and Web sites.
Linux[12] Multiple	1sof Version 4.40 and prior	A buffer overflow condition exists that allows a local user to gain root access.	Patch available at: ftp://vic.cc.purdue.edu/pub/tools/unix/lsof/patches/4.40/arg.c.patch	1sof buffer overflow	**High**	Bug discussed in newsgroups. **Numerous exploit scripts posted to newsgroups and Web sites.**
Linux[13] 2.0.34	Operating System (rpcbind)	Unauthorized remote user can insert and delete files by spoofing source address.	Filter 127.0.0.1 and local nets at border routers.	Rpcbind	Medium	Bug discussed in newsgroups.
Microsoft[14]	Internet Explorer (IE) 4	Information in the clipboard is not accessible to IE unless IE was the program that placed the information in the clipboard. A simple java script will allow IE to access data in the clipboard even if IE does not own the information	Microsoft stated that a patch will be available with the next service pack release	IE4 vulnerability: The clipboard again	Low/ Medium	Bug discussed in newsgroups. Exploit script posted to newsgroups and Web sites.

Operating System	Software Name	Vulnerability/ Impact	Patches/ Workarounds/Alerts	Common Name	Risk*	Attacks/ Scripts
Microsoft Windows[15]	Explorer	If a user right clicks in explorer or attempts to delete a file longer than 217 characters explorer will crash.	No workarounds or patches known at time of publishing.	Windows Explorer 217 character limitation	Low	Bug discussed in news-groups and Web sites. Exploit script not required for the exploit.
Microsoft Win-dows[16] 95/98/NT	Windows Resource Kit (Taskpad)	Hostile Web site can run executables on the user's computer without the user's knowledge.	Patch is available at: ftp://ftp.microsoft.com/reskit	Taskpad Scripting Vulner-ability	Med-ium/ High	Bug discussed in news-groups and Web sites.
Microsoft Win-dows[17] 95, 98, and NT[18] **(change in risk level from Cy-berNotes #3-99 and #4-99)**	IIS and File Transfer Protocol (FTP)	Unauthorized remote user can log in (anonymous access included) and create a DoS condition. This may be a result of a stack overflow or bss (static pointers) overflow.	Patch available for various versions at: ftp://ftp.microsoft.com/bussys/iss	IIS DoS FTP	**High** (Mili-tary CERTs report this attack as re-sult-ing in system access.)	Explana-tion of exploit available in news-groups. Exploit script has been published.

Operating System	Software Name	Vulnerability/ Impact	Patches/ Workarounds/Alerts	Common Name	Risk*	Attacks/ Scripts
Microsoft Windows[19] NT[20]	Operating System (KnownDLL)	All users can read and write to the KnownDLLs list. A user can add malicious DLLs with the same name as system DLLs to gain higher privileges or execute other actions. Vulnerability is restricted to machines that the malicious user is interactively logged onto.	Microsoft is currently working on a patch. Strong protection is possible if the following is add to the registry key: HKEY_LOCAL_MACHINE\System\Current ControlSet\Control\Session Manager: Name: Protection Mode Type: REG_DWORD Value: 1	Windows NOT KnownDLL List Vulnerability	Medium	Bug discussed in newsgroups and Web sites.
NcFTP[21]	NcFTPd	An authorized remote user can execute arbitrary code. The extent of the possible compromise is limited by the program's character restriction. Further exploitation may be possible but has not been verified.	Patch available at: http://www.ncftp/com/down load Vendor reports that only one byte may be changed with this overflow	NcFTPd limited buffer overflow	Low	Bug discussed in newsgroups. Exploit script posted to newsgroups and Web sites.
NFR[22]	Network Flight Recorder (NFR) 2.0.2-Research	Unauthorized user can gain root access.	Patch available from the NFR web site at: http://www.nfr.com	Stack overflow in NFR Web Server	**High**	Bug discussed in newsgroups and Web sites.

Operating System	Software Name	Vulnerability/ Impact	Patches/ Workarounds/Alerts	Common Name	Risk*	Attacks/ Scripts
Seattle Lab[23]	SLMail 3.1 or 3.2 with Remote Administration Service enabled	SLMail allows changes to mail services using HTTP protocol over TCP port 180. An unauthorized user can change mail accounts and services. Other attacks are possible.	Remote administration should be disabled if possible. Other workarounds have been posted but their effectiveness has not been completely tested.	SLMail service change over port 180	Medium	Bug discussed in newsgroups. Exploit script posted to newsgroups and Web sites.
Smart-Max[24]	Mail-Max SMTP server for Windows	Buffer overflow condition exists that will allow an unauthorized user to execute code resulting in root compromise.	No workarounds or patches known at time of publishing.	Mail-Max remote buffer overflow	**High**	Bug discussed in newsgroups. Exploit script posted to newsgroups and Web sites.
Sun Solaris[25] 2.6	Operating System (rpcbind)	Unauthorized remote user can insert and delete files by spoofing source address.	Filter 127.0.0.1 and local nets at border routers.	Rpcbind	Medium/ High	Bug discussed in newsgroups. Exploit script posted to newsgroups and Web sites. **Hackers known to be using this attack.**

Operating System	Software Name	Vulnerability/ Impact	Patches/ Workarounds/Alerts	Common Name	Risk*	Attacks/ Scripts
Trend Micro[26]	InterScan VirusWall for Solaris	HTML request with two "GET" commands combined in the same request will result in the second file not being checked for viruses. Both Netscape Communicator and Microsoft Internet Explorer are known to formulate these types of requests during normal use.	**Vendor stated that they were originally unable to recreate this problem.** Patch posted at: http://www.antivirus.com	InterScan Viruswall second file failure	Medium	Normal browser use may result in virus infection.
Triactive[27] Remote Manage-ment Soft-ware	If Basic authentica-tion is used	the username and password for remote machines is stored in HKLM\SOFTWARE\TriActive\RemoteManager\Username.	Vendor recommends that users opt for the Challenge and Response authentication	Triactive's remote manger software plaintext password	Low/ Medium (high priority targets for hackers)	Bug discussed in news-groups and Web sites.

*Risk is defined in the following manner.

High—A vulnerability that will allow an intruder to immediately gain privileged access (e.g., sysadmin, root) to the system. An example of this would be a vulnerability in which a sequence of instructions is sent to a machine by an unauthorized user and the machine responds with a command prompt.

Medium—Any vulnerability that will allow an intruder immediate access to the system that is not privileged access. This allows the intruder the opportunity to continue the attempt to gain root access. An example would be a configuration error that allows an intruder to capture the password file.

Low—A vulnerability that provides information to an intruder that could lead to further compromise attempts or a Denial-of-Service (DoS) attack. The reader should note that while the DoS attack is deemed low from a threat potential, the frequency of this type of attack is very high. DoS attacks against mission-critical nodes are not included in this rating and any attack of this nature should instead be considered as a "High" threat.

1BUGTRAQ, February 25, 1999.

[2]BUGTRAQ, February 21, 1999.

[3]Reuters, Intel says probing alleged Pentium flaw, February 24, 1999.

[4]BUGTRAQ, February 20, 1999.

[5]BUGTRAQ, February 19, 1999.

[6]BUGTRAQ, February 20,1999.

[7]BUGTRAQ, February 18, 1999.

[8]BUGTRAQ, February 26, 1999.

[9]BUGTRAQ, February 17, 1999.

[10]BUGTRAQ, February 19, 1999.

[11]Debian GNU/Linux Security, February 15, 1999.

[12]HERT advisory # 002.

[13]BUGTRAQ, February 12, 1999.

[14]BUGTRAQ, February 22, 1999.

[15]BUGTRAQ, February 13, 1999.

[16]Microsoft Security Bulletin, MS99-007.

[17]Eye Digital Security Team, Advisory Code - IISEO1.

[18]BUGTRAQ, January 25, 1999.

[19]L0pht Security Advisory, February 18, 1999.

[20]Microsoft Security Bulletin, MS99-006.

[21]Proof of Concept Security Advisory, February 23, 1999.

[22]Network Associates, Inc., Security Advisory, February 16, 1999.

[23]NTBUGTRAQ, February 25, 1999.

[24]BUGTRAQ, February 14, 1999.

[25]BUGTRAQ, February 12, 1999.

[26]Backhats Security Advisory, February 22, 1999.

[27]BUGTRAQ, February 19, 1999.

Recent Exploit Scripts

The table below contains a representative sample of exploit scripts, identified between February 12 and February 26, 1999, listed by date of script, script name, script description, and comments. **Items listed in boldface/red (if any) are attack scripts for which vendors, security vulnerability listservs, or Computer Emergency Response Teams (CERTs) have not published workarounds or patches, or which represent scripts that hackers/crackers are utilizing.** During this period, 65 scripts, programs, and net-news messages containing holes or exploits were identified.

Date of Script (Reverse Chronological Order)	Script Name	Script Description	Comments
Feb 26, 1999	SDI-super.c	Exploit code for buffer overflow vulnerability that exists in the "super" program. See Debian Linux "Super" buffer overflow in "Bugs, Holes & Patches" above.	
Feb 25, 1999	Qwcrash.pl	Denial-of-Service program designed to disrupt quakeworld servers.	
Feb 25, 1999	pbomb.pl	Program that attempt to cause a Denial-of-Service condition by opening multiple connections to a given machine.	
Feb 25, 1999	Novell_login	Fake login screen for Novell that captures usernames and passwords in a file called os31337.sys.	
Feb 25, 1999	netdog.c	General hacker tool that includes scanner, IRC nukes, mailbomber and other tools.	
Feb 24, 1999	RAT	Remote administration tool similar to but less powerful than Back Orifice.	
Feb 24, 1999	blackhole.c	A small backdoor program.	
Feb 23, 1999	9x_c1sco.zip	Program that claims to kill all Cisco 7XX routers running IOS/700 v4.1(x).	
Feb 23, 1999	9x logn zip	A logind Trojan horse.	

Date of Script (Reverse Chronological Order)	Script Name	Script Description	Comments
Feb 23, 1999	Gammaprog 1.50	Bruteforce password cracker for web based e-mail systems and Post Office Protocol (POP) 3 servers.	
Feb 23, 1999	icmpush22.tgz.	Program that gathers information on a remote system by sending ICMP error packages.	
Feb 23, 1999	ie4.clipboard	Exploit code that allows remote attacker to view user's clipboard.	
Feb 23, 1999	Net-RawIP v0.06	Perl module that manipulates raw Internet Protocol (IP) packets and Ethernet headers. The version is ported to Perl 5.005 and BSD and includes the oshare script (causes Microsoft Windows 98 machines to lock).	
Feb 22, 1999	SDI-1sof.c.	Exploit code for the lsof buffer overflow vulnerability.	
Feb 22, 1999	Arcserve.nt.agents.txt	Explanation and exploit code for the ARCserver NT agents username and password vulnerability.	
Feb 22, 1999	Gammaink.zip	Graphic Front-end for the Gammaprog bruteforce password cracker.	
Feb 22, 1999	iss. exploit.c	Exploit code for the tmp-symlink problem in Internet Security Scanner (ISS) that can be used for Denial-of-Service or root compromise.	

Date of Script (Reverse Chronological Order)	Script Name	Script Description	Comments
Feb 22, 1999	1sof-xploit.c	Exploit code for the 1sof buffer overflow.	
Feb 22, 1999	Nmap.tcl	Script that facilitates Port scanning of IRC users.	
Feb 22, 1999	Sm4cl.c.	Program that claims to add three backdoors on machines that it is executed on (operating system not specified).	
Feb 21, 1999	Humpdee2.tgz	Improved version of exploit for the Linux rpc.mountd vulnerability.	
Feb 21, 1999	Imapx.c	Remote exploit for Linux imap vulnerability.	
Feb 20, 1999	Bncx.c	Exploit for Linux BNC.	
Feb 20, 1999	Humpdee.tgz	Exploit for Linux rpc.mountd vulnerability.	
Feb 20, 1999	vcu10.exe	A suite of password cracking utilities.	
Feb 20, 1999	Windows98.pingflood	Exploit code for Windows 98 pingflood Denial-of Service vulnerability.	
Feb 19, 1999	ADMsnmp.0.1.tgz	Tool for remote data gathering, brute force cracking and remote command execution.	
Feb 19, 1999	ADM-spoof-NEW.tgz	Packet spoofing program that includes Send/Spoof IP, GGP Echo packets, connection resetting, and SYN flood.	
Feb 19, 1999	cgiscan2.c	Scanner that checks for various common vulnerabilities in CGI scripts.	

Date of Script (Reverse Chronological Order)	Script Name	Script Description	Comments
Feb 19, 1999	jester.tgz	Backdoor server and spoofer.	
Feb 19, 1999	NetBusPro v2.0	Trojan horse program with a Graphical User Interface (GUI), proxy support, file manager, webcam capture, registry manager, hosts scheduler, application redirect, chat and many other features.	
Feb 19, 1999	Net-RawIP v0.05e.tar.gz	Perl module that manipulates raw Internet Protocol (IP) packets and Ethernet headers. The version is ported to Perl 5.005 and BSD, and includes the oshare script (causes Microsoft Windows 98 machines to lock).	
Feb 19, 1999	NT 4.0 DLL hack	Explanation and exploit script for NT 4.0 DLL cache permissions.	
Feb 19, 1999	Qps.gz	QPOP/UCB/SCO scanner.	
Feb 19, 1999	server.c	Backdoor server.	
Feb 19, 1999	xtyscreen.suse6	Program that can be used to overwrite files on SuSE6.	
Feb 18, 1999	ctontab-backdoor.sh	Shell script that binds a root shell to a selected port for a specific amount of time (crontab controlled).	
Feb 18, 1999	junk.tar.gz	Exploit scripts for the mail.local security hole.	

Date of Script (Reverse Chronological Order)	Script Name	Script Description	Comments
Feb 18, 1999	listerine.tar.gz	Script that tests for the NcFTPd security hole.	
Feb 18, 1999	mailfrm.tar.gz	Exploit scripts for the mail.local security hole.	
Feb 18, 1999	nbtscan.zip	Automates the running of "nbtstat–A" on a set of IP addresses.	
Feb 18, 1999	Net-RawIP v0.05d	See description above	
Feb 18, 1999	Netscape.window.spoof	Explanation and exploit scripts of the Netscape communicator window spoofing bug using HTML and JavaScript.	
Feb 18, 1999	nfr.sof	Explanation and exploit scripts that allows a remote user to gain system management privileges of a Network Flight Recorder v2.0.2 in default configuration.	
Feb 18, 1999	Nmap-2.08.tgz	Network-scanning tool that has a variety of scanning modes, including stealth, Xmas, and Null stealth. This release adds more operating system fingerprinting and several bug fixes. Note: This tool continues to be used by hackers. A number of systems become unstable when scanned if patches are not applied.	

Date of Script (Reverse Chronological Order)	Script Name	Script Description	Comments
Feb 18, 1999	tcplogd.0.0.tar.gz	Stealth-scan detector.	
Feb 18, 1999	Topdesk.passwd	Description of decryption technique for TOPdesk.	
Feb 18, 1999	vprll.zip	Unix password cracker for use on Microsoft Windows machines.	
Feb 17, 1999	1sof	Exploit program for the 1sof vulnerability. See 1sof buffer overflow in "Bugs, Holes & Patches" above.	
Feb 17, 1999	Websitepro	Exploit program that allows remote user to replace files on some websites.	
Feb 16, 1999	Cain v1.0	Password utility that recovers Windows 95/98 passwords including logon, Share, screen savers, dialup and link passwords.	
Feb 16, 1999	inetd.DoS.c	Denial-of-Service program that exploits the inetd buffer overflow.	
Feb 15, 1999	defunct.cpp	Program that attempts to circumvent security measures by translating IP address to raw addresses (identifier). Overrides restrictions set by proxies.	
Feb 15, 1999	exscan-0.4.tar.gz	Port scanner with a strobe scanning capability.	

Date of Script (Reverse Chronological Order)	Script Name	Script Description	Comments
Feb 15, 1999	mailmaxbof.c	Exploit code for the Mial-Max SMTP server buffer overflow.	
Feb 15, 1999	Thetaprog.tgz	Bruteforce password cracker for Hotmail password reminder.	
Feb 15, 1999	tracerouteflood.c	Exploit code that will allow any user to use traceroute as a UDP or ICMP flooder.	
Feb 12, 1999	ActiveX.file.system.object	Exploit code that uses the FileSystemObject vulnerability to modify files on an Active Server Page (ASP) web server.	
Feb 12, 1999	dsu_sat	Exploit code that causes a binary overflow resulting in a terminated DSU satellite communication.	
Feb 12, 1999	Fakebo.bof.c	Exploit code that uses a buffer overflow in fakebo to execute code that results in root access.	
Feb 12, 1999	NT explorer DoS	Exploit code for Windows NT explorer long filename vulnerability.	
Feb 12, 1999	s10scan.pl	Port scanning code that adds address of false machines in an effort to mislead logging programs.	
Feb 12, 1999	sbounce.pl	Perl coded program that functions as an IRC bounce.	

Date of Script (Reverse Chronological Order)	Script Name	Script Description	Comments
Feb 12, 1999	Serve-U.DoS	Exploit code for Serve-U FTP software that results in a Denial-of-Service condition.	
Feb 12, 1999	Ultraprog.zip	Brute force cracker for the MailCity password reminder.	

Script Analysis

This section will supply a short description of scripts that have been analyzed by various security professionals and organizations. If you or your organization wish to contribute, please send e-mail to nipc@fbi.gov with the subject line "CyberNotes Script Analysis." While this section will list only short descriptions, contributors are requested to include a full technical analysis of the script along with release instructions. The release categories are: releasable to anyone; limited releasability (originator-defined list of organizations); or provided for NIPC only. If you would like to receive a copy of the full technical analysis version of any summarized analysis, please send an e-mail listing the script name and requesting the full technical analysis. A member of the CyberNotes editorial team will contact you. All contributions will be credited to the contributing individual or organization unless otherwise requested.

Trends

Trends for this two week period remain the same those trends two weeks ago.

1. Several hackers/hacker groups appear to be using coordinated scans and probes from different sites.

2. Large numbers of scans and attacks continue to be directed at machines running the Linux operating system.

3. Scanning for Internet Message Access Protocol (IMAP) and POP continues.

4. Significant increase in reports of NetBus and Back Orifice scanning.

5. Significant increase in the number of scans directed specifically against Domain Name Servers.

6. Viruses are now being written to capture and transmit information.

Viruses

A list of the top ten viruses infecting two or more sites as reported to various anti-virus vendors has been categorized into the two tables below. The first table list macro viruses, and the second table lists other viruses. Macro viruses have, historically, spread fastest due to their ability to be transferred by e-mail.

For the purposes of collecting and collating data, infections involving multiple systems at a single location are considered a single infection. The tables list the viruses by: ranking (number of sites affected), common virus name, type of virus (i.e., boot, file, macro, multi-partite), trends (based on number of infections during the last three months reported), and approximate date first found.

Note: Virus reporting is normally 6 to 8 weeks behind the first discovery of infection. With the number of viruses that appear each month, it is possible that a new virus will become widely distributed before the next edition of this publication. To limit the possibility of infection, readers are reminded to update their anti-virus packages, as updates become available.

The viruses listed in the virus table have infected over 510 sites in January, which is a slight increase in the number of reported infections from the last prevalence table. The number 1 ranked virus for January/February accounted for 36 infected sites, and the last virus listed in the tables infected 19 sites. A total of 467 distinct viruses were reported this month, infecting over 2,000 sites.

Table 1–Macro viruses:

Ranking	Common Virus Name	Type of Virus	Trends	Date
1	CAP	Macro	Steady	April 1997
2	Class	Macro	Increasing	September 1998
3	ColdApe	Macro	Increasing	December 1998
4	Laroux	Macro	Increasing	July 1997
5	Temple	Macro	Increasing	December 1998
6	Npad	Macro	Steady	December 1996
7	Concept	Macro	Decreasing	December 1996
8	Hark	Macro	Decreasing	December 1998
9	Munch	Macro	Increasing	October 1998
10	Wazzu	Macro	Steady	December 1996

Table 2–Other viruses:

Ranking	Common Virus Name	Type of Virus	Trends	Date
1	One half	Multi	Steady	October 1995
2	Form	Boot	Steady	September 1991
3	Junkie	Multi	Steady	July 1994
4	W95/CIH	File	Increasing	July 1998
5	AntiCMOS	Boot	Steady	October 1995
6	Parity_Boot	Boot	Steady	September 1993
7	AntiEXE	Boot	Decreasing	September 1994
8	Ripper	Boot	Steady	March 1994
9	Empire.Monkey	Boot	Steady	July 1994
10	DelCMOS.B	Boot	Increasing	January 1999

Seagate Backup Exec Warning—Users of Seagate Backup Exec software should be aware that the virus data file downloaded between January 23rd and February 9th contains an incorrect virus signature. This virus data file will cause all Visual Basic class files to be identified as containing a virus. Normal operation of the Network Associates VirusScan API as integrated with Seagate's Backup Exec is to first attempt to clean an infected file. If the file(s) can not be cleaned prior to backup they are deleted from the system.

CHAPTER 33

CRITICAL INFRASTRUCTURE PROTECTION AND INFORMATION WARFARE ISSUES

Statement for the Record of
Michael A. Vatis
Director, National Infrastructure Protection Center
Federal Bureau of Investigation

before the
Senate Armed Service Committee,
Subcommittee on Emerging Threats and Capabilities

Washington, D. C.
March 16, 1999

Introduction

Mr. Chairman, Senator Bingaman, and Members of the Subcommittee: Thank you for inviting me here today to discuss critical infrastructure protection and information warfare issues. My brief remarks will focus on two areas: the role of the NIPC under Presidential Decision Directive-63 (PDD- 63), and current impediments to critical infrastructure protection.

NIPC and PDD-63

PDD-63 creates an unprecedented set of intra-governmental as well as public-private cooperative structures for the vital mission of critical infrastructure protection. Let me begin by reviewing the roles assigned to the NIPC and the other key players in infrastructure protection.
PDD-63 authorized the expansion of the FBI's former organization, the Computer Investigations and Infrastructure Threat Assessment Center, into a full-scale National Infrastructure Protection Center. The PDD states that the NIPC "[s]hall serve as a national critical infrastructure threat assessment, warning, vulnerability, and law enforcement investigation and response entity." It further states that the mission of the NIPC "will include providing timely warnings of intentional threats, comprehensive analyses and law enforcement investigation and response."

Thus, the PDD places the NIPC at the core of the government's warning, threat investigation, and response system for threats to, or attacks on, the nation's critical infrastructures. The NIPC is the focal point for gathering information on threats to the infrastructures as well as "facilitating and coordinating the federal government's response to an incident." The NIPC is also responsible for "mitigating attacks, investigating threats and monitoring reconstitution efforts." The PDD further specifies that the NIPC should include "elements responsible for warning, analysis, computer investigation, coordinating emergency response, training, outreach, and development and application of technical tools."

The NIPC has a vital role in collecting and disseminating information from all relevant sources. Thus, the PDD directs the NIPC to "sanitize law enforcement and intelligence information for inclusion into analyses and reports that it will provide, in appropriate form, to relevant federal, state, and local agencies; the relevant owners and operators of critical infrastructures; and to any private sector information sharing and analysis entity." The NIPC is also charged with issuing "attack warnings or alerts to increases in threat condition to any private sector information sharing and analysis entity and to the owners and operators."

In order to perform its role, the NIPC is establishing a network of relationships with a wide range of entities in both the government and the private sector. The PDD provides for this in several ways. First, it states that the Center will "include representatives from the FBI, U.S. Secret Service, and other investigators experienced in computer crimes and infrastructure protection, as well as representatives detailed from the Department of Defense, Intelligence Community and lead agencies."[1] Second, the NIPC will be "linked electronically to the rest of the government, including warning and operations centers as well as any private sector information sharing centers." Third, all executive departments and agencies are mandated to "cooperate with the NIPC and provide it assistance, information, and advice that the NIPC may request, to the extent permitted by law." Fourth, all executive departments are also mandated to "share with the NIPC information about threats and warning of attacks and actual attacks on critical government and private sector infrastructures, to the extent permitted by law." To ensure that the flow of information is unimpeded—which is imperative when dealing with cyber attacks—the PDD authorizes the NIPC to "establish its own relations directly with others in the private sector and with any information sharing and analysis entity that the private sector may create."

---

1 The lead agencies are: Commerce for information and communications; Treasury for banking and finance; EPA for water supply; Transportation for aviation, highways, mass transit, pipelines, rail, waterborne commerce; Justice/FBI for emergency law enforcement services; Federal Emergency Management Agency for emergency fire service and continuity of government; Health and Human Services for public health services. The lead agencies for special functions are: State for foreign affairs, CIA for intelligence, Defense for national defense, and Justice/FBI for law enforcement and internal security. The NIPC is performing the lead agency and special functions roles specified for "Justice/FBI" in the PDD.

Let me address briefly why the NIPC is located at the FBI. First, as you know, the FBI has had existing programs and authorities to investigate computer crimes and to prevent and investigate acts of espionage and terrorism. These programs and authorities naturally support and mesh with the infrastructure protection mission. Second, in most cyber attacks, the identity, location, and objective of the perpetrator are not immediately apparent. Nor is the scope of his attack—i.e., whether an intrusion is isolated or part of a broader pattern affecting numerous targets. This means it is often impossible to determine at the outset if an intrusion is an act of vandalism, organized crime, domestic or foreign terrorism, economic or traditional espionage, or some form of strategic military attack. The only way to determine the source, nature, and scope of the incident is to gather information from the victim sites and intermediate sites such as Internet Service Providers and telecommunications carriers. Under our constitutional system, gathering such information usually requires some form of legal authority—either criminal investigative or foreign counterintelligence. Thus, the NIPC is housed in the FBI to enable it to utilize the appropriate authorities to gather and retain the necessary information and to act on it. Now, this does not mean that the ultimate response to a cyber attack is limited to criminal investigation and prosecution. The response will be determined by the facts that are uncovered. Thus, for instance, if it is determined that a cyber intrusion is part of a strategic military attack, the President may determine that a military response is called for. But no such determination can be made without adequate factual foundation, and the NIPC's role is to coordinate the process for gathering the facts, analyzing them and making determinations about what is going on, and determining what responses are appropriate.

This role clearly requires the involvement and expertise of many agencies other than the FBI. This is why the NIPC, though housed at the FBI, is an interagency center that brings together personnel from all the relevant agencies. Thus, the Deputy Director is a civilian detailee from the Department of Defense; the Chief of our Analysis and Warning Section is a senior CIA analyst; and managers, investigators, analysts, and computer scientists within the Center come from across the defense, intelligence, and law enforcement communities. In addition, we are seeking infrastructure and technical experts from each of the infrastructure sectors to enhance our ability to understand and coordinate with the owners and operators of the infrastructures. Currently, the NIPC has representatives from multiple government agencies, including FBI, DOD, NSA, DOE, and CIA as well as federal and state law enforcement, including the U.S. Secret Service, the U.S. Postal Service, and, until recently, the Oregon State Police. Private sector representatives are also being sought. In fact, just yesterday the Attorney General and the Information Technology Association of America announced a set of initiatives as part of a "Cybercitizens Partnership" between the government and the information technology (IT) industry. One initiative involves providing IT industry representatives to serve in the NIPC to enhance our technical expertise and our understanding of the information and communications infrastructure. This interagency, public-private composition will ensure that we are able to obtain information necessary to our mission from all relevant sources—criminal investigations, intelligence sources, open sources, automated intrusion detection systems, and private sector contacts— and that we are poised to coordinate closely with the other agencies that may need to participate in the response to an incident.

Other entities are also created by the PDD. The National Coordinator for Security, Infrastructure Protection, and Counter-Terrorism is responsible for overall policy implementation of the PDD. In this capacity he chairs the interagency Critical Infrastructure Coordinating Group. The PDD also created a National Planning Staff (renamed the Critical Infrastructure Assurance Office, or CIAO) to assist the National Coordinator in this policy function by coordinating the drafting of a "national plan" and the implementation of a national education and awareness program. The national plan is currently in the drafting process and is the subject of ongoing interagency discussions.

The PDD also designates certain agencies as the "lead agencies" for each infrastructure sector. These agencies (listed in footnote 1 on page 2) are charged with working with their respective Sectors (via a "Sector Coordinator" chosen to represent the sector) to: assess sector vulnerabilities and develop a plan to eliminate the significant ones; propose a system for identifying and preventing attempted major attacks; and develop a plan for alerting, containing and rebuffing an attack in progress and then reconstituting minimum essential capabilities in the aftermath of an attack. Given its roles in the areas of vulnerability, warning, response, and reconstitution monitoring, the NIPC needs to work closely with the Sector Coordinators and Liaisons in the development, implementation, and testing of their plans.

Finally, under the PDD the federal government is encouraging the creation of one or more Information Sharing and Analysis Centers (ISACs) by the private sector. As envisioned, the ISAC(s) "could serve as a mechanism for gathering, analyzing, appropriately sanitizing and disseminating private sector information to both industry and the NIPC." ISACs could also serve to further disseminate NIPC information to industry. The provision of timely and complete information to the NIPC is critical for the success of its mission, and the PDD states that the ISACs are "not to interfere with direct information exchanges between companies and the government." As the government and private sector consider possible models for an ISAC, it is critical that nothing be created that would impede or delay the flow of incident and threat information to and from the NIPC. Rather, any ISAC should be designed to expedite the flow of information to enable real-time detection, analysis, and response by the NIPC.

Status of the NIPC and its Implementation of the PDD.

To accomplish its goals under the PDD, the NIPC is organized into three sections:

- The Computer Investigations and Operations Section (CIOS) is the operational and response arm of the Center. It program manages computer intrusion investigations conducted by FBI Field Offices throughout the country; provides subject matter experts, equipment, and technical support to cyber investigators in federal, state, and local government agencies involved in critical infrastructure protection; and provides a cyber emergency response capability to help resolve a cyber incident.
- The Analysis and Warning Section (AWS) serves as the indications and warning arm of the NIPC, provides analytical support during computer intrusion investigations, and performs long- term analyses of vulnerability and threat trends. When appropriate, it distributes tactical warnings and analyses to all the relevant partners, informing them of potential vulnerabilities and threats and long-term trends. It also reviews numerous government and private sector databases, media, and other sources daily to gather information that may be relevant to any aspect of our mission, including the gathering of indications of a possible attack.
- The Training, Administration, and Outreach Section (TAOS) coordinates the training and education of cyber investigators within the FBI Field Offices and other federal, state and local law enforcement agencies. It also coordinates our outreach to private sector companies, state and local governments, other government agencies, and the FBI's field offices. In addition, this section manages our collection and cataloging of information concerning "key assets"—i.e., critical individual components within each infrastructure sector, such as specific power grids, telecommunications switch nodes, or financial systems—across the country.

The NIPC is also developing its threat assessment, analytical, and warning capabilities. NIPC assessments form the basis for a variety of products, including alerts and advisories, an Infrastructure Protection Digest, a Y2K Report, a weekly update, CyberNotes, and topical electronic reports. These products are designed for tiered distribution to both government and private sector entities consistent with applicable law through the NIPC Watch and Warning Unit. For example, the Infrastructure Protection Digest is a quarterly publication for sharing analysis and information on critical infrastructure issues. The Digest provides analytical insights into major trends and events affecting the nation's critical infrastructures. It is published in a classified format and reaches national security and civilian government agency officials. CyberNotes is another NIPC publication designed to provide security and information system professionals with timely information on cyber vulnerabilities, hacker exploit scripts, hacker trends, virus information, and other critical infrastructure-related best practices. It is published twice a month on our website and disseminated hardcopy to government and private sector audiences.

In addition, the NIPC is developing processes to ensure that we get relevant information in real time or near real time from all relevant sources, including: the US Intelligence Community, FBI criminal investigations, the private sector, other federal agencies, emerging intrusion detection systems, and open sources. This information is quickly analyzed to determine if a broad scale attack is underway. If we determine an attack is underway, we can issue warnings using an array of mechanisms, and send out sanitized and unsantized warnings to the appropriate parties in Federal Government and the private sector so they can take immediate protective steps. This is a difficult process requiring the design of both procedures for reporting and sanitization, and collection and distribution mechanisms. The NIPC is currently working on these procedures and mechanisms. The long-term goal is to develop a comprehensive "indications and warning" system. This will require participation by the Intelligence Community, DOD, the sector lead agencies, other government agencies, federal, State and local law enforcement, and the private sector owners and operators of the infrastructure. Currently, the NIPC is focusing on developing and implementing a methodology and system for detecting and warning of attacks on the federal government and the national telecommunications and electric power sectors.

Response is central to the NIPC mission. To facilitate our ability to investigate and respond to attacks, the FBI has created a National Infrastructure Protection and Computer Intrusion Program in the 56 FBI field offices across the country. Under this program, managed by the NIPC at FBIHQ, full "NIPCI" squads or smaller teams have been created in each field office to conduct computer intrusion investigations, respond to threats, and collect information on "key assets" within each sector. There are currently 10 full NIPCI squads in Washington DC, New York, San Francisco, Chicago, Dallas, Los Angeles, Atlanta, Charlotte, Boston, and Seattle. The other field offices have smaller teams. The 10 squads have regional responsibilities, assisting the smaller teams in other offices when an incident exceeds the smaller team's resources or capabilities. Ultimately, we need to create a full squad in each field office. During the first nine months of 1998 the NIPCI squads and teams opened 377 new cases, closed 304 cases and had a pending caseload of 526 matters. Currently, there are 680 pending investigations of computer intrusion matters. The pending caseload is expected to markedly increase in the coming years.

The program to protect and respond to *physical* attacks on the US critical infrastructure are handled by the FBI's counter-terrorism program. The NIPC supports this initiative through its management of the Key Asset Program (KAP). A key asset can be defined as an organization, group of organizations, system, or group of systems, or physical plant the loss of which would have widespread and dire economic or social impact on a national, regional, or local basis. The KAP initially will involve determining which assets are key within the jurisdiction of each FBI field office, obtaining 24-hour points of contact at each asset in cases of emergency. Eventually, if resources permit, the Program would include the development of contingency plans to respond to attacks on each asset, exercises to test response plans, and modeling to determine the effects of an attack on particular assets. FBI Field Offices will be responsible for developing a list of the assets within their respective jurisdictions, while the NIPC will maintain the national database. This program will be developed in coordination with DOD and other agencies. This program serves the critical needs of developing lists of the key assets within each critical infrastructure and also of developing the communications and liaison links necessary for the collection of information and the dissemination of warnings to the infrastructure owners and operators.

The FBI, in conjunction with the private sector, has also developed an initiative called "InfraGard" to expand direct contacts with the private sector infrastructure owners and operators and to share information about cyber intrusions, exploited vulnerabilities, and physical infrastructure threats. The initiative encourages the exchange of information by government and private sector members through the formation of local InfraGard chapters within the jurisdiction of each Field Office. Chapter membership includes representatives from the FBI, private industry, other government agencies, State and local law enforcement, and the academic community. The initiative provides four basic services to its members: an intrusion alert network using encrypted e-mail; a secure website; local chapter activities; and a help desk for questions. The critical component of InfraGard is the ability of industry to provide information on intrusions to the NIPC and local FBI field office using secure communications in both a "sanitized" and detailed format. The local FBI Field Offices can, if appropriate, use the detailed version to initiate an investigation; while the

NIPC can analyze that information in conjunction with other law enforcement, intelligence, or industry information to determine if the intrusion is part of a broader attack on numerous sites. The NIPC can simultaneously use the *sanitized* version to inform other members of the intrusion without compromising the confidentiality of the reporting company. InfraGard, which began as a pilot program in the Cleveland, Cincinnati, and Indianapolis field offices, will be expanded to 14 additional offices this month, and to the rest of the country later this year.

The NIPC also serves as the U.S. government lead agency for the Emergency Law Enforcement Services Sector. As Sector Liaison for law enforcement, the NIPC and a Sector Coordinator representing the law enforcement sector are formulating a plan to reduce vulnerabilities of state and local law enforcement to attack and developing methods and procedures to share information within the sector. The NIPC and the FBI Field Offices are also working with the State and local law enforcement agencies to raise awareness with regard to vulnerabilities in this sector.

The NIPC has also been very active in training. Training FBI and other agencies' investigators is critical if we hope to keep pace with the rapidly changing technology and be able to respond quickly and effectively to computer intrusions. The NIPC trained 170 FBI agents and 17 representatives from other law enforcement agencies in 1998. We currently plan to train over 1000 law enforcement personnel in 1999 at the federal, state, and local levels. Additional training initiatives include specialized courses in information security developed by the private sector. Together, these efforts will help place us at the cutting edge of law enforcement and national security in the 21st Century.

Policy and Statutory Impediments to Combating Threats to the Critical Information Infrastructure

There are several policy and statutory impediments to our being able to fully address the threats to the critical information infrastructure.

Hiring sufficient personnel for the National Infrastructure Protection Center and for the nationwide National Infrastructure and Computer Intrusion Program continues to be a major concern. The prevention, detection, analysis, warning, and response missions assigned to the NIPC and FBI field offices all require a large number of skilled personnel. Currently, we believe there are far more intrusions occurring than we know about or can investigate. Additional personnel are therefore a vital need if we are to learn about, investigate, and respond to attacks on our infrastructures. As use of the Internet continues to increase dramatically, the number of intrusions will grow even more, and our capability must keep pace.

I should note that some of the shortfall could be met with detailees from other agencies. Congress has prohibited us, however, from reimbursing other agencies for detailees in FY 99, which has naturally made it somewhat more difficult for other agencies to devote scarce resources to our common mission at the NIPC.

There are a number of statutory issues related to protecting the infrastructure. Fortunately, a number of agencies are focused on identifying these concerns with an aim towards working with the Congress to consider legislative fixes. The NIPC is coordinating in this regard with, among others, the Computer Crime and Intellectual Property Section of the Department of Justice's Criminal Division, the CIAO and the Security Policy Board.

Examples of some of the issues the NIPC or other members of the infrastructure protection community are concerned with include:

- the updating of federal trap and trace and pen register authorities in order to take account of new information technologies;
- the need for multi-jurisdictional pen register and trap and trace orders rather than multiple orders each covering one jurisdiction;
- the need to address sentencing issues regarding minors who commit computer crimes;
- the need to criminalize unauthorized computer access to sensitive computer and information networks when it is difficult to put a dollar value on the harm (since jurisdiction over many types of computer crimes currently attaches only at the $5,000 mark);
- the need to create criminal forfeiture provisions for violations of the Computer Fraud and Abuse Act, so we can seize computers of convicted computer criminals; and
- the need to clarify current law to unambiguously permit the United States to conduct domestic investigations and prosecutions when a United States computer is not itself the target of a computer crime but is used as a conduit to attack systems abroad.

Conclusion

PDD-63 established the NIPC as the operational linchpin of our efforts to protect America's critical infrastructures in the 21st century. Ours is a national mission to combine the inputs from the government lead agencies and the private sector in order to provide analyses and warnings and to respond to an intrusion incident. But the NIPC can perform this mission only if has the necessary resources, interagency support, and information from multiple sources. I believe we have made significant progress in the first year of our existence in establishing the foundation for an effective system for preventing, detecting, and responding to cyber attacks. In just this past year, we have brought on board over 100 personnel from many agencies at NIPC HQ; established a national program for computer investigations in every FBI field office; developed and delivered advanced training in network investigations to nearly 200 FBI and other government agency investigators; developed several mechanisms and programs to share information with the private sector; begun a program to protect "key assets" with each infrastructure sector from cyber attack; and coordinated several national-level investigations involving numerous agencies and FBI field offices. While much as been accomplished, however, much work remains in developing our detection, prevention, warning, and response capabilities. I look forward to working with this Subcommittee and the Congress in protecting our national security against this difficult challenge.

Thank you.

CHAPTER 34

NATIONAL INFRASTRUCTURE PROTECTION CENTER CYBERNOTES

March 17, 1999

CyberNotes is published every two weeks by the National Infrastructure Protection Center (NIPC). Its mission is to support security and information system professionals with timely information on cyber vulnerabilities, hacker exploit scripts, hacker trends, virus information, and other critical infrastructure-related best practices.

You are encouraged to share this publication with colleagues in the information and infrastructure Protection field. Electronic copies are available on the NIPC Web site at http://www.nipc.gov.

Please direct any inquiries regarding this publications to the Editor-CyberNotes, National Infrastructure Protection Center, FBI Building, Room 11719, 935 Pennsylvania Avenue, NW, Washington, DC, 20535.

Bugs, Holes & Patches

The following table provides a summary of software vulnerabilities identified between February 27 and March 12, 1999. The table provides the operating system, software name, potential vulnerability/impact, identified patches/workarounds/alerts, common name of the vulnerability, potential risk, and an indication of whether attacks have utilized this vulnerability or an exploit script is known to exist. Software versions are identified if known. **This information is presented only as a summary; complete details are available from the source of the patch/workaround/alert, indicated in the footnote or linked site.** Please note that even if the method of attack has not been utilized or an exploit script is not currently widely available on the Internet, a potential vulnerability has been identified.

Operating System	Software Name	Vulnerability / Impact	Patches/ Workarounds/ Alerts	Common Name	Risk*	Attacks/ Scripts
Bay Networks[1]	Baystack 350T (SW: V1.2.0.10 and SW: V2.0.0.15)	System contains default passwords.	No workarounds or patches known at time of publishing.	Bay Networks default password	**High**	Default passwords available on Web sites and newsgroups.

Operating System	Software Name	Vulnerability / Impact	Patches/ Workarounds/ Alerts	Common Name	Risk*	Attacks/ Scripts
CISCO[2]	CISCO 700 series routers	Remote attacker can cause CISCO 700 routers to panic and reboot. This attack can be executed continuously. This family of router is commonly used in small businesses with ISDN (BRI).	Users with support contracts can download updates from: http://www.cisco.com/cgibin/tablebuild.p1/760 or CISCO Technical assistance Center at tac@cisco.com.	CISCO 700 denial-of-Service	Medium/ **High**	Explana-tion of exploit available in news-groups. Exploit script has been published.
CISCO[3]	CISCO 700 series routers (Clickstart)	Unauthorized remote users may issue certain command to the route and gain information for future attacks via Clickstart.	Users with support contracts can download updates from: http://www.cisco.com/cgibin/tablebuild.pl/760 or CISCO Technical assistance Center at tac@cisco.com.	CISCO Clickstart vulner-ability.	**High**	Explana-tion of exploit available in news-groups.
Efnet[4]	IRCD	Due to certain character translations, it is possible for a new user to join a channel without current user's knowledge.	Workaround is to edit/src/match.c.	IRCD hidden channel joining	Low	Explana-tion of exploit available in news-groups.

Operating System	Software Name	Vulnerability / Impact	Patches/ Workarounds/ Alerts	Common Name	Risk*	Attacks/ Scripts
IP-SWITCH[5]	Imail	The encrypted password can be easily decrypted allowing an unauthorized individual to gain access to all users' mail.	No workarounds or patches known at time of publishing.	IMAIL password recovery	Medium	Explanation of exploit available in news-groups.
IP-SWITCH[6]	Imail (Imapd)	At login, if two long sets of characters are sent the imapd service will crash.	No workarounds or patches known at time of publishing.	Imail Imapd crash	Medium	Bug discussed in news-groups and Web sites.
IP-SWITCH[7]	Imail (LDAP)	Sending over 2,375 characters along with a specific number of "enters" will cause system resourcing to spike and remain at 90%.	No workarounds or patches known at time of publishing.	Imail LDAP resourcing use	Medium	Bug discussed in news-groups and Web sites.
IP-SWITCH[8]	Imail (Imonitor)	Sending over 2,045 characters along with a specific number of "enters" will cause Imonitor to crash with an overflow.	No workarounds or patches known at time of publishing.	**Imail Imonitor crash**	Medium	Bug discussed in news-groups and Web sites.

Operating System	Software Name	Vulnerability / Impact	Patches/ Workarounds/ Alerts	Common Name	Risk*	Attacks/ Scripts
IP-SWITCH[9]	Imail (Web Service)	Sending over 3,000 characters with "GET" to the Web Service results in a buffer overflow that may be exploitable by unauthorized individual to gain root access.	No workarounds or patches known at time of publishing.	Imail Web Service buffer overflow	**High**	Bug discussed in news-groups and Web sites.
IP-SWITCH[10]	Imail (Whois32 Daemon)	Sending 1,000 characters when telneting to port 43 will cause a buffer overflow.	No workarounds or patches known at time of publishing.	Imail Whois32 buffer overflow	Medium/ **High**	Bug discussed in news-groups and Web sites.
Linux[11]	Operating System Below 2.0.36 (TCP)	Unauthorized remote user can pass data to the application layer without an established connection.	Red hat users can obtain an updated kernel at: http://www.redhat.com/support/docs/errata.html	Linux Blind TCP Spoofing	Low/ Medium	Explanation of exploit available in newsgroups. Exploit script has been published.
Linux[12]	Xcmail (long subject line with autoquote on)	Buffer overflow condition exists that will allow an unauthorized user to execute code possibly resulting in root compromise.	No workarounds or patches known at time of publishing.	Xcmail buffer overflow	**High**	Explanation of exploit available in newsgroups. Exploit script has been published.

Operating System	Software Name	Vulnerability / Impact	Patches/ Workarounds/ Alerts	Common Name	Risk*	Attacks/ Scripts
Linux[13] – S.u.S.E.	Operating System (usr/bin/ gnuplot)	Buffer overflow condition exists that will allow an unauthorized user to execute code resulting in root compromise.	Workaround is to: Chmod -s /usr/bin/ gnuplot	Linux gnuplot buffer overflow	**High**	Explanation of exploit available in newsgroups. Exploit script has been published.
Macromedia[14]	Shockwave 7	Auto-update feature can transmit password and hard disk information back to Macromedia.	Workaround is to disable Autodate.	Shockwave 7 data transfer security hole	Medium/ **High**	Bug discussed in newsgroups.
Microsoft Windows[15] NT 3.51 all, 4.0 SP1, and 5.0 betas	Operating System (Screen saver)	Unauthorized user can gain full system access.	Microsoft Windows NT 4.0 SP 2, 3, and 4 partially correct the problem and prevent the current exploit script from functioning.	Windows NT screen saver system rights problem	Medium/ **High**	Explanation of exploit available in newsgroups. Exploit script for SP1 machines has been published.

Operating System	Software Name	Vulnerability / Impact	Patches/ Workarounds/ Alerts	Common Name	Risk*	Attacks/ Scripts
Microsoft Windows[16] NT[17] (**PATCH UPDATE**[18])	Operating System (KnownDLL)	All users can read and write to the KnownDLLs list. A user can add malicious DLLs with the same name as system DLLs to gain higher privileges or execute other actions. Vulnerability is restricted to machines that the malicious user is interactively logged onto.	Strong protection is possible if the following is add to the registry key: HKEY_ LOCAL_ MACHINE\System\ Current ControlSet\ Control\ Session Manger. Name: Protection Mode Type: REG_DWORD Value: 1 **Patch available at:** ftp://ftp.microsoft.com/bussys/winnt/winnt-public/fixes/usa	Windows NT KnownDLL List Vulnerability	Medium	Bug discussed in newsgroups and Web sites.
Microsoft Windows[19] NT 4.0 SP4	SAMBA running on NT	Under certain conditions if the Primary Domain Controller (PDC) fails or is reset the SAMBA server will take over as PDC. In order to bring the original PDC back on-line the SAMBA server must be closed.	No workarounds or patches known at time of publishing.	NT Domain DoS and Security Exploit with SAMBA server	Medium	Bug discussed in newsgroups and Web sites. Exploit script not required for the exploit.

Operating System	Software Name	Vulnerability / Impact	Patches/ Workarounds/ Alerts	Common Name	Risk*	Attacks/ Scripts
Netscape[20]	Netscape Communicator 4.5	Hostile site can: browse user directories, read user's cache, read content of HTML page that the users see using JavaScript find() function.	Workaround is to Disable Javascript.	Netscape Communicator 4.5 JavaScript find() vulnerability	Medium	Explanation of exploit available in newsgroups. Exploit script has been published.
Oracle 8.0.3[21] (problem has been reviewed in security posting previously)	Oracle Database Assistant v1.0	A log file is created whenever a database is created. This log file stores the password used to create the database as plaintext. The log file has permissions set so anyone can view it.	No workarounds or patches known at time of publishing.	Oracle plaintext password	Medium/ **High**	Bug discussed in newsgroups and Web sites. Exploit script not required for the exploit.
Seapine[22]	TestTrack	If an individual telnet's to port 99 and then disconnects without typing anything, CPU usage goes to 100% and remains there until the TestTrack server is killed.	No workarounds or patches known at time of publishing.	TestTrack Telnet port 99 Denial-of-Service	Low	Bug discussed in newsgroups and Web sites. Exploit script not required for the exploit.

Operating System	Software Name	Vulnerability / Impact	Patches/ Workarounds/ Alerts	Common Name	Risk*	Attacks/ Scripts
Seapine[23]	TestTrack	A log file is created in the/scripts directory that contains username and password in plaintext.	No workarounds or patches known at time of publishing.	TestTrack plaintext password log file	Medium	Bug discussed in newsgroups and Web sites.
Seat-tleLabs[24] **(Patch information update)**	SLMail 3.1	A DoS and a buffer overflow condition exist, if the "helo" command is sent to port 27 followed by 819 to 849 characters. A potential DoS condition occurs, if 855 to 2,041 characters are sent, a buffer overflow occurs.	SeattleLabs representatives report that they are working on a fix and it will be included in Slmail 3.2 maintenance release. **Patch is currently undergoing Alpha texting.**	SLMail Vulner-abilities	Low/ Medium	Bug discussed in newsgroups. Exploit script not required for the exploit.
Silicon Graphics[25]-IRIX	X server (installed by default on all IRIX machines)	Buffer overflow condition exists that will allow an unauthorized local user to execute code resulting in root compromise.	Patches available at: http://www.sgi.com/Support/security/patches.html	X server font path buffer overflow	**High**	Explanation of exploit available in news-groups. Exploit script has been published.

Operating System	Software Name	Vulnerability / Impact	Patches/ Workarounds/ Alerts	Common Name	Risk*	Attacks/ Scripts
Sun Solaris[26] 2.6	Operating System (noexec_user_stack)	This protection method can be bypassed using a buffer overflow. This will allow an unauthorized user to gain root access.	No workarounds or patches known at time of publishing.	Defeating Solaris Non-Executable Stack Protection	**High**	Explanation of exploit available in newsgroups. Exploit script has been published.
Sun Solaris[27] All versions	Operating System (cancel)	All versions contain this buffer overflow, but version 2.6 is setuid root when the overflow occurs. On both the SPARC and i386 versions 2.6 an unauthorized user can gain root access.	No workarounds or patches known at time of publishing. Sun indicates that a patch should be available in 4 weeks.	Sun Solaris cancel buffer overflow	Medium (most versions) **High (version 2.6)**	Explanation of exploit available in newsgroups. Exploit script has been published for both SPARC and i386 versions.
Sun Solaris[28] 2.7, 2.6, and 2.6 x86	Operating System (/usr/bin/ write)	Buffer overflow condition exists that will allow an unauthorized user to execute code with sgid TTY. Further exploit may be possible.	No workarounds or patches known at time of publishing.	Solaris "/usr/bin/ write" bug	Low / Medium	Explanation of exploit available in newsgroups. Exploit script has been published.
Unix - Santa Cruz Operations[29] (SCO) 5.0.4	Operating System	It is possible to erase/ overwrite file using startup scripts.	No workarounds or patches known at time of publishing.	Startup script problem	Medium/ **High**	Explanation of exploit available in newsgroups. Exploit script has been published.

*Risk is defined in the following manner:

High - A vulnerability that will allow an intruder to immediately gain privileged access (e.g., sysadmin, root) to the system. An example of this would be a vulnerability in which a sequence of instructions is sent to a machine by an unauthorized user and the machine responds with a command prompt.

Medium - Any vulnerability that will allow an intruder immediate access to the system that is not privileged access. This allows the intruder the opportunity to continue the attempt to gain root access. An example would be a configuration error that allows an intruder to capture the password file.

Low - A vulnerability that provides information to an intruder that could lead to further compromise attempts or a Denial-of-Service (DoS) attack. The reader should note that while the DoS attack is deemed low from a threat potential, the frequency of this type of attack is very high. DoS attacks against mission-critical nodes are not included in this rating and any attack of this nature should instead be considered as a "High" threat.

1Anonymous, February 27, 1999.

[2]ISS Security Advisory, March 11, 1999.

[3]Ibid.

[4]BUGTRAQ, March 7, 1999.

[5]BUGTRAQ, March 4, 1999.

[6]eEye Digital Security Team, Advisory Code AD03011999.

[7]eEye Digital Security Team, Advisory Code AD03011999.

[8]eEye Digital Security Team, Advisory Code AD03011999.

[9]eEye Digital Security Team, Advisory Code AD03011999.

10eEye Digital Security Team, Advisory Code AD03011999

11Network Associates, Inc., Security Advisory March 9, 1999.

[12]BUGTRAQ, March 2, 1999.

[13]BUGTRAQ, March 4, 1999.

[14]Lingo programming list, March 11, 1999.

[15]Cybermedia, March 1999.

[16]L0pht Security Advisory, February 18, 1999.

[17]Microsoft Security Bulletin, MS99-006.

[18]Microsoft Security Bulletin, MS99006 update.

[19]BUGTRAQ, March 2, 1999.

[20]BUGTRAQ, March 8, 1999.

[21]NTBUGTRAQ, March 4, 1999.

[22]NTBUGTRAQ, March 8, 1999.

[23]NTBUGTRAQ, March 8, 1999.

[24]eEye Digital Security Team, Advisory Code AD02041999.

[25]Silicon Graphics Inc. Security Advisory 19990301-01-PX.

[26]BUGTRAQ, March 3, 1999.

[27]BUGTRAQ, March 5, 1999.

[28]BUGTRAQ, March 8, 1999.

[29]BUGTRAQ, March 7, 1999.

Recent Exploit Scripts

The table below contains a representative sample of exploit scripts, identified between February 27 and March 12, 1999, listed by date of script, script name, script description, and comments. **Items listed in boldface/red (if any) are attack scripts for which vendors, security vulnerability listservs, or Computer Emergency Response Teams (CERTs) have not published workarounds or patches, or which represent scripts that hackers/crackers are utilizing.** During this period, 41 scripts, programs, and net-news messages containing holes or exploits were identified.

Date of Script (Reverse Chronological Order)	Script Name	Script Description	Comments
Mar 10, 1999	**Mtr-033.tar.gz**	**A traceroute and ping program that provides link information on all machines between the host and destination computers.**	
Mar 10, 1999	**Lin35.c**	**Exploit program for the Linux TCP Blind Spoofing vulnerability. See "Bugs, Holes & Patches" (Linux[11]).**	
Mar 10, 1999	Winfreeze-sparc.c	The Winfreeze program ported to Solaris. See Winfreez.c (Mar 8, 1999).	
Mar 10, 1999	**Receive.c**	**Exploit program for the Linux TCP Blind Spoofing vulnerability. See "Bugs, Holes & Patches" (Linux[11]).**	
Mar 10, 1999	Solaris.7.procfs	Local Denial-of-service attack against Solaris 7.	

Date of Script (Reverse Chronological Order)	Script Name	Script Description	Comments
Mar 10, 1999	ScrnSave.zip	Exploit program that allows local user to add themselves to the admin group using the Windows NT screen saver system rights problem. See "Bugs, Holes & Patches" (Microsoft Windows[15]).	
Mar 10, 1999	BEADMIN.zip	Exploit program that allows local user to add themselves to the admin group using the Windows NT screen saver system rights problem. See "Bugs, Holes & Patches" (Microsoft Windows[15]).	
Mar 10, 1999	**Cancelex.c**	**Exploit program for the Sun Solaris cancel buffer overflow vulnerability. See "Bugs, Holes & Patches" (Sun Solaris[27]).**	
Mar 9, 1999	**Mtr-032.tar.gz**	**See Mar 10, 1999.**	
Mar 9, 1999	Miffo-check-1.3.1.c	Port scanner that scans class B or Class C IP ranges.	
Mar 8, 1999	NTSweep v1.0	Brute force password cracker that gathers a list of network users. Does not require access to the registry or SAM database.	
Mar 8, 1999	Sco-filewiper.sh	Shell script that will allow non-privileges user to erase any file on SCO OpenServer.	

Date of Script (Reverse Chronological Order)	Script Name	Script Description	Comments
Mar 8, 1999	Winfreez.c	Program that sends a storm of ICMP/redirect packets that cause Microsoft Windows 9x/NT machines to freeze.	
Mar 8, 1999	**Bitch.x**	**Exploit script that allows a user to join mIRC channel without other users knowledge and spy on the channel.**	
Mar 5, 1999	Miffo-check-1.3.c	See Mar 9, 1999.	
Mar 5, 1999	Imail.sh	Script to decrypt Imail server user passwords.	
Mar 5, 1999	InetdDoS-spewfing.tgz	Denial-of-Service exploit script against inetd.	
Mar 5, 1999	Scanpromisc.c	Remote promiscuous mode Ethernet detector.	
Mar 5, 1999	Ego_21.tar.gz	Wordlist Generator that can be customized and invoked with command-line arguments.	
Mar 5, 1999	**picky_10.tar.gz**	**Program that compares wordlist and creates a master list without duplicates.**	
Mar 5, 1999	**Nlog-1.5.1.tar.gz**	**A set of Perl scripts that assist in the databasing and report generation of Nmap logs.**	
Mar 4, 1999	**xnec_plot.c**	**Exploit script for a buffer overflow in Linux gnuplot. See "Bugs, Holes & Patches" (Linux[13]).**	
Mar 4, 1999	Count.cgi.1.c	Exploit script for count.cgi.	

Date of Script (Reverse Chronological Order)	Script Name	Script Description	Comments
Mar 3, 1999	**Lpstatex.c**	**Buffer overflow exploit against Solaris 2.6 lpstat. See "Bugs, Holes & Patches" (CyberNotes -March 3, 1999, Sun Solaris[29]).**	
Mar 3, 1999	**Rdistex.c**	**Buffer overflow exploit against Solaris 2.6 rdist. See "Bugs, Holes & Patches" (December 10, 1998, Sun Solaris[18]).** Note: "Bugs, Holes & Patches" was the name of this publication before January 1999.	
Mar 3, 1999	**Dexpand.zip**	**Program to expand wordlist by adding plurals and possessive versions. It will also create verb endings.**	
Mar 3, 1999	Samgrab.zip	Newest version of a tool that extracts SAM databases from NTFS volumes and is effective against server and workstations.	
Mar 3, 1999	Dmerge.zip	Program that mergers two wordlists deletes duplicates and outputs result in a list separated by CRLF.	
Mar 3, 1999	**Hintcrack.zip**	**Program that attempts to crack hotmail hints.**	
Mar 3, 1999	**Httpdtype-0.22.tar.gz**	**Program that attempts to identify the type of web server running on a remote host.**	
Mar 3, 1999	Nlog-1.5.tgz	See Mar 5, 1999.	

Date of Script (Reverse Chronological Order)	Script Name	Script Description	Comments
Mar 3, 1999	P3psn.tar.gz	Program that identifies the PIII serial number.	
Mar 3, 1999	Pgpcrack99.tgz	Program that attempts to brute force decrypt a file encrypted by PGP.	
Mar 3, 1999	winscan.zip	Searches a Class C IP range and identifies Wingate proxies.	
Mar 3, 1999	Xcmail_exp.c	Explanation and exploit script for xcmail vulnerability. See "Bugs, Holes & Patches" (Linux[12]).	
Mar 2, 1999	Httpservertype-0.01.tar.gz	See Httptype-0.22.tar.gz (Mar 3, 1999).	
Mar 2, 1999	**Net::Nessus::Client Perl module v0.05**	**Perl-based application that can be used as a non-GUI replacement for other Nessus clients.**	
Mar 2, 1999	**Net-RawIP v0.06b.tar.gz**	**The latest version ported to Perl 5.005 and BSD, and including the oshare script (causes Microsoft Windows 98 machines to lock) with additional bug fixes from version 0.06.**	
Mar 2, 1999	Sbouncer004b.c	Latest version of program that can bounce using wingate and socks proxies.	

Date of Script (Reverse Chronological Order)	Script Name	Script Description	Comments
Mar 1, 1999	Gammaprog151.tgz	Bruteforce password cracker for web based e-mail systems and Post Office Protocol (POP) 3 servers. Latest version includes a configurable cgi cracking program.	
Mar 1, 1999	Spike.sh	Shell script front-end for a number of Denial-of-Service attacks.	

[11]Network Associates, Inc., Security Advisory March 9, 1999.

[12]BUGTRAQ, March 2, 1999.

[13]BUGTRAQ, March 4, 1999.

[15]Cybermedia, March 1999.

[18]Microsoft Security Bulletin, MS99006 update.

[27]BUGTRAQ, March 5, 1999.

[29]BUGTRAQ, March 7, 1999.

Script Analysis

This section will supply a short description of scripts that have been analyzed by various security professionals and organizations. If you or your organization wish to contribute, please send e-mail to nipc@fbi.gov with the subject line "CyberNotes Script Analysis." While this section will list only short descriptions, contributors are requested to include a full technical analysis of the script along with release instructions. The release categories are: releasable to anyone; limited releasability (originator-defined list of organizations); or provided for NIPC only. If you would like to receive a copy of the full technical analysis version of any summarized analysis, please send an e-mail listing the script name and requesting the full technical analysis. A member of the CyberNotes editorial team will contact you. All contributions will be credited to the contributing individual or organization unless otherwise requested.

Trends

Trends for this two week period include all the trends two weeks ago and several additional trends.

1. Several hackers/hacker groups appear to be using coordinated scans and probes from different sites.

2. Large numbers of scans and attacks continue to be directed at machines running the Linux operating system.

3. Scanning for Internet Message Access Protocol (IMAP), POP and Domain Name Serves (DNS) continues.

4. Significant increase in reports of NetBus and Back Orifice scanning.

5. For the quarter the following are the ports targeted by hackers in decreasing order: IMAP (143), DNS (53), Echo (7), POP (110), WWW (80), Sun RPC (111), Finger (79) and Telnet (23).

6. Viruses are now being written to capture and transmit information.

7. Large numbers of SMTP servers are being scanned for common user names, most likely in an effort to obtain names for spam attacks.[1]

Viruses

Anti-virus vendors recommend that their users update the anti-virus definition packages on a regular basis (every two weeks or sooner).

WM97.Marker - This virus collects information on the user from Word including the user's name, address, and date/time of infection. On the first of the month, this information is transmitted via ftp to the site codebreakers.org. The virus will only upload the information once. This is accomplished by tracking the success of the upload using a registry key. The key is "HKEY_CURRENT-USER\Software\Miscrsoft\MS Setup (ACME)\User Info" with the addition of "Logfile". If "Logfile" is TRUE then the virus has upload the information. It also randomly names a temporary text file while infecting. This file is named "HSFxxxx.SYS"where xxxx is a randomly generated number. Because of this, the virus is also called WM97.HSFX. There are currently 4 known variants of this virus.

WM97.Sattelite(sic)– This encrypts and decrypts itself on-the-fly in an effort to make detection more difficult. The current version modifies the registered owner of Microsoft Windows 95/98 to "ThE wEiRd GeNiUs." This virus is reported as being widespread in Northern Europe.

---

1 A number of states have enacted legislation making spamming illegal.

CHAPTER 35

NATIONAL INFRASTRUCTURE PROTECTION CENTER CYBERNOTES

March 31, 1999

CyberNotes is published everytwo weeks by the National Infrastructure Protection Center (NIPC). Its mission is to support security and information system professionals with timely information on cyber vulnerabilities, hacker exploit scripts, hacker trends, virus information, and other critical infrastructure-related best practices.

You are encouraged to share this publication with colleagues in the information and infrastructure protection field. Electronic copies are available on the NIPC Web site at http://www.nipc.gov.

Please direct any inquiries regarding this publication to the Editor-CyberNotes, National Infrastructure Protection Center, FBI Building, Room 11719, 935 Pennsylvania Avenue, NW, Washington, DC, 20535.

Bugs, Holes & Patches

The following table provides a summary of software vulnerabilities identified between March 12 and March 26, 1999. The table provides the hardware/operating system, equipment/software name, potential vulnerability/impact, identified patches/workarounds/alerts, common name of the vulnerability, potential risk, and an indication of whether attacks have utilized this vulnerability or an exploit script is known to exist. Software versions are identified if known. **This information is presented only as a summary; complete details are available from the source of the patch/workaround/alert, indicated in the footnote or linked site.** Please note that even if the method of attack has not been utilized or an exploit script is not currently widely available on the Internet, a potential vulnerability has been identified.**Updates from previous issues of CyberNotes are listed in bold.**

Hardware/ Operating System	Equipment/ Software Name	Vulnerability/ Impact	Patches/ Workarounds/ Alerts	Common Name	Risk*	Attacks/ Scripts
Bay Networks[1] **(UPDATED with additional versions affected)**	Baystack 350T (SW:VI.2.0.10, SW:V2.0.0.15, **SW:V2.02.1, and SW:VI.03) Baystack 350T-HD SW:V2.02.1**	System contains default passwords	Partial Workarounds is to turn off telnet access or limit IP addresses which can telnet in. This can be accomplished in the "TELNET Configuration" menu.	Bay Networks default password	High	Default passwords available on Web sites and newsgroups.
CISCO Catayst[2] 5XXX and some 29XX switches	Operating System (supervisor software	A remote user can reload the switches causing a Denial-of-Service condition.	Vendor suggest that customers with contracts update their software thorough normal channels, and customer without contracts contact CISCO at 1-800-553-2447.	CISCO Catalyst Supervisor Remote Reload	Low/ Medium	Bug discussed in newsgroups and Web sites.
Linux[3] S.u.S.E. with Netscape Communicat or	Talkback	Anyone on the system can create/ overwrite files of another user if Communicator crashers.	Vendor's solution is to disable Talkback.	S.u.S.E. with Netscape communicator Talkback failure	Medium	Bug discussed in newsgroups and Web sites.

Hardware/ Operating System	Equipment/ Software Name	Vulnerability/ Impact	Patches/ Workarounds/ Alerts	Common Name	Risk*	Attacks/ Scripts
Linux[4] - S.u.S.E.[5] **(Update to Cyber Notes #6-99 Linux[11] with new vendor)**	Operating System Below 2.0.36 (/dev/kmem)	Unauthorized remote user can pass data to the application layer without an established connection. If other programs contain the vulnerability this may lead to a root compromise.	Vendor recommends updating to V 2.0.36 or a 2.2.x kernel.	Linux Blind TCP Spoofing	Low/ Medium **High** (if other programs are not patched)	Explanation of exploit a available in newsgroups. Exploit script has been published.
Linux[6] - Slackware V 3.6 and prior	Operating System	If installed with the "net.i"or if a network enabled kernel is used to install, an unauthorized user can gain root access during installation. The time period that the system is vulnerable varies and depends on human interactions during log-in.	Vendor solution is to install post v3.6	Kernel installation root compromise	**High** (limited time only)	Bug discussed in newsgroups and Web sites.

Hardware/ Operating System	Equipment/ Software Name	Vulner-ability/ Impact	Patches/ Workarounds/ Alerts	Common Name	Risk*	Attacks/ Scripts
Lotus[7] Notes V4.5 possibly others	Notes client[8]	Due to differences in Window and Unix path character "/"vs. "/", messages sent to the "Sent Mail" folder of a Notes server may be sent unencrypted over the network.	Workaround is to enable encrypt saved mail" in global preferences.	Lotus Notes saved mail	Low/ Medium	Bug discussed in newsgroups and Web sites.
Lynx[9] V2.8.1 pre.9	Operating System	A IMG tag longer than 250 characters causes the system to crash due to a buffer overflow.	No workarounds or patches known at time of publishing.	Lynx buffer overflow	Medium	Explanation of exploit a available in newsgroups. Exploit script has been published.
Microsoft Exchange[10] V5.5	LDAP[11]	Buffer overflow in LDAP: Bind can lead to a Denial-of-Service condition or potentially lead to root compromise.	Patch is available from: ftp://ftp.microsoft.com/bussys/exchange/exchange-public/fixes/Eng/Exchg5.5/PostSP2/DIR-fix/PSP2DIRI.EXE	Microsoft Exchange Mal-formed Bind Request	Medium	Bug discussed in newsgroups and Web sites. Denial-of-Service exploit script has been published.

Hardware/ Operating System	Equipment/ Software Name	Vulnerability/ Impact	Patches/ Workarounds/ Alerts	Common Name	Risk*	Attacks/ Scripts
Microsoft Internet Explorer[12] V5.XX	Internet Explorer	Any screen saver is disabled during installation. If during the installation the process is aborted then the screen remains disabled.	No workarounds or patches known at time of publishing .	MSIE screen saver disable problem	Medium	Bug discussed in news-groups and Web sites.
Microsoft[13] Internet Information Service (IIS) v4.0	IIS (SMTP service)	The SMTP service will attempt to resend mail immediately if it receives a 4XX error. The resend will occur regardless of the retry time set on the server. This may result in a Denial-of-Service against the recipient's mail.	Microsoft's SMTP service program manager solution is to have the receiver issue a 5XX code instead of a 4XX code.	IIS SMTP service flooding	Low/ Medium	Bug discussed in news-groups and Web sites.

Hardware/ Operating System	Equipment/ Software Name	Vulner-ability/ Impact	Patches/ Workarounds/ Alerts	Common Name	Risk*	Attacks/ Scripts
Microsoft Windows 95/98/NT	Operating System (time and date)	Using the calendar function from the Control Panel can accidentally change Date/time even if the cancel button is used. This may result in passwords expiring and other certifications expiring.	No workarounds or patches known at time of publishing.	Microsoft Windows time advance problem	Low	Explanation of exploit available in newsgroups.
Microsoft Windows 98	Office 97	Every Word and Excel document contains a unique identifier number. This has led to concerns over privacy	Vendor has issued a patch and a program designed to remove the numbers. Both are available at: http://officeupdate.microsoft.com/	Office 97 privacy concerns	Medium	Issue discussed in newsgroups and press.

Hardware/ Operating System	Equipment/ Software Name	Vulner-ability/ Impact	Patches/ Workarounds/ Alerts	Common Name	Risk*	Attacks/ Scripts
Microsoft Windows[14] 98	RegWiz	A Web site may obtain information on a user's computer without the user's knowledge through the hardware identification number (HWID) compiled by RegWiz during product registration with Microsoft.	Microsoft expects to have a utility that that will remove the HWID shortly. Instructions for manually removing the HWID can be found at: http:www.winmag.com/web/regwizoff.htm	Microsoft Windows 98 RegWiz	Low	Bug discussed in newsgroups and Web sites.
Microsoft Windows[15] NT	Operating System	It is possible for a user with write permissions to the "/??"Object directory to gain administrator access.	This vulnerability exists on a limited number of configurations. Machines configured with Steve Sutton's NSA guide consistently block this exploit.	Windows NT Case Sensitivity and Symbolic link problem	Medium/ High (rating due to limited success)	Bug discussed in newsgroups and Web sites.
Microsoft Windows[16] NT 3.51 all, 4.0 SPI, and 5.0 betas **(Vendor patch update[17])**	Operating System (Screen saver)	Unauthorized user can gain full system access.	Patch is available from: ftp://ftp.microsoft.com/bussys/winnt/winntpublic/fixes/usa/NT40/hotfixes-postSP4/Scrn Sav-fix	Windows NT screen saver system rights problem	Medium/ **High**	Explanation of exploit available in newsgroups. Exploit script for SP1 machines has bee published.

Hardware/ Operating System	Equipment/ Software Name	Vulnerability/ Impact	Patches/ Workarounds/ Alerts	Common Name	Risk*	Attacks/ Scripts
Microsoft Windows[18] NT	WINMSD.exe	This program functions as intended but supplies information that may be useful to a hacker. On networked NT machines, currently logged-in users' information including username is provided. This information is useful, as input to other known hacker programs.	This behavior is inherent to the program and represents a misuse of information by hackers.	WINMSD.exe Security exploit	Low	Bug discussed in newsgroups and Web sites. Exploit script not required for the exploit.
Open SSL project[19]	SSLeay and OpenSSL	Under certain circumstances it is possible for a specially written SSL client to obtain an SSL Connection that requires access controls to reuse previous session controls.	Vendor recommended solution is to download OpenSSL v 0.9.2b from: http://www.openssl.org	Open SSL/ SSLeay session reuse	Medium	Bug discussed in newsgroups and Web sites. No known exploits.

Hardware/ Operating System	Equipment/ Software Name	Vulnerability/ Impact	Patches/ Workarounds/ Alerts	Common Name	Risk*	Attacks/ Scripts
Qualcomm[20]	Eudora v4.1	Two long e-mail messages containing the same first 231 characters will cause the system to crash. It may be possible to modify the return address to point to any code the user wishes.	No workarounds or patches known at time of publishing.	Eudora 231 character buffer overflow	Medium/ High	Bug discussed in newsgroups and Web sites. Exploit script not required for the first portion of the exploit.
Sun Solaris[21] 7	Operating System	Non privileged user can cause the system to crash by entering in the command "/usr/xpg4/bin/more "while in "/proc/self/psinfo".	Vendor is currently working on a patch.	Solaris procfs bug	Low	Bug discussed in newsgroups and Web sites. Exploit script not required for the exploit.
Unix[22] HP-UXV10.20	Operating System (hpterm)	PHSS_13560 introduced an access problem into hpterm. Users can increase privileges on the system.	Vendor solution is to install patch PHSS_17830	HP-UX hpterm privilege increase	Medium/ High	Bug discussed in newsgroups and Web sites.

Hardware/ Operating System	Equipment/ Software Name	Vulner-ability/ Impact	Patches/ Workarounds/ Alerts	Common Name	Risk*	Attacks/ Scripts
Unix[23] NetBSD	Operating System (noexec mount flag)	A user is able to execute binaries in a sub-tree even if the "noexec" option is on.	Patch is available at: ftp://ftp. NetBSDORG/ pub/ NetBSD/misc/ security/ patches/1999031 7-mount	NetBSD noexec mount problem	Medium	Bug discussed in news-groups and Web sites. Exploit script not required for the exploit.
Unix[24] NetBSDV1.3. 3 and prior	Operating System (umapfs)	Due to insufficient checking a local user can change their user id to root.	Patch is available at: ftp://ft[/ NetBSD.ORG/ pub/ NetBSD/misc/ security/ patches/ 19990311- umapfs	NetBSD umapfs privilege increase	**High**	Bug discussed in news-groups and Web sites.
Unix[25] NetBSD	Operating System (doscmd)	Buffer overflow exists in the program. This overflow occurs after permissions have been reset to the user's permission.	No workarounds or patches known at time of publishing.	NetBSD doscmd buffer overflow	Low	Bug discussed in news-groups and Web sites.

*Risk is defined in the following manner:

High- A vulnerability that will allow an intruder to immediately gain privileged access (e.g.,sysadmin, root) to the system. An example of this would be a vulnerability in which a sequence of instructions is sent to a machine by an unauthorized user and the machine responds with a command prompt.

Medium- A vulnerability that will allow an intruder immediate access to the system that is not privileged access. This allows the intruder the opportunity to continue the attempt to gain root access. An example would be a configuration error that allows an intruder to capture the password file.

Low- A vulnerability that provides information to an intruder that could lead to further compromise attempts or a Denial-of-Service (DoS) attack. The reader should note that while the DoS attack is deemed low from a threat potential, the frequency of this type of attack is very high. DoS attacks against missioncritical nodes are not included in this rating and any attack of this nature should instead be considered as a "High"threat.

1Anonymous, February 27, 1999,

[2]CISCO software bug announcement (CSCdi74333), March 24, 1999.

[3]S.u.S.E. Security Announcement, March 18, 1999.

[4]Network Associates, Inc., Security Advisory, March 9, 1999.

[5]Security Announcement, March 19, 1999.

[6]Microsoft Security Bulletin, MS99-009.

[7]ISS Security Advisory, March 17, 1999.

[8]Martin Bartosch Security Advisory, March 23, 1999

[9]BUGTRAQ, March 23, 1999

[10]BUGTRAQ, March 16, 1999.

[11]ISS Security Advisory, March 15, 1999

[12]Microsoft Security Bulletin, MS99-009.

[13]BUGTRAQ, March 24, 1999

[14]BUGTRAQ, March 14, 1999.

[15]Windows Magazine, March 12, 1999

[16]NTSD Team Alert, March 12, 1999.

[17]Cybermedia, March 1999.

[18]Microsoft Security Bulletin, MS99-008.

[19]NTSECURITY, March 10, 1999.

[20]OpenSSL and SSLeay Security Alert, March 23, 1999.

[21]BUGTRAQ, March 20, 1999.

[22]BUGTRAQ, March 10, 1999.

[23]HP Security Bulletin #00093, March 18, 1999.

[24]NetBSD Security Advisory 1999-007.

[25]BUGTRAQ, March 18, 1999.

[26]BUGTRAQ, March 17, 1999.

[27]BUGTRAQ, March 18, 1999.)

[28]NTSD Team Alert, March 12, 1999)

Recent Exploit Scripts

The table below contains a representative sample of exploit scripts, identified between March 13 and March 26, 1999, listed by date of script, script name, script description, and comments. **Items listed in boldface/red (if any) are attack scripts for which vendors, security vulnerability listservs, or Computer Emergency Response Teams (CERTs) have not published workarounds or patches, or which represent scripts that hackers are utilizing.** During this period, 28 scripts, programs, and net-news messages containing holes or exploits were identified.

Date of Script (Reverse Chronological Order)	Script Name	Script Description	Comments
March 25, 1999	MSIE5.dntml	Script allow public access to the clipboard on Microsoft Internet Explorer 5.	
March 25, 1999	Quick Hacks & Tricks	Compilation of many simple hacks into one package.	
March 25, 1999	Sesquipedalian.c	Denial-Of-Service against Linux 2.1.89-2.2.3 using the zero length fragment bug.	
March 25, 1999	**Yahoo.c**	**Code that attempts to discover the userid/password of yahoo accounts.**	
March 23, 1999	Httptunnel-2.0.tar.gz	**Allows a user to tunnel through a restrictive firewall that has an HTTP proxy.**	
March 22, 1999	Gammaprog 152.tgz	Bruteforce password cracker for web based e-mail systems and Post Office Protocol (POP) 3 servers. Latest version includes a configurable CGI cracking program and targets angelfire.com and yahoo.com.	
March 22, 1999	Novell95.zip	Trojan horse log-in program that erases the original log-in program.	
March 22, 1999	**Wu-2.4.2-academ[BETA-18].c**	**Exploit code for the FTPD vulnerability.**	
March 21, 1999	**Eudora.bof.c**	**Denial-of-Service code for the Eudora program.**	

Date of Script (Reverse Chronological Order)	Script Name	Script Description	Comments
March 21, 1999	Httpdtype-0.05.tar.gz	Utility that attempts to identify what type of web server a remote host is running	
March 21, 1999	**Promail.1.21.trojan**	**Shareware mail program that e-mails userid/password and server to hacker address.**	
March 19, 1999	Forhack.exe	Program that changes the Fortress password and bypasses security .	
March 18, 1999	Cgichk-11b.c	CGI exploit scanner.	
March 18, 1999	**Domtools1.4.0.tar.g.z**	**Tool that automates traversing DNS domain hierarchies, lists all Domains, converts host names to IP addresses and more .**	
March 18, 1999	Httptunnel-1.1.02.tar.gz	See entry for March 23, 1999	
March 17, 1999	Httptunnel-1.1.01.tar.gz	See entry for March 23, 1999	
March 17, 1999	Netbsd.umapfs.txt	Explanation of the umapfs vulnerability. See See "Bugs, Holes & Patches" (CyberNotes -March 31, 1999, Unix[24]).	
March 16, 1999	Httptunnel-1.99.tar.gz	See entry for March 23, 1999.	
March 15, 1999	Califax.tar.gz	MS-DOS/Linux virus that is placed in every program that is compiled with gcc.	
March 15, 1999	Dip2.c	Buffer overflow in dip-33.3.7 that gives local user root access.	

Date of Script (Reverse Chronological Order)	Script Name	Script Description	Comments
March 15 1999	Module .c	Virus source code that infects Linux kernels. March 15, 1999 My_login.c Trojan horse that captures userid/password combinations.	
March 15, 1999	VLP_I.c	Demonstration virus that infects ELF-executables.	
March 15, 1999	Vsrc.tar.gz	Unix/Linux demonstration virus.	
March 14, 1999	Humpdee2.tar.gz	Exploit for Linux rpc.mountd vulnerability that spoofs the udp packet address of the attacker.	
March 13, 1999	BeSysAdm.zip	Exploit code for the Microsoft Windows NT Case Sensitivity and Symbolic link problem. See See "Bugs, Holes & Patches"(CyberNotes -March 31, 1999, microsoft[15]).	
March 13, 1999	Guideonvl.exe	Program the removes the Globally Unique Identifier (GUID) in Microsoft office applications or allows the user to insert any string in the GUID*.	
March 13, 1999	Httptunnel-1.97.tar.gz	See entry for March 23, 1999	

*** Note this may call into question the ability to solely rely on the GUID as a basis for identification. This method has been suggested as a potential means for identifying malicious code writer in recent press reports.**

Script Analysis

This section will supply a short description of scripts that have been analyzed by various security professionals and organizations. If you or your organization wish to contribute, please send e-mail to nipc@fbi.gov with the subject line "CyberNotes Script Analysis." While this section will list only short descriptions, contributors are requested to include a full technical analysis of the script along with release instructions. The release categories are: releasable to anyone; limited reliability (originator-defined list of organizations); or provided for NIPC only. If you would like to receive a copy of the full technical analysis version of any summarized analysis, please send an e-mail listing the script name and requesting the full technical analysis. A member of the CyberNotes editorial team will contact you. All contributions will be credited to the contributing individual or organization unless otherwise requested.

Trends

Trends for this week period:

1. Several hackers/hacker groups appear to be using coordinated scans and probes from different sites.

2. Large numbers of scans and attacks continue to be directed at machines running the Linux or Sun Solaris operating systems.

3. Increased scanning for Internet Message Access Protocol **(IMAP) (143), POP (110),** Telnet (23) and Domain Name Servers (DNS) (53). Please note that this activity has increased dramatically.

4. Scans have also been reported against the following in decreasing order: Echo (7), WWW (80), Sun RPC (111), and Finger (79).

5. Viruses are now being written to capture and transmit information or spam other Internet sites.

6. Large number of SMTP servers are being scanned for common user names, most likely in an effort to obtain names for spam attacks.[1]

7. Large number of ICMP packets are being sent to source ports just above 1024. The source of these packets are normal dialup accounts or shell accounts.

Viruses

A list of the top ten viruses infecting two or more sites as reported to various anti-virus vendors has been categorized into the two tables below. The first table list macro viruses, and the second table lists other viruses. Macro viruses have, historically, spread fastest due to their ability to be transferred by e-mail.

---

1 A number of states have enacted legislation making spamming illegal.

For the purposes of collecting and collating data, infections involving multiple systems at a single location are considered a single infection. The tables list the viruses by: ranking (number of sites affected),common virus name, type of virus (i.e., boot, file,macro,multi-partite), trends (based on number of infections during the last three months reported), and approximate date first found.

Note: Virus reporting is normally 6 to 8 weeks behind the first discovery of infection. With the number of viruses that appear each month, it is possible that a new virus will become widely distributed before the next edition of this publication. To limit the possibility of infection, readers are reminded to update their anti-virus packages, as updates become available.

The viruses listed in the virus table infected over 1800 machines in January, which represents an increase in the number of reported infections from the last prevalence table. The number 1 ranked virus for February/March accounted for 450 infections, and the last virus listed in the tables infected 22. A total of 465 distinct viruses were reported this month, infecting over 2,000 sites. **Readers should note that CyberNotes has changed the way it keeps statistics on virus infections. Infections rates are now based on** number of machines infected, not number of sites. This change reflects the way statistics are being gathered and reported by a number of anti-virus vendors.

TABLE 1 - Macro viruses:

Ranking	Common Virus Name	Type of Virus	Trends	Date First Reported
1	CAP	Macro	Steady	April 1997
2	Class	Macro	Increasing	September 1998
3	Marker	Macro	Increasing	February 1999
4	ColdApe	Macro	Increasing	December 1998
5	Laroux	Macro	Increasing	July 1997
6	Temple	Macro	Increasing	December 1998
7	Npad	Macro	Steady	December 1996
8	Concept	Macro	Steady	December 1996
9	Munch	Macro	Increasing	October 1998
10	Wazzu	Macro	Steady	December 1996

Table 2 - Other virus:

Ranking	Common Virus Name	Type of Virus	Trends	Date First Reported
1	Form	Boot	Steady	September 1991
2	W95/CIH	File	Increasing	July 1998
3	Parity_Boot	Boot	Steady	September 1993
4	AntiEXE	Boot	Steady	September 1994
5	Sampo	Boot	Increasing	January 1995
6	Stat	Boot	Increasing	February 1998
7	AntiCMOS	Boot	Steady	October 1995
8	Ripper	Boot	Steady	March 1994
9	Empire.Monkey	Boot	Steady	July 1994
10	DelCMOS.B	Boot	Increasing	January 1999

W97M.Melissa - The first reported discovery of this virus occurred on Friday, March 26, 1999. Since the discovery, hundreds of sites have been infected including several Fortune 50 companies. One company reported over 60,000 machines infected with this virus. Most infections have occurred via e-mail. The e-mail contains the subject line "Important Message from (user name)"and the Message body contains: "Here is the document you asked for . . . don't show anyone else;-), "The user must open the document to become infected. If macro warnings are enabled in Microsoft Word 97, the user will receive a warning when he/she attempts to open the infected document. If the infected machine uses Microsoft Outlook R. then the virus will execute a number of Visual Basic instructions to read the user's personal address book and send e-mail to 50 recipients in the address book. Individuals can check for infection by looking in the registry (HKEY_CURRENT_USER/Software/Microsoft/Office) for the term "Melissa?) which indicates the presence of the virus. If the key has a value of "Kwyjibo,"then the machine has already mailed the virus out. It should be noted that several variants have started to appear and readers should contact their anti-virus vendors for an update. Most anti-virus vendors have isssued updates that will detect and remove this virus.

X97M.PAPA - This worm has been reported in the wild as of March 29, 1999. The worm is similar to the Melissa virus and is considered by some anti-virus vendors to be a "copy-cat" of Melissa. This worm affects Excel spreadsheet and is transmitted via Microsoft Outlook. The worm sends e-mail to 60 recipients in the address book. The e-mail contains the subject: "Fwd: Workbook from all.net. and Fred Cohen" and the message body: "Urgent info inside. Disregard macro warning." The workbook is currently named pass.wls. The payload executes a shell which executes ping.exe and attempts to ping two IP addresses. Readers should contact their anti-virus vendors for an update.

CHAPTER 36

NATIONAL INFRASTRUCTURE PROTECTION CENTER CYBERNOTES

April 14, 1999

CyberNotes is published every two weeks by the National Infrastructure Protection Center (NIPC).

Its mission is to support security and information system professionals with timely information on cyber vulnerabilities, hacker exploit scripts, hacker trends, virus information, and other critical infrastructure-related best practices.

You are encouraged to share this publication with colleagues in the information and infrastructure protection field. Electronic copies are available on the NIPC Web site at http://www.nipc.gov.

Please direct any inquiries regarding this publication to the Editor-CyberNotes, National Infrastructure Protection Center, FBI Building, Room 11719, 935 Pennsylvania Avenue, NW, Washington, DC, 20535.

Bugs, Holes & Patches

The following table provides a summary of software vulnerabilities identified between March 27 and April 9, 1999. The table provides the hardware/operating system, equipment/software name, potential vulnerability/impact, identified patches/workarounds/alerts, common name of the vulnerability, potential risk, and an indication of whether attacks have utilized this vulnerability or an exploit script is known to exist. Software versions are identified if known. **This information is presented only as a summary; complete details are available from the source of the patch/workaround/alert, indicated in the footnote or linked site.** Please note that even if the method of attack has not been utilized or an exploit script is not currently widely available on the Internet, a potential vulnerability has been identified.

Updates from previous issues of CyberNotes are listed in bold.

Hardware/ Opeating System	Equipment/ Software Name	Vulner-ability/ Impact	Patches/ Workarounds/ Alerts	Common Name	Risk*	Attacks/ Scripts
HP[1]-UX	Domian Enterprise Server Management System (DESMS)	A number of packages have an extra Domain Management process running in background. Users can gain increase privileges as a result.	Patches and a list of affected programs are available from HP.	HP - UX DESMS extra Domain management process	Med-ium/ **High**	Bug discussed in news-groups and Web sites.
HP[2] -UX	Operating System (Service Guard and LockManager)	Improper implementation of restricted SAM functionality can lead to users gaining increased privileges.	Patches are available from HP.	HP - UX ServiceGaurd and LockManager implementation	Med-ium/ **High**	Bug discussed in news-groups and Web sites.
Internet Se-curity Sys-tems, Inc[3]. (ISS)	Internet Scanner (IS)	If IS is configured to use the suid bit to permit "root privileges," an unauthorized user can execute arbitrary commands.	Workaround is to disable the suid bit. The default configuration IS is not vulnerable to this condition.	IIS Internet Scanner suid bit overflow	Med-ium/ **High**	Bug discussed in news-groups and Web sites. Exploit script has been published.

Hardware/ Opeating System	Equipment/ Software Name	Vulner-ability/ Impact	Patches/ Workarounds/ Alerts	Common Name	Risk*	Attacks/ Scripts
Linux[4] - Debian 2.1	Apache Web server (apache_1.3.3 -7.deb)	The current configuration will allow anyone on the Net to discover the exact version of software packages installed.	The maintainer of the Debian Apache package indicated that this configuration is required by the Debian packaging policy and if user wishes a change, the policy must be changed.	Apache on Debian	Low/ Medium **Can provide into for further attacks.**	Bug discussed in news-groups and Web sites.
Microsoft[5]	Internet Explorer 5.0	Vulnerability in the DHTML edit control allows a remote site to obtain and/or read files on the local machine. Note that this is similar to the "%01 security bug" discovered in January.	Possible workaround is to disable the "allow paste via scripts" option. Note that this workaround has not been tested.	IE 5.0 reading and sending of files (a.k.a %01 bug again)	Med-ium/ **High**	Bug discussed in news-groups and Web sites. Exploit script has been published.

Hardware/ Opeating System	Equipment/ Software Name	Vulner-ability/ Impact	Patches/ Workarounds/ Alerts	Common Name	Risk*	Attacks/ Scripts
Microsoft[6]	Personal Web Server 4.0 (Windows 95 and Windows versions) and FrontPage Web Server (same editions)	A network user can read files on the server if the file's name is known. A file request with a non-standard URL will bypass the server's normal file access controls.	Patches are available for the Personal Web Server at: http://support.microsoft.com/download/support.mslfiles/Pwssecup.exe Patches are available for the FrontPage Web Server at: http://officeupdate.microsoft.com/download Deails.fppws98.htm	Personal Web Server File viewing	Low	Bug discussed in news-groups and Web sites.
Mirabilis[7]	ICQ99a (build 1700 v2.13)	Turning on the "Activate my home page" option starts web server which allows anything in the "program files\icq\homepage\root\YOUR#\files" to be requested.	This information is provided for awareness.	ICQ web server	Low	Bug discussed in news-groups and Web sites.

Hardware/ Opeating System	Equipment/ Software Name	Vulner-ability/ Impact	Patches/ Workarounds/ Alerts	Common Name	Risk*	Attacks/ Scripts
Ramp Net-works[8]	WebRamp Internet access device	Two Denial-of-Service vulnerabilities have been discovered. An attacker can (1) send a specific string of characters to the HTTP causing the device to hang, or (2) send a specific UDP pack to port 5335 changing the local IP address.	Firmware upgrades can be obtained at: http://www.rampnet.com/up grades	WebRamp Denial-of-Service	Medium	Bug discussed in news-groups and Web sites. Exploit script has been published.
Sun Java[9]	JDK 1.1, 1.1.X, 1.2 and 2 (multiple platforms) Netscape 4.5 (FreeBSD and NT) Netscape 4.02 (Solaris) Netscape 4.07 (Linux) Symantec Visual Café V3	Hostile Web site can gain full control of the user's machine	Patch available at: Http//www.sun.com	Java 2 byte code verifier	**High**	Bug discussed in news-groups and Web sites. Exploit script has been tested in Uni-versity Labs.

Hardware/ Opeating System	Equipment/ Software Name	Vulner-ability/ Impact	Patches/ Workarounds/ Alerts	Common Name	Risk*	Attacks/ Scripts
Sun Solaris[10] V 7	Operating System (ff.core)	It is possible for a user to overwrite arbitrary files include "/etc/shadow" using a bug in ff.core.	No workarounds or patches known at time of publishing. Patches on Solaris 2.6 and 2.5.1 exist for this vulnerability.	Solaris 7 ff.core bug	Med-ium/ **High**	Bug discussed in news-groups and Web sites. Exploit scripts for earlier versions of the operating system have been published.
Symantec[11]	PCAnywhere V8.0 upgrade B	This upgrade will disable the NT Screensaver even if SP 4 is reinstalled after the PCAnywhere upgrade.	No workarounds or patches known at time of publishing.	PCAnywhere V8.0 and NT Screensaver	Low/ Medium / **High** (security policy issue)	Bug discussed in news-groups and Web sites.
Unix[12] - Multiple systems	Xfree 86 V3.3.3-5 (X11 -unix)	If an unauthorized user creates a symlink with the name ".X11-unix" and point it to any directory (e.g., root), the permissions of the directory are set to 777. This vulnerability occurs a result of a flaw in the way the program handles the "/tmp" directory.	Vendor recommends upgrading to latest version of Xfree86.	Xf86 directory permissions	Med-ium/ **High**	Explana-tion of exploit available in news-groups. Exploit script has been published.

Hardware/ Opeating System	Equipment/ Software Name	Vulnerability/ Impact	Patches/ Workarounds/ Alerts	Common Name	Risk*	Attacks/ Scripts
Unix[13] - Multiple Systems	rsync	The permissions of your home directory may be changed to the permissions of the directory you are transferring to, if you transfer an empty directory into a non-existent directory.	Patches available at: ftp://rsync.samba.org/pub/rsync	Rsync ghost directory	Medium	Bug discussed in news-groups and Web sites.
Unix[14] Multiple Systems	Xfree86 (Xfs)	If an unauthorized user creates a symlink with the name ".font-unix" and point it to "/etc.shadow" the permissions of "/etc/shadow" are set to 777. This vulnerability occurs a result of a flaw in the way the program handles the "/tmp" directory.	Various non-vendor patches are available but not tested.	Xfree86 xfs bug	**High**	Bug discussed in news-groups and Web sites. Exploit script has been published.

Hardware/ Opeating System	Equipment/ Software Name	Vulnerability/ Impact	Patches/ Workarounds/ Alerts	Common Name	Risk*	Attacks/ Scripts
Xylan OmniSwitch[15]	Operating System 3.1.8 and 3.1.9	Denial-of-Service condition results if an unauthorized user telnets to the interface and uses any username and the password "<ctrl>d."	Vendor recommends upgrading to later version. Later versions have been tested and are not affected by this attack.	OmniSwitch telnet "<ctrl> d" problem	Low	Bug discussed in newsgroups and Web sites.

*Risk is defined in the following manner:

High - A vulnerability that will allow an intruder to immediately gain privileged access (e.g., sysadmin, root) to the system. An example of this would be a vulnerability in which a sequence of instructions is sent to a machine by an unauthorized user and the machine responds with a command prompt.

Medium - A vulnerability that will allow an intruder immediate access to the system that is not privileged access. This allows the intruder the opportunity to continue the attempt to gain root access. An example would be a configuration error that allows an intruder to capture the password file.

Low - A vulnerability that provides information to an intruder that could lead to further compromise attempts or a Denial-of-Service (DoS) attack. The reader should note that while the DoS attack is deemed low from a threat potential, the frequency of this type of attack is very high. DoS attacks against mission-critical nodes are not included in this rating any attack of this nature should instead be considered as a "High" threat.

---

[1] Hewlett-Packard Security Bulletin #095, March 31, 1999.

[2] Hewlett-Packard Daily Security Bulletins Digest, March 31, 1999.

[3] Brazilian Information Security Team, SDI.03-99.iss—scanner.

[4] BUGTRAQ, April 5, 1999.

[5] NTBUGTRAQ, March 30, 1999.

[6] Microsoft Security Bulletin, MS99-010.

[7] BUGTRAQ, March 29, 1999.

[8] Internet Security System Security Advisory - WebRamp Denial of Service Attacks, March 31, 1999.

[9] BUGTRAQ, April 5, 1999.

[10] BUGTRAQ, April 7, 1999.

[11] NTSECURITY, March 30, 1999.

[12] S.u.S.E. Security Announcement, March 28, 1999.

[13]BUGTRAQ, April 7, 1999.

[14]BUGTRAQ, April 30, 1999.

[15]BUGTRAQ, March 31, 1999.

Recent Exploit Scripts

The table below contains a representative sample of exploit scripts, identified between March 27 and April 9, 1999, listed by date of script name, script description, and comments. **Items listed in boldface/red (if any) are attack scripts for which vendors, security vulnerability listservs, or Computer Emergency Response Teams (CERTs) have not published workarounds or patches, or which represent scripts that hackers/crackers are utilizing.** During this period, 30, scripts, programs, and net-news messages containing holes or exploits were identified.

Date of Script (Reverse Chronological Order)	Script Name	Script Description	Comments
April 8, 1999	Subseven 1.3	Trojan Horse program that is similar to netbus/Back Orifice.	
April 8, 1999	Nlog-15.3tar.gz	A set of Perl scripts that assist in the databasing and report generation of Nmap logs.	
April 8, 1999	NetBus Pro v2.01	**Trojan horse program with a Graphical User Interface (GUI), proxy support, file manager, web-cam capture, registry manager, hosts scheduler, application redirect, chat and many other features.**	
April 8, 1999	Gspot121.tar.gz	Control panel for Back Orifice that uses GTK+ widget set and provides cut-and-paste and more descriptive list of the commands.	

Date of Script (Reverse Chronological Order)	Script Name	Script Description	Comments
April 8, 1999	Crackpipe.c	Code designed to allow users to tunnel IP's over "normal TCP/UDP connections" in an effort to go around firewalls.	
April 7, 1999	Proc.c	Sample exploit code for the procmail vulnerability that allows local user to dump contents of any file to the screen.	
April 7, 1999	Nessus-alpha2-fix2.tgz	Network security auditing tool that checks for 208 vulnerabilities, detects remote services, portscans and more.	
April 7, 1999	Gspot.patch	See entry for (Gspot121.tar.gz) April 8, 1999.	
April 6, 1999	Wingate.3.0.txt	Explanation of exploit against WinGate 3.0.	
April 6, 1999	Smdos.c	Denial-of-service exploit against sendmail (up to version 8.9.3) using queue filling.	
April 6, 1999	SDI-iss.c	Sample exploit code for the ISS Internet Scanner buffer overflow. See "Bugs, Holes & Patches" CyberNotes - April 14, 1999, Internet Security Systems, Inc[16]).	

Date of Script (Reverse Chronological Order)	Script Name	Script Description	Comments
April 6, 1999	Rat20.exe	Remote administration tool similar to but less powerful than Back Orifice. Can receive commands from the net via UDP. This version can be used with netcat and IRC.	
April 6, 1999	Nessus-alpha2-fix1.tgz	See entry for April 7, 1999.	
April 5, 1999	Shmkill.c	Local Denial-of-Service exploit code for Linux 2.X IPC and mmap vulnerabilities.	
April 5, 1999	**Nmap-2.12.tgz**	**Network-scanning tool that has a variety of scanning modes, including stealth, Xmas, and Null stealth. This release adds more operating system fingerprinting and several bug fixes. Note: This tool continues to be used by hackers. A number of systems become unstable when scanned if patches are not applied.**	
April 5, 1999	Lock-exploit.c	Denial-of-Service exploit against Washington.EDU ipop3d 4.xx using the lock vulnerability.	
April 4, 1999	Nessus-alpha2.tgz	See entry for April 7, 1999.	
April 3, 1999	Wipe-0.58.3.tar.bz2	Program to overwrite hard disk making recovery of data difficult/impossible.	

Date of Script (Reverse Chronological Order)	Script Name	Script Description	Comments
April 3, 1999	Nmap-2.11.tgz	See entry for April 5, 1999.	
April 2, 1999	Net-RawIP-0.06c.tar.gz	**Perl module that manipulates raw Internet Protocol (IP) packets and Ethernet headers. Also includes the oshare script that cause Microsoft Windows to stop functioning.**	
April 2, 1999	Nessus-alpha2-pre1.tzg	See entry for April 7, 1999.	
April 1, 1999	xfsx.sh	Shell script that exploits the xfs security hole. See "Bugs, Holes & Patches" (CyberNotes April 14, 1999, Xfree86 xfs bug).	
April 1, 1999	Vengine.zip	Program that allows any user to create Microsoft Word polymorphic macro viruses. Melissa and Poyssa viruses' code are included in the sample.	
April 1, 1999	Msie5.0.local	Exploit script that allows a remote user to read and grab files on Microsoft Internet Explorer 5.0.	
April 1, 1999	Insmod	Exploit script that allows a local attacker to gain root access.	
April 1, 1999	Imp211-auto.exe	Self-extracting file that contains a Net Ware password cracking utility with a Microsoft Windows GUI.	
March 31, 1999	Ftpscan.c	Ftp port scanner.	

Date of Script (Reverse Chronological Order)	Script Name	Script Description	Comments
March 30, 1999	smbgrinder.zip	Program that removes duplication from sniffer dumps and facilitates the targeting of a specific user for L0phtcrack.	
March 30, 1999	Tobbv1.zip	Claims to bind Trojan horse or virus programs to any file specified. This is done in an effort to defeat anti-virus programs.	
March 30, 1999	Fortrv1r.exe	Allows an attacker to obtain the passwords for Fortres v4.0.	

[16]Brazilian Information Security Team, SDI.03-99.iss-scanner.

Script Analysis

This section will supply a short description of scripts that have been analyzed by various security professionals and organizations. If you or your organization wish to contribute, please send e-mail to nipc@fbi.gov with the subject line "CyberNotes Script Analysis." While this section will list only short descriptions, contributors are requested to include a full technical analysis of the script along with release instructions. The release categories are: releasable to anyone; limited releasability (originator-defined list of organizations); or provided for NIPC only. If you would like to receive a copy of the full technical analysis version of any summarized analysis, please send an e-mail listing the script name and requesting the full technical analysis. A member of the CyberNotes editorial team will contact you. All contributions will be credited to the contributing individual or organization unless otherwise requested.

Trends

Trends for this two week period:

1. Large numbers of scans and attacks continue to be directed at machines running the Linux or Sun Solaris operating systems.

2. Increased scanning for Internet Message Access Protocol (IMAP) (143), POP (110), Telnet (23) and Domain Name Servers (DNS) (53). Please note that this activity has increased dramatically.

3. Scans have also been reported against the following in decreasing order: Echo (7), WWW (80), Sun RPC (111), and Finger (79).

4. Viruses continue to being written to capture and transmit information or spam other Internet sites.

5. Large numbers of SMTP servers are being scanned for common user names, most likely in an effort to obtain names for spam attacks.[1]

6. A company (Network ICE) has reported observing three "Internet-wide" scans from the same site. Two of the scans were for Back Orifice and one was an SNMP "public" scan.

Viruses

False Positive - Recently, Network Associates Incorporated (NAI) McAfee Anti-Virus product with data file 4.0.4017 has reported that Adobe Reader contains the NetBus Trojan horse. Adobe has indicated that this is a false positive with this data file and NAI corrected this with data file 4.0.4019.

W95.CIH - In the last few weeks a number of anti-virus vendors have issued alerts regarding CIH and its variants. Many of the variants are programmed to execute their payload on **April 26**. Versions of this virus attempt to overwrite the MBR and Boot sector and write to FLASH BIOS on some chip sets. All major anti-virus vendors identify most versions of CIH. Anti-virus vendors recommend that users update their virus definitions at least every two weeks. IBM has issued a warning that several thousand Aptivas manufactured between March 5-17, 1999 and sold in the US may have the virus.

W97M.Melissa - The first reported discovery of this virus occurred on Friday, March 26, 1999. Since the discovery, hundreds of sites have been infected including several Fortune 50 companies. One company reported over 60,000 machines infected with this virus. Most infections have occurred via e-mail. The e-mail contains the subject line: "Important Message from (user name)" and the Message body contains: "Here is the document you asked for . . . don't show anyone else;-)." The user must open the document to become infected. If macro warnings are enabled in Microsoft Word 97, the user will reeive a warning when he/she attempts to open the infected document. If the infected machine uses Microsoft Outlook®, then the virus will execute a number of Visual Basic instructions to read the user's personal address book and send e-mail to 50 recipients in the ad-

[1]A number of states have enacted legislation making spamming illegal.

(HKEY_CURRENT_USER\Software/Microsoft/Office) for the term "Melissa?" which indicates the presence of the virus. If the key has a value of "Kwyjibo," then the machine has already mailed the virus out. It should be noted that several variants have started to appear and readers should contact their anti-virus vendors for an update. Most anti-virus vendors have issued updates that will detect and remove this virus.

X97M.PAPA.B - This worm is a working version of the PAPA virus originally discovered on March 29, 1999. The worm is similar to the Melissa virus and is considered by some anti-virus vendors to be a "copy-cat" of Melissa. This worm affects Excel spreadsheet and is transmitted via Microsoft Outlook. The worm sends e-mail to 60 recipients in the address book. The e-mail contains the subject: "Fwd: Workbook from all.net and Fred Cohen": and the message body: "Urgent info inside. Disregard macro warning." The workbook is currently named pass.xls,. The payload executes a shell, which executes ping.exe and attempts to ping two IP addresses. Readers should contact their anti-virus vendors for an update.

CHAPTER 37

NATIONAL INFRASTRUCTURE PROTECTION CENTER CYBERNOTES

April 28, 1999

CyberNotes is published every two weeks by the National Infrastructure Protection Center (NIPC).

Its mission is to support security and information system professionals with timely information on cyber vulnerabilities, hacker exploit scripts, hacker trends, virus information, and other critical infrastructure-related best practices.

You are encouraged to share this publication with colleagues in the information and infrastructure protection field. Electronic copies are available on the NIPC Web site at http://www.nipc.gov.

Please direct any inquiries regarding this publication to the Editor-CyberNotes, National Infrastructure Protection Center, FBI Building, Room 11719, 935 Pennsylvania Avenue, NW, Washington, DC, 20535.

Bugs, Holes & Patches

The following table provides a summary of software vulnerabilities identified between April 10 and April 24, 1999. The table provides the hardware/operating system, equipment/software name, potential vulnerability/impact, identified patches/workarounds/alerts, common name of the vulnerability, potential risk, and an indication of whether attacks have utilized this vulnerability or an exploit script is known to exist. Software versions are identified if known. **This information is presented only as a summary; complete details are available from the source of the patch/workaround/alert, indicated in the footnote or linked site.** Please note that even if the method of attack has not been utilized or an exploit script is not currently widely available on the Internet, a potential vulnerability has been identified. **Updates from previous issues of CyberNotes are listed bold.**

Hardware/ Operating System	Equipment/ Software Name	Vulnerability/ Impact	Patches/ Workarounds /Alerts	Common Name	Risk*	Attacks/ Scripts
Allaire[1] (Updated vulnerability data)[2]	**ColdFusion Application Server**	Machines leaded with the Expression Evaluator (sample application) will allow authorized individuals to read and delete files on the server. *Unauthorized web users can upload files (possibly including executable files) to the server.*	**Remove all sample applications and code from production servers and obtain patch for Expression Evaluator at: http://www.allaire.com**	ColdFusion Expression Evaluator unauthorized access	High	Bug discussed in newsgroups and Web sites. Exploit script not required for the exploit. Other related hacker tools can be utilized to gain privileged access.
BMC Software[3]	PATROL management software	It is possible to replay the session password enabling an unauthorized user to gain a root shell	No workarounds or patches known at time of publishing.	PATROL session replay	High	Bug discussed in newsgroups and Web sites.
BMC Software[4]	PATROL management software	A Denial-of-Service attack is possible using UDP.	No workarounds or patches known at time of publishing.	PATROL session replay	Low	Bug discussed in newsgroups and Web sites.

Hardware/ Operating System	Equipment/ Software Name	Vulnerability/ Impact	Patches/ Workarounds /Alerts	Common Name	Risk*	Attacks/ Scripts
CISCO[5]	Cisco router (Running version 12.x to 12.0(4)X)	If NAT and the input access list are on the same router interface it is possible for packets to be "leaked." A user could then circumvent security filters/policies without knowing the fact.	Patches and full details of IOS's effected can be obtained at: http://www.cisco.com/warp/public/770/iosnataclpub.shtml	CISCO NAT problem	Low / Medium	
eBay	System problem	A malicious user can use JavaScript embedded in HTML to obtain another user's username and password. With this, the malicious user can retract bids, change password and conduct other operations.	No official patch known at time of publishing.	eBayla	Medium; **High**	Bug discussed in newsgroups and Web sites. Exploit script has been published.

Hardware/ Operating System	Equipment/ Software Name	Vulner- ability/ Impact	Patches/ Workarounds /Alerts	Common Name	Risk*	Attacks/ Scripts
Flowpoint Router	aSDL routers	Many companies ship their routers with no password or easily guessed passwords.	User should change passwords upon receiving the router.	Default passwords	High	The use of default password(s) continues to be the number one exploitable vulnerability.
HP- UX[6] (HP3000)	Operating System (MPE /iX debug)	User can gain increase privileges due to improper handling of commands by debug.	Patches now available from Hewlett-Packard (HP). For: MPE/ix 5.0 - MPEKXM1 MPE/ix 5.5 - MPEKXM1B	HP-UX MPE/iX debug problem	Medium/ **High**	No activity on this vulnerability found.
HP - UX[7] HP9000 se- ries 700/800	Operating System (Sendmail)	Users can create a Denial-of-Service condition.	Patches available: HP-UX 10.20 - PHNE_1713 5 HP-UX 11.00 - PHNE_ 12190	HP-UX Sendmail DoS	Low	Bug discussed in newsgroups and Web sites.
Internet Se- curity Sys- tems, Inc.[8] (ISS)	Internet Scanner (IS) (version prior to 5.6.3)	When checking for IP Spoofing, some check may run against random addresses. Some of these addresses may not be associated with the user of IS.	ISS has made available version 5.6.3 at: http://www. iss.net	IIS Internet Scanner out of range check	**High**	Bug discussed in newsgroups and Web sites. Originator may be identified as an intruder on other systems.

Hardware/ Operating System	Equipment/ Software Name	Vulner-ability/ Impact	Patches/ Workarounds /Alerts	Common Name	Risk*	Attacks/ Scripts
Microsoft	Internet Explorer (IE) V 5.0	Web site owner is notified when an end user adds the web site to "Favorites."	No official patch known at time of publishing.	IE 5.0 favorites problem	Low	Bug discussed in newsgroups and Web sites.
Microsoft	Internet Explorer (IE) V 5.0	Conducting a web search for a specific file "favicon.ico" will reveal sites that may expose CGI scripts, including passwords, to viewing.	No official patch known at time of publishing.	IE 5.0 favorites problem	**High**	Bug discussed in newsgroups and Web sites. This problem may also assist hacker in identifying mis-configured sites.

Hardware/ Operating System	Equipment/ Software Name	Vulnerability/ Impact	Patches/ Workarounds /Alerts	Common Name	Risk*	Attacks/ Scripts
Microsoft[9] (Vendor patch update)	Internet Explorer (IE) V 5.0	A hostile web site owner may be able to determine information about a visiting computer through "IMG SRC" tags. A hostile web site owner may be able to execute a Java Script on the visiting computer with trusted privileges. A hostile web site owner may be able to paste contents into the visiting computer clipboard.	Patches available at: http://www.microsoft.com/windows/ie/security/mshtml.asp	IE 5.0 MSHTML problem	Medium	Bug discussed in newsgroups and Web sites. Exploit script has been published.

Hardware/ Operating System	Equipment/ Software Name	Vulner-ability/ Impact	Patches/ Workarounds /Alerts	Common Name	Risk*	Attacks/ Scripts
Microsoft[10]	Internet Explorer 5.0	Vulnerability in the DHTML edit control allows a remote site to obtain and/or read files on the local machine. Note that this is similar to the "%01 security bug" discovered in January.	Patches available at: http://www.microsoft.com/windows/ie/security/dhtml edit.asp	IE 5.0 reading and sending of files (a.k.a. %01 bug again, DHTML edit)	Medium/ **High**	Bug discussed in newsgroups and Web sites. Exploit script has been published.
Microsoft[11] Windows 95/98	Operating system	When a machine receives multiple ARP requests it may result in the user of that machine having to click "OK" for each request, effectively causing a Denial-of-Service condition.	No official patch known at time of published.	Windows ARP Denial-of-service	Low	Bug discussed in nesgroups and Web sites. Exploit script has been published.

Hardware/ Operating System	Equipment/ Software Name	Vulner-ability/ Impact	Patches/ Workarounds /Alerts	Common Name	Risk*	Attacks/ Scripts
Mirabilis[12]	ICQ99a (build 1700 v2.13)	Telnetting to Port 80 and sending "Quit<LF>" will cause the webserver to crash. This bug appears to be intermittent.	No workarounds or patches known at time of publishing.	ICQ web server crash	Low	Bug discussed in newsgroups and Web sites.
Novell Net-Ware[13] (V 4.1X) (Note that this attack was first discovered in August 1998.)	Pandora V 3	An attacker can spoof NCP calls over the IPX protocol. This will enable the attacker to grant Supervisor rights to a guest account or to create a Denial-of-Service condition	Vendor recommends: Install support pack 5B and set the server signature level to 3 (the highest level) and the client packet signature to 1 (1 is the default setting for both). Note that several people have indicated that setting NCP packet signature to Level 3 will prevent Microsoft's client for Netware from logging in[14].	Pandora Hack	Medium/ **High**	Bug discussed in newsgroups and Web sites.

Hardware/ Operating System	Equipment/ Software Name	Vulner-ability/ Impact	Patches/ Workarounds /Alerts	Common Name	Risk*	Attacks/ Scripts
RealNet-works[15] Inc.	Real Media Server	The program stores the password in cleartext. On Linux machines it is store with a 644 permission setting and on NT machines it is stored with file permissions set to full access for everyone.	No official patch known at time of publishing.	Real Media Server password access problem	Medium/ **High**	Bug discussed in newsgroups and Web sites.
Sun Micro-systems	JAVA JDK 1.1 and Java 2	A suspected flaw in the "byte code verifer" will allow code that has not been checked to run. This may enable an attacker to gain full control of the system by running the attacker's code.	Patch available at: http://java.sun.com	JAVA 2 sandbox problem (a.k.a. Byte code verifer)	Medium	Bug discussed in newsgroups and Web sites.

Hardware/ Operating System	Equipment/ Software Name	Vulnerability/ Impact	Patches/ Workarounds /Alerts	Common Name	Risk*	Attacks/ Scripts
Unix - Digital Unix[16] 4.0 (*prior to 4.0E[17]*) **Update to versions affected**	Operating System (inc)	Buffer overflow condition exists in the **/usr/bin/mh/ inc file. This condition can allow an unauthorized user to gain root access.**	No work-arounds or patches known at time of publishing.	Digital Unix inc buffer overflow	High	Exploit Script now available. Explanation of exploit available in news-groups.
Unix - Digital Unix[18] 4.0 (*prior to 4.0E[19]*) **Update to versions affected**	Operating System (at)	Buffer overflow condition exists in the "at" program that can allow an unauthorized user to gain root access.	Upgrade to Digital Unix 4.0D or obtain appropriate patch at: ftp://ftp.service.digital.**com/public/ dunix**	Digital Unix at buffer overflow	High	Exploit Script now available in news-groups.
Unix[20] - NetBSD	Operating System	Unprivileged users can cause a file-system locking error. This occurs during certain kernel operations when a path name ends with a symbolic link ending with one or more "/".	Patch for NetBSD 1.3.3 is available at: ftp://ftp.NetBSD.ORG/pub/NetBSD/misc/security/patches/19990412-vfs_lookup	NetBSD symbolic link kernel panic	Medium	Bug discussed in newsgroups and Web sites.

Hardware/ Operating System	Equipment/ Software Name	Vulnerability/ Impact	Patches/ Workarounds /Alerts	Common Name	Risk*	Attacks/ Scripts
Unix - NetBSD[21]	Operating System (shell script "SVR4_MAKE DEV")	An unprivileged user can read and write data on the first IDE disk.	Patch is available at: ftp://ftp.NetBSD.ORG/pub/NetBSD/misc/security/patches/19990419-SVR4_MAKEDEV	SVR4_ MAKE DEV script vulnerability	**High**	Explanation of exploit available in news-groups.
Webcom[22] (not Webcom.com)	CGI Guestbook	An unauthorized individual may read any text file if the path to file is known. On NT machines, Anonymous Internet Accounts must have read access, but on Windows 95/98 all text files are readable.	No work-arounds or patches known at time of publishing.	Webcom's CGI Guestbook	Medium	Exploration of exploit available in nesgroups.
Winroute[23]	Winroute Pro 3.04	Unauthorized user can gain access to software configuration without authentication.	No official patch known at time of publishing.	Winroute Pro configuration access	**Medium**	Bug discussed in newsgroups and Web sites.

*Risk is defined in the following manner:

High - A vulnerability that will allow an intruder to immediately again privileged access (e.g., sysadmin, root) to the system. An example of this would be a vulnerability in which a sequence of instructions is sent to a machine by an unauthorized user and the machine responds with a command prompt.

Medium - A vulnerability that will allow an intruder immediate access to the system that is not privileged access. This allows the intruder the opportunity to continue the attempt to again root access. c example would be a configuration error that allows an intruder to capture the password file.

Low - Avulnerability that provides information to an intruder that could lead to further compromise attempts or a Denial-of-Service (DoS) attack. The reader should note that while the DoS attacks against mission-critical nodes are not includes in this rating and attack of this nature should instead be considered as a "High" threat.

1Allaire Security Bulletin, ASB99-01.

[2]BUGTRAQ, April 9, 1999.

[3]CISCO web site.

[4]Hewlett-Packard Daily Security Bulletin Digest, April 13, 1999.

[5]Hewlett-Packard Daily Security Digest, April 20, 1999.

[6]ISSforum, April 12, 1999.

[7]Microsoft Security Bulletin MS99-012, April 21, 1999.

[8]NTBUGTRAQ, March 30, 1999.

[9]BUGTRAQ, April 12, 1999.

[10]BUGTRAQ, April 8, 1999.

[11]Novell Technical Information, 2941119, last modified April 9, 1999

[12]BUGTRAQ, April 13, 1999.

[13]BUGTRAQ, April 14, 1999.

[14]BUGTRAQ, January 25, 1999.

[15]Compaq Software Security Team Advisory, April 6, 1999.

[16]BUGTRAQ, January 25, 1999.

[17]Compaq Software Security Team Advisory, April 6, 1999.

[18]NetBSD Security Advisory 1999-008, April 13, 1999.

[19]NetBSD Security Advisory 1999-009, April 19, 1999.

[20]NTBUGTRAQ, April 9, 1999.

[21]BUGTRAQ, April 9, 1999.

Recent Exploit Scripts

The table below contains a representative sample of exploit scripts, identified between April 10 and April 24, 1999, listed by date of script name, script description, and comments. **Items listed in boldfaced/red (if any) are attack scripts for which vendors, security vulnerability listservs, or Computer Emergency Response Teams (CERTs) have not published worarounds or patches, or which represent scripts that hackers/crackers are utilizing.** During this period, 35 scripts, programs, and net-new messages containing holes or exploits were identified.

Date of Script (Reverse Chronological Order	Script Name	Script Description	Comments
April 23, 1999	**Coldscan.c**	**A scanner that checks for Cold Fusion application Server vulnerabilities.**	
April 23, 1999	Cgiv3.c	ACGI scanner that checks for 43 different vulnerabilities.	
April 23, 1999	**Httptunnel-2.3.tar.gz**	**Allows a user to tunnel through a restrictive firewall that has an HTTP proxy.**	
April 23, 1999	ICQ.webserver.txt	Detailed instructions of to execute a number of attacks against ICQ99a Webservers.	
April 22, 1999	**Nessus-alpha2-fix4.tgz**	**Network security auditing tool that checks for 208 vulnerabilities, detects remote services portscans and more. This version fixes a number of segmentation faults and increases performance.**	
April 21, 1999	Ipps-1.0.tgz	Port scanner	

Date of Script (Reverse Chrono-logical Order	Script Name	Script Description	Comments
April 21, 1999	mc-kill.c	Exploit that allows local user to obtain root on a machine running Midnight Commander 4.XX.	
April 21, 1999	**Mole.cfm**	**Cold Fusion template that allows remote attacker full access to files on a Cold Fusion application server.**	
April 21, 1999.	Shopping cart	Exploit description for use against most commercial and freeware shopping carts.	
April 20, 1999	Mars	Java application the monitors SMTP, POP3, HTTP, and FTP.	
April 18, 1999	**Httpdtype-0.07.tar.gz**	**Program that attempts to identify the type of web server running on a remote host.**	
April 18, 1999	KKI.dos	Denial-of-Service script against several RPC implementations.	
April 17, 1999	**Abelkiller.zip**	**Program that removes both the Cain and Able programs.**	
April 17, 1999	Nessus-alpha2-fix3.tgz	See entry for April 22, 1999.	
April 16, 1999	**Abel v1.1**	**Trojan horse that accesses user passwords on Microsoft Windows 9X machine. It can be used in conjuction with the "Cain" tool.**	

Date of Script (Reverse Chrono-logical Order	Script Name	Script Description	Comments
April 16, 1999	**Cain v1.51**	**Password cracker/recovery tool for use on Microsoft Windows 95/98 systems.**	
April 16, 1999	Nmap/nlog tool set	Collection of scripts and utilities that enhance nmap and/or nlog.	
April 16, 1999	Snort v0.99rc6	Packet logger.	
April 16, 1999	Snort-0.99rc6-lib	Sample rule set for Snort.	
April 15, 1999	Mac	Exploit script for a buffer overflow in the Responder.cgi script for MacHTTP.	
April 15, 1999	Miffo-check-1.4.tar.gz	Port scanner that scans class B Class C IP ranges.	
April 15, 1999	**Nwrap**	**Nmap add-on that will randomize host/port scanning combination in an effort to remain stealthy.**	
April 15, 1999	Scrl.41.zip	Source code for a Windows version of Back Orifice client.	
April 14, 1999	**Cgichk1.3c**	**Program that checks for 41 Common Gateway Interface (CGI) vulnerabilities and, if any are found, will attempt to exploit the most common.**	
April 14, 1999	Tcpflow-0.11.tar.gz	Program that captures data transmittted as part of TCP connections. This program will reconstruct packet sequence.	

Date of Script (Reverse Chronological Order	Script Name	Script Description	Comments
April 12, 1999	Exdump-0.2.tar.gz	Packet logger that can monitor content and direct output to a user-defined file.	
April 12, 1999	**Sdi-wu.c**	**Exploit script that creates a backdoor on systems with the WU-ftpd long directory names vulnerability. This program also includes a worm that will scan for other vulnerable systems and attempt to exploit those systems.**	
April 11, 1999	Hunt-1.3.tgz	Connection hijacking tool.	
April 11, 1999	Netbsd.symlink	Code that causes a file-system locking error on NetBSD systems. This results in a system panic or hang.	
April 11, 1999	NT-BSOD.c	Denial-of-Service code against Microsoft Windows NT. Exploits the Windows NT PE loader vulnerability resulting in "Blue Screen of Death."	
April 11, 1999	**Poink.c**	**Denial-of-Service code against Microsoft Windows 95/98/NT. Exploits handling of ARP packets.**	
April 10, 1999	Explpine.c	Exploit code that allows local user to elevate privileges.	

Date of Script (Reverse Chronological Order	Script Name	Script Description	Comments
April 10, 1999	Netscape.4.5.passwd.c	Exploit code that retrieves passwords from Netscape 4.5 ~user/.netscape/liprefs.js files and decrypts the passwords.	
April 10, 1999	Ratc.c	UNIX/Linux source code for the RAT Trojan horse.	
April 10, 1999	Remote.zip	Remote Novell NLM password decrypting program.	

Script Analysis

This section will supply a short description of scripts that have been analyzed by various security professionals and organizations. If you or your organization wish to contribute, please send e-mail to nipc@fbi.gov with the subject line "CyberNotes Script Analysis." While this section will list only short descriptions, contributors are requested to include a full technical analysis of the script along with release instructions. The release categories are: releasable to anyone; limited releasability (originator-defined list of organizations); or provided for NIPC only. If you would like to receive a copy of the full technical analysis version of any summarized analysis, please send an e-mail listing the script name and requesting the full technical analysis. A member of the CyberNotes editorial team will contact you. All contributions will be credited to the contributing individual or organization unless otherwise requested.

Trends

Trends for this two week period:

1. Large numbers of scans and attacks continue to be directed at machines running the Linux or Sun Solaris operating systems.

2. Viruses continue to be written to capture and transmit information or spam other Internet sites.

3. Large numbers of SMTP servers are being scanned for common user names, most likely in an effort to obtain names for spam attacks.[1]

4. Probes to port 1800 and 1945 have been observed recently.

---

1 A number of states have enacted legislation making spamming illegal.

5. Probes continue to look for machines with Back Orifice installed.

Viruses

A list of the top ten viruses infecting two or more sites as reported to various anti-virus vendors has been categorized into the two tables below. The first table list macro viruses, and the second table lists other viruses. Macro viruses have, historically, spread fastest due to their ability to be transferred by e-mail.

For the purposes of collecting and collating data, infections involving multiple systems at a single location are considered a single infection. The tables list the viruses by: ranking (number of sites affected), common virus name, type of virus (i.e.., boot, file, macro, multi-partite), trends (based on number of infections during the last three months reported), and approximate date first found.

Note: Virus reporting is normally 6 to 8 weeks behind the first discovery of infection. With the number of viruses that appear each month, it is possible that a new virus will become widely distributed before the next edition of this publication. To limit the possibility of infection, readers are reminded to update their anti-virus packages, as soon as updates become available.

The viruses listed in the virus table infected over 84,000 machines in March/April, which represents an increase in the number of reported infections from the last prevalence table. When the Melissa virus is removed from the count, the number of reported infections for the March/April timeframe continues to indicate an increase in viruses discovered over the last reporting period. The number 1 ranked virus (Melissa) for March/April accounted for 81,000 reported infections, and the last virus listed in the tables infected 11. A total of 493 distinct viruses were reported this month, infecting over 3,000 sites. **Infection rates are based on number of machines infected, not number of sites.** This method reflects the way statistic are being gathered and by a number of anti-virus vedors.

Table 1 - Macro Viruses

Ranking	Common Virus Name	Type of Virus	Date First Reported
1	Melissa	Macro	March 1999
2	ColdApe	Macro	December 1998
3	Class	Macro	September 1998
4	Laroux	Macro	July 1997
5	Ethan	Macro	February 1999
6	CAP	Macro	April 1997
7	Npad	Macro	December 1996

8	Concept	Macro	December 1996
9	Temple	Macro	December 1998
10	Marker	Macro	February 1999

Table 2 - Other Viruses

Ranking	Common Virus Name	Type of Virus	Date First Reported
1	W95/CIH	File	July 1998
2	AntiEXE	Boot	September 1994
3	AntiCMOS	Boot	October 1995
4	Form	Boot	September 1991
5	Junkie	Multi	July 1994
6	Empire.Monkey	Boot	July 1994
7	Parity_Boot	Boot	September 1993
8	Sampo	Boot	January 1995
9	Stoned	Boot	September 1994
10	DelCMOS.B	Boot	January 1999

Virus shipped out with newsletter - An Internet freeware/shareware software distribution newsletter has indicated that the program "TimePiece 1.0" may contain a virus. Anyone who downloaded this software before April 20, 1999 should obtain a new copy of this program.

CHAPTER 38

NATIONAL INFRASTRUCTURE PROTECTION CENTER CYBERNOTES

May 12, 1999

CyberNotes is published every two weeks by the National Infrastructure Protection Center (NIPC). Its mission is to support security and information system professionals with timely information on cyber vulnerabilities, hacker exploit scripts, hacker trends, virus information, and other critical infrastructure-related best practices.

You are encouraged to share this publication with colleagues in the information and infrastructure protection field. Electronic copies are available on the NIPC Web site at http://www.nipc.gov.

Please direct any inquiries regarding this publication to the Editor-CyberNotes, National Infrastructure Protection Center, FBI Building, Room 11719, 935 Pennsylvania Avenue, NW, Washington, DC, 20535.

Bugs, Holes & Patches

The following table provides a summary of software vulnerabilities identified between April 24 and May 9, 1999. The table provides the hardware/operating system, equipment/software name, potential vulnerability/impact, identified patches/workarounds/alerts, common name of the vulnerability, potential risk, and an indication of whether attacks have utilized this vulnerability or an exploit script is known to exist. Software versions are identified if known. **This information is presented only as a summary; complete details are available from the source of the patch/workaround/alert, indicated in the footnote or linked site.** Please note that even if the method of attack has not been utilized or an exploit script is not currently widely available on the Internet, a potential vulnerability has been identified. **Updates from previous issues of CyberNotes are listed in bold.**

Hardware/ Operating System	Equipment/ Software Name	Vulnerability/Impact	Patches/ Workarounds /Alerts	Common Name	Risk*	Attacks/ Scripts
America Online[1]	AIM	Sending a specific hyperlink in the form of aim:addbuddy?=screenname me and having the receiver click on the hyperlink will cause the client to crash.	No work-arounds or patches known at time of publishing.	AIM Denial-of-Service	Low	Bug discussed in newsgroups and Web sites. Exploit script has been published. Web site has been set up to execute this vulnerability.
Computer Software manufaktur[2]	CSMMail	Buffer overflow conditions exist in the "VRFY" and "RCPT TO:" commands that will allow an unauthorized user to execute arbitrary commands.	No work-arounds or patches known at time of publishing.	CSMMail remote buffer overflow	**High**	Bug discussed in newsgroups and Web sites. Exploit script has been published.
Dis-cusWare[3]	Discus	The software creates a directory with the permissions 666. In this directory are the files users.txt and admin.txt which contain passwords.	Workaround is to check permissions and make sure the directory has correct permissions (not readable from the web).	Discus directory permissions	**High** (large number of sites potentially vulnerable)	Bug discussed in newsgroups and Web sites. Exploit script not required, web search may reveal vulnerable sites.

Hardware/ Operating System	Equipment/ Software Name	Vulnerability/Impact	Patches/ Workarounds /Alerts	Common Name	Risk*	Attacks/ Scripts
FTP Serv-U[4]	FTP Serv-U (v2.5)	FTP Serv-U will crash if 155 or more characters are sent via a command that accepts parameters.	New version of FTP Serv-U corrects this problem. The author has posted reportedly post the fix at: ftp://ftp.catsoft.com/beta/ but the editor was unable to connect to this site.	FTP Serv-U 155 character problem	Low	Bug discussed in newsgroups and Web sites. Exploit script not required.
Humming-bird[5]	Exceed (V5)	Exceed X allows inbound TCP connection on port 6000. If a SDM host is use to telnet to port 6000 the X server will lock-up.	Version 6.1 appears to correct this problem.	Exceed Denial-of-Service	Low	Bug discussed in newsgroups and Web sites. Exploit script not required.
JDEdwards Enterprise software[6]		Applications have hardcoded passwords. The applications in question appear to install as SECOFR on AS/400s and ADMIN on NT/UNIX systems.	No work-arounds or patches known at time of publishing.	JDEdwards hardcoded passwords	**High**	Bug discussed in newsgroups and Web sites. Exploit script has been published.

Hardware/ Operating System	Equipment/ Software Name	Vulnerability/Impact	Patches/ Workarounds /Alerts	Common Name	Risk*	Attacks/ Scripts
Microsoft Windows 95/98[7]	Operating System	A malicious Java program can cause the system to continue to open threads resulting in a Denial-of-Service condition.	Microsoft spokesperson indicated that in order to correct the problem, the relevant code would require "a major overhaul."	Windows 95/98 Java thread problem	Low	Bug discussed in newsgroups and Web sites.
Network Associates[8]	VirusScan NT (v4.0.2)	Due to a suspected race condition, VirusScan may not update virus signatures. The application log will indicate that the file has been updated.	Users can check the About boxes "Created on" to see if the date of the file has changed. Upgrading to version 4.0.3a corrects the problem.	VirusScan update problem	**High**	Bug discussed in newsgroups and Web sites.
Novell[9]	Rconsole utility	Anyone with access to the sys:system\autoexec.nsf, sys:etc\netinfo.cy or sys:system\ldremote.ncf files can crack the rconsole password and gain remote access to the system.	Novell has no fixes planned and suggests that users remove the Remote Netware Loadable Module.	Rconsole password crack	Medium/ **High**	Bug discussed in newsgroups and Web sites. Exploit script has been published.

Hardware/ Operating System	Equipment/ Software Name	Vulnerability/Impact	Patches/ Workarounds /Alerts	Common Name	Risk*	Attacks/ Scripts
Oracle[10]	Oracle Intelligent Agent	An unauthorized user can run have Tcl commands run as root. It is then possible to fully compromise the system.	Workaround on current release is to *chown* the ORACLE_HOME/bin/ oratc 1sh to oracle or other userid.	Oracle Intelligent Agent Tcl problem	**High**	Bug discussed in newsgroups and Web sites.
UNIX	BASH[11]	A number of vulnerabilities have recently been commented on in Listservs including a Buffer overflow in "make_cmd.c" and a directory problem. All these have been reported on older versions.	Updating to a new version corrects these vulner- abilities.	BASH overflow	**High** (if system is not updated to newer version of BASH)	Bug discussed in newsgroups and Web sites. Exploit script not required.

*Risk is defined in the following manner:

High - A vulnerability that will allow an intruder to immediately gain privileged access (e.g., sysadmin, root) to the system. An example of this would be a vulnerability in which a sequence of instructions is sent to a machine by an unauthorized user and the machine responds with a command prompt.

Medium - A vulnerability that will allow an intruder immediate access to the system that is not privileged access. This allows the intruder the opportunity to continue the attempt to gain root access. An example would be a configuration error that allows an intruder to capture the password file.

Low - A vulnerability that provides information to an intruder that could lead to further compromise attempts or a Denial-of-Service (DoS) attack. The reader should note that while the DoS attack is deemed low from a threat potential, the frequency of this type of attack is very high. DoS attacks against mission-critical nodes are not included in this rating and any attack of this nature should instead be considered as a "High" threat.

1BUGTRAQ, April 19, 1999.

[2]BUGTRAQ, April 27, 1999

[3]BUGTRAQ, April 23, 1999

[4]BUGTRAQ, May 4, 1999.

[5]BUGTRAQ, April 27, 1999.

[6]BUGTRAQ, May 3, 1999.

[7]CNET News.com, "Java bug crashes Windows 95,98," April 27, 1999.

[8]Nomad Mobile Research Centre Advisory, May 5, 1999.

[9]Infoworld, "Novell's Remote password falls victim to weak security measures," April 26, 1999.

[10]BUGTRAQ, April 30, 1999

[11]BUGTRAQ, April 19, 1999

Special Note - There have been many reports of Shopping Cart systems exposing personal Credit Card numbers, Orders and other data. Many listservs and newsgroups have also discussed this problem. To date, the software vendors' response has been to refer users to "the manuals which describe in detail the steps to take to avoid" this type of mis-configuration problem. If you run one of these programs, you may wish to recheck the configuration to ensure that information is not available to unauthorized individuals.

Recent Exploit Scripts

The table below contains a representative sample of exploit scripts, identified between April 24 and May 9, 1999, listed by date of script, script name, script description, and comments. **Items listed in boldface/red (if any) are attack scripts for which vendors, security vulnerability listservs, or Computer Emergency Response Teams (CERTs) have not published workarounds or patches , or which represent scripts that hackers/crackers are utilizing.** During this period, 40 scripts, programs, and net-news messages containing holes or exploit were identified.

Date of Script (Reverse Chronological Order)	Script Name	Script Description	Comments
May 4, 1999	Saint-1.3.7.tar.gz	Gathers information on remote hosts by looking at network services.	
May 4, 1999	Tcpblast-19990504.tar.gz	Tool to probe networks and estimate network bandwidth.	
May 4, 1999	**Cgichk1.33.c**	**Program that checks for 53 Common Gateway Interface (CGI) vulnerabilities and, if any are found, will attempt to exploit the most common.**	
May 4, 1999	Gatescan.c	A class B/C network scanner with a number of command lines.	
May 4, 1999	Fav.c	Denial-of-Service exploit code used against Microsoft Internet Explore 5.0 favicon.ico bug.	
May 4, 1999	Edwards.txt	List of hard-coded administrative passwords for a number of JDEdwards applications.	
May 4, 1999	W00f.c	Exploit for obtaining root against machines running wu-ftpd. This code exploits the realpath() overflow.	
May 3, 1999	**NessusJ-alpha2.tar.gz**	**Java client for Nessus.**	
May 3, 1999	Uin2ip.pl	Perl script that resolves ICQ UINs to IP addressees.	

Date of Script (Reverse Chronological Order)	Script Name	Script Description	Comments
May 2, 1999	**Nmap-2.2-BETA3.tgz**	**Network-scanning tool that has a variety of scanning modes, including stealth, Xmas, and Null stealth. This version includes an optional GTK interface.**	
May 2, 1999	**Mns-v.75beta.tar.gz**	**Network scanner similar to nmap and scan.**	
May 2, 1999	Icqget.pl	Program that allows an attacker to retrieve any file from a vulnerable ICQ-Webserver.	
May 1, 1999	Ethereal-0.6.1.tar.gz	Network protocol analyzer (sniffer).	
May 1, 1999	Fizzbounce.tar.gz	Bounces the connection of any HTTP proxy.	
May 1, 1999	Snoop2.c	Packet sniffer for SGI IRIX.	
May 1, 1999	Backdoor.c	Backdoor program that includes security features to insure backdoor is more difficult to discover and activate.	
May 1, 1999	Cgichk1.32.c	See entry for May 4, 1999	
April 30, 1999	Snort-1.0.tar.gz	Packet sniffer that places data in directories based on IP addresses of the captured packets.	
April 30, 1999	Mns-v.69.beta.tar.gz	See entry for May 2, 1999	
April 29, 1999	Netxmon_0.6.tgz	Network sniffer with an X interface.	
April 29, 1999	Net-RawIP-0.06d.tar.gz	Perl module that manipulates raw Internet Protocol (IP) packets and Ethernet headers.	

Date of Script (Reverse Chronological Order)	Script Name	Script Description	Comments
April 29, 1999	**Scrack15.zip**	**Password cracker.**	
April 28, 1999	csmmail.c	Exploit code for CSMMail SMTP server resulting in root compromise.	
April 27, 1999	**Leetscan.tar.gz**	**Single port scanner that is very fast.**	
April 27, 1999	Firewalk-0.99beta.tar.gz	Portscanner that attempts to identify what protocols are used on a given gateway.	
April 27, 1999	Mns-v.67beta.tar.gz	See entry for May 2, 1999.	
April 27, 1999	**Fawx.c**	**Program that sends oversized fragmented IGMP packets that cause systems to freeze pr significantly degrade performance.**	
April 27, 1999	Xxploit.c	Exploit code the Xfree sysmlink compromise.	
April 27, 1999	**ConFusion.zip**	**Win32 program that exploits vulnerabilities in Cold Fusion software.**	
April 27, 1999	Zone	This program exploits a vulnerability in Microsoft Outlook Express that results in unauthorized access to information (number of unread e-mails, etc.).	

Date of Script (Reverse Chronological Order)	Script Name	Script Description	Comments
April 26, 1999	**Lamescan-1.401b.tar.gz**	**Portscanner that conducts random scans and attempts to defeat warning programs by including a username with a port open request.**	
April 26, 1999	Detect-scans-0.80.tar.gz	Detects and logs portscans and a limited number of Denial-of-Service attacks.	
April 25, 1999	B4b0 #7	Electronic hacker magazine that includes a number of exploit scripts including: bouncer.c, GHCgi.c, FreeBSD rootkit and others.	
April 25, 1999	Netconfig.zip	Netware Trojan that presents the user with a failed log-in screen in an effort to have the user enter their password again.	
April 24, 1999	Egghack.tar.gz	Eggdrop userfile password cracker.	
April 24, 1999	Nskan-0.61b.tgz	Network scanner.	
April 24, 1999	Mailing.list.txt	Explanation of exploit that allows anyone to approve messages for a moderated mailing list that uses eGroups web site.	
April 24, 1999	**Qpop242.c**	**Exploit code for FreeBSD 2.2.5 and BSDi 2.1.**	
April 24, 1999	**Bsdi-imapd.c**	**Exploit code for imap2bis.**	
April 24, 1999	**Aixinfod.c**	**Exploit code for the AIX infod vulnerability leading to local root compromise.**	

Date of Script (Reverse Chronological Order)	Script Name	Script Description	Comments
April 24, 1999	NBTscan	This program lists the IP address, NetBIOS name, logged-in user name and MAC address.	

Script analysis

This section will supply a short description of scripts that have been analyzed by various security professionals and organizations. If you or your organization wish to contribute, please send e-mail to nipc@fbi.gov with the subject line "CyberNotes Script Analysis". While this section will list only short descriptions, contributors are requested to include a full technical analysis of the script along with release instructions. The release categories are: releasable to anyone; limited releasability (originator-defined list of organizations); or provided for NIPC only. If you would like to receive a copy of the full technical analysis version of any summarized analysis, please send an e-mail listing the script name and requesting the full technical analysis. A member of the CyberNotes editorial team will contact you. All contributions will be credited to the contributing individual or organization unless otherwise requested.

Trends

Trends for this two week period:

1. Large number of probes for sites with ezmall2000 software installed.

2. Probes to port 1800 and 1945 continue.

3. Large numbers of web sites running the Cold Fusion Software have been hacked recently.

4. Probes looking for Cold Fusion sites continue.

5. Probes continue to look for machines with Back Orifice installed.

6. DoD has reported that the following methods have been used to compromise system in the last quarter (listed by number of successful pentrations): compromised passwords, vulnerability associated with Microsoft Frontpage and Windows NT, Post Office Protocol (POP), Common Gateway Interface (CGI), Telnet, CGI PHF, Internet Control Message Protocol (ICMP), and Internet Message Access Protocol (IMAP)

Viruses

A list of the top ten viruses infecting two or more sites as reported to various anti-virus vendors has been categorized into the two tables below.

The first table list macro viruses, and the second table lists other viruses. Macro viruses have, historically, spread fastest due to their ability to be transferred by e-mail.

For the purposes of collecting and collating data, infections involving multiple systems at a single location are considered a single infection. The tables list the viruses by: ranking (number of sites affected), common virus name, type of virus (i.e., boot, file, macro, multi-partite), trends (based on number of infections during the last three months reported), and approximate date first found.

Note: Virus reporting is normally 6 to 8 weeks behind the first discovery of infection. With the number of viruses that appear each month, it is possible that a new virus will become widely distributed before the next edition of this publication. To limit the possibility of infection, readers are reminded to update their anti-virus packages, as soon as updates become available.

The viruses listed in the virus table infected over 300,000 machines in April, which represents an increase in the number of reported infections from the last prevalence table. When the CIH and Melissa viruses are removed from the count, the number of reported infections for the April timeframe continues to indicate an increase in viruses discovered over the last reporting period. The number 1 ranked virus (CIH) for April accounted for 300,000 reported infections worldwide, and the last virus listed in the tables infected 11. A total of 463 distinct viruses were reported this month, infecting over 2,000 sites (slight decrease from last months reporting). **Infection rates are based on number of machines infected, not number of sites.** This method reflects the way statistics are being gathered and reported by a number of anti-virus vendors.

Table 1 - Macro viruses:

Ranking	Common Virus Name	Type of Virus	Date First Reported
1	Melissa	Macro	March 1999
2	ColdApe	Macro	December 1998
3	CAP	Macro	April 1997
4	Class	Macro	September 1998
5	Ethan	Macro	February 1999
6	Npad	Macro	December 1996
7	Appender	Macro	May 1997
8	Concept	Macro	December 1996
9	Laroux	Macro	July 1997
10	Marker	Macro	February 1999

Table 2 - Other viruses:

Ranking	Common Virus Name	Type of Virus	Date First Reported
1	W95/CIH	File	July 1998
2	Form	Boot	September 1991
3	Parity_Boot	Boot	September 1993
4	AntiEXE	Boot	September 1994
5	AntiCMOS	Boot	October 1995
6	Junkie	Multi	July 1994
7	Empire	Monkey	July 1994
8	Sampo	Boot	January 1995
9	Stoned	Boot	September 1994
10	DelCMOS.B	Boot	January 1999

Win32/Ska (a.k.a HAPPY99) - This Trojan Horse continues to be reported to anti-virus vendors.

This program has been distributed via e-mail, newsgroups and Web software distributions sites.

Happy99.exe is the most common name of this Trojan horse, but reports have identified the use of other files names. When run, the program displays fireworks graphics and installs two files (ska.exe and ska.dll) in the System directory. The program then attempts to spam e-mail recipients and newsgroups to which the infected machine has sent or received messages. Sites with an e-mail filtering capability can filter messages with the header "X-Spanska: Yes." Most major anti-virus vendors have included code in their latest data file updates to detect this Trojan horse program.

CHAPTER 39

NATIONAL INFRASTRUCTURE PROTECTION CENTER CYBER NOTES

May 26, 1999

CyberNotes is published every two weeks by the National Infrastructure Protection Center (NIPC).

Its mission is to support security and information system professionals with timely information on cyber vulnerabilities, hacker exploit scripts, hacker trends, virus information, and other critical infrastructure-related best practices.

You are encouraged to share this publication with colleagues in the information and infrastructure protection field. Electronic copies are available on the NIPC Web site at http://www.nipc.gov.

Please direct any inquiries regarding this publication to the Editor-CyberNotes, National Infrastructure Protection Center, FBI Building, Room 11719, 935 Pennsylvania Avenue, NW, Washington, DC, 20535.

Bugs, Holes & Patches

The following table provides a summary of software vulnerabilities identified between May 9 and May 21, 1999. The table provides the hardware/operating system, equipment/software name, potential vulnerability/impact, identified patches/workarounds/alerts, common name of the vulnerability, potential risk, and an indication of whether attacks have utilized this vulnerability or an exploit script is known to exist. Software versions are identified if known. **This information is presented only as a summary; complete details are available from the source of the patch/workaround/alert, indicated in the footnote or linked site.** Please note that even if the method of attack has not been utilized or an exploit script is not currently widely available on the Internet, a potential vulnerability has been identified. **Updates from previous issues of CyberNotes are listed in bold.**

Hardware/ Operating System	Equipment/ Software Name	Vulnerability/ Impact	Patches/ Workarounds /Alerts	Common Name	Risk*	Attacks/ Scripts
Apple[1]	At Ease 5.0	A user can open Netscape Communicator and enter a command string that allows them to access any user's volume on the server.	Currently no patch available from Apple. However, 3rd party extension may be available.	At Ease 5.0 File Viewing Problem	**Low (Risk may be higher if confidential files are maintained on the system.)**	Bug discussed in newsgroups and websites. No exploit script required. Explanation of exploit discussed.
Computer Associates[2]	InoculateIT v4. 53 for Windows NT Build 169	The InoculateIT Real-Time Scanner only scans the Inbox folder tree. The Inbox Rules Wizard stores the user's rules on the Exchange Server which moves a message to a specific folder without the message ever being placed in a user's inbox. This causes it to completely by-pass the InoculateIT Real-Time Scanner.	No work-arounds or patches known at time of publishing.	Mailbox Virus Vulnerability	Low	Bug discussed in newsgroups and websites.

Hardware/ Operating System	Equipment/ Software Name	Vulnerability/ Impact	Patches/ Workarounds /Alerts	Common Name	Risk*	Attacks/ Scripts
FreeBSD 2.2.7, 2.2.8, 3.1, OpenBSD[3]	Library Calls	A problem in adjusting pointers after realloc() affects a number of programs including 'du' and 'find.' As a result of this problem, it may be possible to construct a directory tree that will result in arbitrary code being executed as a root.	No workaround or patch for affected version supported by the vendor. Currently supported versions of FreeBSD haven't been tested for this problem.	FreeBSD realloc() Pointer Problem	**High**	Bug discussed in newsgroups and websites.
FreeBSD-3.0[4]	Operating system (domain protocol)	Security problem with sockets in FreeBSD's implementation of UNIX-domain protocol family. May cause the system to crash.	FreeBSD 3.1 does not appear to be vulnerable.	FreeBSD Domain Protocol Problem.	**High**	Bug discussed in newsgroups and websites. Exploit script has been published.

Hardware/ Operating System	Equipment/ Software Name	Vulnerability/ Impact	Patches/ Workarounds /Alerts	Common Name	Risk*	Attacks/ Scripts
Intel Experimental Technologies Department[5]	Iparty Conferencing Program	A small voice conferencing program that is used for quick Internet voice chat can be killed by sending a large amount of expended characters to the server port without being logged. This causes the system to crash.	No workarounds or patches known at time of publishing.	Conferencing Program Vulnerability	Low	Bug discussed in newsgroups and websites. Exploit script has been published.
Internet Software Consortium[6]	INN 2.0 and higher (INNDSTART)	Potential for any local user to execute arbitrary code as root by modifying the pathrun parameter in the inn.conf file.	Workaround: Modify the source file innd/ inndstart.c to use a hard-coded pathrun.	INN Pathrun Configuration	**High**	Bug discussed in newsgroups and websites.

Hardware/ Operating System	Equipment/ Software Name	Vulnerability/ Impact	Patches/ Workarounds /Alerts	Common Name	Risk*	Attacks/ Scripts
Internet Software Consortium[7]	INN 2.0 and higher (INND-START) INN 1.7.2. or lower not affected.	Potential exists for a local user to set INNCONF affecting INND-START, which may lead to arbitrary code execution resulting in root access.	Workaround: Install INNDSTART in a directory with 0700 permissions owned by user news.	INNCONF Environmental Variable Trust Problem	**High**	Bug discussed in newsgroups and websites.
Microsoft[8]	Internet Explorer 5.0	A bug in the way Internet Explorer 5.0 cache handles sites that do not employ standards-based HTTP cache controls may result in other users gaining access to password-protected sites without entering username or password.	Workaround is to manually clear the cache at the end of each session.	Internet Explorer 5.0 Site Cache Problem	Low / Medium	Bug discussed in newsgroups and websites.

Hardware/ Operating System	Equipment/ Software Name	Vulnerability/ Impact	Patches/ Workarounds /Alerts	Common Name	Risk*	Attacks/ Scripts
Microsoft[9]	Microsoft Site Server 3.0	If the AdSamples directory is installed (demonstration files for Ad Server) it is possible for an unauthorized individual to gain access to the site configuration file. The configuration file (SITE.CSC) contains username and passwords to access the SQL server database.	Vendor recommends that you remove this sample directory from production machines.	Site Server 3.0 sample directory problem	**High**	Bug discussed in newsgroups and websites. No exploit script required.

Hardware/ Operating System	Equipment/ Software Name	Vulnerability/ Impact	Patches/ Workarounds /Alerts	Common Name	Risk*	Attacks/ Scripts
Microsoft NT 4.0[10]	Microsoft Site Server 3.0; Microsoft BackOffice Server 4.0, 4.5; IIS 4.0 Web Server (sample files)	Web users can view ASP source code and other sensitive files on the web server. The sample files that lead to this problem are installed by default under Site Server but must be explicitly installed under IIS.	Patches available at: ISS: ftp://ftp, microsoft.com /bussys/iis/ iis-public/ fixes/usa/ View code-fix/ Site Server: ftp//ftp. microsoft. com/ buss ys/ sitesrv/ siterv-public/ fixes/usa/si teserver3/ hotfizes- postsp2/ View code-fix/ Vendor recommends that you remove the following files unless they are specifically required on the web site: ViewCode. asp, ShowCode. asp, CodeBrws. asp and Winmsdp.exe.	File Viewer Vulnerability	Low/ Medium	Bug discussed in newsgroups and websites.

Hardware/ Operating System	Equipment/ Software Name	Vulnerability/ Impact	Patches/ Workarounds /Alerts	Common Name	Risk*	Attacks/ Scripts
Microsoft Windows 95[11]	Internet Explorer 5.0; Netscape Communicator 4.x	If a user bookmarks a site containing a specially-coded JavaScript and then views a local file through the web browser, the JavaScript will execute in the security context of the bookmarked file.	Workaround: Disable JavaScript or do not bookmark untrusted pages.	Bookmarks and JavaScript Vulnerability	Medium	Bug discussed in newsgroups and websites. Explanation of exploit discussed.
Novell NetWare 4.x[12]	Transaction Tracking System	It is possible to overflow the Transaction Tracking System (TTS) built into Novell NetWare and possibly crash multiple servers. TTS can also be disabled, which is a form of Denial of Service.	Upgrade to NetWare 5 or apply the latest service pack for NetWare 4.x located at: http://support.com.novell/cgi-bin/search/tidfinder.cgi?290 8153	Transaction Tracking System Server Vulnerability	**High**	Bug discussed in newsgroups and websites

Hardware/ Operating System	Equipment/ Software Name	Vulnerability/ Impact	Patches/ Workarounds /Alerts	Common Name	Risk*	Attacks/ Scripts
Nullsoft[13]	Win AMP 2.x	When opening a file location which is over 256 bytes long, a general protection fault occurs in Windows 95/98. Windows NT will open the files but not play them.	No work-arounds or patches known at time of publishing	WinAMP Long File Name Problem	Low	Bug discussed in newsgroups and websites.
Open-Linux 2.2[14]	OpenLinux 2.2 (installation process)	The LISA installation leaves an account on the system (help) with root access. This account doesn't require a password.	Only the LISA (old style interface) is affected. Quick fix is to remove (after installation) the lines starting with "help" form/ect/passwd and /etc/shadow.	OpenLinux 2.2 LISA Account Problem	**High**	Bug discussed in newsgroups and websites. Affected account with account name identified.
SunOS[15]	SunOS 5.7	Root privileges can be obtained by inserting a floppy/cdrom that has a setuid shell and a volcheck command.	Workaround: Add the following lines to your /etc/rmmount.conf: Mount hsfs-0 nosuid Mount ufs-0 nosuid	Rmmount Vulner-ability	**High**	Bug discussed in newsgroups and websites.

Hardware/ Operating System	Equipment/ Software Name	Vulnerability/ Impact	Patches/ Workarounds /Alerts	Common Name	Risk*	Attacks/ Scripts
Windows 98 Outlook Express[16]	Outlook Express 4.72.3110.1 and 4.72.3120.0	A mail message with a single dot by itself on a line may be sent in a manner that causes Outlook Express to interpret the dot as an End of Message. Any text received after this may be misinterpreted as the response from a POP3 server. This will cause the session to hang and may require the sysadmin to remove the Message manually.	Upgrade to MSIE 5.0; no patch available at this time.	Outlook Express Dot Problem	Low	Bug discussed in newsgroups and websites.

*Risk is defined in the following manner:

High - A vulnerability that will allow an intruder to immediately gain privileged access (e.g., sysadmin, and root) to the system. An example of this would be a vulnerability in which a sequence of instructions is sent to a machine by an unauthorized user and the machine responds with a command prompt.

Medium - A vulnerability that will allow an intruder immediate access to the system that is not privileged access. This allows the intruder the opportunity to continue the attempt to gain root access. An example would be a configuration error that allows an intruder to capture the password file.

Low - A vulnerability that provides information to an intruder that could lead to further compromise attempts or a Denial-of-Service (DoS) attack. The reader should note that while the DoS attack is deemed low from a threat potential, the frequency of this type of attack is very high. DoS attacks against mission-critical nodes are not included in this rating and any attack of this nature should instead be considered as a "High" threat.

1Bugtraq, May 13, 1999.

[2]NTBugtraq, May 12, 199.

[3]Bugtraq, May 12, 1999.

[4]Bugtraq, Many 6, 1999.

[5]Bugtraq, May 8, 1999.

[6]Bugtraq, May 11, 1999.

[7]Bugtraq, May 11, 1999.

[8]Infor World, "Bug Circumvents IE 5 Passwords," May 10, 1999.

[9]NTSEC, May 11, 1999.

[10]LOpht Advisory, May 7, 1999.

[11]Bugtraq, May 9, 1999.

[12]Bugtraq, May 12, 1999.

[13]Bugtraq, May 12, 1999

[14]bugtraq, May 9, 1999.

[15]Bugtraq, May 10, 1999.

16Bugtraq, May 11, 1999.

NOTE: Discussions on a popular security listserve have indicated that by design Microsoft IIS does not log FTP DELE (delete) commands. This has not been verified to date by Microsoft or independent testing.

Recent Exploit Scripts

The table below contains a representative sample of exploit scripts, identified between May 9 and May 21, 1999, listed by date of script name, script description, and comments. **Items listed in boldface/red (if any) are attack scripts for which vendors, security vulnerability listservs, or Computer Emergency Response Teams (CERT's) have not published workarounds or patches, or which represent scripts that hackers/crackers are utilizing.** During this period, 69 scripts, programs., and net-news messages containing holes or exploits were identified.

Date of Script (Reverse Chronological Order)	Script Name	Script Description	Comments
May 18, 1999	Roolover.c	Code snippet that uses ptrace() to intercept and modify the return value of a system call.	

Date of Script (Reverse Chronological Order)	Script Name	Script Description	Comments
May 18, 1999	**Mns-v.91beta.tar.gz**	**Multifunctional network scanner. Network auditing and vulnerability logging package comparable to nmap and sscan. Uses the latest methods of stealthing, os detection and vulnerability checking.**	
May 18, 1999	**Heh.pl**	**Creates a user specified number of rootshells in /tmp, disguises itself, monitors the clone rootshells and logs out all root terminals and disables console when desired (which gives you time to clean logs and make a quick exit).**	
May 18, 1999	Bisonware.ftp.txt	Security vulnerabilities in Bison Ware FTP Server 3.5 allow remote attacker to execute Denial of Service attacks, browse through any directories, and access account passwords, resulting in possible root compromise.	
May 17, 1999	Yaps 12.zip	Windows95/NT intelligent IP scanner,. It can quickly analyze a single computer or identify all computers on a specified network.	

Date of Script (Reverse Chronological Order)	Script Name	Script Description	Comments
May 17, 1999	netscape.bookmarks.title.js.txt	Netscape Communicator v.4.x security vulnerability involving improper handling of special bookmarks with JavaScript code in the title allows malicious webmaster to read user bookmarks, browse user directors, read use files.	
May 17, 1999	**Mns-v.90beta.tar.gz**	**See entry mns-v.91.beta.tar.gz.**	
May 17, 1999	Jammer 17f.exe	Monitors all services, ports, and protocols and also functions as a low-level network sniffer and real-time packet analyzer.	
May 17, 1999	Iputils-ss990417.tar.gz	Suite of programs, optimized for Linux, which are normally used to diagnose network problems. Includes: ping, ping6, traceroute6, rdisc, clockdiff, tracepath, tracepath6, and arping.	
May 17, 1999	Gammaprog 153.tgz	Bruteforce password cracker for web based e-mail addresses (angelifire.com, usa.net, and yahoo.com) and regular POP3 accounts.	

Date of Script (Reverse Chrono-logical Order)	Script Name	Script Description	Comments
May 17, 1999	Firewalk-linus-elf-static-0.99.gz	Network auditing tool that attempts to determine what transport protocol a given gateway will pass.	
May 17, 1999	Cheops-0.60pre3.tar.gz	Network "swiss army knife." A combination of a variety of network tools to provide system administrators and users with a simple interface to managing and accessing their networks.	
May 17, 1999	**Cgichk1.35.c**	**CGI vulnerability scanner that checks a remote host for 65 CGI security holes.**	
May 16, 199	Winfingerprint.zip	Enumerates NetBIOS shares.	
May 16, 1999	Ftpbounce-1.25.tar.gz	Uses the PORT/PASV command in the FTP protocol to bounce ftp files to third parties.	
May 16, 1999	**Cgichk1.34.7.c**	**See entry for cigchkl.35.c.**	
May 16, 1999	**Cgichkl.34.2.c**	**See entry for cgichkl.35.c.**	
May 16, 1999	Banner.zip	Banner-beta is a connect() port scanner and banner gatherer for the Windows platform.	
May 12, 1999	Sshdx.c	Trojan exploit script that creates a new user with uid0/gid0 and no password on local machine.	

Date of Script (Reverse Chronological Order)	Script Name	Script Description	Comments
May 12, 1999	Shockwave.zip	Denial of Service program that sends random data to random ports on a remote host at a very fast rate.	
May 12, 1999	Quickie.c	Confusion exploit scanner. Scans entire subnets and is much faster than other Cold Fusion exploit scanners.	
May 12, 1999	**Mns-v.83beta.tar.gz**	**See entry mns-v.91.beta.tar.gz.**	
May 12, 1999	Ex_sdtcm_convert.c	Solaris Sparc machine local root exploit for buffer overflow condition in sdtcm_convert.	
May 12, 1999	Ex_lpset86.c	Solaris x86 machines local root exploit code for buffer overflow in lpset.	
May 12, 1999	Ex_lpset.c	Updated version of the local root compromise exploit code for buffer overflow condition in lpset for Solaris 2.6 and 2.7x86 machines.	
May 12, 1999	Ex_dtprintinfosp.c	Sparc port of the exploit code for the dtprintinfo stack buffer overflow present in Solaris 2.6 and 2.7 for Sparc.	
May 12, 1999	Ex_admintool.c	Solaris Sparc machines admintool local root exploit.	
May 12, 1999	E2.tgz	Whitepaper about EXT2-weaknesses in Linux 2.0 kernels. "file-hiding" exploit code included.	

Date of Script (Reverse Chronological Order)	Script Name	Script Description	Comments
May 12, 1999	**Cheops-0.60pre2.tar.gz**	**See entry for Cheops-0.60pre3.tar.gz.**	
May 12, 1999	Check.pl	Runs through all of the files and directories that it is given as arguments and determines the permissions. It then sends a list of 'dangerous' files to stdout, which can be redirected to a file.	
May 11, 1999	**Pingscan-1.2.1.tgz**	**Scans networks via ping for reachable hosts. It does DNS lookup and checks for forward/reverse entries. Can enter network addresses in CDIR notation to give start and end addresses.**	
May 10, 1999	**Tds-0.02.tar.gz**	**Toplevel domain scanner that scans through DNS records. Allows you to plug scanned data into security software that checks your networks for holes, or look for weaknesses in other networks.**	
May 10, 1999	**Mns-v.81beta.tar.gz**	**See entry mns-91.beta.tar.gz.**	
May 10, 1999	Ex_dtprintinfo.c	Exploits a stack buffer overflow present in x86 versions of Solaris 2.6 and 2.7 Local root compromise.	
May 8, 1999	WinDump95.exe	TCP dump dor Windows 95/98. It is a network capture program.	

Date of Script (Reverse Chronological Order)	Script Name	Script Description	Comments
May 8, 1999	WinDump.exe	TCP dump for Windows NT. It is a network capture program.	
May 8, 1999	Webster.zip	Dictionary wordlist for use with any password cracking program.	
May 8, 1999	**TopV4.exe**	**Telflon Oil Patch v4.1 program used to bind trojans to any files you specify, defeating virus/trojan detection programs in most cases.**	
May 8, 1999	Tbg.zip	Scans and monitors your network for trojans.	
May 8, 1999	Showcode.tar.gz	Vulnerability scanner that checks for the Microsoft IIS 4,9 showcode.asp vulnerability.	
May 8, 1999	Pgppass.zip	Dictionary-based attack (password cracking) program for use against PCP secret key rings.	
May 8, 1999	PacketNT.exe	Network capture driver required for use with WinDump: TCP dump for Windows NT.	
May 8, 1999	Packet95.exe	Network capture driver required for use with WinDump: TCPdump for Windows 95/98.	
May 8, 1999	**OpenSEC Exploit & Vulnerability Archive**	**Over 1,300 of the most recent and useful exploits and security vulnerabilities, organized by OS and network service.**	

Date of Script (Reverse Chronological Order)	Script Name	Script Description	Comments
May 8, 1999	Nmap-2.2-BETA4.tgz	Utility for network exploration or security auditing.	
May 8, 1999	Netbuf.c	Exploit Code for FreeBSD-2.x and IRIX TCP buffer overflow mbug panic that results in system carash.	
May 8, 1999	**Mns-v.80beta.tar.gz**	**See entry mns-v.91.beta.tar.gz.**	
May 8, 1999	Lowkill2.1zip	Windows9x/NT port of the modem escape character Denial of Service exploit (+++ATHO).	
May 8, 1999	Kki.freebsd.socket.c	Security vulnerability related to FreeBSD implementation of sockets allows any local user to crash any version of FreeBSD.	
May 8, 1999	Ippooper.sh	Exploit script for the iParty daemon Denial of Service vulnerability.	
May 8, 1999	Hypno.zip	Brute force POP3 password cracker that connects to a given mail server and attempts to crack the password of any specified account, using a directory file.	
May 8, 1999	Hotmail.browser.trust .txt	Exploit code and description of another new Hotmail password-grabbing security hole.	
May 8, 1999	Cfscan.c	Faster, better confusion exploit scanner.	

Date of Script (Reverse Chronological Order)	Script Name	Script Description	Comments
May 8, 1999	**ADMgates-v0.2.tgz**	**ADM Linux-based Wingate scanner that allows you to scan entire zones.**	
May 6, 1999	Xip-1.2tar.gz	Acts like tcpfump(8) but with the possibility of changing packet values, creating packets and sending them.	
May 6, 1999	Winperl.zip	Winperl files needed for Windows 9x/NT version of Viper v1.2. (puts the files in windows/system directory).	
May 6, 1999	Viper12.zip	Windows 9x/NT password cracker that goes through every combination of characters you define until it finds a match.	
May 6, 1999	Viper12.tar.gz	UNIX/inux password cracker that goes through every combination of characters you define until it finds a match.	
May 6, 1999	Saint-1.3.7.tar.gz	Security Administrator's Integrated Network Tool that gathers as much information about remote hosts and networks as possible by examining all network services and potential security flaws.	
May 6, 1999	Lsekure.vl-alpha3.fts.tgz	Linux security auditing tool that checks for several local security holes.	

Date of Script (Reverse Chronological Order)	Script Name	Script Description	Comments
May 6, 1999	Gatescan20.c	Wingate scanner with a built in port scanner and logging options.	
May 6, 1999	Gateprobe.c	Wingate scanner that includes A/B/C scanning, IP resolving, View option, File save, etc.	
May 6, 1999	Fts-rvscan.v2-b3.tgz	Remote vulnerability scanner that determines the operating system and finds common vulnerabilities.	
May 6, 1999	Fehmalfvl.tgz	Denial of Service suite that utilizes back orifice servers to launch "spoofed" smurf-like attacks with malformed packets and a amplification ratio.	
May 6, 1999	Cracker.tgz	Cracker that randomly generates class A, B, C, portscans for 111; does quasi OS check for Linux and attempts to exploit it.	
May 6, 1999	**Cgichk1.34.c**	**See entry for cigchkl.35.c.**	
May 6, 1999	Apache.c	Trojan horse that creates a local account with no password.	

Script Analysis

This section will supply a short description of scripts that have been analyzed by various security professionals and organizations. If you or your organization wish to contribute, please send e-mail to nipc@fbi.gov with the subject line "CyberNotes Script Analysis." While this section will list only short descriptions, contributors are requested to include a full technical analysis of the script along with release instructions. The release categories are: releasable to anyone; limited releasability (originator-defined list of organizations); or provided for NIPC only. If you would like to receive a copy of the full technical analysis version of any summarized analysis, please send an e-mail listing the script name and requesting the full technical analysis. A member of the CyberNotes editorial team will contact you. All contributions will be credited to the contributing individual or organization unless otherwise requested.

Trends

Trends for this two week period:

1. Hackers are currently testing methods to defeat detection systems. These include experimentation with timing thresholds and scanning methodologies.

2. Probes to port 1800 and 1945 continue.

3. Large numbers of web sites running the Cold Fusion Software have been hacked recently.

4. Probes looking for Cold Fusion sites continue.

Viruses

CS-Galadriel: *This is the first Corel script virus. This virus attempts to check the system date to determine if it is the* 6th of June of any year. If it is the 6th of June, the payload is activated. This displays a message and attempts to infect the first Corel script. Due to an error in programming, the date check function appears to be faulty in the first sample of this virus. Therefore, it is unlikely that the payload will activate. It is possible that this error may be corrected and released as a variant of CS-Galadriel.

W97M.Melissa.A (NEW VARIANT)Ð The first reported discovery of the MELISSA virus occurred on Friday, March 26, 1999. According to Symantec Corporation, there exists recently discovered variant of the virus that **may not be detected by virus scanning software packages** using the configuration check program files only. This is not a new virus. The file is sent out as an email attachment with the extension .RTF but is actually a word document with the .RTF extension. A suggested solution for this is to include .RTF files in the list of files search by your anti-virus program.

The following table is a list assembled by Joakim von Braun of Trojan horse programs known to him and the default ports that they utilize. Although several of the trojans could use any port, this list provides a starting point for determining Trojan horse activity.

Port No.	Trojan
21	Blade Runner, Doly Trojan, Fore, Invisible FTP, WebEX, WinCrash
23	Tiny Telnet Server

Port No.	Trojan
25	Antigen, Email Password Sender, Haebu Coceda, Shtrilitz Stealth, Terminator, WinPC, WinSpy
31	Hackers Paradise
80	Executor
456	Hackers Paradise
555	Ini-Killer
666	Santanz Backdoor
1001	Silencer, WebEx
1011	Doly Trojan
1170	Psyber Stream Server, Voice
1234	Ultors Trojan
1245	VooDoo Doll
1492	FTP99CMP
1600	Shivka-Burka
1807	SpySender
1981	Shockrave
1999	BackDoor
2001	Trojan Cow
2023	Ripper
2115	Bugs
2140	Deep Throat, The Invasor
2801	Phineas Phucker
3024	WinCrash
3129	Masters Paradise
3150	Deep Throat
3700	Portal of Doom
4092	WinCrash
4590	ICQTrojan
5000	Sockets de Troie
5001	Sockets de Troie
5321	Firehotcker
5400	Blade Runner
5401	Blade Runner
5402	Blade Runner
5569	Robo-Hack
5742	WinCrash
6670	Deep Throat
6771	Deep Throat
6969	GateCrasher, Priority
7000	Remote Grab

Port No.	Trojan
7300	NetMonitor
7301	NetMonitor
7306	NetMonitor
7307	NetMonitor
7308	NetMonitor
7789	ICKiller
9872	Portal of Doom
9873	Portal of Doom
9874	Portal of Doom
9875	Portal of Doom
9989	iNi-Killer
10067	Portal of Doom
10167	Portal of Doom
11000	Senna Spy
11223	Progenic Trojan
12223	Hack'99 KeyLogger
12345	GabanBus
12346	GahanBus, NetBus
12361	Whack-a-mole
12362	Whack-a-mole
16969	Priority
20001	Millennium
20034	NetBus 2 Pro
21544	GirlFriend
22222	Prosiak
23456	Evil FTP, Ugly FTP
26274	Delta
31337	Back Orifice
31338	Back Orifice, DeepBO
31339	NetSpy DK
31666	BOWhack
33333	Prosiak
34324	Big Gluck, TN
40412	The Spy
40421	Masters Paradise
40422	Masters Paradise
40423	Masters Paradise
40426	Masters Paradise
47262	Delta

Port No.	Trojan
50505	Sockets de Troie
50766	Fore
53001	Remote Windows Shutdown
61466	Telecommando
65000	Devil